MATHEMATICS
FOR
BUSINESS STUDIES

MATHEMATICS FOR BUSINESS STUDIES

[*A Text book for B. Com. (Hons.) Semester III, Paper DSC-3.1 University of Delhi under NEP*]

Dr. J.K. Thukral
Former Associate Professor
Department of Mathematics
Shri Ram College of Commerce
University of Delhi-110007

Mayur Paperbacks

Mayur Paperbacks
A-95, Sector-5,
NOIDA–201301

•

Sales Office :
K. L. Malik and Sons Pvt. Ltd.
23, Darya Ganj, New Delhi-110002

•

First Edition : 1993
Twenty Sixth Revised Edition : 2025

•

Printed in India at
Nice Printing Press
Sahibabad, Ghaziabad, U.P.

•

ISBN : 81-7198-124-0

Dedicated

to

Ranjna, Sahil and Sanchit

B.COM. (HONS.)

Discipline Specific Course- 3.1(DSC-3.1)

Business Mathematics

Unit 1: Matrices and Determinants **(9 hours)**

Overview of Matrices. Solution of a system of linear equations (having a unique solution and involving not more than three variables) using matrix inversion method and Cramer's Rule

Leontief Input Output Model (Open Model Only).

Unit 2: Calculus-I **(6 hours)**

Concepts and rules of differentiation. Concept of Marginal Analysis: Marginal Revenue, Marginal Cost. Concept of Elasticity of demand and supply.

Application of Maxima and Minima problems: Revenue, Cost, Profit, Economic Order Quantity, Optimal trade in time.

Unit 3: Calculus-II **(12 hours)**

Partial Differentiation: Partial derivatives up to second order. Homogeneity of a function and Euler's theorem. Production Function: Returns to factor, Returns to scale. MRTS and Elasticity of Substitution.

Application of Maxima and Minima problems involving two independent variables.

Integration: Nature of commodities and partial elasticity of demand, Applications of marginal analysis, Consumer Surplus and Producer Surplus.

Unit 4: Mathematics of Finance **(9 hours)**

Rates of interest: nominal, effective and their inter-relationships in different compounding situations.

Compounding a sum using different types of rates. Applications relating to Depreciation of assets and average due date.

Types of annuities: ordinary, due, and deferred - Discrete and continuous. Perpetuity. Determination of future and present values using different types of rates of interest. Applications relating to Capital Expenditure and Leasing.

Unit 5: Linear Programming **(9 hours)**

Formulation and Assumptions of LPP, Solution by Simplex Method- maximization and minimization cases. Shadow prices of the resources. Special Cases: Identification of unique and multiple optimal solutions, unbounded solution, infeasibility and degeneracy.

Practical Exercises: **30 hours**

The learners are required to:

1. Assess the use of matrices in evaluating competing alternatives.
2. Apply differential calculus to solve hypothetical business problems.
3. Evaluate business problems as an application of linear programming.
4. Gather information about various deposit and loan schemes of banks to find out interest rate differentials, and compounded value.
5. Gather information about annuity schemes in the investment markets like periodic home mortgage payments, insurance payments and pension payments, life insurance products as an annuity.
6. Identify the decision-making variables and assess their functional relationship with other variables affecting the decision in a hypothetical business and economic situation.
7. Develop programming for hypothetical business problems involving constrained optimisation.

PREFACE OF THE FIRST EDITION

This book is an outcome of the lectures delivered by the author during the past thirteen years to the students of commerce and economics at different levels. The book incorporates mathematical topics essential for the students of commerce, economics and management. The material in this book is designed to provide maximum flexibility of use for teachers and students alike.

A look at the contents of the book will give the reader a clear idea of the variety and range of topics discussed and analysed. The book has been organised in the form of twelve chapters and an appendix giving table of e^x *and* e^{-x}. *Each chapter begins with a brief introduction which describes its main theme in general terms. An attempt has been made to introduce new ideas is a manner which will give the reader a good intuitive feeling and understanding for the subject.*

Special care has been taken to design and devise the book in such manner as to enable the reader to fully appreciate the application of Mathematics to business and economics.

The book has been written in a simple language, lucid and clear style, so as to enable the students to understand the topics easily. Each topic is illustrated with solved examples which have been designed to clarify the principles involved.

Designed to further enhance the understanding and given a deeper insight into the concepts involved, exercises to well-graded problems are given at appropriate places in the text. Answers to all the exercises are given at the back of the book. The careful organisation and explanations help the student learn and comprehend difficult concepts.

The author is indebted to many of his teachers, colleagues and students for direct and indirect ideas in the production of this volume. I owe a great deal to my esteemed teachers Prof. B.S. Yadav, Dr. S.C. Arora and Prof. (Mrs.) Ajit Iqbal Singh of the Mathematics Department, University of Delhi for the numerous conversations we have had about the book, its contents and its approach. The author gratefully acknowledges the advice and criticism of Prof. Vidya Ratna, Dr. R.N. Aggarwal and Dr. C.B. Gupta who read the manuscript in its intermediate stages and offered numerous suggestions to make it more comprehensive. Others who deserve specific mention are Mr. V.K. Kapoor, Mr. J.R. Monga, Mr. R.C. Bhardwaj, Mrs. Savita Bhagi, Mrs. Misha Govil, Ms. Pushp Lata Jain, Dr. Ramesh Kumar, Mr. Sushil Azad and Mrs. Neelam Malhotra. Above all, I thank Dr. D.C. Sancheti, Principal, Shri Ram College of Commerce, for the many incisive suggestions and observations that he has made about the presentation, both as to its style as well as to its contents. Finally, a special word of thanks is due to Mr. S.K. Malik, the Production Editor and to the entire National Publishing House Staff for their efficient, competent assistance and enthusiastic cooperation.

Every effort has been made to avoid printing errors, though some might have crept in inadvertently. I shall be obliged if such errors are brought to my notice. Suggestions from fellow teachers and students for further improvement in the book are invited and will be most welcome.

Delhi **J.K. THUKRAL**

October 16, 1993

PREFACE TO THE TWENTY SIXTH EDITION

*The author takes great pleasure in presenting to the readers the twenty sixth completely revised edition of the book. The twenty sixth edition of **"Mathematics for Business Studies"** has been designed strictly in accordance with the **latest syllabus prescribed by the University of Delhi for B.Com. (Hons.) Semester III, Paper DSC-3.1 as per Undergraduate Curriculum Framework (UGCF) 2022 based on NEP 2020.***

In preparing this twenty sixth edition, the author is immensely grateful to all friends and discreet readers who were kind enough to give constructive criticism and extended valuable suggestions for the improvement of the book. Any suggestion to ensure further improvement of the book will be greatly acknowledged.

Some of the Special Features of this Book are :

- *Written in a lucid style and simple language.*
- *Written with a view to present a qualitative understanding of the subject.*
- *Comprehensive step-by-step explanations for easier understanding of the subject*
- *Each chapter starts with chapter objectives which state specifically what the reader should be able to do upon completion of the chapter.*
- *Numerous illustrations and careful page layout are designed to make the material as easy to follows as possible.*
- *A large number of solved examples and unsolved problems have been drawn from recent examination papers of B. Com. (H) Delhi University.*
- *Answers to all the problems in each exercise are given immediately after the exercise.*
- *Delhi University Examination paper of **Business Mathematics** for the year 2024 has been appended at the end of the book with complete solutions.*

I again wish to thank the entire staff of Mayur Paperbacks, Paperback division of K.L. Malik and Sons Pvt. Ltd. for their assistance, cooperation and support during the production of this edition.

Mobile : 9810587661

E-mail : *jitendrathukral@gmail.com*

DR. J.K. THUKRAL

CONTENTS

CHAPTER

1

MATRICES AND DETERMINANTS: AN OVERVIEW

OBJECTIVES

After studying the material in this chapter, you should be able to:

- *Identify various types of matrices.*
- *Add, subtract and multiply matrices.*
- *Distinguish between matrix and determinant.*
- *Compute the inverse of a matrix using adjoint.*
- *Solve systems of linear equations by matrix inversion method.*
- *Solve systems of linear equations by Cramer's rule (Determinant method).*

1.1 MATRICES

Definition . A rectangular arrangement of *mn* numbers (real or complex) into *m* horizontal rows and *n* vertical columns enclosed by a pair of brackets [ ], such as

$$\begin{bmatrix} a_{11} & a_{12} & \cdots & a_{1n} \\ a_{21} & a_{22} & \cdots & a_{2n} \\ \cdot & \cdot & & \cdot \\ \cdot & \cdot & & \cdot \\ \cdot & \cdot & & \cdot \\ a_{m1} & a_{m1} & \cdots & a_{mn} \end{bmatrix}$$

is called an $m \times n$ (read "*m* by *n*") **matrix** or **a matrix of order** $m \times n$. Parentheses ( ) are also commonly used to enclose numbers constituting matrices.

The numbers forming a matrix are called **elements** or **entries** of the matrix. For the entry a_{ij}, the first subscript i specifies the row and the second subscript j, the column in which the entry appears. It must be noted that even though an $m \times n$ matrix contains mn elements, the entire matrix should be considered as a single entity. In keeping with this point of view, matrices are denoted by single capital letters such as **A, B, C** and so on.

***Example* 1.** What is the order of each of the following matrices?

$$\mathbf{A} = \begin{bmatrix} 1 & 5 & 2 \\ 0 & 3 & 6 \end{bmatrix} \qquad \mathbf{B} = \begin{bmatrix} 1+i & -1 \\ 2i & 3+5i \end{bmatrix} \qquad \mathbf{C} = \begin{bmatrix} 5 & 7 \\ -1 & 3 \\ 6 & 4 \end{bmatrix}$$

***Solution* .** The matrix **A** has 2 rows and 3 columns, so **A** has order 2 × 3, where the number of rows is specified first.

The matrix **B** has 2 rows and 2 columns, so **B** has order 2 × 2.

The matrix **C** has 3 rows and 2 columns, so **C** has order 3 × 2.

It is customary to abbreviate the matrix

$$A = \begin{bmatrix} a_{11} & a_{12} & \cdots & a_{1n} \\ a_{21} & a_{22} & \cdots & a_{2n} \\ \cdot & \cdot & & \cdot \\ \cdot & \cdot & & \cdot \\ \cdot & \cdot & & \cdot \\ a_{m1} & a_{m1} & \cdots & a_{mn} \end{bmatrix}$$

by the symbol $(a_{ij})_{m \times n}$ or more simply (a_{ij}). This notation merely indicates what type of symbols we are using to denote the general entry.

Equality of Matrices

Definition. *Two matrices **A** and **B** are said to be **equal,** written **A** = **B**, if they are of the same order and if all corresponding entries are equal.* For example,

$$\begin{bmatrix} 5 & 1 & 0 \\ 2 & 3 & 4 \end{bmatrix} = \begin{bmatrix} 2+3 & 1 & 0 \\ 2 & 3 & 2\times 2 \end{bmatrix} \quad \text{but} \quad \begin{bmatrix} 2 & -2 \end{bmatrix} \neq \begin{bmatrix} 2 \\ -2 \end{bmatrix}$$

Special Types of Matrices

Certain types of matrices, which play important roles in matrix theory, are now considered.

Row Matrix. *A matrix that has exactly one row is called a **row matrix.*** For example, the matrix

$$\mathbf{A} = \begin{bmatrix} 5 & 2 & -1 & 4 \end{bmatrix}$$

is a row matrix of order 1 × 4.

Column Matrix *. A matrix consisting of a single column is called a* **column matrix.** For example, the matrix

$$\mathbf{B} = \begin{bmatrix} 3 \\ 1 \\ 4 \end{bmatrix}$$

is a 3 × 1 column matrix.

Zero or Null Matrix *. An m × n matrix whose entries are all 0 is called the m × n* ***zero or null matrix.*** *It is usually denoted by* $\mathbf{O}_{m \times n}$ *or more simply by* ***O.*** For example,

$$\mathbf{O} = \begin{bmatrix} 0 & 0 & 0 & 0 \\ 0 & 0 & 0 & 0 \end{bmatrix}$$

is a 2 × 4 zero matrix.

Square Matrix *. An m × n matrix is said to be a* ***square matrix*** *of order n if m = n. That is, if it has the same number of columns as rows.* For example,

$$\begin{bmatrix} -3 & 4 & 6 \\ 2 & 1 & 3 \\ 5 & 2 & -1 \end{bmatrix} \text{ and } \begin{bmatrix} 2 & -1 \\ 5 & 6 \end{bmatrix}$$

are square matrices of order 3 and 2 respectively.

In a square matrix $\mathbf{A} = (a_{ij})$ of order n, the entires $a_{11}, a_{22}, \ldots, a_{nn}$ which lie on the diagonal extending from the left upper corner to the lower right corner are called the **main diagonal** entries, or more simply the **main diagonal.** Thus in the matrix

$$\begin{bmatrix} 3 & 2 & 4 \\ 1 & 6 & 0 \\ 5 & 1 & 8 \end{bmatrix}$$

the entries $a_{11} = 3$, $a_{22} = 6$ and $a_{33} = 8$ constitute the main diagonal.

Identity Matrix or Unit Matrix *. A square matrix is said to be* ***identity matrix or unit matrix*** *if all its main diagonal entries are 1's and all other entries are 0's. An identity matrix of order n is denoted by* $\mathbf{I}_n$ *or more simply by* ***I.*** For example,

$$\mathbf{I}_3 = \begin{bmatrix} 1 & 0 & 0 \\ 0 & 1 & 0 \\ 0 & 0 & 1 \end{bmatrix}$$

is identity matrix of order 3.

1.2 ALGEBRA OF MATRICES

Addition of Matrices. *Let **A** and **B** be two matrices of the same order. Then the addition of **A** and **B**, denoted by **A + B**, is the matrix obtained by adding corresponding entries of **A** and **B**.*

Thus, if $\mathbf{A} = (a_{ij})_{m \times n}$ *and* $\mathbf{B} = (b_{ij})_{m \times n}$, *then* $\mathbf{A} + \mathbf{B} = (a_{ij} + b_{ij})_{m \times n}$

Remark. *Notice that we can add two matrices only if they are of the same order. If they are, we say they are **conformable** for addition. Also, the order of the sum of two matrices is same as that of the two original matrices.*

Example 2. $$\begin{bmatrix} 2 & 1 & 0 \\ 5 & 7 & -1 \end{bmatrix} + \begin{bmatrix} 5 & -3 & 2 \\ 1 & 4 & 6 \end{bmatrix} = \begin{bmatrix} 7 & -2 & 2 \\ 6 & 11 & 5 \end{bmatrix}.$$

Negative of a Matrix. *If **A** is any matrix, the **negative** of **A**, denoted by **–A**, is the matrix obtained by replacing each entry in **A** by its negative.* For example, if

$$\mathbf{A} = \begin{bmatrix} 2 & -1 \\ 5 & 4 \\ -6 & 0 \end{bmatrix}, \quad \text{then} \quad -\mathbf{A} = \begin{bmatrix} -2 & 1 \\ -5 & -4 \\ 6 & 0 \end{bmatrix}$$

Note. *The zero matrix plays the same role in matrix addition as the number zero does in addition of numbers.*

Subtraction of Matrices . *Let **A** and **B** be two matrices of the same order. Then by **A – B,** we mean **A + (– B).** In other words, to find **A – B** we subtract each entry of **B** from the corresponding entry of **A.***

Example **3.** Let $\mathbf{A} = \begin{bmatrix} 4 & -1 \\ 2 & 3 \\ 5 & -7 \end{bmatrix}$ and $\mathbf{B} = \begin{bmatrix} 0 & 2 \\ 5 & -2 \\ 6 & 1 \end{bmatrix}$. Then

$$\mathbf{A} - \mathbf{B} = \begin{bmatrix} 4-0 & -1-2 \\ 2-5 & 3-(-2) \\ 5-6 & -7-1 \end{bmatrix} = \begin{bmatrix} 4 & -3 \\ -3 & 5 \\ -1 & -8 \end{bmatrix}$$

Multiplication of a Matrix by a Scalar . *Let **A** be an m × n matrix and k be a real or a complex number (called a **scalar**). Then the multiplication of **A** by k, denoted by k **A,** is the m × n matrix obtained by multiplying each entry of **A** by k. This operation is called **scalar multiplication.***

Example 4 . *If* $\mathbf{A} = \begin{bmatrix} 2 & -3 & -1 \\ 5 & 6 & 4 \end{bmatrix}$, *then* $4\mathbf{A} = 4\begin{bmatrix} 2 & -3 & -1 \\ 5 & 6 & 4 \end{bmatrix} = \begin{bmatrix} 8 & -12 & -4 \\ 20 & 24 & 16 \end{bmatrix}.$

Multiplication of Matrices* .** *While the operations of matrix addition and scalar multiplication are fairly straight forward, the product* **AB** *of matrices* **A** *and* **B** *can be defined under the condition that the number of columns of* **A** *must be equal to the number of rows of* **B.** *If the number of columns in the matrix* **A** *does equal the number of rows in the matrix* **B,** *we say that the matrices are* ***conformable for the product AB.

***Definition* .** *Let* $\mathbf{A} = (a_{ij})$ *be an* $m \times n$ *matrix and* $\mathbf{B} = (b_{jk})$ *be an* $n \times p$ *matrix. Then the product* **AB** *is the* $m \times p$ *matrix defined by* $\mathbf{AB} = (c_{ik})$, *where*

$$c_{ik} = a_{i1}\, b_{1k} + a_{i2}\, b_{2k} + \ldots + a_{in}\, b_{nk} = \sum_{j=1}^{n} a_{ij}\, b_{jk}$$

Thus the product **AB** is the $m \times p$ matrix, where each entry c_{ik} of **AB** is obtained by multiplying corresponding entries of the *i*th row of **A** by those of the *k*th column of **B** and then finding the sum of the results.

Remark* .** *The definition refers to the product* **AB,** *in that order,* **A** *is the left factor called* ***prefactor *and* **B** *is the right factor called* ***postfactor.***

The definition of multiplication of matrices is best illustrated by an example.

***Example* 5 .** Find the product *AB* if $\mathbf{A} = \begin{bmatrix} 1 & -4 \\ 5 & 3 \\ 0 & 2 \end{bmatrix}$ and $\mathbf{B} = \begin{bmatrix} -2 & 4 & 1 & 6 \\ 2 & 7 & 3 & 8 \end{bmatrix}$.

***Solution* .** Since the number of columns of **A** is equal to the number of rows of **B,** the product **AB** is defined. Since **A** is 3×2 and **B** is 2×4, the product **AB** will be 3×4 :

$$\mathbf{AB} = \begin{bmatrix} c_{11} & c_{12} & c_{13} & c_{14} \\ c_{21} & c_{22} & c_{23} & c_{24} \\ c_{31} & c_{32} & c_{33} & c_{34} \end{bmatrix}$$

The entry c_{11} is obtained by summing the products of each entry in row 1 of **A** by the corresponding entry in column 1 of **B,** *i.e.*

$$c_{11} = (1)(-2) + (-4)(2) = -10$$

Similarly, for c_{21} we use the entries in row 2 of **A** and those in column 1 of **B** :

$$c_{21} = (5)(-2) + (3)(2) = -4$$

Also,

$$c_{12} = (1)(4) + (-4)(7) = -24$$

$$c_{13} = (1)(1) + (-4)(3) = -11$$

$$c_{14} = (1)(6) + (-4)(8) = -26$$
$$c_{22} = (5)(4) + (3)(7) = 41$$
$$c_{23} = (5)(1) + (3)(3) = 14$$
$$c_{24} = (5)(6) + (3)(8) = 54$$
$$c_{31} = (0)(-2) + (2)(2) = 4$$
$$c_{32} = (0)(4) + (2)(7) = 14$$
$$c_{33} = (0)(1) + (2)(3) = 6$$
$$c_{34} = (0)(6) + (2)(8) = 16$$

Thus $$\mathbf{AB} = \begin{bmatrix} -10 & -24 & -11 & -26 \\ -4 & 41 & 14 & 54 \\ 4 & 14 & 6 & 16 \end{bmatrix}$$

Observe that the product **BA** is not defined since the number of columns of **B** is not equal to the number of rows of **A. This shows that matrix multiplication is not commutative.** That is, for any two matrices **A** and **B,** it is usually the case that **AB** ≠ **BA** (even if both products are defined).

Positive Integral Power of A Square Martix

*Let **A** be a square matrix of order n. We define A^2 = **AA.** In general, for any positive integer m, A^m is defined to be*

$$\mathbf{A}^m = \mathbf{AA} \ldots \mathbf{A} \ (m \text{ times})$$

***Example* 6 .** Show that the matrix $\mathbf{A} = \begin{bmatrix} 2 & 2 & 0 \\ 2 & 1 & 1 \\ -7 & 2 & -3 \end{bmatrix}$ satisfies the equation $\mathbf{A}^3 - 13\mathbf{A} + 12\mathbf{I} = \mathbf{O}$.

***Solution* .** We have

$$\mathbf{A}^2 = \mathbf{AA} = \begin{bmatrix} 2 & 2 & 0 \\ 2 & 1 & 1 \\ -7 & 2 & -3 \end{bmatrix} \begin{bmatrix} 2 & 2 & 0 \\ 2 & 1 & 1 \\ -7 & 2 & -3 \end{bmatrix} = \begin{bmatrix} 8 & 6 & 2 \\ -1 & 7 & -2 \\ 11 & -18 & 11 \end{bmatrix}$$

$$\mathbf{A}^3 = \mathbf{A}^2\mathbf{A} = \begin{bmatrix} 8 & 6 & 2 \\ -1 & 7 & -2 \\ 11 & -18 & 11 \end{bmatrix} \begin{bmatrix} 2 & 2 & 0 \\ 2 & 1 & 1 \\ -7 & 2 & -3 \end{bmatrix} = \begin{bmatrix} 14 & 26 & 0 \\ 26 & 1 & 13 \\ -91 & 26 & -51 \end{bmatrix}$$

Thus

$$A^3 - 13A + 12I = \begin{bmatrix} 14 & 26 & 0 \\ 26 & 1 & 13 \\ -91 & 26 & -51 \end{bmatrix} - \begin{bmatrix} 26 & 26 & 0 \\ 26 & 13 & 13 \\ -91 & 26 & -39 \end{bmatrix} + \begin{bmatrix} 12 & 0 & 0 \\ 0 & 12 & 0 \\ 0 & 0 & 12 \end{bmatrix} = \begin{bmatrix} 0 & 0 & 0 \\ 0 & 0 & 0 \\ 0 & 0 & 0 \end{bmatrix}$$

1.3 DETERMINANTS

We interrupt our discussion of matrices to introduce the concept of the determinant function. Remember that a matrix is simply an ordered arrangement of elements, it is meaningless to assign a single numerical value to a matrix.

However, if **A** is a square matrix, then the determinant function associates with **A** exactly one numerical value called the **determinant** of **A.** Denoting the determinant of **A** by $|\mathbf{A}|$ (that is, using vertical bars) we can think of the determinant function as a correspondence :

$$\underset{\text{square matrix}}{\mathbf{A}} \rightarrow \underset{\text{determinant of A}}{|\mathbf{A}|}$$

The determinant has a varicty of uses, but first we need to find out how determinant may be evaluated.

***Definition*. (*Determinant of order 1*)** *Let* $\mathbf{A} = [a_{11}]$ *be a square matrix of order 1. Then determinant of A is defined as the number* a_{11} *itself. That is,* $|a_{11}| = a_{11}$.

***Example* 7 .** $|3| = 3$, $|-5| = -5$ and $|0| = 0$.

***Definition*. (*Determinant of order 2*)** *Let* $\mathbf{A} = \begin{bmatrix} a_{11} & a_{12} \\ a_{21} & a_{22} \end{bmatrix}$ be a 2×2 *matrix, then we define*

$$|\mathbf{A}| = a_{11}\, a_{22} - a_{12}\, a_{21}.$$

That is, the determinant of a 2 × 2 matrix is obtained by taking the product of the entries in the main diagonal and subtracting from it the product of the entries in the other diagonal.

***Example* 8 .** Evaluate each of the following determinants :

(i) $\begin{vmatrix} 3 & 4 \\ 9 & -7 \end{vmatrix}$ *(ii)* $\begin{vmatrix} -5 & 2 \\ -3 & 1 \end{vmatrix}$

***Solution* .**

(i) $\begin{vmatrix} 3 & 4 \\ 9 & -7 \end{vmatrix} = 3 \times (-7) - 9 \times 4 = -21 - 36 = -57$

(ii) $\begin{vmatrix} -5 & 2 \\ -3 & 1 \end{vmatrix} = -5 \times 1 - (-3) \times 2 = -5 + 6 = 1$

To define the determinant of a square matrix **A** of order n ($n > 2$), we need the concepts of the minor and the cofactor of an element.

1.4 MINORS AND COFACTORS

Minor of an element of a determinant**.** *Let $|\mathbf{A}| = |a_{ij}|$ be a determinant of order n. The* ***minor*** *of a_{ij}, the element in the ith row and jth column of $|\mathbf{A}|$, is the determinant that is left by deleting the ith row and the jth column. It is denoted by M_{ij}.*

For example, given the 3 × 3 determinant

$$\begin{vmatrix} a_{11} & a_{12} & a_{13} \\ a_{21} & a_{22} & a_{23} \\ a_{31} & a_{32} & a_{33} \end{vmatrix}$$

The minor of a_{11} is $M_{11} = \begin{vmatrix} a_{22} & a_{23} \\ a_{32} & a_{33} \end{vmatrix}$, obtained by deleting the first row and first column.

The minor of a_{12} is $M_{12} = \begin{vmatrix} a_{21} & a_{23} \\ a_{31} & a_{33} \end{vmatrix}$, obtained by deleting the first row and second column and so on.

Cofactor of an element of a determinant**.** *Let $|\mathbf{A}| = |a_{ij}|$ be a determinant of order n. The* ***cofactor*** *of a_{ij}, denoted C_{ij} or A_{ij}, is defined as $(-1)^{i+j} M_{ij}$, where i + j is the sum of the row number i and column number j in which the entry lies. Thus,*

$$C_{ij} = \begin{cases} M_{ij} & \text{,if } i+j \text{ is even} \\ -M_{ij} & \text{,if } i+j \text{ is odd} \end{cases}$$

For example, the cofactor of a_{12} in the 3 × 3 determinant $\begin{vmatrix} a_{11} & a_{12} & a_{13} \\ a_{21} & a_{22} & a_{23} \\ a_{31} & a_{32} & a_{33} \end{vmatrix}$ is

$$C_{12} = (-1)^{1+2} \begin{vmatrix} a_{21} & a_{23} \\ a_{31} & a_{33} \end{vmatrix} = -\begin{vmatrix} a_{21} & a_{23} \\ a_{31} & a_{33} \end{vmatrix}$$

Definition **. (*****Determinant of order n*****).** *If A is a square matrix of order n (n > 2), then its determinant may be calculated by multiplying the entries of any row (or column) by their cofactors and summing the resulting products.*

Remark **.** *It is a fact that determinant of a matrix is unique and does not depend on the row or column chosen for its evaluation.*

A Particular Case : Determinant of order 3

Let $\mathbf{A} = \begin{bmatrix} a_{11} & a_{12} & a_{13} \\ a_{21} & a_{22} & a_{23} \\ a_{31} & a_{32} & a_{33} \end{bmatrix}$ be a 3 × 3 matrix. Then, expanding along the first row, determinant

of **A** is given by
$$
\begin{aligned}
|\mathbf{A}| &= a_{11}C_{11} + a_{12}C_{12} + a_{13}C_{13} \\
&= a_{11}M_{11} - a_{12}M_{12} + a_{13}M_{13} \\
&= a_{11}\begin{vmatrix} a_{22} & a_{23} \\ a_{32} & a_{33} \end{vmatrix} - a_{12}\begin{vmatrix} a_{21} & a_{23} \\ a_{31} & a_{33} \end{vmatrix} + a_{13}\begin{vmatrix} a_{21} & a_{22} \\ a_{31} & a_{32} \end{vmatrix} \\
&= a_{11}(a_{22}a_{33} - a_{32}a_{23}) - a_{12}(a_{21}a_{33} - a_{31}a_{23}) + a_{13}(a_{21}a_{32} - a_{31}a_{22}) \\
&= a_{11}a_{22}a_{33} - a_{11}a_{32}a_{23} - a_{12}a_{21}a_{33} + a_{12}a_{31}a_{23} + a_{13}a_{21}a_{32} - a_{13}a_{31}a_{22}
\end{aligned}
$$

If we had expanded along the first column, then
$$
\begin{aligned}
|\mathbf{A}| &= a_{11}C_{11} + a_{21}C_{21} + a_{31}C_{31} \\
&= a_{11}M_{11} - a_{21}M_{21} + a_{31}M_{31} \\
&= a_{11}\begin{vmatrix} a_{22} & a_{23} \\ a_{32} & a_{33} \end{vmatrix} - a_{21}\begin{vmatrix} a_{12} & a_{13} \\ a_{32} & a_{33} \end{vmatrix} + a_{31}\begin{vmatrix} a_{12} & a_{13} \\ a_{22} & a_{23} \end{vmatrix} \\
&= a_{11}(a_{22}a_{33} - a_{32}a_{23}) - a_{21}(a_{12}a_{33} - a_{32}a_{13}) + a_{31}(a_{12}a_{23} - a_{22}a_{13}) \\
&= a_{11}a_{22}a_{33} - a_{11}a_{32}a_{23} - a_{21}a_{12}a_{33} + a_{21}a_{32}a_{13} + a_{31}a_{12}a_{23} - a_{31}a_{22}a_{13}
\end{aligned}
$$

which is same as before.

***Example* 9 .** Find the value of $\begin{vmatrix} 1 & -3 & 4 \\ 0 & 2 & 5 \\ -2 & 6 & 3 \end{vmatrix}$.

***Solution* .** Choose a given row or column. Let us arbitrarily select the first row. Then
$$
\begin{vmatrix} 1 & -3 & 4 \\ 0 & 2 & 5 \\ -2 & 6 & 3 \end{vmatrix} = (1)\begin{vmatrix} 2 & 5 \\ 6 & 3 \end{vmatrix} + (-3)\,(-1)\begin{vmatrix} 0 & 5 \\ -2 & 3 \end{vmatrix} + 4\begin{vmatrix} 0 & 2 \\ -2 & 6 \end{vmatrix}
$$
$$
= 1\,(6-30) + 3\,(0+10) + 4\,(0+4) = -24 + 30 + 16 = 22.
$$

If we had expanded along the first column, then
$$
\begin{vmatrix} 1 & -3 & 4 \\ 0 & 2 & 5 \\ -2 & 6 & 3 \end{vmatrix} = (1)\begin{vmatrix} 2 & 5 \\ 6 & 3 \end{vmatrix} + 0 + (-2)\begin{vmatrix} -3 & 4 \\ 2 & 5 \end{vmatrix}
$$
$$
= 1\,(6-30) - 2\,(-15-8) = -24 + 46 = 22, \text{ as before.}
$$

***Example* 10 .** Find the value of $|\mathbf{A}| = \begin{vmatrix} 1 & 2 & 0 & -1 \\ 3 & -1 & 4 & 1 \\ -2 & 0 & -3 & 3 \\ 4 & 3 & 1 & 2 \end{vmatrix}$.

Solution **.** Expanding along first row, we have

$$\begin{aligned}|\mathbf{A}| &= a_{11}C_{11} + a_{12}C_{12} + a_{13}C_{13} + a_{14}C_{14}\\ &= a_{11}M_{11} - a_{12}M_{12} + a_{13}M_{13} - a_{14}M_{14}\\ &= (1)\begin{vmatrix} -1 & 4 & 1\\ 0 & -3 & 3\\ 3 & 1 & 2\end{vmatrix} - 2\begin{vmatrix} 3 & 4 & 1\\ -2 & -3 & 3\\ 4 & 1 & 2\end{vmatrix} + 0 + (1)\begin{vmatrix} 3 & -1 & 4\\ -2 & 0 & -3\\ 4 & 3 & 1\end{vmatrix}\\ &= 54 - 94 + 13 = -27.\end{aligned}$$

Remarks **.** *The expansion of a determinant of order n (n > 2) can be algorithmically given as follows :*

Assign a_{11} *the + sign and going along the first row, assign alternate signs, i.e. assign* a_{12} *the – sign,* a_{13} *the + sign and so on. Make this assignment through the columns as well, i.e., assign* a_{21} *the – sign,* a_{31} *the + sign and so on. To expand the determinant along any row or column, write the elements of the row or column with assigned signs in order and juxtapose each of them with corresponding minors, i.e., determinants of order (n – 1) obtained by deleting the row and column in which the element lies.* For example, *consider the determinant of order* 3:

$$\begin{vmatrix} a_{11} & a_{12} & a_{13}\\ a_{21} & a_{22} & a_{23}\\ a_{31} & a_{32} & a_{33}\end{vmatrix}$$

The assignment of sign is as follows :

$$\begin{vmatrix} + & - & +\\ - & + & -\\ + & - & +\end{vmatrix}$$

Expanding along the second row, for instance, the determinant is :

$$-a_{21}\begin{vmatrix} a_{12} & a_{13}\\ a_{32} & a_{33}\end{vmatrix} + a_{22}\begin{vmatrix} a_{11} & a_{13}\\ a_{31} & a_{33}\end{vmatrix} - a_{23}\begin{vmatrix} a_{11} & a_{12}\\ a_{31} & a_{32}\end{vmatrix}$$

1.5 TRANSPOSE OF A MATRIX

We now return our discussion to matrices.

Definition **.** *Let* **A** *be an m × n matrix. The* **transpose** *of* **A***, denoted by* **A'** *or* $\mathbf{A}^T$*, is the n × m matrix obtained from* **A** *by interchanging the rows and columns of* **A***. Thus the first row of* **A** *is the first column of* **A'** *the second row of* **A** *is the second column of* **A'** *and so on.*

Example **11 .** If $\mathbf{A} = \begin{bmatrix} 2 & -4 & 6\\ 3 & 1 & 4\end{bmatrix}$, find **A'**.

***Solution*.** Column 1 of **A** becomes Row 1 of **A'**. Column 2 becomes Row 2 and Column 3 becomes Row 3. Thus

$$A' = \begin{bmatrix} 2 & 3 \\ -4 & 1 \\ 6 & 4 \end{bmatrix}$$

1.6 ADJOINT OF A SQUARE MATRIX

Definition*.** *Let* **A** *= (a_{ij}) be a square matrix of order n and let C_{ij} be the cofactor of a_{ij} in the determinant* $|\mathbf{A}|$. *Then the* ***adjoint *of A, denoted by adj A, is defined as the transpose of the cofactor matrix (C_{ij}).*

Thus the adjoint of a square matrix **A** is obtained on replacing each (i, j)th element of **A** by the cofactor of the (j, i)th element in $|\mathbf{A}|$. In particular, if $\mathbf{A} = \begin{bmatrix} a_{11} & a_{12} & a_{13} \\ a_{21} & a_{22} & a_{23} \\ a_{31} & a_{32} & a_{33} \end{bmatrix}$ is a square matrix of order 3, then the adjoint of **A** is given by

$$\text{adj } \mathbf{A} = \begin{bmatrix} C_{11} & C_{21} & C_{31} \\ C_{12} & C_{22} & C_{32} \\ C_{13} & C_{23} & C_{33} \end{bmatrix}.$$

***Example* 12.** Find the adjoint of matrix $\mathbf{A} = \begin{bmatrix} 1 & 2 & 3 \\ -1 & 0 & 1 \\ 4 & 3 & 2 \end{bmatrix}$.

***Solution*.** We have

$$\begin{array}{lll} C_{11} = -3 & C_{12} = 6 & C_{13} = -3 \\ C_{21} = 5 & C_{22} = -10 & C_{23} = 5 \\ C_{31} = 2 & C_{32} = -4 & C_{33} = 2 \end{array}$$

Thus,

$$\text{adj } \mathbf{A} = \begin{bmatrix} C_{11} & C_{21} & C_{31} \\ C_{12} & C_{22} & C_{32} \\ C_{13} & C_{23} & C_{33} \end{bmatrix} = \begin{bmatrix} -3 & 5 & 2 \\ 6 & -10 & -4 \\ -3 & 5 & 2 \end{bmatrix}.$$

***Remark*.** *If A is a square matrix of order 2, then its adjoint can be obtained directly by simply interchanging the diagonal elements and changing the signs of non-diagonal elements as is illustrated below:*

***Example* 13.** If $\mathbf{A} = \begin{bmatrix} a_{11} & a_{12} \\ a_{21} & a_{22} \end{bmatrix}$, show that adj $\mathbf{A} = \begin{bmatrix} a_{22} & -a_{12} \\ -a_{21} & a_{11} \end{bmatrix}$.

***Solution*.** We have

$$C_{11} = a_{22} \qquad C_{12} = -a_{21}$$
$$C_{21} = -a_{12} \qquad C_{22} = a_{11}$$

$$\therefore \qquad \text{adj } \mathbf{A} = \begin{bmatrix} C_{11} & C_{21} \\ C_{12} & C_{22} \end{bmatrix} = \begin{bmatrix} a_{22} & -a_{12} \\ -a_{21} & a_{11} \end{bmatrix}.$$

1.7 INVERSE OF A SQUARE MATRIX

***Definition*.** *Let **A** be a square matrix of order n. Then a square matrix **B** of order n, if it exists, is called an **inverse** of A if **AB** = **BA** = I_n.*

A matrix **A** having an inverse is called an **invertible** matrix. It may easily be seen that if a matrix **A** is invertible, its inverse is unique. The inverse of an invertible matrix **A** is denoted by $\mathbf{A}^{-1}$.

Does every square matrix possess an inverse? To answer this let us consider the matrix

$$\mathbf{A} = \begin{bmatrix} 0 & 0 \\ 0 & 0 \end{bmatrix}$$

If **B** is any square matrix of order 2, we find that **AB = BA = O.**

We thus see that there cannot be any matrix **B** for which **AB** and **BA** both are equal to $\mathbf{I}_2$. Therefore, **A** is not invertible. Hence, we conclude that a square matrix may fail to have an inverse.

However, if **A** is a square matrix such that $|\mathbf{A}| \neq 0$, then **A** is invertible and $\mathbf{A}^{-1} = \dfrac{1}{|\mathbf{A}|}$ adj **A.**

***Singular and Non-Singular Matrices*.** *A square matrix **A** is said to be **singular** if* $|\mathbf{A}| = 0$, *and it is called **non-singular** if* $|\mathbf{A}| \neq 0$. For example, if

$$\mathbf{A} = \begin{bmatrix} 3 & 1 & 2 \\ 2 & -3 & -1 \\ 1 & 2 & 1 \end{bmatrix},$$

then $|\mathbf{A}| = 3(-3+2) - 1(2+1) + 2(4+3) = 8$. Since $|\mathbf{A}| \neq 0$, **A** is non-singular. Further, if

$$\mathbf{B} = \begin{bmatrix} 2 & 0 & 0 \\ 3 & -1 & 4 \\ 5 & -2 & 8 \end{bmatrix},$$

then $|\mathbf{B}| = 2(-8+8) + 0 + 0 = 0$. Since $|\mathbf{B}| = 0$, **B** is singular.

***Example* 14.** Find the inverse of the matrix $\mathbf{A} = \begin{bmatrix} 4 & -2 & 1 \\ 7 & 3 & 3 \\ 2 & 0 & 1 \end{bmatrix}$.

***Solution* .** We have

$|\mathbf{A}| = (4)(3) - (-2)(1) + 1(-6) = 8 \neq 0$. Thus $\mathbf{A}^{-1}$ exists and is given by $\mathbf{A}^{-1} = \dfrac{1}{|\mathbf{A}|}$ adj **A.**

To find adj **A,** let $\mathbf{C}_{ij}$ denote the cofactor of a_{ij}, the element in the ith row and jth column of $|\mathbf{A}|$.

Then

$$\begin{array}{lll} C_{11} = 3 & C_{12} = -1 & C_{13} = -6 \\ C_{21} = 2 & C_{22} = 2 & C_{23} = -4 \\ C_{31} = -9 & C_{32} = -5 & C_{33} = 26 \end{array}$$

$$\therefore \quad \text{adj}\,\mathbf{A} = \begin{bmatrix} 3 & 2 & -9 \\ -1 & 2 & -5 \\ -6 & -4 & 26 \end{bmatrix} \Rightarrow \mathbf{A}^{-1} = \frac{1}{|\mathbf{A}|}\text{adj}\,\mathbf{A} = \frac{1}{8}\begin{bmatrix} 3 & 2 & -9 \\ -1 & 2 & -5 \\ -6 & -4 & 26 \end{bmatrix}.$$

1.8 SYSTEM OF LINEAR EQUATIONS

Consider the system of n linear equations in n unknowns $x_1, x_2, ..., x_n$:

$$\left.\begin{array}{l} a_{11}x_1 + a_{12}x_2 + \ldots + a_{1n}x_n = b_1 \\ a_{21}x_1 + a_{22}x_2 + \ldots + a_{2n}x_n = b_2 \\ \quad\vdots \\ a_{n1}x_1 + a_{n2}x_2 + \ldots + a_{nn}x_n = b_n \end{array}\right\} \qquad \ldots\ldots\ldots(1)$$

Solution* .** *By a* ***solution *of the system (1), we mean a set of values of unknowns $x_1, x_2, ..., x_n$ which simultaneously satisfy the n equations given by (1). If $x_1, x_2, ..., x_n$ are all zero, then the solution is said to be a* ***trivial*** *solution, otherwise the solution is said to be* ***non-trivial.***

Consistent System* .** *A system of linear equations is said to be* ***consistent *if it possesses at least one solution.*

Inconsistent System* .** *A system of linear equations which is not consistent is said to be* ***inconsistent.

Remark. *A system of linear equations can have (i) no solution, (ii) a unique solution, or (iii) an infinitely many solutions.*

Note. In this chapter we shall discuss only those systems of linear equations which possess unique solutions.

Matrix Representation of a System of Equations

Setting

$$\mathbf{A} = \begin{bmatrix} a_{11} & a_{12} & \cdots & a_{1n} \\ a_{21} & a_{22} & \cdots & a_{2n} \\ \cdot & \cdot & & \cdot \\ \cdot & \cdot & & \cdot \\ \cdot & \cdot & & \cdot \\ a_{n1} & a_{n2} & \cdots & a_{nn} \end{bmatrix}, \mathbf{X} = \begin{bmatrix} x_1 \\ x_2 \\ \cdot \\ \cdot \\ \cdot \\ x_n \end{bmatrix} \text{ and } \mathbf{B} = \begin{bmatrix} b_1 \\ b_2 \\ \cdot \\ \cdot \\ \cdot \\ b_n \end{bmatrix}$$

The given system of equations (1) may be written as

$$\mathbf{AX} = \mathbf{B} \qquad \text{........(2)}$$

which is known as the *matrix form* or the *matrix representation* of the given system of equations (1). The matrix **A** is called the *coefficient matrix* of the system.

We shall discuss the following methods of solving a system of linear equations :

(*i*) *Matrix Inverse Method, and*

(*ii*) *Cramer's Rule (or Determinant Method).*

Matrix Inverse Method

We shall first discuss the method of solving the system of n linear equations in n unknowns using inverse of the coefficient matrix.

Consider the system of n linear equations in n unknowns :

$$\left.\begin{aligned} a_{11}x_1 + a_{12}x_2 + \ldots + a_{1n}x_n &= b_1 \\ a_{21}x_1 + a_{22}x_2 + \ldots + a_{2n}x_n &= b_2 \\ \cdot \quad \cdot \quad & \quad \cdot \\ \cdot \quad \cdot \quad & \quad \cdot \\ \cdot \quad \cdot \quad & \quad \cdot \\ a_{n1}x_1 + a_{n2}x_2 + \ldots + a_{nn}x_n &= b_n \end{aligned}\right\} \qquad \text{........(1)}$$

This system can be represented by the single matrix equation

$$\mathbf{AX} = \mathbf{B} \qquad \text{........(2)}$$

where **A** is the coefficient matrix of the system. If **A** is non-singular, *i.e.*, $|\mathbf{A}| \neq 0$, so that $\mathbf{A}^{-1}$ exists, then from equation (2), we obtain

$$\mathbf{X} = \mathbf{A}^{-1}\mathbf{B} \qquad \text{......(3)}$$

which gives the required solution of the given system. Since $\mathbf{A}^{-1}$ is unique, the solution of the system of equations (1) as given by (3) is unique.

We shall now illustrate the matrix inversion method with the help of following example :

***Example* 15.** Using matrix inversion method, solve the following system of equations for x, y and z.

$$x + y + z = 5; \quad 2x + y - z = 2; \quad 2x - y + z = 2$$

Solution . The given system can be expressed in the matrix form as

$$\mathbf{AX} = \mathbf{B},$$

where $\mathbf{A} = \begin{bmatrix} 1 & 1 & 1 \\ 2 & 1 & -1 \\ 2 & -1 & 1 \end{bmatrix}$, $\mathbf{X} = \begin{bmatrix} x \\ y \\ z \end{bmatrix}$ and $\mathbf{B} = \begin{bmatrix} 5 \\ 2 \\ 2 \end{bmatrix}$

Now $|\mathbf{A}| = 1\,(1 - 1) - 1\,(2 + 2) + 1\,(-2 - 2) = 0 - 4 - 4 = -8 \neq 0.$

Since $|\mathbf{A}| \neq 0$, $\mathbf{A}^{-1}$ exists and the system has a unique solution given by $\mathbf{X} = \mathbf{A}^{-1}\mathbf{B}$. It can be easily verified that

$$\mathbf{A}^{-1} = -\frac{1}{8}\begin{bmatrix} 0 & -2 & -2 \\ -4 & -1 & 3 \\ -4 & 3 & -1 \end{bmatrix}$$

Thus, $$\mathbf{X} = \mathbf{A}^{-1}\mathbf{B} = -\frac{1}{8}\begin{bmatrix} 0 & -2 & -2 \\ -4 & -1 & 3 \\ -4 & 3 & -1 \end{bmatrix}\begin{bmatrix} 5 \\ 2 \\ 2 \end{bmatrix} = -\frac{1}{8}\begin{bmatrix} -8 \\ -16 \\ -16 \end{bmatrix} = \begin{bmatrix} 1 \\ 2 \\ 2 \end{bmatrix}.$$

Hence $x = 1$, $y = 2$ and $z = 2$.

Cramer's Rule (Determinant Method)

Determinants can also be applied for solving a system of n linear equations in n variables. The method for solving the system of equations by making use of determinant was given by Swiss mathematician Gabriel Cramer and is referred to as *Cramer's rule.*

Let the system of n linear equations in n unknowns $x_1, x_2 \ldots, x_n$ be given by

$$\left.\begin{aligned} a_{11}x_1 + a_{12}x_2 + \ldots + a_{1n}x_n &= b_1 \\ a_{21}x_1 + a_{22}x_2 + \ldots + a_{2n}x_n &= b_2 \\ \vdots \qquad\quad \vdots \qquad\quad & \;\;\vdots \\ a_{n1}x_1 + a_{n2}x_2 + \ldots + a_{nn}x_n &= b_n \end{aligned}\right\}$$

Let

$$D = \begin{vmatrix} a_{11} & a_{12} & \cdots & a_{1n} \\ a_{21} & a_{22} & \cdots & a_{2n} \\ \cdot & \cdot & & \cdot \\ \cdot & \cdot & & \cdot \\ \cdot & \cdot & & \cdot \\ a_{n1} & a_{n2} & \cdots & a_{nn} \end{vmatrix}$$

That is, D is the determinant of the coefficient matrix. Then, if $D \neq 0$, the system has a unique solution given by

$$x_1 = \frac{D_1}{D}, \qquad x_2 = \frac{D_2}{D}, \ldots, \quad x_n = \frac{D_n}{D},$$

where D_i is the determinant obtained from D by replacing the ith column of D by the elements of matrix **B.**

***Example* 16 .** Solve the following system of equations by Cramer's rule :

$$\begin{aligned} x - 4y - z &= 11 \\ 2x - 5y + 2z &= 39 \\ -3x + 2y + z &= 1 \end{aligned}$$

***Solution* .** We have $D = \begin{vmatrix} 1 & -4 & -1 \\ 2 & -5 & 2 \\ -3 & 2 & 1 \end{vmatrix} = 1\,(-5-4) + 4\,(2+6) - 1\,(4-15) = 34.$

Since $D \neq 0$, therefore by Cramer's rule, the solution is given by

$$x = \frac{D_1}{D}, \qquad y = \frac{D_2}{D}, \qquad z = \frac{D_3}{D},$$

where $D_1 = \begin{vmatrix} 11 & -4 & -1 \\ 39 & -5 & 2 \\ 1 & 2 & 1 \end{vmatrix} = 11\,(-5-4) + 4\,(39-2) - 1\,(78+5) = -34$

$$D_2 = \begin{vmatrix} 1 & 11 & -1 \\ 2 & 39 & 2 \\ -3 & 1 & 1 \end{vmatrix} = 1\,(39-2) - 11\,(2+6) - 1\,(2+117) = -170$$

$$D_3 = \begin{vmatrix} 1 & -4 & 11 \\ 2 & -5 & 39 \\ -3 & 2 & 1 \end{vmatrix} = 1\,(-5-78) + 4\,(2+117) + 11\,(4-15) = 272$$

$$\therefore \quad x = \frac{D_1}{D} = \frac{-34}{34} = -1; \quad y = \frac{D_2}{D} = \frac{-170}{34} = -5; \quad z = \frac{D_3}{D} = \frac{272}{34} = 8$$

Hence the solution is : $x = -1$, $y = -5$ and $z = 8$.

EXERCISE 1.1

1. Solve the following systems of equations using matrix inversion method :

(*i*) $x+2y-3z=6$; $3x+2y-2z=3$; $2x-y+z=2$

(*ii*) $3x+2y-z=4$; $x+y+z=6$; $2x-y+z=3$

(*iii*) $x+y-z=1$; $3x+y-2z=3$; $x-y-z=-1$

(*iv*) $2x-y-z=7$; $3x+y-z=7$; $x+y-z=3$

2. Solve the following systems of equations using Cramer's rule :

(*i*) $5x-7y+z=11$; $6x-8y-z=15$; $3x+2y-6z=7$

(*ii*) $x+2y-2z=-7$; $2x-y+z=6$; $x-y-3z=-3$

(*iii*) $6x+y-3z=5$; $x+3y-2z=5$; $2x+y+4z=8$

(*iv*) $2x-3y-4z=29$; $-2x+5y-z=-15$; $3x-y+5z=-11$

ANSWERS

1. (*i*) $x=1$, $y=-5$, $z=-5$ (*ii*) $x=1$, $y=2$, $z=3$

(*iii*) $x=2$, $y=1$, $z=2$ (*iv*) $x=2$, $y=-1$, $z=-2$

2. (*i*) $x=1$, $y=-1$, $z=-1$ (*ii*) $x=1$, $y=-2$, $z=2$

(*iii*) $x=1$, $y=2$, $z=1$ (*iv*) $x=2$, $y=-3$, $z=-4$

CHAPTER

2

APPLICATIONS OF MATRICES TO BUSINESS AND ECONOMICS

OBJECTIVES

After studying the material in this chapter, you should be able to:

- *Understand and appreciate the various methods of representing large quantities of data in matrix form.*
- *Use matrix algebra in solving word problems.*
- *Discuss common methods of solving systems of linear equations that result from applied business and economics problems.*
- *Understand the concept of Leontief input-output model.*

2.1 INTRODUCTION

One of the major methods by which data are tabulated, stored and manipulated makes use of matrices. Matrices have proved their usefulness in the analysis and decision-making processes in business, economics, sociology, chemistry-in fact, in any field that has a quantitative side to it. Another application of matrices deals with the interrelationships that exist between various sectors of an economy and is known as *input-output analysis.*

2.2 MATRIX REPRESENTATION OF DATA

A matrix provides a very convenient and compact notation for representing data in many business and economic situations. This is illustrated with a few examples :

1. Annual production relating to two branches and three types of items may be represented in the following manner :

$$\begin{array}{c} \textit{Branch} \\ \\ I \\ II \end{array} \quad \begin{array}{ccc} \textit{Item A} & \textit{Item B} & \textit{Item C} \\ \\ \end{array}$$
$$\begin{array}{c} I \\ II \end{array} \begin{bmatrix} 2000 & 5000 & 6000 \\ 3000 & 4500 & 7000 \end{bmatrix}$$

2. Vitamin content of two types of foods and three types of vitamins may be represented in the following matrix :

Food	*Vitamin A*	*Vitamin B*	*Vitamin C*
I	5	10	7
II	4	8	9

3. Annual sale of three products in two markets may be represented as follows :

Product	*Market I*	*Market II*
A	5000	6000
B	7000	6000
C	4000	3000

4. Number of staff in the office may be represented by the matrix

2	*Peon*
4	*Clerk*
2	*Typist*
1	*Head Clerk*
1	*Office Superintendent*

5. Processing time of three products in two different production departments may be represented by the matrix

Product	*Department I*	*Department II*
P	3	2
Q	4	5
R	1	2

6. The unit cost of transportation of an item from each of the three factories to each of the four warehouses can be represented by the matrix as follows :

Factory	*Warehouse* W_1	W_2	W_3	W_4
I	15	17	16	13
II	20	16	18	12
III	16	14	15	20

2.3 MATRIX ADDITION AND SUBTRACTION – APPLICATIONS

The following examples illustrate applications of matrix addition and subtraction.

***Example* 1 .** An automobile dealer sells two car models, standard and deluxe. Each is available in one of two colours, white and red. His sales for the months of January and February are given by the matrices:

$$\mathbf{A} = \begin{array}{c} \\ \textit{White} \\ \textit{Red} \end{array} \begin{array}{c} \begin{array}{cc} \textit{Standard} & \textit{Deluxe} \end{array} \\ \begin{bmatrix} 2 & 1 \\ 3 & 4 \end{bmatrix} \end{array}, \qquad \mathbf{B} = \begin{bmatrix} 3 & 1 \\ 2 & 3 \end{bmatrix}$$

Find his total sales for each model and colour for both months.

***Solution* .** The total sales for each model and colour for both months can be obtained by adding the two matrices **A** and **B** :

$$\mathbf{A} + \mathbf{B} = \begin{bmatrix} 2 & 1 \\ 3 & 4 \end{bmatrix} + \begin{bmatrix} 3 & 1 \\ 2 & 3 \end{bmatrix} = \begin{bmatrix} 5 & 2 \\ 5 & 7 \end{bmatrix}$$

***Example* 2 .** The sales figure for two car dealers during January showed that Dealer A sold 5 deluxe, 3 premium and 4 standard cars, while Dealer B sold 7 deluxe, 2 premium and 3 standard cars. Total sales over the 2-month period of January-February revealed that Dealer A sold 8 deluxe, 7 premium and 6 standard cars. In the same 2-month period, Dealer B sold 10 deluxe, 5 premium and 7 standard cars.

Write 2×3 matrices summarizing sales data for January and the 2-month period for each dealer. Hence find the sales in February for each dealer.

***Solution* .** The sales for the month of January can be represented by the matrix

$$\mathbf{A} = \begin{array}{c} \\ \textit{Dealer A} \\ \textit{Dealer B} \end{array} \begin{array}{c} \begin{array}{ccc} \textit{Deluxe} & \textit{Premium} & \textit{Standard} \end{array} \\ \begin{bmatrix} 5 & 3 & 4 \\ 7 & 2 & 3 \end{bmatrix} \end{array}$$

Each row of **A** gives the number of each model sold by a given dealer. The sales for the 2-month period can be represented by the matrix

$$\mathbf{B} = \begin{bmatrix} 8 & 7 & 6 \\ 10 & 5 & 7 \end{bmatrix}$$

The sales for the month of February is equal to the sales for the 2-month period of January-February minus the sales for the month of January. Thus the sales in February is

$$\mathbf{B} - \mathbf{A} = \begin{bmatrix} 8 & 7 & 6 \\ 10 & 5 & 7 \end{bmatrix} - \begin{bmatrix} 5 & 3 & 4 \\ 7 & 2 & 3 \end{bmatrix} = \begin{bmatrix} 3 & 4 & 2 \\ 3 & 3 & 4 \end{bmatrix}.$$

2.4 SCALAR MULTIPLICATION – AN APPLICATION

The following example illustrates an application of scalar multiplication.

***Example* 3 .** In a certain city there are 30 colleges. Each college has 15 peons, 6 clerks, 2 cashiers, 1 typist and 1 section officer. Express the given information as a column matrix.

Using scalar multiplication, find the total number of posts of each kind in all the colleges.

Solution . The number of posts of each kind in a college can be represented by the 5 × 1 column matrix:

$$\mathbf{A} = \begin{bmatrix} 15 \\ 6 \\ 2 \\ 1 \\ 1 \end{bmatrix}$$

Using scalar multiplication, the total number of posts of each kind in all the colleges is

$$30\mathbf{A} = 30 \begin{bmatrix} 15 \\ 6 \\ 2 \\ 1 \\ 1 \end{bmatrix} = \begin{bmatrix} 450 \\ 180 \\ 60 \\ 30 \\ 30 \end{bmatrix}.$$

2.5 MULTIPLICATION OF MATRICES – APPLICATIONS

In the following we give several examples to illustrate applications of multiplication of matrices.

***Example* 4** . Three firms *A*, *B* and *C* supplied 40, 35 and 25 truck loads of stones and 10, 5, 8 truck loads of sand respectively to a contractor. If the costs of stone and sand are ₹ 1200 and ₹ 500 per truck load respectively, find the total amount paid by the contractor to each of these firms, by using matrix method.

Solution . The amount of stone and sand supplied to a contractor by three firms *A*, *B* and *C* can be represented by the matrix

$$\mathbf{X} = \begin{matrix} A \\ B \\ C \end{matrix} \overset{\begin{matrix} Stone & \quad Sand \end{matrix}}{\begin{bmatrix} 40 & 10 \\ 35 & 5 \\ 25 & 8 \end{bmatrix}}$$

The costs (per truck load) of stone and sand can be represented by the column matrix

$$\mathbf{Y} = \begin{matrix} Stone \\ Sand \end{matrix} \begin{bmatrix} 1200 \\ 500 \end{bmatrix}$$

Thus the total amount paid by the contractor to each of these firms is given by the matrix product

$$\mathbf{XY} = \begin{bmatrix} 40 & 10 \\ 35 & 5 \\ 25 & 8 \end{bmatrix} \begin{bmatrix} 1200 \\ 500 \end{bmatrix} = \begin{bmatrix} 53{,}000 \\ 44{,}500 \\ 34{,}000 \end{bmatrix} \begin{matrix} A \\ B \\ C \end{matrix}$$

Hence the amount paid to firm A– ₹ 53,000; B– ₹ 44,500 and C– ₹ 34,000.

***Example* 5.** A transport company uses 3 types of trucks T_1, T_2 and T_3, to transport 3 types of vehicles V_1, V_2 and V_3. The capacity of each truck in terms of 3 types of vehicles is given below :

	V_1	V_2	V_3
T_1	1	3	2
T_2	2	2	3
T_3	3	2	2

Using matrix method, find the number of vehicles of each type which can be transported if company has 10, 20 and 30 trucks of each type respectively.

[*Delhi Univ. B.Com. (H) 1999, 2006, 2019*]

Solution. The capacity of each truck in terms of 3 types of vehicles can be represented by the matrix

$$\mathbf{X} = \begin{matrix} V_1 \\ V_2 \\ V_3 \end{matrix} \overset{\begin{matrix} T_1 & T_2 & T_3 \end{matrix}}{\begin{bmatrix} 1 & 2 & 3 \\ 3 & 2 & 2 \\ 2 & 3 & 2 \end{bmatrix}}$$

The number of trucks of each type used by the company can be represented by the column matrix

$$\mathbf{Y} = \begin{matrix} T_1 \\ T_2 \\ T_3 \end{matrix} \begin{bmatrix} 10 \\ 20 \\ 30 \end{bmatrix}$$

The total number of vehicles of each type which can be transported is given by the matrix product

$$\mathbf{XY} = \begin{bmatrix} 1 & 2 & 3 \\ 3 & 2 & 2 \\ 2 & 3 & 2 \end{bmatrix} \begin{bmatrix} 10 \\ 20 \\ 30 \end{bmatrix} = \begin{bmatrix} 140 \\ 130 \\ 140 \end{bmatrix}$$

Thus the number of 3 types of vehicles V_1, V_2 and V_3 which can be transported are :

$$V_1 = 140, \quad V_2 = 130 \quad \text{and} \quad V_3 = 140$$

***Example* 6 .** There are two Families A and B. There are 4 men, 6 women and 2 children in Family A and 2 men, 2 women, and 4 children in Family B. The recommended daily requirement for calories is : Man : 2400, Woman : 1900, child : 1800 and for proteins is: Man : 55 gram, Woman : 45 gram and Child 33 gram.

Represent the above information by matrices. Using matrix multiplication, calculate the total requirements of calories and proteins for each of the two families.

***Solution* .** The members of the two families can be represented by the 2 × 3 matrix:

$$\mathbf{F} = \begin{array}{c} \\ A \\ B \end{array} \begin{array}{c} \begin{matrix} M & W & C \end{matrix} \\ \begin{bmatrix} 4 & 6 & 2 \\ 2 & 2 & 4 \end{bmatrix} \end{array}$$

and the recommended daily requirements of calories and proteins for each member can be represented by the 3 × 2 matrix

$$\mathbf{R} = \begin{array}{c} \\ M \\ W \\ C \end{array} \begin{array}{c} \begin{matrix} \text{Calories} & \text{Proteins} \end{matrix} \\ \begin{bmatrix} 2400 & 55 \\ 1900 & 45 \\ 1800 & 33 \end{bmatrix} \end{array}$$

The total requirements of calories and proteins for each of the two families is given by the matrix multiplication :

$$\mathbf{FR} = \begin{bmatrix} 4 & 6 & 2 \\ 2 & 2 & 4 \end{bmatrix} \begin{bmatrix} 2400 & 55 \\ 1900 & 45 \\ 1800 & 33 \end{bmatrix} = \begin{array}{c} A \\ B \end{array} \begin{bmatrix} 24600 & 556 \\ 15800 & 332 \end{bmatrix}$$

Hence Family A requires 24,600 calories and 556 gm proteins and Family B requires 15,800 calories and 332 gm proteins.

***Example* 7 .** The following matrix gives the number of units of three products (P, Q and R) that can be processed per hour on three machines (A, B and C)

$$\begin{array}{c} \\ P \\ Q \\ R \end{array} \begin{array}{c} \begin{matrix} A & B & C \end{matrix} \\ \begin{bmatrix} 10 & 12 & 15 \\ 13 & 11 & 20 \\ 16 & 18 & 14 \end{bmatrix} \end{array}$$

Determine by using matrix algebra, how many units of each product can be produced, if the hours available on machines A, B and C are 54, 46 and 48 respectively.

[*Delhi Univ. B.Com. (H) 1992*]

***Solution* .** The number of hours available on machines A, B and C can be represented by 3 × 1 column matrix :

$$\begin{matrix} A \\ B \\ C \end{matrix}\begin{bmatrix} 54 \\ 46 \\ 48 \end{bmatrix}$$

Using matrix multiplication, the number of units of each product that can be produced is given by

$$\begin{bmatrix} 10 & 12 & 15 \\ 13 & 11 & 20 \\ 16 & 18 & 14 \end{bmatrix}\begin{bmatrix} 54 \\ 46 \\ 48 \end{bmatrix} = \begin{bmatrix} 1812 \\ 2168 \\ 2364 \end{bmatrix}$$

Thus Product P = 1812 units, Product Q = 2168 units and Product R = 2364 units.

***Example* 8.** Mr. *A* went to a market to purchase 3 kg of sugar, 10 kg of wheat and 1 kg of salt. In a shop near to Mr. *A* residence, these commodities are priced at ₹ 20, ₹ 10 and ₹ 8 per kg whereas in the local market these commodities are priced at ₹ 15, ₹ 8 and ₹ 6 per kg respectively. If cost of travelling to local market is ₹ 25, find the net savings of Mr. *A* using matrix multiplication. **[*Delhi Univ. B.Com. (H) 2011*]**

Solution. The quantity of sugar, wheat and salt purchased by Mr. *A* can be represented by the matrix

$$\mathbf{Q} = A\begin{matrix} \textit{Sugar} & \textit{Wheat} & \textit{Salt} \\ [\ 3 & 10 & 1\ ] \end{matrix}$$

The prices of these commodities in the nearby shop and the local market can be represented by the matrix

$$\mathbf{P} = \begin{matrix} \textit{Sugar} \\ \textit{Wheat} \\ \textit{Salt} \end{matrix}\begin{bmatrix} \textit{Shop} & \textit{Local Market} \\ 20 & 15 \\ 10 & 8 \\ 8 & 6 \end{bmatrix}$$

The total cost of purchasing at two different places is given by

$$\mathbf{QP} = [3 \ \ 10 \ \ 1]\begin{bmatrix} 20 & 15 \\ 10 & 8 \\ 8 & 6 \end{bmatrix} = [168 \ \ 131]$$

$\therefore$ Cost of purchasing from a nearby shop = ₹ 168

and cost of purchasing from the local market = ₹ 131

However, cost of travelling to local market = ₹ 25

$\therefore$ Net savings of Mr. A = 168 – 131 – 25 = ₹12 .

***Example* 9.** A student has 4 places where he can eat lunch. The college canteen charges ₹ 8 for a Dosa, ₹ 3 for French fries and ₹ 5 for a soft drink. The campus coffee house charges ₹ 10 for a Dosa, ₹ 2 for French Fries and ₹ 4.50 for a soft drink. A fast food place charges ₹ 8 for a Dosa, ₹ 4 for French fries and ₹ 5 for a soft drink. A nearby restaurant serves Dosa for ₹ 12, French fries for ₹ 5 and a free soft drink for any order.

Express the above information in a 4 × 3 matrix. The student wishes to buy 1 Dosa, 2 orders of French fries and a soft drink. Find, using matrix algebra, the cost of lunch at each place. Assuming that the student has no preference for any of the places, decide where he should eat to spend the least amount of money.

Solution . The information in this problem may be expressed in a 4 × 3 matrix using rows to represent the eating places and columns to represent the food purchases :

$$\begin{array}{l} \\ \textit{College canteen} \\ \textit{Coffee house} \\ \textit{Fast food place} \\ \textit{Restaurant} \end{array} \begin{array}{c} \begin{array}{ccc} \textit{Dosa} & \textit{French fries} & \textit{Soft drink} \end{array} \\ \begin{bmatrix} 8 & 3 & 5 \\ 10 & 2 & 4.50 \\ 8 & 4 & 5 \\ 12 & 5 & 0 \end{bmatrix} \end{array}$$

His order can be represented by the column matrix

$$\begin{array}{l} \textit{Dosa} \\ \textit{French fries} \\ \textit{Soft drink} \end{array} \begin{bmatrix} 1 \\ 2 \\ 1 \end{bmatrix}$$

Using matrix multiplication, the cost of lunch at each place can be determined as follows :

$$\begin{bmatrix} 8 & 3 & 5 \\ 10 & 2 & 4.50 \\ 8 & 4 & 5 \\ 12 & 5 & 0 \end{bmatrix} \begin{bmatrix} 1 \\ 2 \\ 1 \end{bmatrix} = \begin{bmatrix} ₹\,19 \\ ₹\,18.50 \\ ₹\,21 \\ ₹\,22 \end{bmatrix} \begin{array}{l} \textit{College canteen} \\ \textit{Coffee house} \\ \textit{Fast food place} \\ \textit{Restaurant} \end{array}$$

Hence the student would spend the least amount of money by eating at campus coffee house.

***Example* 10 .** A firm produces three products A, B and C which it sells in two markets. Annual sales in units are given as follows :

Market	*Units Sold*		
	A	*B*	*C*
I	8000	4000	16,000
II	7000	18,000	9000

If the prices per unit of A, B and C are ₹ 2.50, ₹ 1.25 and ₹ 1.50 respectively and the costs per unit are ₹ 1.70, ₹ 1.20 and ₹ 0.80 respectively, find the total profit in each market by using matrix algebra. **[*Delhi Univ. B.Com. (H) 1990*]**

Solution . The profits per unit of A, B and C can be obtained as follows :

$$\begin{bmatrix} 2.50 \\ 1.25 \\ 1.50 \end{bmatrix} - \begin{bmatrix} 1.70 \\ 1.20 \\ 0.80 \end{bmatrix} = \begin{bmatrix} 0.80 \\ 0.05 \\ 0.70 \end{bmatrix}$$

The total profit in each market is given by the matrix product

$$\begin{bmatrix} 8000 & 4000 & 16,000 \\ 7000 & 18,000 & 9000 \end{bmatrix} \begin{bmatrix} 0.80 \\ 0.05 \\ 0.70 \end{bmatrix} = \begin{bmatrix} 17,800 \\ 12,800 \end{bmatrix}$$

Thus the profit in Market I is ₹ 17,800 and the profit in Market II is ₹ 12,800.

***Example* 11 .** A manufacturer produces three products X, Y and Z which he sells in three markets I, II and III. Monthly sales and sale prices are given as follows :

Market	*Items Sold*			*Sale Price (in ₹)*		
	X	*Y*	*Z*	*X*	*Y*	*Z*
I	1000	1500	2000	2.00	3.00	4.00
II	2000	2800	3000	2.50	2.80	3.70
III	3000	3500	4000	2.30	3.10	4.20

If the costs per item for three products X, Y and Z are ₹ 1.80, ₹ 2.50 and ₹ 3.40 respectively, find by matrix algebra his total profit. **[*Delhi Univ. B.Com. (H) 2006, 2017*]**

Solution . The total profit is given by

$$[1000 \quad 1500 \quad 2000] \begin{bmatrix} 2.00-1.80 \\ 3.00-2.50 \\ 4.00-3.40 \end{bmatrix} + [2000 \quad 2800 \quad 3000] \begin{bmatrix} 2.50-1.80 \\ 2.80-2.50 \\ 3.70-3.40 \end{bmatrix} + [3000 \quad 3500 \quad 4000] \begin{bmatrix} 2.30-1.80 \\ 3.10-2.50 \\ 4.20-3.40 \end{bmatrix}$$

$$= [1000 \quad 1500 \quad 2000] \begin{bmatrix} 0.20 \\ 0.50 \\ 0.60 \end{bmatrix} + [2000 \quad 2800 \quad 3000] \begin{bmatrix} 0.70 \\ 0.30 \\ 0.30 \end{bmatrix} + [3000 \quad 3500 \quad 4000] \begin{bmatrix} 0.50 \\ 0.60 \\ 0.80 \end{bmatrix}$$

$$= [2150] + [3140] + [6800] = [12,090].$$

Thus the total profit of the manufacturer is ₹ 12,090.

***Example* 12** .In a certain city there are 50 colleges and 400 schools. Each school and college has 18 peons, 5 clerks and 1 cashier. Each college, in addition, has 1 section officer and 1 librarian. The monthly salary of each of them is as follows :

Peon– ₹ 300, Clerk– ₹ 500, Cashier– ₹ 600, Section Officer– ₹ 700 and Librarian– ₹ 900.

Using matrix notation, find (*i*) total number of posts of each kind in schools and colleges taken together, (*ii*) the total monthly salary bill of all the schools and colleges taken together.

[*Delhi Univ. B.Com. (H) 1983*]

***Solution* .** The number of schools and colleges can be represented by the matrix

$$\mathbf{A} = \begin{matrix} \textit{School} & \textit{College} \\ [\ 400 & 50\ ] \end{matrix}$$

The number of posts of each kind in each school and college can be represented by the matrix

$$\mathbf{B} = \begin{matrix} & \textit{Peon} & \textit{Clerk} & \textit{Cashier} & \textit{S.officer} & \textit{Librarian} \\ \textit{School} & 18 & 5 & 1 & 0 & 0 \\ \textit{College} & 18 & 5 & 1 & 1 & 1 \end{matrix}$$

(*i*) The total number of posts of each kind in schools and colleges taken together can be obtained by the matrix product

$$\begin{matrix} \textit{School} & \textit{College} \\ [\ 400 & 50\ ] \end{matrix} \begin{matrix} \textit{Peon} & \textit{Clerk} & \textit{Cashier} & \textit{S.officer} & \textit{Librarian} & \\ 18 & 5 & 1 & 0 & 0 & \textit{School} \\ 18 & 5 & 1 & 1 & 1 & \textit{College} \end{matrix}$$

$$= \begin{matrix} \textit{Peon} & \textit{Clerk} & \textit{Cashier} & \textit{S.officer} & \textit{Librarian} \\ [\ 8100 & 2250 & 450 & 50 & 50\ ] \end{matrix}$$

Thus there are 8100 peons, 2250 clerks, 450 cashiers, 50 section officers and 50 librarians in all schools and colleges taken together.

(*ii*) The total monthly salary bill of all the schools and colleges taken together is given by

$$[8100 \quad 2250 \quad 450 \quad 50 \quad 50] \begin{bmatrix} 300 \\ 500 \\ 600 \\ 700 \\ 900 \end{bmatrix} = [39{,}05{,}000]$$

Hence total monthly salary bill = ₹ 39,05,000.

***Example* 13** . A firm produces three products P_1, P_2 and P_3 requiring the mix-up of three materials M_1, M_2 and M_3. The per unit requirement of each product for each material is as follows :

$$\mathbf{A} = \begin{array}{c} \\ P_1 \\ P_2 \\ P_3 \end{array} \begin{array}{c} \begin{array}{ccc} M_1 & M_2 & M_3 \end{array} \\ \begin{bmatrix} 2 & 3 & 1 \\ 4 & 2 & 5 \\ 2 & 4 & 2 \end{bmatrix} \end{array}$$

Using matrix notations, find

(*i*) The total requirement of each material if the firm produces 100 units of each product.

(*ii*) The per unit cost of production of each product if the per unit costs of materials M_1, M_2 and M_3 are ₹ 5, ₹ 10 and ₹ 5 respectively, and

(*iii*) The total cost of production if the firm produces 200 units of each product.

[*Delhi Univ. B.Com. (H) 1998*]

Solution **.** (*i*) The total requirement of each material if the firm produces 100 units of each product is given by the matrix product

$$\begin{array}{c} \\ M_1 \\ M_2 \\ M_3 \end{array} \begin{array}{c} \begin{array}{ccc} P_1 & P_2 & P_3 \end{array} \\ \begin{bmatrix} 2 & 4 & 2 \\ 3 & 2 & 4 \\ 1 & 5 & 2 \end{bmatrix} \end{array} \begin{bmatrix} 100 \\ 100 \\ 100 \end{bmatrix} \begin{array}{c} P_1 \\ P_2 \\ P_3 \end{array} = \begin{array}{c} M_1 \\ M_2 \\ M_3 \end{array} \begin{bmatrix} 800 \\ 900 \\ 800 \end{bmatrix}$$

Thus the firm requires 800 units of material M_1 , 900 units of material M_2 and 800 units of material M_3 to produce 100 units of each product.

(*ii*) The per unit cost of production of each product is given by the matrix product

$$\begin{array}{c} \\ P_1 \\ P_2 \\ P_3 \end{array} \begin{array}{c} \begin{array}{ccc} M_1 & M_2 & M_3 \end{array} \\ \begin{bmatrix} 2 & 3 & 1 \\ 4 & 2 & 5 \\ 2 & 4 & 2 \end{bmatrix} \end{array} \begin{bmatrix} 5 \\ 10 \\ 5 \end{bmatrix} \begin{array}{c} M_1 \\ M_2 \\ M_3 \end{array} = \begin{array}{c} P_1 \\ P_2 \\ P_3 \end{array} \begin{bmatrix} 45 \\ 65 \\ 60 \end{bmatrix}$$

Thus the per unit costs of three products P_1, P_2 and P_3 are ₹ 45, ₹ 65 and ₹ 60 respectively.

(*iii*) The total cost of production if the firm produces 200 units of each product is given by

$$\begin{bmatrix} 200 & 200 & 200 \end{bmatrix} \begin{bmatrix} 45 \\ 65 \\ 60 \end{bmatrix} = [34{,}000]$$

Thus the total cost of production is ₹ 34,000.

Example **14 .** The following matrix gives the proportionate mix of constituents used for the three fertilizers:

		Constituent			
		A	*B*	*C*	*D*
	I	0.5	0	0.5	0
Fertilizer	*II*	0.2	0.3	0	0.5
	III	0.2	0.2	0.1	0.5

(*i*) If sales are 1000 tins (of one kilogram) per week, 20% being fertilizer *I*, 30% being fertilizer *II* and 50% being fertilizer *III*, how much of each constituent is used ?

(*ii*) If the cost of each constituent is 50 paise, 60 paise, 75 paise and 100 paise per 100 grams respectively, how much does one kilogram tin of each fertilizer cost ?

(*iii*) What is the total cost per week ? **[Delhi Univ. B.Com. (H) 2008, 2009, 2018]**

Solution . The proportionate mix of constituents used for three fertilizers can be represented by the 4 × 3 matrix :

$$\mathbf{P} = \begin{matrix} & I & II & III \\ A & \\ B & \\ C & \\ D & \end{matrix}\begin{bmatrix} 0.5 & 0.2 & 0.2 \\ 0 & 0.3 & 0.2 \\ 0.5 & 0 & 0.1 \\ 0 & 0.5 & 0.5 \end{bmatrix}$$

The weekly sales of three fertilizers can be represented by the 3 × 1 matrix :

$$\mathbf{S} = 1000\begin{bmatrix} 0.2 \\ 0.3 \\ 0.5 \end{bmatrix}\begin{matrix} I \\ II \\ III \end{matrix} = \begin{bmatrix} 200 \\ 300 \\ 500 \end{bmatrix}\begin{matrix} I \\ II \\ III \end{matrix}$$

(*i*) The total quantity of each constituent used is given by the matrix product :

$$\mathbf{PS} = \begin{matrix} A \\ B \\ C \\ D \end{matrix}\overset{\begin{matrix} I & II & III \end{matrix}}{\begin{bmatrix} 0.5 & 0.2 & 0.2 \\ 0 & 0.3 & 0.2 \\ 0.5 & 0 & 0.1 \\ 0 & 0.5 & 0.5 \end{bmatrix}}\begin{bmatrix} 200 \\ 300 \\ 500 \end{bmatrix}\begin{matrix} I \\ II \\ III \end{matrix} = \begin{bmatrix} 260 \\ 190 \\ 150 \\ 400 \end{bmatrix}\begin{matrix} A \\ B \\ C \\ D \end{matrix}$$

(*ii*) The cost (in ₹) of each constituent (per kg) can be represented by 4 × 1 column matrix:

$$\begin{matrix} \\ A \\ B \\ C \\ D \end{matrix}\begin{matrix} Cost \\ \begin{bmatrix} 5 \\ 6 \\ 7.5 \\ 10 \end{bmatrix} \end{matrix}$$

Thus the cost of 1 kg tin of each fertilizer is given by the matrix product :

$$\begin{array}{c} \\ I \\ II \\ III \end{array}\overset{\begin{array}{cccc} A & B & C & D \end{array}}{\begin{bmatrix} 0.5 & 0 & 0.5 & 0 \\ 0.2 & 0.3 & 0 & 0.5 \\ 0.2 & 0.2 & 0.1 & 0.5 \end{bmatrix}} \begin{bmatrix} 5 \\ 6 \\ 7.5 \\ 10 \end{bmatrix}\begin{array}{c} A \\ B \\ C \\ D \end{array} = \begin{array}{c} I \\ II \\ III \end{array}\begin{bmatrix} 6.25 \\ 7.80 \\ 7.95 \end{bmatrix}$$

(*iii*) The total cost of fertilizers, when 1000 tins (of one kilogram) are sold per week, is given by

$$\begin{bmatrix} 200 & 300 & 500 \end{bmatrix}\begin{bmatrix} 6.25 \\ 7.80 \\ 7.95 \end{bmatrix} = \begin{bmatrix} 7565 \end{bmatrix}$$

Thus the total cost per week is ₹ 7565.

***Example* 15 .** The manager of a car rental company is planning its maintenance programme for the next year and wants to determine the needs for certain repair parts and the expected costs. The company rents out three types of cars: large, medium and small. The matrix **N** gives the number of each size of cars available for renting in the four regions.

$$\mathbf{N} = \overset{\begin{array}{c} \textit{Region} \\ \begin{array}{cccc} I & II & III & IV \end{array} \end{array}}{\begin{bmatrix} 160 & 150 & 100 & 120 \\ 400 & 300 & 100 & 400 \\ 500 & 200 & 150 & 300 \end{bmatrix}}\begin{array}{l} \textit{Large} \\ \textit{Medium} \\ \textit{Small} \end{array}$$

Because of cost and frequency of replacement, the manager is concentrating on four repair parts in particular : fan belts, spark plugs, batteries and tyres. On the basis of past records, the average number of repair parts needed per car during a year are given in matrix **R**. Further, the matrix **C** gives the cost (in ₹) per unit for the four parts.

$$\mathbf{R} = \begin{bmatrix} 1.7 & 1.6 & 1.5 \\ 12.0 & 8.0 & 5.0 \\ 0.9 & 0.75 & 0.5 \\ 4.0 & 6.5 & 6.0 \end{bmatrix}\begin{array}{l} \textit{Fan belts} \\ \textit{Spark Plugs} \\ \textit{Batteries} \\ \textit{Tyres} \end{array} \qquad \mathbf{C} = \begin{bmatrix} 125 & 80 & 300 & 350 \end{bmatrix}$$

Using matrix operations, find :

(*i*) the total demand for each type of car,

(*ii*) total demand for each repair part needed for the fleet, and

(*iii*) total combined cost for all repair parts. **[*Delhi Univ. B.Com. (H) 2000*]**

Solution. (*i*) Total demand for each type of car is :

$$\begin{bmatrix} 160 & 150 & 100 & 120 \\ 400 & 300 & 100 & 400 \\ 500 & 200 & 150 & 300 \end{bmatrix} \begin{bmatrix} 1 \\ 1 \\ 1 \\ 1 \end{bmatrix} = \begin{bmatrix} 530 \\ 1200 \\ 1150 \end{bmatrix}$$

Thus the total demand for each type of car is : Large - 530, Medium - 1200 and Small - 1150.

(*ii*) Total demand for each repair part is :

$$\begin{bmatrix} 1.7 & 1.6 & 1.5 \\ 12 & 8 & 5 \\ 0.9 & 0.75 & 0.5 \\ 4 & 6.5 & 6.0 \end{bmatrix} \begin{bmatrix} 530 \\ 1200 \\ 1150 \end{bmatrix} = \begin{bmatrix} 4546 \\ 21,710 \\ 1952 \\ 16,820 \end{bmatrix}$$

Thus the total demand for each repair part is as follows :

Fan belts - 4546, Spark plugs - 21,710, Batteries - 1952, Tyres - 16,820.

(*iii*) Total combined cost for all repair parts is :

$$\begin{bmatrix} 125 & 80 & 300 & 350 \end{bmatrix} \begin{bmatrix} 4546 \\ 21,710 \\ 1952 \\ 16,820 \end{bmatrix} = \begin{bmatrix} 87,77,650 \end{bmatrix}$$

Thus the combined cost for all repair parts is ₹ 87,77,650.

***Example* 16 .** A baker makes bread, sweet patties and biscuits. He requires flour, egg, sugar, milk and yeast for his preparations. The requirement of these basic items for making the bread, sweet patty and biscuit is as follows :

	Bread	*Sweet Patty*	*Biscuit*
Flour	250 g	50 g	8 g
Egg	0	1/2	1/4
Sugar	125 g	100 g	25 g
Milk	0.05 litre	0.02 litre	0.01 litre
Yeast	1/4 cake	1/8 cake	0

The baker buys the flour for ₹ 2.20 per kg and sugar for ₹ 2.40 per kg at controlled prices. An egg costs him ₹ 0.25, a litre of milk ₹ 1.80 and a cake of yeast ₹ 0.80. What is the cost of making a bread, a sweet patty and a biscuit ? Use matrix system.

[*Delhi Univ. B.Com. (H) 2001*]

Solution . The requirement of flour, egg, sugar, milk and yeast for making the bread, sweet patty and biscuit can be represented by 3 × 5 matrix :

$$\mathbf{A} = \begin{matrix} \\ \textit{bread} \\ \textit{sweet patty} \\ \textit{biscuit} \end{matrix} \begin{matrix} \begin{matrix} \textit{flour} & \textit{egg} & \textit{sugar} & \textit{milk} & \textit{yeast} \end{matrix} \\ \begin{bmatrix} 0.25 & 0 & 0.125 & 0.05 & 0.25 \\ 0.05 & 0.5 & 0.1 & 0.02 & 0.125 \\ 0.008 & 0.25 & 0.025 & 0.01 & 0 \end{bmatrix} \end{matrix}$$

The cost of the basic items can be represented by 5 × 1 column matrix :

$$\mathbf{B} = \begin{matrix} \textit{flour} \\ \textit{egg} \\ \textit{sugar} \\ \textit{milk} \\ \textit{yeast} \end{matrix} \begin{bmatrix} 2.20 \\ 0.25 \\ 2.40 \\ 1.80 \\ 0.80 \end{bmatrix}$$

The cost of making a bread, a sweet patty and a biscuit is given by the matrix multiplication:

$$\mathbf{AB} = \begin{bmatrix} 0.25 & 0 & 0.125 & 0.05 & 0.25 \\ 0.05 & 0.5 & 0.1 & 0.02 & 0.125 \\ 0.008 & 0.25 & 0.025 & 0.01 & 0 \end{bmatrix} \begin{bmatrix} 2.20 \\ 0.25 \\ 2.40 \\ 1.80 \\ 0.80 \end{bmatrix} = \begin{bmatrix} 1.14 \\ 0.611 \\ 0.1581 \end{bmatrix}.$$

***Example* 17 .** A firm produces chairs, tables and cupboards, each requiring three types of raw material– timber, nails and varnish. You are given below, the units of different raw materials required for producing one unit of each product :

Product	*Timber (c.ft.)*	*Nails (dozen)*	*Varnish (litres)*
Chair	0.7	2	1
Table	1	4	1.5
Cupboard	3.2	6	2

If the firm produces 300 units of each product, find the quantity of each raw material using matrix algebra. ***[Delhi Univ. B.Com. (H) 2002]***

Solution . The quantity of each raw material required if the firm produces 300 units of each product is given by the matrix product :

$$\begin{matrix} \\ \textit{Timber} \\ \textit{Nails} \\ \textit{Varnish} \end{matrix} \begin{matrix} \begin{matrix} \textit{Chair} & \textit{Table} & \textit{Cupboard} \end{matrix} \\ \begin{bmatrix} 0.7 & 1 & 3.2 \\ 2 & 4 & 6 \\ 1 & 1.5 & 2 \end{bmatrix} \end{matrix} \begin{bmatrix} 300 \\ 300 \\ 300 \end{bmatrix} \begin{matrix} \textit{Chair} \\ \textit{Table} \\ \textit{Cupboard} \end{matrix} = \begin{matrix} \textit{Timber} \\ \textit{Nails} \\ \textit{Varnish} \end{matrix} \begin{bmatrix} 1470 \\ 3600 \\ 1350 \end{bmatrix}$$

Thus the firm requires 1470 c.ft. of timber, 3600 dozens of nails and 1350 litres of varnish to produce 300 units of each product.

***Example* 18.** A firm purchases two machines at a cost of ₹ 31,000 and ₹ 35,000 respectively. Both machines have a useful life of 4 years and have zero salvage values at the end of 4 years. Find the depreciation of each machine for each year using matrix notation if:

(*i*) Both machines are depreciated by the straight-line method,

(*ii*) Both machines are depreciated by the sum of years digit method,

(*iii*) First is depreciated by straight-line method and second by the sum of years digit method.

[*Delhi Univ. B.Com.(H) 2022*]

Solution. (*i*) According to the straight line method, machines are depreciated by equal amounts each year. In other words, the annual depreciation rate for each year would be constant 1/4. Thus, the annual depreciation of each machine for each year is given by the matrix product:

$$\begin{matrix} M_1 \\ M_2 \end{matrix}\begin{bmatrix} 31{,}000 \\ 35{,}000 \end{bmatrix}\overset{\begin{matrix}1 & 2 & 3 & 4\end{matrix}}{\begin{bmatrix} \frac{1}{4} & \frac{1}{4} & \frac{1}{4} & \frac{1}{4} \end{bmatrix}} = \begin{bmatrix} 7{,}750 & 7{,}750 & 7{,}750 & 7{,}750 \\ 8{,}750 & 8{,}750 & 8{,}750 & 8{,}750 \end{bmatrix}\begin{matrix} M_1 \\ M_2 \end{matrix}$$

(*ii*) According to the sum of the year's digit method, the annual depreciation of each machine for each year for the first four years would be in the ratio 4 : 3 : 2 : 1 or 4/10, 3/10, 2/10 and 1/10 respectively. Thus if we define

$$\text{Total depreciation} \qquad\qquad \begin{matrix}1 & 2 & 3 & 4\end{matrix}$$

$$\boldsymbol{W} = \begin{bmatrix} 31{,}000 \\ 35{,}000 \end{bmatrix}\begin{matrix} M_1 \\ M_2 \end{matrix} \quad \text{and} \quad \boldsymbol{R} = \begin{bmatrix} 4/10 & 3/10 & 2/10 & 1/10 \end{bmatrix}$$

Then the annual depreciation of each machine for each year is given by the matrix product :

$$\boldsymbol{WR} = \begin{bmatrix} 31{,}000 \\ 35{,}000 \end{bmatrix}\begin{bmatrix} 4/10 & 3/10 & 2/10 & 1/10 \end{bmatrix}$$

$$= \overset{\begin{matrix}1 & 2 & 3 & 4\end{matrix}}{\begin{bmatrix} 12{,}400 & 9{,}300 & 6{,}200 & 3{,}100 \\ 14{,}000 & 10{,}500 & 7{,}000 & 3{,}500 \end{bmatrix}}\begin{matrix} M_1 \\ M_2 \end{matrix}$$

(*iii*) According to the straight line method, machines are depreciated by equal amounts each year. In other words, the annual depreciation rate for each year would be constant 1/4. Thus if first machine is depreciated by straight line method and second by sum of the year's digit method, then the annual depreciation of each machine for each year can be calculated by

$$\begin{bmatrix} 31{,}000 & 0 \\ 0 & 35{,}000 \end{bmatrix}\begin{bmatrix} 1/4 & 1/4 & 1/4 & 1/4 \\ 4/10 & 3/10 & 2/10 & 1/10 \end{bmatrix}$$

$$= \begin{bmatrix} 7{,}750 & 7{,}750 & 7{,}750 & 7{,}750 \\ 14{,}000 & 10{,}500 & 7{,}000 & 3{,}500 \end{bmatrix}\begin{matrix} M_1 \\ M_2 \end{matrix}$$

EXERCISE 2.1

1. A bicycle company has two branches whose daily productions are given below:

	3 speed	*5 speed*	*10 speed*
Branch 1	40	120	150
Branch 2	60	175	100

Express the given information as a matrix. Using multiplication of a scalar times a matrix, find the new daily productions if production is to double.

2. A man buys 8 dozen mangoes, 10 dozen of apples and 4 dozen of bananas. Mangoes cost ₹ 18 per dozen, apples ₹ 15 per dozen and banana ₹ 12 per dozen. Represent the quantities bought by a row matrix and prices by a column matrix. Using matrix multiplication, determine the total amount spent by the man.

3. The prices of three commodities A, B and C in a shop are ₹ 5, ₹ 6 and ₹ 10 respectively. Customer X buy 8 units of A, 7 units of B and 8 units of C. Customer Y buys 6 units of A, 7 units of B and 8 units of C. Show, in matrix notation, the prices of the commodities, quantities bought and the amount spent.

4. A company has two plants. Plant 1 is capable of producing 5 items of A, 10 items of B and 3 items of C per one hour of operation. Plant 2 is capable of producing 5 items of A, 6 items of B and 6 items of C per one hour of operation. Express this information in a 3×2 matrix. Using matrix multiplication, determine the total number of items A, B and C produced if Plant 1 is operated for 10 hours and Plant 2 is operated for 5 hours.

5. A company is considering which of the three methods of production it should use in producing three goods A, B and C. The amount of each goods produced by each method is shown in the matrix :

$$\begin{array}{cc} & \begin{array}{ccc} A & B & C \end{array} \\ \text{Method} \begin{array}{c} I \\ II \\ III \end{array} & \begin{bmatrix} 4 & 8 & 2 \\ 5 & 7 & 1 \\ 5 & 3 & 9 \end{bmatrix} \end{array}$$

The matrix vector $\begin{bmatrix} 10 & 4 & 6 \end{bmatrix}$ represents the profit per unit for the goods A, B, C in that order. Using matrix multiplication, find which method maximizes total profit.

6. A company is to employ 60 labourers from either of the Party A and B, comprising of persons in different age groups as under :

Category	*I* (20 -25 years)	*II* (26 -30 years)	*III* (31- 40 years)
Party A	25	20	15
Party B	20	30	10

Rate of labour applicable to categories *I*, *II* and *III* are ₹ 1200, ₹ 1000 and ₹ 600 respectively. Using matrix notations, find, which party is economically preferable over the other.

7. A factory produces 3 types of portable radio sets called Audio 1, Audio 2 and Audio 3. Audio 1 contains 1 transistor, 10 resistors and 5 capacitors, while Audio 2 contains 2 transistors, 18 resistors and 7 capacitors, and Audio 3 contains 3 transistors, 24 resistors and 10 capacitors.

Represent these information in the form of a matrix. Using matrix multiplication, calculate the factory's weekly consumption of transistors, resistors and capacitors, if weekly output of sales is 100 of Audio 1, 250 of Audio 2 and 80 of Audio 3.

8. Three persons buy cold drinks of different brands *A*, *B* and *C*. The first person buys 12 bottles of *A*, 5 bottles of B and 3 bottles of *C*. The second person buys 4 bottles of *A*, 6 bottles of *B* and 10 bottles of C. The third person buys 6 bottles of A, 7 bottles of *B* and 9 bottles of *C*.

Represent these information in the form of a matrix. If each bottle of brand *A* costs ₹ 4, each bottle of *B* costs ₹ 5 and each bottle of *C* costs ₹ 6, then using matrix operations, find the total sum of money spent individually by the three persons for the purchase of cold drinks.

9. The annual sale volumes of three products *X*, *Y*, *Z* whose sale prices per unit are ₹ 3.50, ₹ 2.75, ₹ 1.50 respectively, in two different markets *I* and *II* are shown below :

	Product		
Market	*X*	*Y*	*Z*
I	6000	9000	13,000
II	12,000	6000	17,000

Find the total revenue in each market with the help of matrices.

10. There are two families *X* and *Y*. Family *X* has 2 men, 3 women and one child, while family *Y* has one man, one woman and two children. Their individual daily requirements are as follows :

Man : 2400 calories and 55 gms protein; Woman : 1900 calories and 45 gms protein; Child : 1800 calories and 33 gms protein.

Present the above information in the form of matrices. Using matrix multiplication, calculate the total daily requirements of calories and protein for each of the two families.

11. A student has 4 places where he can have his lunch. The college canteen charges ₹ 9 for a cold drink, ₹ 6 for a cutlet and ₹ 5 for a sandwich. The coffee house charges ₹ 10, ₹ 8 and ₹ 9 for the same items, while fast food joint charges ₹ 12, ₹ 15, ₹ 15 and the restaurant charges ₹ 15, ₹ 25 and ₹ 20 for the above items respectively. The student wants to have one cold drink, two cutlets and one sandwich. Where should he have his lunch so that the lunch costs him the least ? (Use of Matrix-Algebra is required).

12. In a certain city, there are 5 colleges and 20 schools. Each school has 3 peons, 1 clerk and 1 head clerk, whereas a college has 5 peons, 3 clerks, 1 head clerk and an additional staff as a caretaker. The monthly salary of each of them is as follows :

Peon– ₹ 1100, Clerk– ₹ 1700, Head Clerk– ₹ 3000 and Caretaker– ₹ 2500

Using matrix method, find

(*i*) the total number of posts of each kind in schools and colleges taken together,

(*ii*) the total monthly salary bill of each school and college, and

(*iii*) the total monthly salary bill of all the schools and colleges taken together.

13. In a certain city there are 25 colleges and 100 schools. Each school and college has 5 peons, 2 clerks and 1 cashier. Each college, in addition, has 1 accountant and 1 head clerk. The monthly salary of each of them is as follows :

Peon – ₹ 625, Clerk – ₹ 740, Cashier – ₹ 760, Accountant – ₹ 1200, Head clerk – ₹ 1250

Using matrix method, find

(*i*) the total number of posts of each kind in schools and colleges taken together,

(*ii*) the total monthly salary bill of all the schools and colleges taken together.

14. There are three electric dealers in a pure competition market who sell Radios, Tape recorders and Television sets. *A* sells weekly 48 radios, 20 tape recorders and 12 television sets; *B* sells weekly 57 radios, 17 tape recorders and 15 television sets, and *C* sells weekly 35 radios, 22 tape recorders and 18 television sets. Using matrices, calculate their individual profits, if the net revenue on a radio is ₹ 50, on a tape recorder ₹ 75 and on a television set ₹ 250.

15. Two television companies TV_1 and TV_2, both televise documentary programmes and variety programmes. TV_1 has two transmitting stations and TV_2 has three transmitting stations, all stations transmit different programmes. On an average, each TV_1 station broadcasts 1 hour of documentary and 3 hours of variety programmes each day, whereas each TV_2 station broadcasts 2 hours of documentary and $1\frac{1}{2}$ hours of variety programmes each day. The transmission of documentary and variety programmes costs approximately ₹ 50 and ₹ 200 per hour respectively. Express, in matrix form and hence evaluate :

(*i*) the daily cost of transmission from each TV_1 and each TV_2 station.

(*ii*) the total number of hours which are devoted daily to documentary and to variety programmes by both companies.

(*iii*) the total daily cost of transmission incurred by both companies.

16. A firm produces three products P_1, P_2 and P_3 requiring the mix-up of three materials M_1, M_2 and M_3. The per unit requirement of each product for each material is as follows :

$$\mathbf{A} = \begin{array}{c} \\ P_1 \\ P_2 \\ P_3 \end{array}\begin{array}{c} \begin{array}{ccc} M_1 & M_2 & M_3 \end{array} \\ \begin{bmatrix} 2 & 4 & 5 \\ 3 & 2 & 4 \\ 1 & 3 & 2 \end{bmatrix} \end{array}$$

Using matrix notations, find

(*i*) The total requirement of each material if the firm produces 100, 200 and 300 units of three products respectively.

(*ii*) The per unit cost of production of each product if the per unit costs of materials M_1, M_2 and M_3 are ₹ 10, ₹ 15 and ₹ 12 respectively, and

(*iii*) The total cost of production. **[*Delhi Univ. B.Com. (H) 2013*]**

17. A firm produces three products P_1, P_2 and P_3 requiring the mixup of four materials M_1, M_2, M_3 and M_4. The matrix below gives the amount of material needed for each product :

$$\mathbf{A} = \begin{matrix} & M_1 & M_2 & M_3 & M_4 \\ P_1 & 2 & 3 & 1 & 12 \\ P_2 & 7 & 9 & 5 & 20 \\ P_3 & 8 & 12 & 6 & 15 \end{matrix}$$

Using matrix notations, find

(*i*) the total requirement of each material if the firm produces 70 units of P_1, 120 units of P_2 and 50 units of P_3.

(*ii*) the per unit cost of production of each product if the per unit costs of materials M_1, M_2, M_3 and M_4 are ₹ 10, ₹ 12, ₹ 15 and ₹ 20 respectively, and

(*iii*) the total cost of production.

18. A firm purchases two machines at a cost of ₹ 10,000 and ₹ 20,000 respectively. Both machines have a life time of four years and have zero salvage values at the end of the four years. Find depreciation of each machine for each year using matrix notation if

(*i*) both are depreciated by sum of the years digit method,

(*ii*) both are depreciated by straight line method, and

(*iii*) first is depreciated by sum of the years digit method and second by straight line method.

19. A firm has two machines M_1 and M_2 costing ₹ 45,000 and ₹ 30,000. Each has 5 years life with scrap value nil. Find depreciation of each machine for each year using matrix notation if (*i*) both are depreciated by sum of the year's digit method, (*ii*) first is depreciated by sum of the year's digit method and second by straight line method. **[Delhi Univ. B.Com.(H) 1997, 2011]**

20. A transport company uses 3 types of trucks T_1, T_2 and T_3 to transport 3 types of vehicles V_1, V_2 and V_3. The capacity of each truck in terms of 3 types of vehicles is given below :

	V_1	V_2	V_3
T_1	2	3	6
T_2	6	7	6
T_3	9	12	8

Using matrix method, find the number of vehicles of each type which can be transported if company has 15, 10 and 20 trucks of each type respectively.

ANSWERS

1. $\begin{bmatrix} 80 & 240 & 300 \\ 120 & 350 & 200 \end{bmatrix}$ **2.** ₹ 342 **3.** $\begin{bmatrix} 5 \\ 6 \\ 10 \end{bmatrix}, \begin{bmatrix} 8 & 7 & 8 \\ 6 & 7 & 8 \end{bmatrix}, \begin{bmatrix} 162 \\ 152 \end{bmatrix}$

4. $A = 75$ units, $B = 130$ units, $C = 60$ units **5.** Method *III*

6. Party *A* is more economical as compared to Party *B*.

7. Transistors : 840, Registors : 7420, Capacitors : 3050

8. ₹ 91, ₹ 106, ₹ 113 **9.** Market I : ₹ 65,250, Market II : ₹ 84,000

10. Family X : 12,300 calories and 278 gms protein;
Family Y : 7900 calories and 166 gms protein.

11. College canteen

12. (*i*) Peons : 85, Clerks : 35, Head clerks : 25, Caretaker : 5

(*ii*) Monthly salary bill for each school : ₹ 8000 and for each college : ₹ 16,100

(*iii*) ₹ 2,40,500

13. (*i*) Peons : 625, Clerks : 250, Cashiers : 125, Accountants : 25, Head Clerks : 25

(*ii*) ₹ 7,31,875

14. ₹ 6900, ₹ 7875, ₹ 7900

15. (*i*) ₹ 650, ₹ 400 (*ii*) Documentary : 8 hours, Variety : 10.5 hours (*iii*) ₹ 2500

16. (*i*) M_1 : 1100, M_2 : 1700, M_3 : 1900

(*ii*) ₹ 140, ₹ 108, ₹ 79 (*iii*) ₹ 59,300

17. (*i*) M_1 : 1380, M_2 : 1890, M_3 : 970, M_4 : 3990

(*ii*) ₹ 311, ₹ 653, ₹ 614 (*iii*) ₹ 1,30,830

18. (*i*) $\begin{matrix} M_1 \\ M_2 \end{matrix}\begin{bmatrix} 4000 & 3000 & 2000 & 1000 \\ 8000 & 6000 & 4000 & 2000 \end{bmatrix}$ (*ii*) $\begin{matrix} M_1 \\ M_2 \end{matrix}\begin{bmatrix} 2500 & 2500 & 2500 & 2500 \\ 5000 & 5000 & 5000 & 5000 \end{bmatrix}$

(*iii*) $\begin{matrix} M_1 \\ M_2 \end{matrix}\begin{bmatrix} 4000 & 3000 & 2000 & 1000 \\ 5000 & 5000 & 5000 & 5000 \end{bmatrix}$

19. (*i*) $\begin{matrix} M_1 \\ M_2 \end{matrix}\begin{bmatrix} 15,000 & 12,000 & 9,000 & 6,000 & 3,000 \\ 10,000 & 8,000 & 6,000 & 4,000 & 2,000 \end{bmatrix}$

(*ii*) $\begin{matrix} M_1 \\ M_2 \end{matrix}\begin{bmatrix} 15,000 & 12,000 & 9,000 & 6,000 & 3,000 \\ 6,000 & 6,000 & 6,000 & 6,000 & 6,000 \end{bmatrix}$

20. $V_1 = 270, \ V_2 = 355, \ V_3 = 310$

2.6 SYSTEMS OF LINEAR EQUATIONS – APPLICATIONS

In this section we will review and discuss common methods of solving systems of linear equations that result from applied business and economics problems.

***Example* 19** . Given the following equations for two related markets *A* and *B*, find the equilibrium conditions for each market and the price for each market :

$$x_d\,(A) = 82 - 3p_A + p_B \qquad x_s\,(A) = -\,5 + 15p_A$$
$$x_d\,(B) = 92 + 2p_A - 4p_B \qquad x_s\,(B) = -\,6 + 32p_B,$$

where x_d and x_s denote quantity demanded and quantity supplied respectively. Use matrix inversion method. **[*Delhi Univ. B.Com. (H) 1984*]**

***Solution* .** Equilibrium conditions are :

$$x_d(A) = x_s(A) \quad \text{and} \quad x_d(B) = x_s(B)$$

$$\Rightarrow \quad 82 - 3p_A + p_B = -5 + 15p_A \quad \text{and} \quad 92 + 2p_A - 4p_B = -6 + 32p_B$$

$$\Rightarrow \quad \begin{aligned} 18p_A - p_B &= 87 \\ -2p_A + 36p_B &= 98 \end{aligned}$$

These can be written in the matrix form as

$$\underset{\mathbf{A}}{\begin{bmatrix} 18 & -1 \\ -2 & 36 \end{bmatrix}} \underset{\mathbf{X}}{\begin{bmatrix} p_A \\ p_B \end{bmatrix}} = \underset{\mathbf{B}}{\begin{bmatrix} 87 \\ 98 \end{bmatrix}}$$

Now, $|\mathbf{A}| = 646 \neq 0$. Since $|\mathbf{A}| \neq 0$, $\mathbf{A}^{-1}$ exists and the solution is given by $\mathbf{X} = \mathbf{A}^{-1}\mathbf{B}$. It can be readily verified that

$$\mathbf{A}^{-1} = \frac{1}{646}\begin{bmatrix} 36 & 1 \\ 2 & 18 \end{bmatrix} \quad \therefore \quad \mathbf{X} = \mathbf{A}^{-1}\mathbf{B} = \frac{1}{646}\begin{bmatrix} 36 & 1 \\ 2 & 18 \end{bmatrix}\begin{bmatrix} 87 \\ 98 \end{bmatrix} = \begin{bmatrix} 5 \\ 3 \end{bmatrix}$$

Hence the equilibrium prices are: $p_A = 5$ and $p_B = 3$.

***Example* 20 .** Given the following National Income model :

$$C = a + bY \; (a > 0, 0 < b < 1)$$
$$I = d + eY \; (d > 0, 0 < e < 1)$$
$$Y = I + C$$

Solve for the endogeneous variables *C, I* and *Y* using matrix or determinant method.

[*Delhi Univ. B.Com. (H) 1989*]

***Solution* .** We can rewrite the given system of equations as

$$\begin{aligned} C + 0I - bY &= a \\ 0C + I - eY &= d \\ -C - I + Y &= 0 \end{aligned}$$

The system in the matrix form is

$$\underset{\mathbf{A}}{\begin{bmatrix} 1 & 0 & -b \\ 0 & 1 & -e \\ -1 & -1 & 1 \end{bmatrix}} \underset{\mathbf{X}}{\begin{bmatrix} C \\ I \\ Y \end{bmatrix}} = \underset{\mathbf{B}}{\begin{bmatrix} a \\ d \\ 0 \end{bmatrix}}$$

The determinant of the coefficient matrix is : $D = |\mathbf{A}| = 1 - b - e$

Assuming $D \neq 0$, the system possesses a unique solution given by

$$C = \frac{D_1}{D}, \quad I = \frac{D_2}{D}, \quad Y = \frac{D_3}{D}, \text{ where}$$

$$D_1 = \begin{vmatrix} a & 0 & -b \\ d & 1 & -e \\ 0 & -1 & 1 \end{vmatrix} = a - ae + bd$$

$$D_2 = \begin{vmatrix} 1 & a & -b \\ 0 & d & -e \\ -1 & 0 & 1 \end{vmatrix} = d + ea - bd$$

$$D_3 = \begin{vmatrix} 1 & 0 & a \\ 0 & 1 & d \\ -1 & -1 & 0 \end{vmatrix} = a + d$$

$$\therefore \qquad C = \frac{a - ae + bd}{1 - b - e},\ I = \frac{d + ea - bd}{1 - b - e},\quad Y = \frac{a + d}{1 - b - e}.$$

***Example* 21 .** An amount of ₹ 5,000 is put into three investments at the rates of interest of 6%, 7% and 8% per annum respectively. The total annual income is ₹ 358. If the combined income from the first two investments is ₹ 70 more than the income from the third, find the amount of each investment by using matrix algebra.

[*Delhi Univ. B.Com. (H) 1989*]

***Solution* .** Let x, y and z represent the investments (in ₹) at the rates of interest of 6%, 7% and 8% per annum respectively. Then, according to the given information, we have

$$\begin{aligned} x + y + z &= 5000 \\ 0.06x + 0.07y + 0.08z &= 358 \\ 0.06x + 0.07y &= 0.08z + 70 \end{aligned} \quad \Leftrightarrow \quad \begin{aligned} x + y + z &= 5000 \\ 6x + 7y + 8z &= 35{,}800 \\ 6x + 7y - 8z &= 7000 \end{aligned}$$

The above system can be expressed in the matrix form as

$$\underset{\mathbf{A}}{\begin{bmatrix} 1 & 1 & 1 \\ 6 & 7 & 8 \\ 6 & 7 & -8 \end{bmatrix}} \underset{\mathbf{X}}{\begin{bmatrix} x \\ y \\ z \end{bmatrix}} = \underset{\mathbf{B}}{\begin{bmatrix} 5000 \\ 35{,}800 \\ 7000 \end{bmatrix}}$$

The determinant of the coefficient matrix **A** is : $|\mathbf{A}| = -16$

Since $|\mathbf{A}| \neq 0$, $\mathbf{A}^{-1}$ exists and the solution is given by $\mathbf{X} = \mathbf{A}^{-1}\mathbf{B}$. It can be easily verified that

$$\mathbf{A}^{-1} = -\frac{1}{16}\begin{bmatrix} -112 & 15 & 1 \\ 96 & -14 & -2 \\ 0 & -1 & 1 \end{bmatrix}$$

Hence the solution is :

$$\mathbf{X} = \mathbf{A}^{-1}\mathbf{B} = -\frac{1}{16}\begin{bmatrix} -112 & 15 & 1 \\ 96 & -14 & -2 \\ 0 & -1 & 1 \end{bmatrix}\begin{bmatrix} 5000 \\ 35{,}800 \\ 7000 \end{bmatrix} = -\frac{1}{16}\begin{bmatrix} -16{,}000 \\ -35{,}200 \\ -28{,}800 \end{bmatrix} = \begin{bmatrix} 1000 \\ 2200 \\ 1800 \end{bmatrix}$$

Thus $x =$ ₹ 1000, $y =$ ₹ 2200 and $z =$ ₹ 1800.

***Example* 22 .** Mr. X has invested a part of his investment in 10% bond A and a part in 15% bond B. His interest income during first year is ₹ 4000. If he invests 20% more in 10% bond A and 10% more in 15% bond B, his income during second year increases by ₹ 500. Find his initial investments and the new investments in bonds A and B using matrix method.

[*Delhi Univ. B.Com. (H) 1999*]

***Solution* .** Let his initial investments be ₹ x and ₹ y in bond A and bond B respectively. Then according to the given information, we have

$$0.10x + 0.15y = 4000 \qquad ...(i)$$

Since he invests 20% more in bond A and 10% more in bond B in the next year, his new investments are $1.20x$ and $1.10y$. Thus again according to the given information, we have

$$0.10\,(1.20x) + 0.15\,(1.10y) = 4500$$

or,
$$0.12x + 0.165y = 4500 \qquad ....(ii)$$

Now we'll solve Eqs. (*i*) and (*ii*) using matrix method. Expressed in the matrix from, we obtain

$$\underset{\mathbf{A}}{\begin{bmatrix} 0.10 & 0.15 \\ 0.12 & 0.165 \end{bmatrix}} \underset{\mathbf{X}}{\begin{bmatrix} x \\ y \end{bmatrix}} = \underset{\mathbf{B}}{\begin{bmatrix} 4000 \\ 4500 \end{bmatrix}}$$

Now $|\mathbf{A}| = 0.0165 - 0.0180 = -0.0015 \neq 0$. Hence the solution is given by $\mathbf{X} = \mathbf{A}^{-1}\mathbf{B}$.
It can be easily verified that

$$\mathbf{A}^{-1} = -\frac{1}{0.0015}\begin{bmatrix} 0.165 & -0.15 \\ -0.12 & 0.10 \end{bmatrix}$$

$$\therefore \quad \mathbf{X} = \mathbf{A}^{-1}\mathbf{B} = -\frac{1}{0.0015}\begin{bmatrix} 0.165 & -0.15 \\ -0.12 & 0.10 \end{bmatrix}\begin{bmatrix} 4000 \\ 4500 \end{bmatrix} = = -\frac{1}{0.0015}\begin{bmatrix} -15 \\ -30 \end{bmatrix} = \begin{bmatrix} 10{,}000 \\ 20{,}000 \end{bmatrix}$$

Thus $x = 10{,}000$, $y = 20{,}000$. That is, his initial investments are ₹ 10,000 and ₹ 20,000 in bond A and bond B respectively. Substituting these values in $1.20x$ and $1.10y$, we get his new investments as ₹ 12,000 and ₹ 22,000.

***Example* 23 .** A salesman has the following record of sales during three months for three items A, B and C which have different rates of commission.

Months	*Sales of Units*			*Total Commission Drawn (in ₹)*
	A	*B*	*C*	
January	90	100	20	800
February	130	50	40	900
March	60	100	30	850

Find out the rates of commission on items A, B and C. **[*Delhi Univ. B.Com. (H) 2006*]**

Solution . Let x, y and z be the rates of commission (in ₹) on items A, B and C respectively. Then we have

$$\begin{aligned} 90x + 100y + 20z &= 800 \\ 130x + 50y + 40z &= 900 \\ 60x + 100y + 30z &= 850 \end{aligned} \quad \Leftrightarrow \quad \underset{\mathbf{A}}{\begin{bmatrix} 90 & 100 & 20 \\ 130 & 50 & 40 \\ 60 & 100 & 30 \end{bmatrix}} \underset{\mathbf{X}}{\begin{bmatrix} x \\ y \\ z \end{bmatrix}} = \underset{\mathbf{B}}{\begin{bmatrix} 800 \\ 900 \\ 850 \end{bmatrix}}$$

The determinant of the coefficient matrix **A** is

$$D = |\mathbf{A}| = \begin{vmatrix} 90 & 100 & 20 \\ 130 & 50 & 40 \\ 60 & 100 & 30 \end{vmatrix} = 1000 \begin{vmatrix} 9 & 10 & 2 \\ 13 & 5 & 4 \\ 6 & 10 & 3 \end{vmatrix} = 5000 \begin{vmatrix} 9 & 2 & 2 \\ 13 & 1 & 4 \\ 6 & 2 & 3 \end{vmatrix} = -1{,}75{,}000$$

Since $D \neq 0$, the system has a unique solution. Using Cramer's rule, the solution is given by

$$x = \frac{D_1}{D}, \quad y = \frac{D_2}{D} \quad \text{and} \quad z = \frac{D_3}{D}$$

where D_i is the determinant obtained from D by replacing the ith column of D by the elements of the matrix B, $i = 1, 2, 3$.

Thus
$$x = \frac{\begin{vmatrix} 800 & 100 & 20 \\ 900 & 50 & 40 \\ 850 & 100 & 30 \end{vmatrix}}{D} = \frac{1000 \begin{vmatrix} 80 & 10 & 2 \\ 90 & 5 & 4 \\ 85 & 10 & 3 \end{vmatrix}}{-1{,}75{,}000} = \frac{-3{,}50{,}000}{-1{,}75{,}000} = 2,$$

$$y = \frac{\begin{vmatrix} 90 & 800 & 20 \\ 130 & 900 & 40 \\ 60 & 850 & 30 \end{vmatrix}}{D} = \frac{1000 \begin{vmatrix} 9 & 80 & 2 \\ 13 & 90 & 4 \\ 6 & 85 & 3 \end{vmatrix}}{-1{,}75{,}000} = \frac{-7{,}00{,}000}{-1{,}75{,}000} = 4,$$

$$z = \frac{\begin{vmatrix} 90 & 100 & 800 \\ 130 & 50 & 900 \\ 60 & 100 & 850 \end{vmatrix}}{D} = \frac{1000 \begin{vmatrix} 9 & 10 & 80 \\ 13 & 5 & 90 \\ 6 & 10 & 85 \end{vmatrix}}{-1{,}75{,}000} = \frac{-1{,}92{,}5000}{-1{,}75{,}000} = 11.$$

Hence the solution is $x = 2$, $y = 4$ and $z = 11$.

***Example* 24 .** A transport company uses 3 types of trucks T_1, T_2 and T_3, to transport 3 types of vehicles V_1, V_2 and V_3. The capacity of each truck in terms of 3 types of vehicles is given as follows :

	V_1	V_2	V_3
T_1	1	3	2
T_2	2	2	3
T_3	3	2	2

Using matrix method, find the number of trucks of each type required to transport 85, 105 and 110 vehicles of V_1, V_2 and V_3 types respectively.

[*Delhi Univ. B.Com. (H) 1999, 2006, 2019*]

***Solution*.** Let *x, y* and *z* denote the number of trucks required of each type. Then we have

$$x + 2y + 3z = 85$$
$$3x + 2y + 2z = 105$$
$$2x + 3y + 2z = 110$$

The system in the matrix form is :

$$\underset{\mathbf{A}}{\begin{bmatrix} 1 & 2 & 3 \\ 3 & 2 & 2 \\ 2 & 3 & 2 \end{bmatrix}} \underset{\mathbf{X}}{\begin{bmatrix} x \\ y \\ z \end{bmatrix}} = \underset{\mathbf{B}}{\begin{bmatrix} 85 \\ 105 \\ 110 \end{bmatrix}}$$

Now $\quad |\mathbf{A}| = 1(4-6) - 2(6-4) + 3(9-4) = -2 - 4 + 15 = 9$

Since $|\mathbf{A}| \neq 0$, $\mathbf{A}^{-1}$ exists and the solution is given by $\mathbf{X} = \mathbf{A}^{-1}\mathbf{B}$. It can be easily verified that

$$\mathbf{A}^{-1} = \frac{1}{9}\begin{bmatrix} -2 & 5 & -2 \\ -2 & -4 & 7 \\ 5 & 1 & -4 \end{bmatrix}$$

Hence the solution is :

$$\mathbf{X} = \mathbf{A}^{-1}\mathbf{B} = \frac{1}{9}\begin{bmatrix} -2 & 5 & -2 \\ -2 & -4 & 7 \\ 5 & 1 & -4 \end{bmatrix}\begin{bmatrix} 85 \\ 105 \\ 110 \end{bmatrix} = \frac{1}{9}\begin{bmatrix} 135 \\ 180 \\ 90 \end{bmatrix} = \begin{bmatrix} 15 \\ 20 \\ 10 \end{bmatrix}$$

Thus $\quad x = 15, \quad y = 20$ and $z = 10$.

***Example* 25.** One unit of commodity A is produced by combining 1 unit of land, 2 units of labour and 5 units of capital. One unit of B is produced by combining 2 units of land, 3 units of labour and 1 unit of capital, one unit of C is produced by 3 units of land, 1 unit of labour and 2 units of capital. If the price of A, B and C are ₹ 27, ₹ 16 and ₹ 19 respectively, find the rent R, wages W and rate of interest I. **[*Delhi Univ. B.Com. (H) 2018*]**

Solution. According to the given data, we have

$$R + 2W + 5I = 27$$
$$2R + 3W + I = 16$$
$$3R + W + 2I = 19$$

The above system of equations can be written in the matrix form as

$$\begin{bmatrix} 1 & 2 & 5 \\ 2 & 3 & 1 \\ 3 & 1 & 2 \end{bmatrix} \begin{bmatrix} R \\ W \\ I \end{bmatrix} = \begin{bmatrix} 27 \\ 16 \\ 19 \end{bmatrix}$$

$$\quad \mathbf{A} \qquad \mathbf{X} \qquad \mathbf{B}$$

The determinant of the coefficient matrix is:

$$|\mathbf{A}| = 1\,(6-1) - 2(4-3) + 5\,(2-9) = -32$$

Since $|\mathbf{A}| \neq 0$, $\mathbf{A}^{-1}$ exists and the solution is given by $\mathbf{X} = \mathbf{A}^{-1}\mathbf{B}$. It can be easily verified that

$$\mathbf{A}^{-1} = -\frac{1}{32}\begin{bmatrix} 5 & 1 & -13 \\ -1 & -13 & 9 \\ -7 & 5 & -1 \end{bmatrix}$$

$$\therefore \qquad \mathbf{X} = \mathbf{A}^{-1}\mathbf{B} = \frac{1}{32}\begin{bmatrix} 5 & 1 & -13 \\ -1 & -13 & 9 \\ -7 & 5 & -1 \end{bmatrix}\begin{bmatrix} 27 \\ 16 \\ 19 \end{bmatrix} = -\frac{1}{32}\begin{bmatrix} -96 \\ -64 \\ -128 \end{bmatrix} = \begin{bmatrix} 3 \\ 2 \\ 4 \end{bmatrix}$$

$$\therefore \qquad R = 3, \quad W = 2 \quad \text{and } I = 4$$

***Example* 26.** The total sales (S) in thousands of rupees of a firm selling two products X and Y is given by the relationship : $S = a + bX + cY$. Data for the first three months is given as under:

Months	*Total Sales*	*X*	*Y*
1	12	2	3
2	13	6	2
3	15	5	3

Using determinant method, determine the sales in the next month when it sells 4 units of X and 5 units of Y. **[*Delhi Univ. B. Com. (H) 2014*]**

Solution. From the given data, we have

$$a + 2b + 3c = 12$$
$$a + 6b + 2c = 13$$
$$a + 5b + 3c = 15$$

We solve the above system for a, b and c using determinant method. The determinant of the coefficient matrix is :

$$D = \begin{vmatrix} 1 & 2 & 3 \\ 1 & 6 & 2 \\ 1 & 5 & 3 \end{vmatrix} = 3 \neq 0$$

$\therefore$ By Cramer's rule, the solution is given by

$$a = \frac{D_1}{D}, \qquad b = \frac{D_2}{D}, \qquad c = \frac{D_3}{D},$$

where

$$D_1 = \begin{vmatrix} 12 & 2 & 3 \\ 13 & 6 & 2 \\ 15 & 5 & 3 \end{vmatrix} = 12\,(18-10) - 13\,(6-15) + 15\,(4-18) = 3$$

$$D_2 = \begin{vmatrix} 1 & 12 & 3 \\ 1 & 13 & 2 \\ 1 & 15 & 3 \end{vmatrix} = 1\,(39-30) - 12\,(3-2) + 3\,(15-13) = 3$$

$$D_3 = \begin{vmatrix} 1 & 2 & 12 \\ 1 & 6 & 13 \\ 1 & 5 & 15 \end{vmatrix} = 1\,(90-65) - 2\,(15-13) + 12\,(5-6) = 9$$

$$\therefore \qquad a = \frac{3}{3} = 1, \quad b = \frac{3}{3} = 1, \quad c = \frac{9}{3} = 3$$

Hence the total sales (S) is given by $S = 1 + X + 3Y$. In particular, when $X = 4$ and $Y = 5$, the total sales is given by $S = 1 + 4 + 3\,(5) = 20$.

Example* 27.** To control a crop disease it is necessary to use 8 units of chemical A, 14 units of chemical B and 13 units of chemical C. One barrel of spray P contains one unit of A, 2 units of B and 3 units of C. One barrel of spray Q contains 2 units of A, 3 units of B and 2 units of C. One barrel of spray R contains 1 unit of A, 2 units of B and 2 unit of C. Find how many barrels of each spray be used to just meet the requirement. (Solve by Matrix Algebra.) ***[Delhi Univ. B.Com. (H) 2005, 2007, 2022]

Solution. Let x barrels of spray P, y barrels of spray Q and z barrels of spray R be used to just meet the requirements. Then, according to the given information, we have

$$x + 2y + z = 8$$

$$2x + 3y + 2z = 14$$

$$3x + 2y + 2z = 13$$

The above system can be expressed in the matrix form as

$$\underset{\mathbf{A}}{\begin{bmatrix} 1 & 2 & 1 \\ 2 & 3 & 2 \\ 3 & 2 & 2 \end{bmatrix}} \underset{\mathbf{X}}{\begin{bmatrix} x \\ y \\ z \end{bmatrix}} = \underset{\mathbf{B}}{\begin{bmatrix} 8 \\ 14 \\ 13 \end{bmatrix}}$$

The determinant of the coefficient matrix **A** is : $|\mathbf{A}| = 1$

Since $|\mathbf{A}| \neq 0$, $\mathbf{A}^{-1}$ exists and the solution is given by $\mathbf{X} = \mathbf{A}^{-1}\mathbf{B}$. It can be easily verified that

$$\mathbf{A}^{-1} = \begin{bmatrix} 2 & -2 & 1 \\ 2 & -1 & 0 \\ -5 & 4 & -1 \end{bmatrix}$$

Hence the solution is : $\mathbf{X} = \mathbf{A}^{-1}\mathbf{B} = \begin{bmatrix} 2 & -2 & 1 \\ 2 & -1 & 0 \\ -5 & 4 & -1 \end{bmatrix} \begin{bmatrix} 8 \\ 14 \\ 13 \end{bmatrix} = \begin{bmatrix} 1 \\ 2 \\ 3 \end{bmatrix}$

$\Rightarrow$ $x=1,\ y=2$ and $z=3.$

Thus 1 barrel of *P*, 2 barrels of *Q* and 3 barrels of *R* must be used to just meet the requirements.

***Example* 28.** In an engineering workshop there are 10 machines for drilling, 8 machines for turning and 7 machines for grinding. Three types of brackets are made. Type *I* brackets require 0 minutes for drilling, 5 minutes for turning and 4 minutes for grinding. The corresponding times for type *II* and *III* brackets are 3, 3, 2 and 3, 2, 2 minutes respectively. How many brackets of each type should be produced per hour so that all the machines remain fully occupied during an hour ? Solve by matrix algebra.

[*Delhi Univ. B.Com. (H) 2008, 2010*]

***Solution*.** The data of the problem may be summarized in the tabular form as follows :

Brackets	*Drilling*	*Turning*	*Grinding*
Type I	0	5	4
Type II	3	3	2
Type III	3	2	2
Time available (in minutes)	10 × 60 = 600	8 × 60 = 480	7 × 60 = 420

Let *x, y* and *z* denote the number of brackets produced of each type. Then we have

$$\begin{aligned} 0x + 3y + 3z &= 600 \\ 5x + 3y + 2z &= 480 \\ 4x + 2y + 2z &= 420 \end{aligned}$$

The above system can be expressed in the matrix form as

$$\underset{\mathbf{A}}{\begin{bmatrix} 0 & 3 & 3 \\ 5 & 3 & 2 \\ 4 & 2 & 2 \end{bmatrix}} \underset{\mathbf{X}}{\begin{bmatrix} x \\ y \\ z \end{bmatrix}} = \underset{\mathbf{B}}{\begin{bmatrix} 600 \\ 480 \\ 420 \end{bmatrix}}$$

The determinant of the coefficient matrix is : $|\mathbf{A}| = 0 - 3\,(10-8) + 3\,(10-12) = -12 \neq 0$

Since $|\mathbf{A}| \neq 0$, $\mathbf{A}^{-1}$ exists and the solution is given by $\mathbf{X} = \mathbf{A}^{-1}\mathbf{B}$. It can be easily verified that

$$\mathbf{A}^{-1} = -\frac{1}{12}\begin{bmatrix} 2 & 0 & -3 \\ -2 & -12 & 15 \\ -2 & 12 & -15 \end{bmatrix}$$

$$\therefore \quad \mathbf{X} = \mathbf{A}^{-1}\mathbf{B} = -\frac{1}{12}\begin{bmatrix} 2 & 0 & -3 \\ -2 & -12 & 15 \\ -2 & 12 & -15 \end{bmatrix}\begin{bmatrix} 600 \\ 480 \\ 420 \end{bmatrix} = -\frac{1}{12}\begin{bmatrix} -60 \\ -660 \\ -1740 \end{bmatrix} = \begin{bmatrix} 5 \\ 55 \\ 145 \end{bmatrix}$$

Thus $x = 5$, $y = 55$ and $z = 145$.

***Example* 29.** Based on the past marketing trends and his own experience, marketing expert suggested to the concerned company the segments of market for their product as follows : The first segment consisted of lower income class, the second segment that of middle income and the third segment that of high income. The data based on the income of the customers was readily available. During a particular month in a particular area, the agent reported that for the three products of the company the following were the sales : There were 200 customers who bought all the three products, 240 customers bought only products *I* and *III*, 60 customers only products *II* and *III* and 80 customers only products *I* and *II*. There were 120 persons who bought only product *I*; 220 persons only product *II* and 80 persons only *III* regardless of the market segmentation groups. Based on the market segmentation analysis, for product *I*, the percentage for the income groups are given as (40%, 20% and 40%), for product *II* (30%, 20% and 50%), for product *III* (10%, 50%) and 40%). Using matrix method, find out the number of persons in the lower income, middle income and higher income class in the region referred.

Solution. Number of persons purchasing product *I* $= 200+240+80+120 = 640$

Number of persons purchasing product *II* $= 200+60+80+220 = 560$

Number of persons purchasing product *III* $= 200+60+240+80 = 580$

Let x, y and z denote the number of persons in the lower income group, middle income group and higher income group, respectively. Then, according to the given data, we have

$$\begin{aligned} \frac{40}{100}x + \frac{20}{100}y + \frac{40}{100}z &= 640 & \qquad 4x + 2y + 4z &= 6400 \\ \frac{30}{100}x + \frac{20}{100}y + \frac{50}{100}z &= 560 & \Leftrightarrow \qquad 3x + 2y + 5z &= 5600 \\ \frac{10}{100}x + \frac{50}{100}y + \frac{40}{100}z &= 580 & \qquad x + 5y + 4z &= 5800 \end{aligned}$$

The above system can be written in the matrix form as

$$\underset{\mathbf{A}}{\begin{bmatrix} 4 & 2 & 4 \\ 3 & 2 & 5 \\ 1 & 5 & 4 \end{bmatrix}}\underset{\mathbf{X}}{\begin{bmatrix} x \\ y \\ z \end{bmatrix}} = \underset{\mathbf{B}}{\begin{bmatrix} 6400 \\ 5600 \\ 5800 \end{bmatrix}}$$

The determinant of the coefficient matrix **A** is : $|\mathbf{A}| = 4(8-25) - 2(12-5) + 4(15-2) = -30$.

Since $|\mathbf{A}| \neq 0$, $\mathbf{A}^{-1}$ exists and the solution is given by $\mathbf{X} = \mathbf{A}^{-1}\mathbf{B}$. It can be easily verified that

$$\mathbf{A}^{-1} = -\frac{1}{30}\begin{bmatrix} -17 & 12 & 2 \\ -7 & 12 & -8 \\ 13 & -18 & 2 \end{bmatrix}$$

$$\therefore \quad \mathbf{X} = \mathbf{A}^{-1}\mathbf{B} = -\frac{1}{30}\begin{bmatrix} -17 & 12 & 2 \\ -7 & 12 & -8 \\ 13 & -18 & 2 \end{bmatrix}\begin{bmatrix} 6400 \\ 5600 \\ 5800 \end{bmatrix} = \begin{bmatrix} 1000 \\ 800 \\ 200 \end{bmatrix}$$

Hence $x = 1000,\ y = 800$ and $z = 200.$

***Example* 30.** A Company earns *before-tax* profits of ₹ 1,00,000. It is committed to making a donation to the Red Cross 10% of its *after-tax* profits. The central government levies corporate taxes of 50% of profits after deducting charitable donations and any local taxes. The company must also pay local taxes of 10% of its profits less the donation to the Red Cross. Compute how much the company pays in corporate taxes, local taxes and as a donation to the Red Cross, using Cramer's rule.

Solution. Let x = the amount donated to the Red Cross

y = the amount paid in corporate taxes

z = the amount paid in local taxes

Then, according to the given information, we have

$$\begin{aligned} x &= 0.10(1,00,000 - y - z) \\ y &= 0.50(1,00,000 - x - z) \\ z &= 0.10(1,00,000 - x) \end{aligned} \quad \Leftrightarrow \quad \begin{aligned} 10x + y + z &= 1,00,000 \\ x + 2y + z &= 1,00,000 \\ x + 0y + 10z &= 1,00,000 \end{aligned}$$

Now D = the determinant of the coefficient matrix $= \begin{vmatrix} 10 & 1 & 1 \\ 1 & 2 & 1 \\ 1 & 0 & 10 \end{vmatrix} = 189$

Since $D \neq 0$, therefore by Cramer's rule, the solution is given by

$$x = \frac{D_1}{D},\quad y = \frac{D_2}{D},\quad z = \frac{D_3}{D},\ \text{where}$$

$$D_1 = \begin{vmatrix} 1,00,000 & 1 & 1 \\ 1,00,000 & 2 & 1 \\ 1,00,000 & 0 & 10 \end{vmatrix} = 1,00,000\begin{vmatrix} 1 & 1 & 1 \\ 1 & 2 & 1 \\ 1 & 0 & 10 \end{vmatrix} = 1,00,000(9) = 9,00,000$$

$$D_2 = \begin{vmatrix} 10 & 1,00,000 & 1 \\ 1 & 1,00,000 & 1 \\ 1 & 1,00,000 & 10 \end{vmatrix} = 1,00,000\begin{vmatrix} 10 & 1 & 1 \\ 1 & 1 & 1 \\ 1 & 1 & 10 \end{vmatrix} = 1,00,000(81) = 81,00,000$$

$$D_3 = \begin{vmatrix} 10 & 1 & 1,00,000 \\ 1 & 2 & 1,00,000 \\ 1 & 0 & 1,00,000 \end{vmatrix} = 1,00,000\begin{vmatrix} 10 & 1 & 1 \\ 1 & 2 & 1 \\ 1 & 0 & 1 \end{vmatrix} = 1,00,000(18) = 18,00,000$$

$$\therefore \quad x = \frac{D_1}{D} = \frac{9,00,000}{189},\quad y = \frac{D_2}{D} = \frac{81,00,000}{189} \ \text{and}\ z = \frac{D_3}{D} = \frac{18,00,000}{189}$$

***Example* 31.** A firm produces two products P_1 and P_2 passing through two machines M_1 and M_2 before completion. M_1 can produce either 10 units of P_1 or 15 units of P_2 per hour. M_2 can produce 15 units of either product per hour. Find daily production of P_1 and P_2 if time available is 12 hours on machine M_1 and 10 hours on M_2 per day using matrix inversion. **[*Delhi Univ. B.Com. (H) 1997*]**

***Solution* .** Let the daily production of products P_1 and P_2 be x units and y units respectively. Then

$$\left.\begin{aligned}\frac{x}{10}+\frac{y}{15} &= 12\\ \frac{x}{15}+\frac{y}{15} &= 10\end{aligned}\right. \quad\Leftrightarrow\quad \begin{aligned}3x+2y &= 360\\ x+y &= 150\end{aligned}$$

The above system can be expressed in the matrix form as

$$\begin{bmatrix}3 & 2\\ 1 & 1\end{bmatrix}\begin{bmatrix}x\\ y\end{bmatrix}=\begin{bmatrix}360\\ 150\end{bmatrix}$$

$$\mathbf{A}\quad \mathbf{X} = \mathbf{B}$$

The determinant of the coefficient matrix is : $|\mathbf{A}| = 1$.

Since $|\mathbf{A}| \neq 0$, $\mathbf{A}^{-1}$ exists and the solution is given by $\mathbf{X}=\mathbf{A}^{-1}\mathbf{B}$. It can be easily verified that

$$\mathbf{A}^{-1}=\begin{bmatrix}1 & -2\\ -1 & 3\end{bmatrix} \quad\Rightarrow\quad \mathbf{X}=\mathbf{A}^{-1}\mathbf{B}=\begin{bmatrix}1 & -2\\ -1 & 3\end{bmatrix}\begin{bmatrix}360\\ 150\end{bmatrix}=\begin{bmatrix}60\\ 90\end{bmatrix}$$

Thus $\qquad x = 60$ and $y = 90$.

***Example* 32.** A firm produces three products P_1, P_2 and P_3 processed on three machines M_1, M_2, M_3 before completion. M_1 can process 25 units of P_1 or 50 units of P_2 or 75 units of P_3 per hour. M_2 can process 50 units of any one product per hour. M_3 can process 50 or 25 or 100 units per hour of products P_1, P_2 and P_3 respectively. The processing hours available on machines M_1, M_2 and M_3 are 12, 12 and 13 respectively. Using Matrix notations, find:

(*i*) How many units of three products can be produced per day.

(*ii*) The production cost per unit, if costs per hour of operating machines M_1, M_2 and M_3 are ₹ 500, ₹ 1000 and ₹ 1500 respectively.

(*iii*) The total cost of production. **[*Delhi Univ. B.Com. (H) 2009, 2017*]**

***Solution* .** (*i*) Let the daily production of products P_1, P_2 and P_3 be x units, y units and z units respectively. Then

$$\left.\begin{aligned}\frac{x}{25}+\frac{y}{50}+\frac{z}{75} &= 12\\ \frac{x}{50}+\frac{y}{50}+\frac{z}{50} &= 12\\ \frac{x}{50}+\frac{y}{25}+\frac{z}{100} &= 13\end{aligned}\right. \quad\Leftrightarrow\quad \begin{aligned}6x+3y+2z &= 1800\\ x+y+z &= 600\\ 2x+4y+z &= 1300\end{aligned}$$

The above system can be expressed in the matrix form as

$$\underset{\mathbf{A}}{\begin{bmatrix} 6 & 3 & 2 \\ 1 & 1 & 1 \\ 2 & 4 & 1 \end{bmatrix}} \underset{\mathbf{X}}{\begin{bmatrix} x \\ y \\ z \end{bmatrix}} = \underset{\mathbf{B}}{\begin{bmatrix} 1800 \\ 600 \\ 1300 \end{bmatrix}}$$

Now $\quad |\mathbf{A}| = 6(1-4) - 3(1-2) + 2(4-2) = -11$

Since $|\mathbf{A}| \neq 0$, $\mathbf{A}^{-1}$ exists and the solution is given by $\mathbf{X} = \mathbf{A}^{-1}\mathbf{B}$. It can be easily verified that

$$\mathbf{A}^{-1} = -\frac{1}{11}\begin{bmatrix} -3 & 5 & 1 \\ 1 & 2 & -4 \\ 2 & -18 & 3 \end{bmatrix}$$

$$\therefore \quad \begin{matrix} P_1 \\ P_2 \\ P_3 \end{matrix}\ \mathbf{X} = \mathbf{A}^{-1}\mathbf{B} = -\frac{1}{11}\begin{bmatrix} -3 & 5 & 1 \\ 1 & 2 & -4 \\ 2 & -18 & 3 \end{bmatrix}\begin{bmatrix} 1800 \\ 600 \\ 1300 \end{bmatrix} = -\frac{1}{11}\begin{bmatrix} -1100 \\ -2200 \\ -3300 \end{bmatrix} = \begin{bmatrix} 100 \\ 200 \\ 300 \end{bmatrix}$$

$\Rightarrow \quad \underset{P_1}{x = 100}, \quad \underset{P_2}{y = 200} \quad$ and $\quad \underset{P_3}{z = 300}.$

(*ii*) The production cost per unit of each of the three products is given by

$$\begin{matrix} & M_1 & M_2 & M_3 \end{matrix}$$

$$\begin{bmatrix} \frac{1}{25} & \frac{1}{50} & \frac{1}{50} \\ \frac{1}{50} & \frac{1}{50} & \frac{1}{25} \\ \frac{1}{75} & \frac{1}{50} & \frac{1}{100} \end{bmatrix}\begin{bmatrix} 500 \\ 1000 \\ 1500 \end{bmatrix}\begin{matrix} M_1 \\ M_2 \\ M_3 \end{matrix} = \begin{matrix} P_1 \\ P_2 \\ P_3 \end{matrix}\begin{bmatrix} 70 \\ 90 \\ 41.67 \end{bmatrix}$$

(*iii*) Total cost of production is given by

$$\begin{bmatrix} 100 & 200 & 300 \end{bmatrix}\begin{bmatrix} 70 \\ 90 \\ 41.67 \end{bmatrix}\begin{matrix} P_1 \\ P_2 \\ P_3 \end{matrix} = [37{,}500]$$

Thus the total cost of production is ₹ 37,500.

***Example* 33.** A holding company H has two subsidiaries S_1 and S_2. The subsidiaries also own a portion of the share capital of H. The percentage ownership of the group is given as follows :

Owners of Shares	*Percentage of Share Capital Held in Company*		
	H	S_1	S_2
H	0	60	80
S_1	3	0	10
S_2	6	10	0
Outside Shareholders	91	30	10
Total	100	100	100

If the separately earned profits of H, S_1 and S_2 are ₹ 30,000, ₹ 17,500 and ₹ 5000 respectively; find the total net profit of each company (separately earned profit plus share in the profits of the other companies) by using matrix algebra. Also, verify that sum of the profits allocated to the outside shareholders equals the sum of the separately earned profits of all the companies. **[*Delhi Univ. B.Com. (H) 1990*]**

Solution. Let the total net profit of H, S_1 and S_2 be x, y and z respectively. Then from the given data, we obtain the following equations :

$$\begin{aligned} x &= 30{,}000 + 0.6y + 0.8z \\ y &= 17{,}500 + 0.03x + 0.1z \\ z &= 5{,}000 + 0.06x + 0.1y \end{aligned} \quad \Rightarrow \quad \begin{aligned} x - 0.6y - 0.8z &= 30{,}000 \\ -0.03x + y - 0.1z &= 17{,}500 \\ -0.06x - 0.1y + z &= 5{,}000 \end{aligned}$$

These equations can be expressed in the matrix form as :

$$\underset{\mathbf{A}}{\begin{bmatrix} 1 & -0.6 & -0.8 \\ -0.03 & 1 & -0.1 \\ -0.06 & -0.1 & 1 \end{bmatrix}} \underset{\mathbf{X}}{\begin{bmatrix} x \\ y \\ z \end{bmatrix}} = \underset{\mathbf{B}}{\begin{bmatrix} 30{,}000 \\ 17{,}500 \\ 5000 \end{bmatrix}}$$

Now $\quad |\mathbf{A}| = 1\,(1 - 0.01) + 0.6\,(-0.03 - 0.006) - 0.8\,(0.003 + 0.06) = 0.918$

Since $|\mathbf{A}| \neq 0$, $\mathbf{A}^{-1}$ exists and the solution is given by $\mathbf{X} = \mathbf{A}^{-1}\mathbf{B}$. It can be verified that

$$\mathbf{A}^{-1} = \frac{1}{0.918}\begin{bmatrix} 0.99 & 0.68 & 0.86 \\ 0.036 & 0.952 & 0.124 \\ 0.063 & 0.136 & 0.982 \end{bmatrix}$$

$$\therefore \quad \mathbf{X} = \mathbf{A}^{-1}\mathbf{B} = \frac{1}{0.918}\begin{bmatrix} 0.99 & 0.68 & 0.86 \\ 0.036 & 0.952 & 0.124 \\ 0.063 & 0.136 & 0.982 \end{bmatrix}\begin{bmatrix} 30{,}000 \\ 17{,}500 \\ 5000 \end{bmatrix} = \frac{1}{0.918}\begin{bmatrix} 45{,}900 \\ 18{,}360 \\ 9180 \end{bmatrix} = \begin{bmatrix} 50{,}000 \\ 20{,}000 \\ 10{,}000 \end{bmatrix}$$

Thus $x = ₹\ 50{,}000$, $y = ₹\ 20{,}000$ and $z = ₹\ 10{,}000$.

Now sum of the profits allocated to the outside shareholders is given by

$$[0.91 \quad 0.3 \quad 0.1]\begin{bmatrix} 50{,}000 \\ 20{,}000 \\ 10{,}000 \end{bmatrix} = [52{,}500]$$

and sum of the separately earned profits of all the companies is :

$$30{,}000 + 17{,}500 + 5000 = 52{,}500$$

Hence the sum of the profits allocated to the outside shareholders equals the sum of the separately earned profits of all the companies.

***Example* 34.** A firm has three service departments, S_1, S_2, and S_3 and two production departments, P_1 and P_2. The direct cost of each department and the percentage of total cost of each service department allocated to various departments are given as follows :

Department	*Direct cost (in ₹)*	*Percentage allocation of total cost of department*		
		S_1	S_2	S_3
S_1	14,000	0	30	20
S_2	12,000	10	0	10
S_3	45,000	10	10	0
P_1	50,000	50	30	40
P_2	60,000	30	30	30
	Total	100	100	100

Find the total cost (direct plus allocated) of each service department by using matrix algebra. Verify that the sum of the service costs allocated to the production departments equals the sum of the direct costs of the service departments. **[*Delhi Univ. B.Com. (H) 1992*]**

Solution. Let x denote the total cost of the service department S_1, y the total cost of S_2 and z the total cost of S_3. Since the total cost of each department is its direct costs plus its allocated costs, therefore we obtain the following equations :

$$\begin{aligned} x &= 14{,}000 + 0.3y + 0.2z \\ y &= 12{,}000 + 0.1x + 0.1z \\ z &= 45{,}000 + 0.1x + 0.1y \end{aligned} \quad \Leftrightarrow \quad \begin{aligned} x - 0.3y - 0.2z &= 14{,}000 \\ -0.1x + y - 0.1z &= 12{,}000 \\ -0.1x - 0.1y + z &= 45{,}000 \end{aligned}$$

This can be written in the matrix form as

$$\underset{\mathbf{A}}{\begin{bmatrix} 1 & -0.3 & -0.2 \\ -0.1 & 1 & -0.1 \\ -0.1 & -0.1 & 1 \end{bmatrix}} \underset{\mathbf{X}}{\begin{bmatrix} x \\ y \\ z \end{bmatrix}} = \underset{\mathbf{B}}{\begin{bmatrix} 14{,}000 \\ 12{,}000 \\ 45{,}000 \end{bmatrix}}$$

Now, $|\mathbf{A}| = 1\,(1 - 0.01) + 0.3\,(-0.1 - 0.01) - 0.2\,(0.01 + 0.1) = 0.935$

Since $|\mathbf{A}| \neq 0$, $\mathbf{A}^{-1}$ exists and the solution is given by $\mathbf{X} = \mathbf{A}^{-1}\mathbf{B}$. It can be verified that

$$\mathbf{A}^{-1} = \frac{1}{0.935}\begin{bmatrix} 0.99 & 0.32 & 0.23 \\ 0.11 & 0.98 & 0.12 \\ 0.11 & 0.13 & 0.97 \end{bmatrix}$$

$$\therefore \quad \mathbf{X} = \mathbf{A}^{-1}\mathbf{B} = \frac{1}{0.935}\begin{bmatrix} 0.99 & 0.32 & 0.23 \\ 0.11 & 0.98 & 0.12 \\ 0.11 & 0.13 & 0.97 \end{bmatrix}\begin{bmatrix} 14{,}000 \\ 12{,}000 \\ 45{,}000 \end{bmatrix} = \frac{1}{0.935}\begin{bmatrix} 28{,}050 \\ 18{,}700 \\ 46{,}750 \end{bmatrix} = \begin{bmatrix} 30{,}000 \\ 20{,}000 \\ 50{,}000 \end{bmatrix}$$

Thus $x =$ ₹ 30,000, $y =$ ₹ 20,000 and $z =$ ₹ 50,000.

Now, sum of the service costs allocated to the production departments is given by

$$\begin{bmatrix} 0.5 & 0.3 & 0.4 \end{bmatrix}\begin{bmatrix} 30{,}000 \\ 20{,}000 \\ 50{,}000 \end{bmatrix} + \begin{bmatrix} 0.3 & 0.3 & 0.3 \end{bmatrix}\begin{bmatrix} 30{,}000 \\ 20{,}000 \\ 50{,}000 \end{bmatrix} = [41{,}000] + [30{,}000] = [71{,}000],$$

which is same as the sum of the direct costs of the service departments (14,000 + 12,000 + 45,000 = 71,000).

EXERCISE 2.2

1. An amount of ₹ 10,000 is put into three investments at the rate of interest of 10, 12 and 15 per cent per annum. The combined income is ₹ 1310 and the combined income of the first and second investment is ₹ 190 short of the income from the third. Find the investment in each using determinant method (Cramer's rule). ***[Delhi Univ. B.Com. (H) 1984]***

2. A person has ₹35,000 that he has divided into three investments. Part of the money is invested in a savings account with an annual rate of interest of 6%, part in 7% annual yield bonds, and the remainder in a business. In 2015, when he lost 6% of the money that he invested in that business, his net income from all three investments was ₹ 660. If he invested ₹ 3000 more in the business than in the savings account, how much was invested in each ? Using matrix algebra. ***[Delhi Univ. B.Com. (H) 2017]***

3. An amount of ₹ 65,000 is invested in three investments at rates of 6%, 8% and 9% per annum respectively. The total annual income is ₹ 4800. The income from the third investment is ₹ 600 more than the income from the second investment. Using matrix algebra, determine the amount of each investment. ***[Delhi Univ. B.Com. (H) 1995]***

4. An investor has 30,000 that he wants to invest in bank deposits, equity shares and unit trust. In view of the risk involved in buying equity shares, he wants to invest an amount in equity shares equal to 20% of his total investment in bank deposits and unit trust. Because of certain tax

exemptions available to him, he would also like to maintain a 3 : 2 ratio between investments in bank deposits and unit trust. Determine the amount he would invest in each of the three forms of investments, using matrix inversion method. **[*Delhi Univ. B.Com. (H) 2019*]**

5. *A*, *B* and *C* have ₹ 1250, ₹ 1700 and ₹ 2100 respectively. They utilized the amount to purchase three types of shares of prices *x*, *y* and *z* respectively. *A* purchases 20 shares of price *x*; 50 shares of price *y* and 30 shares of price *z*. *B* purchases 44 shares of price *x*, 30 shares of price *y* and 60 shares of price *z*. *C* purchases 12 shares of price *x*, 40 shares of price *y* and 100 shares of price *z*. Find *x*, *y* and *z* using matrix inversion method.

6. The prices of three commodities *X*, *Y* and *Z* are, *x*, *y* and *z* per unit respectively. A purchases 4 units of *Z* and sells 3 units of *X* and 5 units of *Y*. *B* purchases 3 unit of *Y* and sells 2 units of *X* and 1 unit of *Z*. *C* purchases 1 unit of *X* and sells 4 units of *Y* and 6 units of *Z*. In the process, *A*, *B*, *C* earn ₹ 6000, ₹ 5000 and ₹ 13,000 respectively. Using matrices, find the prices per unit of the three commodities (note that selling the units is positive earning and buying the units is negative earning).

7. A company produces three products everyday. Their total production on a certain day is 45 tons. It is found that the production of third product exceeds the production of first product by 8 tons while the total production of first and third product is twice the production of second product. Determine the production level of each product using Cramer's rule.

[*Delhi Univ. B. Com. (H) 1988, 2009*]

8. A business has ₹ 40,000 to spend on advertising in an upcoming sale. The money is to be divided between television (*x*), radio (*y*) and newspapers (*z*). The business manager has decided to spend three times as much money on television as on radio. The manager has also decided to spend ₹ 8000 less on radio advertising than on newspapers. Find out, using matrix inversion method, the total amount divided between television, radio and newspapers.

9. An automobile company uses three types of steel s_1, s_2 and s_3 for producing three types of cars c_1, c_2 and c_3. Steel requirements (in tons) for each type of cars are given below :

		Cars		
		c_1	c_2	c_3
	s_1	2	3	4
Steel	s_2	1	1	2
	s_3	3	2	1

Determine the number of cars of each type which can be produced using 29, 13 and 16 tons of steel of three types respectively. **[*Delhi Univ. B.Com. (H) 1985*]**

10. Three products *X*, *Y* and *Z* are produced after being processed through three departments D_1, D_2 and D_3. The following data are available :

Products	*Hours required for a unit product*		
	in D_1	*in* D_2	*in* D_3
X	2	5	1
Y	1	2	3
Z	2	2	3
Maximum time available in hours	1100	1800	1400

Find by matrix method the number of units produced for each product to have full utilization of capacity. **[*Delhi Univ. B.Com. (H) 1986*]**

11. A salesman has the following record of sales during three weeks for the three items *X*, *Y* and *Z* which have different rates of commission.

Weeks	*Units sold*			*Total Commission (in ₹)*
	X	*Y*	*Z*	
I	40	30	20	270
II	50	50	40	450
III	60	30	10	260

Find out, using matrix method, the rates of commission on items *X*, *Y* and *Z*.

[*Delhi Univ. B.Com. (H) 1988*]

12. The total cost *C* for a manufacturer producing two items *X* and *Y* is given as $C = p + qX + rY$. Data for three weeks is as follows :

Week	*Cost*	*No. of items produced*	
		X	*Y*
1	6950	40	10
2	6725	35	9
3	7100	40	12

Determine *p, q* and *r* using determinant method.

13. The equilibrium condition for three related markets is given by :

$$\begin{aligned} 11p_1 - p_2 - p_3 &= 31 \\ -p_1 + 6p_2 - 2p_3 &= 26 \\ -p_1 - 2p_2 + 7p_3 &= 24 \end{aligned}$$

Using matrix inversion, find the equilibrium price for each market.

[*Delhi Univ. B.Com. (H) 1994*]

14. The prices per ton of wheat and rye are p_1 and p_2 respectively. The market demand for wheat is given by $x_1 = 4 - 10p_1 + 7p_2$ and for rye by $x_2 = 3 + 7p_1 - 5p_2$. The supply of wheat is related to prices by the relation $x_1 = 7 + p_1 - p_2$ and supply of rye by $x_2 = -27 - p_1 + 2p_2$.

(*i*) Write down the relevant simultaneous equations for determining the equilibrium prices in matrix form.

(*ii*) Find the equilibrium prices and quantities using Cramer's rule.

[*Delhi Univ. B.A. Eco. (H.) 1990*]

15. A factory producing leather bags has three types of workers : skilled (Grade *I*), skilled (Grade *II*) and semi-skilled. Average number of bags produced per day by the skilled (Grade *I*), skilled (Grade *II*) and semi-skilled are 8, 6 and 4 respectively, the average daily production being 108 bags. One day due to local strike only half of the skilled (Grade *II*) and semi-skilled and one-fourth of skilled (Grade *I*) workers attended the factory and only 46 bags were produced. Find the number of workers of each type, the total number of workers in the factory being 20 only. (*Use matrix inversion method*)

16. In a certain university, the number of students from 3 different religions aggregate 10,000. The largest religion group is 1000 more than twice as large as the other two groups. If the smallest group has 2000 less than the medium group, then find how many students are there in each group. Use matrix method.

17. The purchasing officer of the Pain-killing Co. informs the production manager that during the month, following supply of three chemicals, Asprin (A), Caffeine (C) and Decongestant (D) used in the production of three types of pain-killing tablets, Paingo (x), X-Prene (y) and Relaxo (z), will be 16, 10 and 16 kg respectively. According to the specifications, each strip of 10 tablets of Paingo requires 2 gm of A, 3gm of C and 1 gm of D. The requirements for other tablets are :

X-Prene : 4gm of A, 1 gm of C and 3 gm of D

Relaxo : 1 gm of A, 2 gm of C and 3 gm of D

Use matrix inversion method to find the number of strips of each type so that raw materials are consumed entirely.

18. A transport company uses 3 types of trucks T_1, T_2 and T_3 to transport 3 types of vehicles V_1, V_2 and V_3. The capacity of each truck in terms of 3 types of vehicles is given below :

	V_1	V_2	V_3
T_1	2	3	6
T_2	6	7	6
T_3	9	12	8

Using matrix method, find :

(*i*) the number of trucks of each type required to transport 58, 75 and 62 vehicles of V_1, V_2 and V_3 types respectively.

(*ii*) the number of vehicles of each type which can be transported if company has 15, 10 and 20 trucks of each type respectively.

19. A firm produces two products P_1 and P_2 passing through two machines M_1 and M_2 before completion. M_1 can produce either 8 units of P_1 or 10 units of P_2 per hour. M_2 can produce 12 units of either product per hour. Using matrix notations, determine:

(*i*) Production of P_1 and P_2 if time available on two machines is 33 hours and 25 hours respectively.

(*ii*) Per unit cost of production if cost of operating per hour on two machines is ₹ 400 and ₹ 300 respectively.

(*iii*) Total cost of production. **[*Delhi Univ. B.Com. (H) 2013*]**

20. A firm produces three products P_1, P_2 and P_3 processed on four machines M_1, M_2, M_3 and M_4. M_1 can process 25 units of P_1 or 50 units of P_2 or 75 units of P_3 per hour. M_2 can process 50 units of any product per hour. M_3 can process 50 or 25 or 100 units per hour and machine M_4 can process 50 or 40 or 50 units per hour of P_1, P_2 and P_3 respectively. The processing hours available on the machines M_1, M_2, M_3 and M_4 are 12, 12, 13 and 13 respectively. Using matrices, find how many units of three products can be produced with the available time fully used? **[*Delhi Univ. B.Com. (H) 2010*]**

21. Three companies X, Y and Z hold shares in each other. The percentage ownership of the three companies is given below :

Owners of Shares	*Percentage of Share Capital held in Company*		
	X	Y	Z
X	0	30	20
Y	30	0	10
Z	10	20	0
Outside Share holders	60	50	70
Total	100	100	100

If the separately earned profits of X, Y and Z are ₹ 53,000, ₹ 4000 and ₹ 1,40,000 respectively, find the total net profit of each company (separately earned profit plus share in the profits of the other companies) by using matrix algebra. Also, verify that sum of the profits allocated to the outside shareholders equals the sum of the separately earned profits of all the companies.

22. Two companies A and B are holding shares in each other. A is holding 20% shares of B and B is holding 10% shares of A. If the separately earned profits of the two companies are ₹ 98,000 and ₹ 49,000 respectively, find the total profit of each company using matrix notations. Also, show that the total of the profits allocated to outside share holders is equal to the total of separately earned profits. **[*Delhi Univ. B.Com. (H) 2004*]**

23. A company has two production departments, P_1 and P_2 and three service departments S_1, S_2 and S_3. The direct cost allocated to each of the departments and the percentage of total cost of each service department apportioned to various departments are given below :

Department	*Direct Cost* (₹)	*Percentage Allocation of Total Cost*		
		S_1	S_2	S_3
P_1	60,000	40	35	25
P_2	74,000	20	40	20
S_1	6000	0	20	35
S_2	8000	15	0	20
S_3	68,500	25	5	0

Determine the total cost (allocated and apportioned) for each production department by using matrix algebra. **[*Delhi Univ. B.Com. (H) 1994*]**

24. A firm has two service departments, S_1, S_2 and three production departments, P_1, P_2 and P_3. The direct cost of each department and the percentage of the total cost of each department allocated to various departments are given below :

Service Department	*Percentages to be Allocated to Departments*				
	S_1	S_2	P_1	P_2	P_3
S_1	0	10	20	40	30
S_2	20	0	50	10	20
Direct Cost (in ₹)	98,000	1,17,600	14,00,000	21,00,000	6,40,000

(*i*) Express the relevant simultaneous equations for determining the total cost (direct plus allocated) of each service department in matrix forms;

(*ii*) Determine the total cost of each service department using matrix method; and

(*iii*) Verify that the sum of the sevice costs allocated to the production departments equals the sum of the direct costs of the service departments.

25. ABC Co. has two service departments S_1 and S_2 and four production departments P_1, P_2, P_3 and P_4. Overhead is allocated to production department for inclusion in the stock valuation. The analysis of benefits received by each department during the last quarter and the direct overhead expenses (in thousand of ₹) incurred by each department were :

	Percentage to be allocated to Departments					
Service Departments	S_1	S_2	P_1	P_2	P_3	P_4
S_1	0	20	30	25	15	10
S_2	30	0	10	35	20	5
Direct Overhead Exp.	20	40	25	30	20	10

(*i*) Express the total overhead of the service department in the form of simultaneous equations.

(*ii*) Express these equations in a matrix form and solve for total overhead of service departments using matrix inverse method.

(*iii*) Determine total overhead to be allocated from each S_1 and S_2 to the production departments.

[*Delhi Univ. B.Com. (H) 2003*]

26. A firm has three service departments S_1, S_2 and S_3 and two production departments P_1 and P_2. The direct cost of each department and the percentage of the total cost of each department allocated to various departments are given below :

Department	*Direct Cost (in ₹)*	*Percentage Allocation of Total Cost*		
		S_1	S_2	S_3
S_1	25,000	0	30	10
S_2	1,24,000	20	0	10
S_3	77,000	10	10	0
P_1	1,00,000	40	30	30
P_2	2,00,000	30	30	50
	Total	100	100	100

Find the total cost (direct plus allocated) of each department by using matrix algebra. Verify that the sum of the service costs allocated to the production departments equals the sum of the direct costs of the service departments.

ANSWERS

1. ₹ 2000, ₹ 3000, ₹ 5000 **2.** Savings : ₹ 10,000, Bonds : ₹ 12,000, Business : ₹ 13,000

3. ₹ 30,000, ₹ 15,000, ₹ 20,000

4. Bank deposits : ₹ 15,000, Equity shares : ₹ 5000, Unit Trust : ₹ 10,000.

5. $x = 10, \quad y = 12, \quad z = 15$ **6.** $x = 3000, \quad y = 1000, \quad z = 2000.$ **7.** 11, 15, 19

8. $x = 19{,}200, \quad y = 6400, \quad z = 14{,}400.$

9. Type $c_1 = 2$, Type $c_2 = 3$ and Type $c_3 = 4$

10. $X = 200, \quad Y = 100, \quad Z = 300$ **11.** Rates of commission : $X = ₹\,2, \quad Y = ₹\,3, \quad Z = ₹\,5$

12. $p = 5000, \quad q = 30, \quad r = 75$ **13.** $p_1 = 4, \quad p_2 = 7, \quad p_3 = 6$

14. (*i*) $\begin{bmatrix} 11 & -8 \\ 8 & -7 \end{bmatrix} \begin{bmatrix} p_1 \\ p_2 \end{bmatrix} = \begin{bmatrix} -3 \\ -30 \end{bmatrix}$ (*ii*) $p_1 = \dfrac{219}{13}, \quad p_2 = \dfrac{306}{13}, \quad x_1 = \dfrac{4}{13}, \quad x_2 = \dfrac{42}{13}$

15. Skilled (Grade *I*) = 4, Skilled (Grade *II*) = 6, Semi-skilled = 10

16. 7000, 2500, 500

17. Paingo : 1000, X-Prene : 3000, Relaxo : 2000

18. (*i*) $T_1 = 2, \quad T_2 = 3, \quad T_3 = 4$ (*ii*) $V_1 = 270, \quad V_2 = 355, \quad V_3 = 310$

19. (*i*) $P_1 = 120, \quad P_2 = 180$ (*ii*) $P_1 = ₹75, \quad P_2 = ₹65$ (*iii*) ₹ 20,700

20. $P_1 = 100, \quad P_2 = 200, \quad P_3 = 300$ **21.** ₹ 1,00,000, ₹ 50,000 ₹ 1,60,000

22. A : ₹ 1,10,000, B : ₹ 60,000 **23.** P_1 : ₹1,06,500, P_2 : ₹1,10,000

24. (*i*) $\begin{bmatrix} 0 & -0.2 \\ -0.1 & 1 \end{bmatrix} \begin{bmatrix} x \\ y \end{bmatrix} = \begin{bmatrix} 98{,}000 \\ 1{,}17{,}600 \end{bmatrix}$ (*ii*) S_1 : ₹1,24,000, S_2 : ₹1,30,000

25. (*i*) $x - 0.3y = 20{,}000$ (*ii*) S_1 : ₹34,043 (*app.*), S_2 : ₹46,809 (*app.*)
$-0.2x + y = 40{,}000$

(*iii*) Total overhead expenses of S_1 allocated to production departments P_1, P_2, P_3 and P_4 :
$P_1 = ₹10{,}213, \quad P_2 = ₹8511, \quad P_3 = ₹5106, \quad P_4 = ₹3404$

Total overhead expenses of S_2 allocated to production departments P_1, P_2, P_3 and P_4 :
$P_1 = ₹4681, \quad P_2 = ₹16{,}383, \quad P_3 = ₹9362, \quad P_4 = ₹2340$

26. $S_1 = ₹80{,}000, \quad S_2 = ₹1{,}50{,}000, \quad S_3 = ₹1{,}00{,}000, \quad P_1 = ₹2{,}07{,}000, \quad P_2 = ₹3{,}19{,}000$

2.7 THE LEONTIEF INPUT-OUTPUT MODEL

The Leontief input-output model in economics is named after Wassily Leontief who received the Nobel Prize in Economics in 1973. This model gives a description of an economy in which it is assumed that whatever is produced is consumed, *i.e.* production equals consumption or, in other words, input equals output. There are two types of Leontief models, *viz.*, closed and open.

Closed Model* . In this model, the entire production is consumed by those participating in the production, *i.e.*, production is consumed internally. The objective in the closed model is to find the relative income of each participant in the system.

Open Model. In this model, some of the production is consumed by those who produce it and the rest of the production is consumed by external bodies such as households,

* **Note.** *The Closed Model has been deleted as per the new syllabus issued by the University of Delhi for B. Com. (H) (Semester III), Paper BCH : DSC-3.1*

government and so on. The objective in the open model is to find the amount of production needed to meet a forecasted demand, when the amount of production needed to meet the current demand is known.

The Open Model

The general open Leontief input-output model can be described as follows :

Suppose there are n industries in the economy. Each industry produces some goods or services, which are partially consumed internally by the n industries, while the rest are used by consumers such as households, government and so on for final use called the *final demand.*

The following table, called **input-output table**, describes the interdependence of industries during some time period:

INPUT-OUTPUT TABLE

	Consumers (input)					
Producers (output)	*Industry* 1	*Industry* 2	....	*Industry* n	*Final Demand*	*Total output*
Industry 1	x_{11}	x_{12}	....	x_{1n}	c_1	x_1
Industry 2	x_{21}	x_{22}	....	x_{2n}	c_2	x_2
.	.	.		.	.	.
.	.	.		.	.	.
.	.	.		.	.	.
Industry n	x_{n1}	x_{n2}	....	x_{nn}	c_n	x_n

All entries in the table are in appropriate units, say, in millions of rupees of output. Each industry appears in a row and column. The rows show the sales, called *outputs,* that each industry makes to the others. The columns show the purchases, called *inputs,* that each industry makes from the others. There are two additional columns. The first represents the final demand, showing the consumption by other agencies such as government or in exports. The last column represents the total output.

Reading across Row 1, we see that of the total output of x_1 units worth of product of Industry 1, x_{11} is used as input by industry 1 itself (for internal use), x_{12} by Industry 2, and so on and the remaining c_1 used directly by the consumer for final use. The other rows interpreted in the same way.

Reading down Column 1, we see that in producing x_1 units' worth of product, Industry 1 requires x_{11} units' worth from itself, x_{21} units' worth from Industry 2, and so on. The other columns are interpreted in the same way.

Input-Output analysis allows us to estimate the total production of each industry if there is a change in final demand as long as the basic structure of the economy remains the same. That is, the amount of input required from each industry to produce one unit worth of output of a given industry must remain fixed.

Now, suppose that the value of final demand changes from c_1 to d_1 for Industry 1, c_2 to d_2 for Industry 2, and so on. To effect this change, we would like to estimate how much should each industry now produce ?

To obtain the solution, we need to determine how much of outputs of each industry is required to produce 1 unit worth of output of a given industry. For example, to obtain x_1 units worth of output, Industry 1 requires x_{11} units' worth from itself, x_{21} units' worth from Industry 2, and so on. Forming the ratios, we find that to produce 1 unit worth of output, Industry 1 requires $\frac{x_{11}}{x_1}$ of Industry 1, $\frac{x_{21}}{x_1}$ of Industry 2, and so on. Continuing in this way, we can give the input requirements per unit worth of output for each industry.

Setting
$$a_{ij} = \frac{x_{ij}}{x_j},$$

we can construct the matrix :
$$\mathbf{A} = \begin{bmatrix} a_{11} & a_{12} & \cdots & a_{1n} \\ a_{21} & a_{22} & \cdots & a_{2n} \\ \cdot & \cdot & \cdot & \cdot \\ \cdot & \cdot & \cdot & \cdot \\ \cdot & \cdot & \cdot & \cdot \\ a_{n1} & a_{n2} & \cdots & a_{nn} \end{bmatrix}$$

The matrix **A** is called the **input-output coefficient matrix** or the **technology matrix.** The entries in the matrix **A** are called the **input-output coefficients or technological coefficients.**

Observe that the technological coefficient a_{ij} represents the amount of input required from ith industry to produce 1 unit's worth of output of jth industry. If

$$\mathbf{X} = \begin{bmatrix} y_1 \\ y_2 \\ \cdot \\ \cdot \\ \cdot \\ y_n \end{bmatrix}$$

is the output matrix representing the production of each industry in the system to meet the final demand

$$\mathbf{D} = \begin{bmatrix} d_1 \\ d_2 \\ \cdot \\ \cdot \\ \cdot \\ d_n \end{bmatrix}$$

then the condition that production equals consumption requires that

Internal consumption + Consumer demand = Total output

This leads to the system of equations, called **balancing equations :**

$$\begin{aligned} a_{11}y_1 + a_{12}y_2 + \dots + a_{1n}y_n + d_1 &= y_1 \\ a_{21}y_1 + a_{22}y_2 + \dots + a_{2n}y_n + d_2 &= y_2 \\ \vdots \\ a_{n1}y_1 + a_{n2}y_2 + \dots + a_{nn}y_n + d_n &= y_n \end{aligned}$$

In terms of the matrix notation, this requirement is equivalent to the equation

$$\mathbf{AX} + \mathbf{D} = \mathbf{X} \quad \Rightarrow \quad (1 - \mathbf{A})\mathbf{X} = \mathbf{D}$$

If $(\mathbf{I} - \mathbf{A})^{-1}$ exists, then the output **X** is given by $\mathbf{X} = (\mathbf{I} - \mathbf{A})^{-1}\mathbf{D}$.

The matrix **I – A** is called the **Leontief matrix.**

Hawkins-Simon Conditions for the Viability of the System. The mere formulation of an input-output system does not guarantee that it has a solution with each output as non-negative. In order that the system does not give negative numbers as outputs, certain conditions have to be satisfied. The following conditions, given by D. Hawkins and H.A. Simon, are presented here which ensure that the system does not give negative outputs :

(*i*) *The determinant* $|\mathbf{I} - \mathbf{A}|$ *of the Leontief matrix must be positive; and*

(*ii*) *The diagonal elements :* $1 - a_{11}, 1 - a_{22}, \dots 1 - a_{nn}$ *of the Leontief matrix* **I – A** *should all be positive, or in other words,* $a_{11}, a_{22}, \dots, a_{nn}$ *should all be less than l.*

***Remark*.** If $a_{ii} \geq 1$ for some *i*, one or more units of *i*th industry would be required to produce one unit of itself and as such no net profit can be secured under such circumstances.

***Example* 35.** For the following transaction matrix, find the gross output for each industry for the final demands 18 and 44 units respectively.

Industry	*Input to* *I*	*II*	*Final Demand*
I	16	20	4
II	8	40	32

[*Delhi Univ. B.Com. (H) 1986, 2004 (Modified)*]

Solution. The total output of industry *I* is the sum of industrial demand and final demand : 16 + 20 + 4 = 40. Similarly, the total output of industry *II* is 8 + 40 + 32 = 80.

To find the input-output coefficient matrix, we divide the industry entries in each industry column by the total value of output for that industry. Thus the input-output coefficient matrix is

$$\mathbf{A} = \begin{bmatrix} \frac{16}{40} & \frac{20}{80} \\ \frac{8}{40} & \frac{40}{80} \end{bmatrix} = \begin{bmatrix} \frac{2}{5} & \frac{1}{4} \\ \frac{1}{5} & \frac{1}{2} \end{bmatrix}$$

If $\mathbf{X} = \begin{bmatrix} x_1 \\ x_2 \end{bmatrix}$ is the output matrix representing the gross output of each industry to meet the final demand $\mathbf{D} = \begin{bmatrix} 18 \\ 44 \end{bmatrix}$, then the condition that production equals consumption implies that

$$\mathbf{X} = \mathbf{AX} + \mathbf{D} \quad \text{or} \quad (\mathbf{I} - \mathbf{A})\mathbf{X} = \mathbf{D}$$

Now $$\mathbf{I} - \mathbf{A} = \begin{bmatrix} 1 & 0 \\ 0 & 1 \end{bmatrix} - \begin{bmatrix} \frac{2}{5} & \frac{1}{4} \\ \frac{1}{5} & \frac{1}{2} \end{bmatrix} = \begin{bmatrix} \frac{3}{5} & -\frac{1}{4} \\ -\frac{1}{5} & \frac{1}{2} \end{bmatrix} \quad \therefore \quad |\mathbf{I} - \mathbf{A}| = \frac{3}{10} - \frac{1}{20} = \frac{5}{20} = \frac{1}{4}$$

Since $|\mathbf{I} - \mathbf{A}| \neq 0$, the output $\mathbf{X}$ is given by $\mathbf{X} = (\mathbf{I} - \mathbf{A})^{-1}\mathbf{D}$. It can be readily seen that

$$(\mathbf{I} - \mathbf{A})^{-1} = 4\begin{bmatrix} \frac{1}{2} & \frac{1}{4} \\ \frac{1}{5} & \frac{3}{5} \end{bmatrix} \quad \therefore \quad \mathbf{X} = (\mathbf{I} - \mathbf{A})^{-1}\mathbf{D} = 4\begin{bmatrix} \frac{1}{2} & \frac{1}{4} \\ \frac{1}{5} & \frac{3}{5} \end{bmatrix}\begin{bmatrix} 18 \\ 44 \end{bmatrix} = \begin{bmatrix} 80 \\ 120 \end{bmatrix}$$

Hence the gross outputs of two industries, *I* and *II*, for the given final demand, are 80 and 120 units respectively.

***Example* 36.** Suppose the inter-industry flow of the products of two industries is given as under:

	Consumption X	Consumption Y	*Domestic demand*	*Gross output*
Production X	30	40	50	120
Production Y	20	10	30	60

Determine the technology matrix and test Hawkins-Simon conditions for the viability of the system. Compute the equilibrium level of output of the products when the domestic demand vector is $\begin{bmatrix} 80 \\ 40 \end{bmatrix}$. **[*Delhi Univ. B.Com. (H) 1988, 2009*]**

Solution. Let **A** be the technology matrix. Then $\mathbf{A} = \begin{bmatrix} \frac{30}{120} & \frac{40}{60} \\ \frac{20}{120} & \frac{10}{60} \end{bmatrix} = \begin{bmatrix} \frac{1}{4} & \frac{2}{3} \\ \frac{1}{6} & \frac{1}{6} \end{bmatrix}$.

To test Hawkins-Simon conditions, we consider

$$\mathbf{I} - \mathbf{A} = \begin{bmatrix} 1 & 0 \\ 0 & 1 \end{bmatrix} - \begin{bmatrix} \frac{1}{4} & \frac{2}{3} \\ \frac{1}{6} & \frac{1}{6} \end{bmatrix} = \begin{bmatrix} \frac{3}{4} & -\frac{2}{3} \\ -\frac{1}{6} & \frac{5}{6} \end{bmatrix} \quad \therefore \quad |\mathbf{I} - \mathbf{A}| = \frac{15}{24} - \frac{2}{18} = \frac{37}{72} > 0.$$

Also, the diagonal elements, $\frac{1}{4}$ and $\frac{1}{6}$, of the technology matrix are both less than 1. Hence Hawkins-Simon conditions are satisfied and the system is viable.

Let $\mathbf{X} = \begin{bmatrix} x \\ y \end{bmatrix}$ denote the total output and $\mathbf{D} = \begin{bmatrix} 80 \\ 40 \end{bmatrix}$ denote the demand vector. Then

$$\mathbf{AX} + \mathbf{D} = \mathbf{X} \quad \Rightarrow \quad (\mathbf{I} - \mathbf{A})\mathbf{X} = \mathbf{D} \quad \Rightarrow \quad \mathbf{X} = (\mathbf{I} - \mathbf{A})^{-1}\mathbf{D}$$

Now, $$(\mathbf{I} - \mathbf{A})^{-1} = \frac{1}{|\mathbf{I} - \mathbf{A}|} \operatorname{adj}(\mathbf{I} - \mathbf{A}) = \frac{72}{37}\begin{bmatrix} \frac{5}{6} & \frac{2}{3} \\ \frac{1}{6} & \frac{3}{4} \end{bmatrix}$$

$\therefore$ $$\mathbf{X} = \frac{72}{37}\begin{bmatrix} \frac{5}{6} & \frac{2}{3} \\ \frac{1}{6} & \frac{3}{4} \end{bmatrix}\begin{bmatrix} 80 \\ 40 \end{bmatrix} = \frac{72}{37}\begin{bmatrix} \frac{280}{3} \\ \frac{130}{3} \end{bmatrix} = \begin{bmatrix} \frac{6720}{37} \\ \frac{3120}{37} \end{bmatrix}$$

Thus the equilibrium level of output of the products is $\frac{6720}{37}$ and $\frac{3120}{37}$ units respectively.

***Example* 37.** Find the demand vector which is consistent with the input-output matrix

$$\mathbf{A} = \begin{bmatrix} 0.2 & 0.3 & 0.2 \\ 0.4 & 0.1 & 0.2 \\ 0.1 & 0.3 & 0.2 \end{bmatrix} \text{ and the output vector } \mathbf{X} = \begin{bmatrix} 25 \\ 21 \\ 18 \end{bmatrix}$$

Also test Hawkins-Simon conditions for the viability of the system.

[Delhi Univ. B.Com. (H) 1985]

Solution . Since production equals consumption, we have

$$\mathbf{AX} + \mathbf{D} = \mathbf{X}, \text{ where } \mathbf{D} = \begin{bmatrix} d_1 \\ d_2 \\ d_3 \end{bmatrix} \text{ is the demand vector.}$$

$\therefore$ $$\mathbf{D} = (\mathbf{I} - \mathbf{A})\mathbf{X}$$

Now $$\mathbf{I} - \mathbf{A} = \begin{bmatrix} 1 & 0 & 0 \\ 0 & 1 & 0 \\ 0 & 0 & 1 \end{bmatrix} - \begin{bmatrix} 0.2 & 0.3 & 0.2 \\ 0.4 & 0.1 & 0.2 \\ 0.1 & 0.3 & 0.2 \end{bmatrix} = \begin{bmatrix} 0.8 & -0.3 & -0.2 \\ -0.4 & 0.9 & -0.2 \\ -0.1 & -0.3 & 0.8 \end{bmatrix}$$

$\Rightarrow$ $$\mathbf{D} = (\mathbf{I} - \mathbf{A})\mathbf{X} = \begin{bmatrix} 0.8 & -0.3 & -0.2 \\ -0.4 & 0.9 & -0.2 \\ -0.1 & -0.3 & 0.8 \end{bmatrix}\begin{bmatrix} 25 \\ 21 \\ 18 \end{bmatrix} = \begin{bmatrix} 10.1 \\ 5.3 \\ 5.6 \end{bmatrix}$$

To test Hawkins-Simon conditions, we have

$$|\mathbf{I}-\mathbf{A}| = 0.384 > 0$$

Also each entry on the main diagonal of the input-output matrix **A** is less than 1. Thus the Hawkins-Simon conditions are satisfied and the system is viable.

Equilbrium Prices

Let us consider a hypothetical economy producing only two goods X and Y. The two commodities serve as intermediate inputs in each others' production. Let

$$\mathbf{A} = \begin{bmatrix} a_{11} & a_{12} \\ a_{21} & a_{22} \end{bmatrix}$$

be the technology matrix for the given economy. This means a_{11} units of X and a_{21} units of Y are needed to produce a unit of X, whereas a_{12} units of X and a_{22} units of Y are needed to produce a unit of Y. In addition to this, suppose labour is used as a primary input and l_1 units and l_2 units of primary input are needed to produce each unit of X and Y respectively. If the price of primary input is w per unit, we would like to find the equilibrium prices of the two goods.

Let equilibrium prices per unit of the two goods be p_1 and p_2. Since each unit of X requires a_{11} units of X, a_{21} units of Y and l_1 units of primary input, therefore it costs $a_{11}\, p_1 + a_{21}\, p_2 + l_1\, w$ to produce a unit of X. Similarly, it costs $a_{12}\, p_1 + a_{22}\, p_2 + l_2\, w$ to produce a unit of Y.

Equilibrium prices must satisfy the following equations :

$$\begin{aligned} a_{11}\, p_1 + a_{21}\, p_2 + l_1\, w &= p_1 \\ a_{12}\, p_1 + a_{22}\, p_2 + l_2\, w &= p_2 \end{aligned} \quad \Leftrightarrow \quad \begin{aligned} (1-a_{11})\, p_1 - a_{21}\, p_2 &= l_1\, w \\ -a_{12}\, p_1 + (1-a_{22})\, p_2 &= l_2\, w \end{aligned}$$

Expressed in matrix form, we have

$$\begin{bmatrix} 1-a_{11} & -a_{21} \\ -a_{12} & 1-a_{22} \end{bmatrix} \begin{bmatrix} p_1 \\ p_2 \end{bmatrix} = \begin{bmatrix} l_1\, w \\ l_2\, w \end{bmatrix}$$

Note that the very first matrix on the left-hand side is transpose of the matrix **(I – A)**. Thus we have

$$(\mathbf{I}-\mathbf{A})' \begin{bmatrix} p_1 \\ p_2 \end{bmatrix} = \begin{bmatrix} l_1\, w \\ l_2\, w \end{bmatrix} \Rightarrow \begin{bmatrix} p_1 \\ p_2 \end{bmatrix} = [(\mathbf{I}-\mathbf{A})']^{-1} \begin{bmatrix} l_1\, w \\ l_2\, w \end{bmatrix} = [(\mathbf{I}-\mathbf{A})^{-1}]' \begin{bmatrix} l_1\, w \\ l_2\, w \end{bmatrix},$$

since transpose and inverse are interchangeable.

***Example* 38.** An economy produces only coal and steel. The two commodities serve as intermediate inputs in each other's production. 0.4 tonne of steel and 0.7 tonne of coal are needed to produce a tonne of steel. Similarly, 0.1 tonne of steel and 0.6 tonne of coal are required to produce a tonne of coal. No capital inputs are needed. Do you think that the system is viable ?

2 and 5 labour days are required to produce a tonne of coal and steel respectively. If the economy needs 100 tonne of coal and 50 tonne of steel, calculate the gross output of the two commodities and total labour required. Also determine the equilibrium prices and the value added, if the wage rate is ₹ 10 per man-day. **[*Delhi Univ. B.Com. (H) 2006*]**

Solution. The input-output coefficients may be summarized in the tabular form as follows :

		Input	
		Steel	*Coal*
Output	*Steel*	0.4	0.1
	Coal	0.7	0.6

Let **A** denote the input-output coefficient matrix. Then $\mathbf{A} = \begin{bmatrix} 0.4 & 0.1 \\ 0.7 & 0.6 \end{bmatrix}$ and the Leontief matrix is :

$$\mathbf{I} - \mathbf{A} = \begin{bmatrix} 1 & 0 \\ 0 & 1 \end{bmatrix} - \begin{bmatrix} 0.4 & 0.1 \\ 0.7 & 0.6 \end{bmatrix} = \begin{bmatrix} 0.6 & -0.1 \\ -0.7 & 0.4 \end{bmatrix}$$

Now $$|\mathbf{I} - \mathbf{A}| = 0.24 - 0.07 = 0.17$$

Since $|\mathbf{I} - \mathbf{A}|$ is positive and all the elements on the main diagonal of **A** are less than 1, Hawkins-Simon conditions are satisfied and the system is viable.

Let the gross output of steel be x_1 units and the gross output of coal be x_2 units to meet the final demand vector $\mathbf{D} = \begin{bmatrix} 50 \\ 100 \end{bmatrix}$. Then the balancing equations are

$$0.4x_1 + 0.1x_2 + 50 = x_1$$
$$0.7x_1 + 0.6x_2 + 100 = x_2$$

In matrix notation, these can be written as

$$\begin{bmatrix} 0.4 & 0.1 \\ 0.7 & 0.6 \end{bmatrix}\begin{bmatrix} x_1 \\ x_2 \end{bmatrix} + \begin{bmatrix} 50 \\ 100 \end{bmatrix} = \begin{bmatrix} x_1 \\ x_2 \end{bmatrix} \quad \textit{i.e.,} \quad \mathbf{AX} + \mathbf{D} = \mathbf{X}, \text{ where } \mathbf{X} = \begin{bmatrix} x_1 \\ x_2 \end{bmatrix}.$$

Thus we have $$\mathbf{X} = (\mathbf{I} - \mathbf{A})^{-1}\mathbf{D} = \frac{1}{0.17}\begin{bmatrix} 0.4 & 0.1 \\ 0.7 & 0.6 \end{bmatrix}\begin{bmatrix} 50 \\ 100 \end{bmatrix} = \frac{1}{0.17}\begin{bmatrix} 30 \\ 95 \end{bmatrix} = \begin{bmatrix} 176.5 \\ 558.8 \end{bmatrix}$$

Hence the gross outputs of steel and coal, for a given final demand, are 176.5 and 558.8 units respectively.

$$\text{Total labour days required} = \begin{bmatrix} l_1 & l_2 \end{bmatrix} \begin{bmatrix} x_1 \\ x_2 \end{bmatrix} = \begin{bmatrix} 5 & 2 \end{bmatrix} \begin{bmatrix} 176.5 \\ 558.8 \end{bmatrix} = [2000]$$

Equilibrium prices, when the wage rate is ₹ 10, are given by

$$\begin{bmatrix} p_1 \\ p_2 \end{bmatrix} = [(1-\mathbf{A})^{-1}]' \begin{bmatrix} l_1 w \\ l_2 w \end{bmatrix} = \frac{1}{0.17} \begin{bmatrix} 0.4 & 0.7 \\ 0.1 & 0.6 \end{bmatrix} \begin{bmatrix} 50 \\ 20 \end{bmatrix} = \frac{1}{0.17} \begin{bmatrix} 34 \\ 17 \end{bmatrix} = \begin{bmatrix} 200 \\ 100 \end{bmatrix}$$

Hence the equilibrium prices per tonne of steel and coal are ₹ 200 and ₹ 100 respectively. Also,

$$\text{Value added} = \begin{bmatrix} x_1 & x_2 \end{bmatrix} \begin{bmatrix} l_1 w \\ l_2 w \end{bmatrix} = \begin{bmatrix} 176.5 & 558.8 \end{bmatrix} \begin{bmatrix} 50 \\ 20 \end{bmatrix} = [20{,}001].$$

***Example* 39.** A country produces only two goods X and Y. The two commodities serve as intermediate inputs in each other's production. 0.1 units of X and 0.55 units of Y are needed to produce a unit of X, whereas 0.4 units of X and 0.2 units of Y arc needed to produce a unit of Y. In additon to this 4 units and 3 units of primary input are needed to produce each unit of X and Y respectively.

(i) If 240 and 140 units of X and Y respectively are needed for final consumption, find gross output levels of the two goods.

(ii) If the price of primary input is ₹ 10 per unit, compute the prices of the goods.

(iii) Also compute total value added. **[*Delhi Univ. B.Com. (H) 2008, 2022 Modified*]**

Solution. *(i)* The input-output coefficient matrix is : $\mathbf{A} = \begin{bmatrix} 0.1 & 0.4 \\ 0.55 & 0.2 \end{bmatrix}$. Let $\mathbf{X} = \begin{bmatrix} x \\ y \end{bmatrix}$ be the gross output vector required to meet the final demand $\boldsymbol{D} = \begin{bmatrix} 240 \\ 140 \end{bmatrix}$. Then

$$\mathbf{X} = \mathbf{AX} + \mathbf{D} \quad \text{or} \quad (\mathbf{I} - \mathbf{A})\,\mathbf{X} = \mathbf{D}$$

Now
$$\mathbf{I} - \mathbf{A} = \begin{bmatrix} 0.9 & -0.4 \\ -0.55 & 0.8 \end{bmatrix} \quad \therefore \quad |\mathbf{I} - \mathbf{A}| = 0.72 - 0.22 = 0.5 \neq 0$$

Since $|\mathbf{I} - \mathbf{A}| \neq 0$, the output $\mathbf{X}$ is given by $\mathbf{X} = (\mathbf{I} - \mathbf{A})^{-1}\,\mathbf{D}$. It can be easily verified that

$$(\mathbf{I} - \mathbf{A})^{-1} = \frac{1}{0.5} \begin{bmatrix} 0.8 & 0.4 \\ 0.55 & 0.9 \end{bmatrix}$$

$$\therefore \quad \mathbf{X} = (\mathbf{I} - \mathbf{A})^{-1}\,\mathbf{D} = \frac{1}{0.5} \begin{bmatrix} 0.8 & 0.4 \\ 0.55 & 0.9 \end{bmatrix} \begin{bmatrix} 240 \\ 140 \end{bmatrix} = \begin{bmatrix} 496 \\ 516 \end{bmatrix}$$

Hence the gross output levels of two goods are 496 and 516 units respectively.

(ii) Equilibrium prices, when the price of primary input is ₹ 10 per unit, are given by

$$\begin{bmatrix} p_1 \\ p_2 \end{bmatrix} = \left[(\mathbf{I}-\mathbf{A})^{-1}\right]' \begin{bmatrix} l_1 w \\ l_2 w \end{bmatrix} = \frac{1}{0.5}\begin{bmatrix} 0.8 & 0.55 \\ 0.4 & 0.9 \end{bmatrix}\begin{bmatrix} 40 \\ 30 \end{bmatrix} = \begin{bmatrix} 97 \\ 86 \end{bmatrix}$$

Hence the equalibrium prices per unit of two goods are ₹ 97 and ₹ 86 respectvely.

(iii) Total value added $= \begin{bmatrix} x & y \end{bmatrix}\begin{bmatrix} l_1 w \\ l_2 w \end{bmatrix} = \begin{bmatrix} 496 & 516 \end{bmatrix}\begin{bmatrix} 40 \\ 30 \end{bmatrix} = [35{,}320].$

***Example* 40.** In a two sector economy, it is known that Industry *I* uses 10 paise worth of its own product and 60 paise worth of commodity *II* to produce a rupee worth of commodity *I*; industry *II* uses 30 paise worth of its own product and 50 paise worth of commodity *I* in producing a rupee worth of commodity *II*; and the final demands are ₹1100 billion worth of commodity *I* and ₹2200 billion worth of commodity *II*.

(i) Write down the input-output matrix for this economy.

(ii) Find the solution output using matrix inverse method.

(iii) Find the total of primary input requirement to produce the solution output if there is only one primary input requirement. **[*Delhi Univ. B.Com. (H) 2007*]**

Solution. *(i)* The input-output matrix is : $\mathbf{A} = \begin{bmatrix} 0.1 & 0.5 \\ 0.6 & 0.3 \end{bmatrix}$

(ii) Let $\mathbf{X} = \begin{bmatrix} x \\ y \end{bmatrix}$ be the output matrix required to meet the final demand $\mathbf{D} = \begin{bmatrix} 1100 \\ 2200 \end{bmatrix}$. Then,

by Leontief Principle, we have

$$\mathbf{X} = \mathbf{AX} + \mathbf{D} \quad \text{or} \quad (\mathbf{I}-\mathbf{A})\,\mathbf{X} = \mathbf{D}$$

Now, $\mathbf{I}-\mathbf{A} = \begin{bmatrix} 1 & 0 \\ 0 & 1 \end{bmatrix} - \begin{bmatrix} 0.1 & 0.5 \\ 0.6 & 0.3 \end{bmatrix} = \begin{bmatrix} 0.9 & -0.5 \\ -0.6 & 0.7 \end{bmatrix}$

$\Rightarrow \quad |\mathbf{I}-\mathbf{A}| = 0.63 - 0.30 = 0.33 \neq 0$

Since $|\mathbf{I}-\mathbf{A}| \neq 0$, the output **X** is given by $\mathbf{X} = (\mathbf{I}-\mathbf{A})^{-1}\mathbf{D}$. It can be easily verified that

$$(\mathbf{I}-\mathbf{A})^{-1} = \frac{1}{0.33}\begin{bmatrix} 0.7 & 0.5 \\ 0.6 & 0.9 \end{bmatrix}$$

$$\therefore \quad \mathbf{X} = (\mathbf{I}-\mathbf{A})^{-1}\mathbf{D} = \frac{1}{0.33}\begin{bmatrix} 0.7 & 0.5 \\ 0.6 & 0.9 \end{bmatrix}\begin{bmatrix} 1100 \\ 2200 \end{bmatrix} = \frac{1}{0.33}\begin{bmatrix} 1870 \\ 2640 \end{bmatrix} = \begin{bmatrix} 5666.67 \\ 8000 \end{bmatrix}$$

Thus, the gross outputs for the two sectors are 5666.67 and 8000 units respectively.

(*iii*) The total of primary input requirement to produce the solution output, if there is only one primary input requirement, is given by

$$\begin{bmatrix} 0.3 & 0.2 \end{bmatrix}\begin{bmatrix} 5666.67 \\ 8000 \end{bmatrix} = [1700.001 + 1600] = [3300.001].$$

***Example* 41.** A three sector economy has the following input-output coefficient matrix :

$$\mathbf{A} = \begin{bmatrix} 0 & 0.5 & 0 \\ 0.2 & 0 & 0.5 \\ 0.4 & 0 & 0 \end{bmatrix}$$

The labour days required per unit of output of the three sectors are 0.4, 0.7 and 1.2 respectively and their consumer output targets are 1000, 5000 and 4000 units respectively. The wage rate is ₹ 10 per labour day. By using matrix algebra, you are required to find:

(*i*) the gross output of each sector,

(*ii*) total labour days required,

(*iii*) equilibrium prices, and

(*iv*) total value added. **[*Delhi Univ. B.Com. (H) 2011, 2018*]**

Solution. (*i*) Let $X = \begin{bmatrix} x_1 \\ x_2 \\ x_3 \end{bmatrix}$ be the gross output vector required to meet the final demand

$$\mathbf{D} = \begin{bmatrix} 1000 \\ 5000 \\ 4000 \end{bmatrix}. \quad \text{Then} \quad \mathbf{X} = \mathbf{AX} + \mathbf{D} \quad \text{or} \quad (\mathbf{I} - \mathbf{A})\,\mathbf{X} = \mathbf{D}$$

$$\text{Now} \quad \mathbf{I} - \mathbf{A} = \begin{bmatrix} 1 & -0.5 & 0 \\ -0.2 & 1 & -0.5 \\ -0.4 & 0 & 1 \end{bmatrix} \Rightarrow |\mathbf{I} - \mathbf{A}| = 0.8 \neq 0$$

Since $|\mathbf{I} - \mathbf{A}| \neq 0$, the output **X** is given by $\mathbf{X} = (\mathbf{I} - \mathbf{A})^{-1}\mathbf{D}$. It can be easily verified that

$$(\mathbf{I} - \mathbf{A})^{-1} = \frac{1}{0.8}\begin{bmatrix} 1 & 0.5 & 0.25 \\ 0.4 & 1 & 0.5 \\ 0.4 & 0.2 & 0.9 \end{bmatrix}$$

$$\therefore \quad \mathbf{X} = (\mathbf{I}-\mathbf{A})^{-1}\mathbf{D} = \frac{1}{0.8}\begin{bmatrix} 1 & 0.5 & 0.25 \\ 0.4 & 1 & 0.5 \\ 0.4 & 0.2 & 0.9 \end{bmatrix}\begin{bmatrix} 1000 \\ 5000 \\ 4000 \end{bmatrix} = \begin{bmatrix} 5625 \\ 9250 \\ 6250 \end{bmatrix}$$

Hence the gross output levels of 3 sectors are 5625, 9250 and 6250 units respectively.

(*ii*) Total labour days required $= [0.4 \quad 0.7 \quad 1.2]\begin{bmatrix} 5625 \\ 9250 \\ 6250 \end{bmatrix} = [16,225]$

(*iii*) Equilibrium prices, when the wage rate is ₹ 10, are given by

$$\begin{bmatrix} p_1 \\ p_2 \\ p_3 \end{bmatrix} = [(\mathbf{I}-\mathbf{A})^{-1}]'\begin{bmatrix} l_1 w \\ l_2 w \\ l_3 w \end{bmatrix} = \frac{1}{0.8}\begin{bmatrix} 1 & 0.4 & 0.4 \\ 0.5 & 1 & 0.2 \\ 0.25 & 0.5 & 0.9 \end{bmatrix}\begin{bmatrix} 4 \\ 7 \\ 12 \end{bmatrix} = \frac{1}{0.8}\begin{bmatrix} 11.6 \\ 11.4 \\ 15.3 \end{bmatrix} = \begin{bmatrix} 14.5 \\ 14.25 \\ 19.125 \end{bmatrix}$$

Hence equilibrium prices are : p_1 = ₹14.5, p_2 = ₹14.25 and p_3 = ₹19.125

(*iv*) Total value added $= [x_1 \quad x_2 \quad x_3]\begin{bmatrix} l_1 w \\ l_2 w \\ l_3 w \end{bmatrix} = [5625 \quad 9250 \quad 6250]\begin{bmatrix} 4 \\ 7 \\ 12 \end{bmatrix} = [1,62,250].$

***Example* 42.** The interrelationships between the production of three industries *R*, *S* and *T* in a given year are given in the following table :

	R	*S*	*T*	*Consumer demand*	*Total output*
R	50	20	40	70	180
S	20	30	20	90	160
T	30	20	20	50	120

If the forecast demand in 2 years be $\mathbf{D} = \begin{bmatrix} 60 \\ 110 \\ 60 \end{bmatrix}$, what should the total output **X** be ?

Solution. The input-output martix is : $\mathbf{A} = \begin{bmatrix} \frac{50}{180} & \frac{20}{160} & \frac{40}{120} \\ \frac{20}{180} & \frac{30}{160} & \frac{20}{120} \\ \frac{30}{180} & \frac{20}{160} & \frac{20}{120} \end{bmatrix} = \begin{bmatrix} \frac{5}{18} & \frac{1}{8} & \frac{1}{3} \\ \frac{1}{9} & \frac{3}{16} & \frac{1}{6} \\ \frac{1}{6} & \frac{1}{8} & \frac{1}{6} \end{bmatrix}$

Let $\mathbf{X} = \begin{bmatrix} x_1 \\ x_2 \\ x_3 \end{bmatrix}$ be the output matrix required to meet the given demand $\mathbf{D} = \begin{bmatrix} 60 \\ 110 \\ 60 \end{bmatrix}$. Then

$$\mathbf{X} = \mathbf{AX} + \mathbf{D} \quad \text{or} \quad (\mathbf{I} - \mathbf{A})\,\mathbf{X} = \mathbf{D}$$

Now $$\mathbf{I} - \mathbf{A} = \begin{bmatrix} \frac{13}{18} & -\frac{1}{8} & -\frac{1}{3} \\ -\frac{1}{9} & \frac{13}{16} & -\frac{1}{6} \\ -\frac{1}{6} & -\frac{1}{8} & \frac{5}{6} \end{bmatrix} \quad \text{and} \quad |\mathbf{I} - \mathbf{A}| = \frac{707}{1728}$$

Since $|\mathbf{I} - \mathbf{A}| \neq 0$, the output $\mathbf{X}$ is given by $\mathbf{X} = (\mathbf{I} - \mathbf{A})^{-1}\,\mathbf{D}$. It can be checked that

$$(\mathbf{I} - \mathbf{A})^{-1} = \frac{1728}{707} \begin{bmatrix} \frac{21}{32} & \frac{7}{48} & \frac{7}{24} \\ \frac{13}{108} & \frac{59}{108} & \frac{17}{108} \\ \frac{43}{288} & \frac{1}{9} & \frac{165}{288} \end{bmatrix}$$

$$\therefore \quad \mathbf{X} = (\mathbf{I} - \mathbf{A})^{-1}\,\mathbf{D} = \frac{1728}{707} \begin{bmatrix} 7000/96 \\ 8290/108 \\ 16{,}000/288 \end{bmatrix} = \begin{bmatrix} 1{,}26{,}000/707 \\ 1{,}32{,}640/707 \\ 96{,}000/707 \end{bmatrix} = \begin{bmatrix} 178.218 \\ 187.610 \\ 135.785 \end{bmatrix}$$

Thus, the total outputs of R, S and T required for the forecast demand **D** are :

$$x_1 = 178.218, \quad x_2 = 187.610 \quad \text{and} \quad x_3 = 135.785.$$

***Example* 43.** Two industries input-output relationships are given below in **A** with final demand vector **B** (in units) :

$$\mathbf{A} = \begin{array}{c} \\ I \\ II \end{array}\!\!\begin{array}{c} \begin{array}{cc} I & II \end{array} \\ \begin{bmatrix} 50 & 75 \\ 100 & 50 \end{bmatrix} \end{array} \qquad \mathbf{B} = \begin{array}{c} I \\ II \end{array}\!\begin{bmatrix} 75 \\ 50 \end{bmatrix}$$

If the gross output increases to $\begin{array}{c} I \\ II \end{array}\!\begin{bmatrix} 400 \\ 600 \end{bmatrix}$, determine the final demand which can be satisfied.

Also test the Hawkins-Simon conditions. **[*Delhi Univ. B.Com. (H) 1998*]**

Solution. The inter-relationship between the production of two industries can be described in the following input-output table :

Producing Industry	*Input to Industry*		*Final Demand*	*Total Output*
	I	*II*		
I	50	75	75	200
II	100	50	50	200

Let **R** denote the input-output coefficient matrix. Then

$$\mathbf{R} = \begin{bmatrix} 50/200 & 75/200 \\ 100/200 & 50/200 \end{bmatrix} = \begin{bmatrix} 1/4 & 3/8 \\ 1/2 & 1/4 \end{bmatrix}$$

Let $\mathbf{D} = \begin{bmatrix} d_1 \\ d_2 \end{bmatrix}$ denote the final demand which can be satisfied, if the gross output increases to

$\mathbf{X} = \begin{bmatrix} 400 \\ 600 \end{bmatrix}$. Since production equals consumption, we have

$$\mathbf{RX} + \mathbf{D} = \mathbf{X} \quad \Rightarrow \quad \mathbf{D} = \mathbf{X} - \mathbf{RX} = (\mathbf{I} - \mathbf{R})\mathbf{X}$$

Now $\quad \mathbf{I} - \mathbf{R} = \begin{bmatrix} 1 & 0 \\ 0 & 1 \end{bmatrix} - \begin{bmatrix} 1/4 & 3/8 \\ 1/2 & 1/4 \end{bmatrix} = \begin{bmatrix} 3/4 & -3/8 \\ -1/2 & 3/4 \end{bmatrix}$

$\therefore \quad \mathbf{D} = (\mathbf{I} - \mathbf{R})\mathbf{X} = \begin{bmatrix} 3/4 & -3/8 \\ -1/2 & 3/4 \end{bmatrix}\begin{bmatrix} 400 \\ 600 \end{bmatrix} = \begin{bmatrix} 75 \\ 250 \end{bmatrix}$

To test Hawkins-Simon conditions, we have $|\mathbf{I} - \mathbf{R}| = 3/8 > 0$. Further each entry on the main diagonal of the input-output matrix **R** is less than 1. Hence Hawkins-Simon conditions are satisfied.

***Example* 44.** For a two sector economy, the input-output coefficient matrix is : $\mathbf{A} = \begin{bmatrix} 0.5 & 0.3 \\ 0.2 & 0.4 \end{bmatrix}$

If the final demands of the two sectors are 10 and 30, find the gross output. Land and labour are used as two primary inputs. Their coefficients for the two sectors are given as

$$\begin{matrix} \textit{Labour} \\ \textit{Land} \end{matrix} \begin{bmatrix} 0.4 & 0.3 \\ 0.5 & 0.4 \end{bmatrix}$$

If the wage rate and rent are ₹ 40 and ₹ 100 respectively, find the equilibrium prices for the two sectors. **[*Delhi Univ. B.Com. (H) 2006, 2011, 2014*]**

Solution. Let $X = \begin{bmatrix} x_1 \\ x_2 \end{bmatrix}$ be the gross output required to meet the final demand $\mathbf{D} = \begin{bmatrix} 10 \\ 30 \end{bmatrix}$.

Since production equals consumption, we have

$$\mathbf{X} = \mathbf{AX} + \mathbf{D} \quad \text{or} \quad (\mathbf{I} - \mathbf{A})\,\mathbf{X} = \mathbf{D}$$

Now $\mathbf{I-A} = \begin{bmatrix} 1 & 0 \\ 0 & 1 \end{bmatrix} - \begin{bmatrix} 0.5 & 0.3 \\ 0.2 & 0.4 \end{bmatrix} = \begin{bmatrix} 0.5 & -0.3 \\ -0.2 & 0.6 \end{bmatrix}$ $\therefore |\mathbf{I-A}| = 0.30 - 0.06 = 0.24$

Since $|\mathbf{I-A}| \neq 0$, the output **X** is given by $\mathbf{X} = (\mathbf{I-A})^{-1}\mathbf{D}$. It can be easily verified that

$$(\mathbf{I-A})^{-1} = \frac{1}{0.24}\begin{bmatrix} 0.6 & 0.3 \\ 0.2 & 0.5 \end{bmatrix}$$

$$\therefore \quad \mathbf{X} = (\mathbf{I-A})^{-1}\mathbf{D} = \frac{1}{0.24}\begin{bmatrix} 0.6 & 0.3 \\ 0.2 & 0.5 \end{bmatrix}\begin{bmatrix} 10 \\ 30 \end{bmatrix} = \frac{1}{0.24}\begin{bmatrix} 15 \\ 17 \end{bmatrix} = \begin{bmatrix} 62.5 \\ 70.83 \end{bmatrix}$$

Thus the gross output for the two sectors are 62.5 and 70.83 respectively.

The cost of primary inputs for each unit of two sectors is given by

$$\begin{bmatrix} 40 & 100 \end{bmatrix}\begin{bmatrix} 0.4 & 0.3 \\ 0.5 & 0.4 \end{bmatrix} = \begin{bmatrix} 66 & 52 \end{bmatrix}$$

If p_1 and p_2 are the equilibrium prices for the two sectors, then

$$\begin{bmatrix} p_1 \\ p_2 \end{bmatrix} = \left[(\mathbf{I-A})^{-1}\right]'\begin{bmatrix} 66 \\ 52 \end{bmatrix} = \frac{1}{0.24}\begin{bmatrix} 0.6 & 0.2 \\ 0.3 & 0.5 \end{bmatrix}\begin{bmatrix} 66 \\ 52 \end{bmatrix} = \frac{1}{0.24}\begin{bmatrix} 50 \\ 45.8 \end{bmatrix} = \begin{bmatrix} 208.33 \\ 190.83 \end{bmatrix}$$

Thus the equilibrium prices are: p_1 : ₹ 208.33 and p_2 : ₹190.83.

***Example* 45.** For two industries, input-output relationship (in units) is as follows:

Industry → ↓	*I*	*II*	*Final Demand*	*Gross Output*
I	20	40	40	100
II	30	50	40	120
Capital (₹)	2000	3000		
Labour day	200	300		

Using matrix notations, determine :

(*i*) Gross output required to satisfy the new final demand of 110 units and 220 units for industry *I* and *II* respectively.

(*ii*) Total requirement of capital and the labour days.

(*iii*) Total value addition, if cost of capital is 10% and the labour cost is ₹ 200 per labour day.

(*iv*) Is the system viable? **[*Delhi Univ. B.Com. (H) 2013, 2022 Modified*]**

Solution. (*i*) The input-output coefficient matrix is :

$$\mathbf{A} = \begin{bmatrix} 20/100 & 40/120 \\ 30/100 & 50/120 \end{bmatrix} = \begin{bmatrix} 1/5 & 1/3 \\ 3/10 & 5/12 \end{bmatrix}.$$

Let $X = \begin{bmatrix} x_1 \\ x_2 \end{bmatrix}$ be the gross output required to meet the final demand $D = \begin{bmatrix} 110 \\ 220 \end{bmatrix}$. Since production equals consumption, we have

$$\mathbf{X} = \mathbf{AX} + \mathbf{D} \quad \text{or} \quad (\mathbf{I} - \mathbf{A})\mathbf{X} = \mathbf{D}$$

Now $$\mathbf{I} - \mathbf{A} = \begin{bmatrix} 1 & 0 \\ 0 & 1 \end{bmatrix} - \begin{bmatrix} 1/5 & 1/3 \\ 3/10 & 5/12 \end{bmatrix} = \begin{bmatrix} 4/5 & -1/3 \\ -3/10 & 7/12 \end{bmatrix}$$

$$\therefore \quad |\mathbf{I} - \mathbf{A}| = \frac{11}{30} \neq 0$$

Since $|\mathbf{I} - \mathbf{A}| \neq 0$, the output **X** is given by $\mathbf{X} = (\mathbf{I} - \mathbf{A})^{-1}\mathbf{D}$. It can be easily verified that

$$(\mathbf{I} - \mathbf{A})^{-1} = \frac{30}{11}\begin{bmatrix} 7/12 & 1/3 \\ 3/10 & 4/5 \end{bmatrix}$$

$$\therefore \quad \mathbf{X} = (\mathbf{I} - \mathbf{A})^{-1}\mathbf{D} = \frac{30}{11}\begin{bmatrix} 7/12 & 1/3 \\ 3/10 & 4/5 \end{bmatrix}\begin{bmatrix} 110 \\ 220 \end{bmatrix} = \begin{bmatrix} 375 \\ 570 \end{bmatrix}$$

Thus the gross output for the two sectors are 375 and 570 respectively.

(*ii*) The total requirement of two primary inputs is given by

$$\begin{bmatrix} 20 & 25 \\ 2 & 2.5 \end{bmatrix}\begin{bmatrix} 375 \\ 570 \end{bmatrix} = \begin{bmatrix} 21{,}750 \\ 2175 \end{bmatrix}$$

(*iii*) The total value addition $= [0.10 \quad 200]\begin{bmatrix} 21{,}750 \\ 2175 \end{bmatrix} = [4{,}37{,}175]$

(*iv*) Yes, because each entry on the main diagonal of the input-output matrix **A** is less than 1 and $|\mathbf{I} - \mathbf{A}| > 0$.

Note. Suppose there are *n* industries in an economy and it is required to find the change in the gross output of each of the industries if the final demand of Industry 1 changes by Δd_1 units, final demand of Industry 2 changes by Δd_2 units and so on. In such a situation, if we let

$\mathbf{\Delta X} = \begin{bmatrix} \Delta x_1 \\ \Delta x_2 \\ . \\ . \\ \Delta x_n \end{bmatrix}$ denote the change in the gross output and $\mathbf{\Delta D} = \begin{bmatrix} \Delta d_1 \\ \Delta d_2 \\ . \\ . \\ \Delta d_n \end{bmatrix}$ denote the change in demand, then $\mathbf{\Delta X}$ is given by $\mathbf{\Delta X} = (\mathbf{I} - \mathbf{A})^{-1} \mathbf{\Delta D}$.

***Example* 46.** A hypothetical economy produces only two commodities *X* and *Y*. The two commodities serve as intermediate inputs in each other's production. To produce a unit of *X*, 0.2 unit of *X* and 0.6 unit of *Y* are needed. To produce a unit of *Y*, 0.4 unit of *X* and 0.3 unit of *Y* are needed. Three and five labour hours are required to produce a unit of *X* and a unit of *Y* respectively. The wage rate is ₹ 20 per labour-hour. If the final demand of *X* increases by 150 units and that of *Y* decreases by 120 units, find

(*i*) Change in the gross outputs of each of the two commodities;

(*ii*) Change in labour requirement;

(*iii*) Change in the value-added in the two producing sectors.

[*Delhi Univ. B.Com. (H) 2010, 2018*]

Solution. The technology matrix is : $\mathbf{A} = \begin{bmatrix} 0.2 & 0.4 \\ 0.6 & 0.3 \end{bmatrix}$

$$\Rightarrow \mathbf{I} - \mathbf{A} = \begin{bmatrix} 0.8 & -0.4 \\ -0.6 & 0.7 \end{bmatrix} \quad \therefore \quad |\mathbf{I} - \mathbf{A}| = 0.56 - 0.24 = 0.32$$

(*i*) Let $\mathbf{\Delta X} = \begin{bmatrix} \Delta x \\ \Delta y \end{bmatrix}$ denote the change in the gross output and $\mathbf{\Delta D} = \begin{bmatrix} 150 \\ -120 \end{bmatrix}$ denote the change in demand. Then

$$\mathbf{\Delta X} = (\mathbf{I} - \mathbf{A})^{-1} \mathbf{\Delta D} = \frac{1}{0.32} \begin{bmatrix} 0.7 & 0.4 \\ 0.6 & 0.8 \end{bmatrix} \begin{bmatrix} 150 \\ -120 \end{bmatrix} = \begin{bmatrix} 178.125 \\ -18.750 \end{bmatrix}$$

Thus the output of *X* increases by 178.125 units and the output of *Y* decreases by 18.750 units.

(*ii*) Change in labour requirement $= [l_1 \quad l_2] \begin{bmatrix} \Delta x \\ \Delta y \end{bmatrix} = [3 \quad 5] \begin{bmatrix} 178.125 \\ -18.750 \end{bmatrix} = [440.625]$

i.e., 440.625 more labour hours are required.

(*iii*) Value added increase in sector of $X = (\Delta x)(l_1 w) = (178.125)(3 \times 20) = ₹\ 10{,}687.50$

Value added decrease in sector of $Y = (\Delta y)(l_2 w) = (18.750)(5 \times 20) = ₹\ 1875.$

***Example* 47.** The following gives technology matrix for a two sector economy :

	Sector I	*Sector II*
Sector I	0.5	0.3
Sector II	0.3	0.2
Labour	0.5	0.4

Final demands for the two sectors are 15 and 20 respectively. Write down the input-output table for the economy. If the total labour available is 20 units, is the solution feasible ?

Solution. The technology matrix is : $\mathbf{A} = \begin{bmatrix} 0.5 & 0.3 \\ 0.3 & 0.2 \end{bmatrix}$

Let $\mathbf{X} = \begin{bmatrix} x_1 \\ x_2 \end{bmatrix}$ be the gross output vector required to meet the final demand $\mathbf{D} = \begin{bmatrix} 15 \\ 20 \end{bmatrix}$. Since production equals consumption, we have $\mathbf{X} = \mathbf{AX} + \mathbf{D}$ or $(\mathbf{I} - \mathbf{A})\,\mathbf{X} = \mathbf{D}$

Now
$$\mathbf{I} - \mathbf{A} = \begin{bmatrix} 0.5 & -0.3 \\ -0.3 & 0.8 \end{bmatrix} \Rightarrow |\mathbf{I} - \mathbf{A}| = 0.31$$

Since $|\mathbf{I} - \mathbf{A}| \neq 0$, the output **X** is given by $\mathbf{X} = (\mathbf{I} - \mathbf{A})^{-1}\,\mathbf{D}$. It can be easily verified that

$$(\mathbf{I} - \mathbf{A})^{-1} = \frac{1}{0.31}\begin{bmatrix} 0.8 & 0.3 \\ 0.3 & 0.5 \end{bmatrix}$$

$$\therefore \quad \mathbf{X} = (\mathbf{I} - \mathbf{A})^{-1}\,\mathbf{D} = \frac{1}{0.31}\begin{bmatrix} 0.8 & 0.3 \\ 0.3 & 0.5 \end{bmatrix}\begin{bmatrix} 15 \\ 20 \end{bmatrix} = \frac{1}{0.31}\begin{bmatrix} 18 \\ 14.5 \end{bmatrix} = \begin{bmatrix} 58.06 \\ 46.77 \end{bmatrix}$$

To construct the input-output table for the given economy, let us compute the amount x_{ij} of the output of industry i used as input in industry j. Using the fact $x_{ij} = a_{ij}\,x_j$, we obtain

$$\begin{aligned} x_{11} &= a_{11}\,x_1 = 0.5 \times 58.06 = 29.03 \\ x_{12} &= a_{12}\,x_2 = 0.3 \times 46.77 = 14.03 \\ x_{21} &= a_{21}\,x_1 = 0.3 \times 58.06 = 17.42 \\ x_{22} &= a_{22}\,x_2 = 0.2 \times 46.77 = 9.35 \end{aligned}$$

Hence the input-output table is :

	Sector I	*Sector II*	*Final Demand*	*Total Output*
Sector I	29.03	14.03	15	58.06
Sector II	17.42	9.35	20	46.77

Finally, total labour required $= [0.5 \;\; 0.4]\begin{bmatrix} 58.06 \\ 46.77 \end{bmatrix} = [47.73]$

However, the total labour available is 20 units, and hence the solution is not feasible.

***Example* 48.** For a three-sector economy, the input-output co-efficients, a_{ij}'s are :

$$a_{11} = 0.5,\ a_{21} = 0.2,\ a_{31} = 0.1,\ a_{12} = 0.1,\ a_{22} = 0.6,\ a_{32} = 0.2,\ a_{13} = 0.1,\ a_{23} = 0.2,\ a_{33} = 0.6$$

(*i*) Using Simon-Hawkins conditions, test whether the system is viable.

(*ii*) Determine the gross output for the three sectors when the final demand values for them are 21,000, 42,000 and 63,000 units respectively.

(*iii*) Determine the new level of output if the final demand increases by 1500, 3000 and 4500 units. **[*Delhi Univ. B.Com. (H) 2000*]**

Solution. (*i*) The input-output coefficient matrix is : $\mathbf{A} = \begin{bmatrix} 0.5 & 0.1 & 0.1 \\ 0.2 & 0.6 & 0.2 \\ 0.1 & 0.2 & 0.6 \end{bmatrix}$

To test Simon-Hawkins conditions, we consider

$$\mathbf{I} - \mathbf{A} = \begin{bmatrix} 1 & 0 & 0 \\ 0 & 1 & 0 \\ 0 & 0 & 1 \end{bmatrix} - \begin{bmatrix} 0.5 & 0.1 & 0.1 \\ 0.2 & 0.6 & 0.2 \\ 0.1 & 0.2 & 0.6 \end{bmatrix} = \begin{bmatrix} 0.5 & -0.1 & -0.1 \\ -0.2 & 0.4 & -0.2 \\ -0.1 & -0.2 & 0.4 \end{bmatrix}$$

$$\therefore \qquad |\mathbf{I} - \mathbf{A}| = 0.5(0.16 - 0.04) + 0.1(-0.08 - 0.02) - 0.1(0.04 + 0.04) = 0.042 > 0$$

Also, all the elements on the main diagonal of matrix **A** are less than 1. Hence Simon-Hawkings conditions are satisfied and the system is viable.

(*ii*) Let $\mathbf{X} = \begin{bmatrix} x_1 \\ x_2 \\ x_3 \end{bmatrix}$ be the output matrix required to meet the final demand $\mathbf{D} = \begin{bmatrix} 21,000 \\ 42,000 \\ 63,000 \end{bmatrix}$.

Since production equals consumption, we have

$$\mathbf{X} = \mathbf{AX} + \mathbf{D} \quad \text{or} \quad (\mathbf{I} - \mathbf{A})\,\mathbf{X} = \mathbf{D} \quad \Rightarrow \quad \mathbf{X} = (\mathbf{I} - \mathbf{A})^{-1}\mathbf{D}$$

It can be verified that

$$(\mathbf{I} - \mathbf{A})^{-1} = \frac{1}{0.042}\begin{bmatrix} 0.12 & 0.06 & 0.06 \\ 0.10 & 0.19 & 0.12 \\ 0.08 & 0.11 & 0.18 \end{bmatrix}$$

$$\therefore \qquad \mathbf{X} = (\mathbf{I} - \mathbf{A})^{-1}\mathbf{D} = \frac{1}{0.042}\begin{bmatrix} 0.12 & 0.06 & 0.06 \\ 0.10 & 0.19 & 0.12 \\ 0.08 & 0.11 & 0.18 \end{bmatrix}\begin{bmatrix} 21,000 \\ 42,000 \\ 63,000 \end{bmatrix} = \begin{bmatrix} 2,10,000 \\ 4,20,000 \\ 4,20,000 \end{bmatrix}$$

(*iii*) If the final demand for the three sectors are increased by 1500, 3000 and 4500 units respectively, then the new level of output is given by

$$(\mathbf{I} - \mathbf{A})^{-1}\begin{bmatrix} 22,500 \\ 45,000 \\ 67,500 \end{bmatrix} = \frac{1}{0.042}\begin{bmatrix} 0.12 & 0.06 & 0.06 \\ 0.10 & 0.19 & 0.12 \\ 0.08 & 0.11 & 0.18 \end{bmatrix}\begin{bmatrix} 22,500 \\ 45,000 \\ 67,500 \end{bmatrix} = \begin{bmatrix} 2,25,000 \\ 4,50,000 \\ 4,50,000 \end{bmatrix}.$$

***Example* 49.** Given the input-output matrix and final demand vector :

$$\mathbf{A} = \begin{bmatrix} 0.05 & 0.25 & 0.34 \\ 0.33 & 0.10 & 0.72 \\ 0.19 & 0.38 & 0.00 \end{bmatrix}, \qquad \mathbf{D} = \begin{bmatrix} 1800 \\ 200 \\ 900 \end{bmatrix}$$

(*i*) Explain the economic meaning of the elements 0.33, 0.00 and 200.

(*ii*) Explain the economic meaning (if any) of the third-column sum.

(*iii*) Explain the economic meaning (if any) of the third-row sum.

[*Delhi Univ. B.Com. (H) 2014*]

Solution. (*i*) In the given model : $0.33 = a_{21}$. This means that industry *I* uses 33 paise worth of the output produced by industry *II* to produce a rupee worth of its product.

$0.00 = a_{33}$. This means that industry *III* does not require its own product for producing its output.

$200 = d_2$. This means that the economy needs 200 units of the output of industry *II*.

(*ii*) The third-column sum : 0.34 + 0.72 + 0.00 = 1.06 shows that ₹ 1.06 worth of output is required from all the industries to produce a rupee worth of output in industry *III*. Thus the system is not feasible.

(*iii*) The third (or any other) row sum has no economic meaning.

EXERCISE 2.3

1. Given the following transaction matrix, find the gross output to meet the final demand of 200 units of Agriculture and 800 units of Industry.

Producing Sector	*Purchasing Sector*		*Final Demand*
	Agriculture	*Industry*	
Agriculture	300	600	100
Industry	400	1200	400

[*Delhi Univ. B.Com. (H) 1982*]

2. Suppose the inter-relationship between the production of two industries *A* and *B* in a given year is :

	Consumers		*Current Demand*	*Total Output*
	A	*B*		
A	14	6	8	28
B	7	18	11	36

If the forecast demand next year is $\mathbf{D} = \begin{bmatrix} 20 \\ 30 \end{bmatrix}$, what should the total output of the industries A and B be ?

3. Two commodities A and B are produced such that 0.4 ton A and 0.7 ton of B are required to produce a ton of A. Similarly, 0.1 ton of A and 0.6 ton of B are needed to produce a ton of B. Write down the technology matrix. If 68 tons of A and 102 tons of B are required, find out the gross production of both of them. **[Delhi Univ. B.Com. (H) 1984]**

4. In an economy of three industries P, Q, R, the following data (all figures in millions of ₹) are available :

Producers	*Users*			*Final Demand*	*Output*
	P	Q	R		
P	80	100	100	40	320
Q	80	200	60	60	400
R	80	100	100	20	300

Determine the output if the final demand changes to 60 for P, 40 for Q and 60 for R.

[Delhi Univ. B.Com. (H) 1993]

5. Given the following technology matrix : $\mathbf{A} = \begin{bmatrix} 0.5 & 0.1 & 0.1 \\ 0.2 & 0.6 & 0.2 \\ 0.1 & 0.2 & 0.6 \end{bmatrix}$. Required :

   (*i*) Test Simon-Hawkins conditions for the viability of the system.

   (*ii*) Determine the gross output vector **X**, given the demand vector $\mathbf{D} = \begin{bmatrix} 2100 \\ 4200 \\ 6300 \end{bmatrix}$.

[Delhi Univ. B.Com. (H) 1995]

6. A three sector economy has the following input output coefficient matrix **A** and final demand vector **F** :

$$\mathbf{A} = \begin{bmatrix} 0.3 & 0.2 & 0.2 \\ 0.2 & 0.1 & 0.5 \\ 0.2 & 0.4 & 0.2 \end{bmatrix}, \qquad \mathbf{F} = \begin{bmatrix} 80 \\ 30 \\ 50 \end{bmatrix}$$

Find the gross output of each sector to meet the final demand.

[Delhi Univ. B.Com. (H) 1996]

7. The input-output technology matrix for three industries is : $\begin{bmatrix} 0.2 & 0.3 & 0.1 \\ 0 & 0.5 & 0.2 \\ 0 & 0 & 0.5 \end{bmatrix}$. If the final demand

for three industries is $\begin{bmatrix} 100 \\ 200 \\ 300 \end{bmatrix}$, find the gross output. **[Delhi Univ. B.Com. (H) 1984]**

8. Suppose that the final demand for steel, coal and electricity in an economy consisting only of three sectors are ₹ 10 crores, ₹ 5 crores and ₹ 6 crores, respectively. A rupee worth of steel requires 20 paise, 40 paise and 10 paise worth of steel, coal and electricity respectively as inputs; a rupee worth of coal requires 30 paise, 10 paise and 30 paise worth of steel, coal and electricity respectively as inputs and a rupee worth of electricity requires 20 paise worth each of steel, coal and electricity respectively as inputs. How much of steel, coal and electricity should be produced to satisfy both final and intermediate demands ?

9. A country produces only two goods X and Y. The input-output coefficient matrix is $\mathbf{A} = \begin{bmatrix} 0.3 & 0.6 \\ 0.5 & 0.0 \end{bmatrix}$ and the primary input coefficients are (0.2, 0.4). The final demand for X is 10 and for Y is 5.

(*i*) Write down the open input-output model as a set of simultaneous linear equations.

(*ii*) Determine the output solution by finding the inverse of the relevant matrix.

[Delhi Univ. B.Com. (H) 2005]

10. You are given the following transaction matrix for a two sector economy :

Sector		*Sales*		*Final demand*	*Gross output*
		I	*II*		
Purchases	*I*	4	3	13	20
	II	5	4	3	12
Primary input		11	5		

(*i*) Write the technology matrix.

(*ii*) Rewrite the new transaction matrix when the final demand for the output of sector *I* increases to 23 units. **[Delhi Univ. B.Com. (H) 2019]**

11. Consider an economy consisting of three sectors: agriculture, manufactures and services. The hypothetical flow of goods and services in physical units is summarized in the following table:

	Agriculture	*Manufactures*	*Services*	*Final Demand*
Agriculture	60	180	0	120
Manufactures	30	60	30	180
Services	0	40	10	100

If the demand changes to [150, 160, 180], what will be the new output ? (Calculate upto two decimals.) **[Delhi Univ. B.Com. (H) 2001]**

12. A hypothetical economy produces only two commodities, X and Y. The two commodities serve as intermediate inputs in each other's production. To produce a unit of X, 0.6 unit of X and 0.1 unit of Y are needed. Similarly, to produce a unit of Y, 0.7 unit of X and 0.4 unit of Y are needed. Two and five labour-days are required to produce a unit of X and a unit of Y respectively. The wage rate is ₹ 70 per man-day. The economy needs 100 units of X and 50 units of Y for final consumption.

You are required to :

(*i*) examine the viability of input-output system using Hawkins-Simon conditions;

(*ii*) calculate the gross output of each of the two commodities;

(*iii*) determine the equilibrium prices; and

(*iv*) find the total value added. **[*Delhi Univ. B.Com. (H) 2002*]**

13. An economy consists of two sectors—manufacturing and agriculture. To produce one unit of manufacturing output, 0.1 unit of manufacturing goods and 0.01 unit of agriculture goods are required as inputs. One unit of agriculture output requires 40 units of manufacturing goods and nil unit of agricultural goods as inputs. A unit of manufacturing goods requires 4 man-hours of labour and a unit of agriculture output requires 100 man-hours of labour. Calculate the total labour requirement if 50 million units of manufacturing goods and 2 million units of agricultural goods need to be produced for final consumption.

Supposing the wage rate as ₹ 10 per man-hour, calculate the equilibrium prices of manufacturing goods and agricultural output. **[*Delhi Univ. B.Com. (H) 2003, 2019*]**

14. A two industry input-output relationships are given below :

Industry → ↓	*I*	*II*	*Final Demand*	*Gross Output*
I	16	20	4	40
II	8	40	32	80
Labour day	80	120		

Using matrix notations, determine :

(*i*) Gross output required to satisfy the new final demand of 18 units and 44 units for industry *I* and *II* respectively.

(*ii*) Total labour days required.

(*iii*) Total value additions, if wage is ₹ 40 per labour day.

(*iv*) Equilibrium prices using the same wage rate.

Also test the Hawkins-Simon conditions. **[*Delhi Univ. B.Com. (H) 2004, 2017*]**

15. Suppose the inter-industry flow of the products of two industries is given as under :

	Consumption		*Domestic Demand*	*Gross Output*
	X	*Y*		
Production X	30	40	50	120
Production Y	20	10	30	60
Labour	40	20		

(*i*) Determine the technology matrix and test Hawkins-Simon conditions for the viability of the system.

(*ii*) Write down the input-output table for the economy, when the final demand targets are 80 and 40 units of X and Y respectively.

(*iii*) If the total labour available is 100 units, is the solution feasible ?

(*iv*) If the wage rate is ₹ 50 per unit, find equilibrium prices.

(*v*) Find the value added in each sector.

16. For a three sector economy, the input-output coefficient matrix is :

$$\mathbf{A} = \begin{bmatrix} 0.2 & 0.3 & 0.2 \\ 0.4 & 0.1 & 0.2 \\ 0.1 & 0.3 & 0.2 \end{bmatrix}$$

If the final demands of the three sectors are 10, 5, 6 respectively, find the gross output consistent with the given demand.

Land and labour are used as two primary inputs. Their coefficients for the three sectors are given as

$$\begin{matrix} \textit{Land} \\ \textit{Labour} \end{matrix} \begin{bmatrix} 0.3 & 0.2 & 0.5 \\ 0.4 & 0.1 & 0.2 \end{bmatrix}$$

If the wage rate and rent are ₹ 50 and ₹ 100 respectively, find the equilibrium prices for the three sectors.

17. Two industries input-output relationship states that industry I requires 20 paisa worth of its own output and 50 paisa worth of indutry II for producing output of ₹ 1. Industry II requires 30 paisa worth of its own output and 40 paisa worth of indutry I for producing output of ₹ 1.

(*i*) Write the technology coefficient matrix.

(*ii*) Give the interpretation of its row sum and column sum, if any.

(*iii*) Determine the gross output required to satisfy the economy's demand of ₹ 180 crores and ₹ 270 crores respectively

(*iv*) Total value addition. **[*Delhi Univ. B.Com. (H) 2013*]**

ANSWERS

1. Agriculture = 2000, Industry = 4000

2. $A = 72$, $B = 96$

3. $A = 220$, $B = 640$

4. $P : 417.391$, $Q : 455.652$, $R : 417.391$

5. 21,000, 42,000, 42,000

6. 241.54, 215, 230.38

7. 440, 640, 600

8. Steel = 24.84, Coal = 20.68, Electricity = 18.36

9. (*i*) $0.7x - 0.6y = 10$; $-0.5x + y = 5$ (*ii*) $x = 32.5$, $y = 21.25$

10. (*i*) $\begin{bmatrix} 1/5 & 1/4 \\ 1/4 & 1/3 \end{bmatrix}$

(*ii*) The new transaction matrix is :

Sector		Sales I	Sales II	Final Demand	Gross Output
Purchases	I	6.83	4.32	23	34.16
	II	8.54	5.77	3	17.31
Primary Inputs		18.77	7.23		

11. Agriculture : 396.14, Manufactures : 300.13, Services : 235.75

12. (*ii*) $X = 558.82$ units, $Y = 176.47$ units (*iii*) $p_1 = ₹\,700$, $p_2 = ₹\,1400$ (*iv*) 1,39,999.3

13. Manufacturing : 260 million units, Agriculture : 4.6million units; Total labour requirement : 1500 million man-hours; $p_1 = ₹\,100$, $p_2 = ₹\,5000$

14. (*i*) Industry $I = 80$ units, Industry $II = 120$ units (*ii*) 340

(*iii*) 13,600 (*iv*) $p_1 = ₹\,208$, $p_2 = ₹\,224$

15. (*i*) $\begin{bmatrix} 1/4 & 2/3 \\ 1/6 & 1/6 \end{bmatrix}$

(*ii*) The input-output table is :

	X	Y	Final Demand	Total Output
X	45.41	56.21	80	181.62
Y	30.27	14.05	40	84.32
Labour	60.54	28.11		

(*iii*) Yes (*iv*) $p_1 = ₹\,32.43$, $p_2 = ₹\,45.94$

(*v*) Value added in Sector $I = 3027$; Value added in Sector $II = 1405$

16. Gross output : 24.84, 20.68 and 18.36

Equilibrium prices : ₹ 140.88, ₹ 121.61 and ₹ 140.62

17. (*i*) $\begin{bmatrix} 0.20 & 0.40 \\ 0.50 & 0.30 \end{bmatrix}$

(*ii*) Column sum implies the cost of intermediate inputs to produce a rupee worth of output and the balance denotes the value addition per rupee – value addition is 30 paisa per rupee for each of the industries; row sum has no economic meaning.

(*iii*) Industry I = 650 units, Industry II = 850 units.

(*iv*) ₹ 450 crores.

CHAPTER

3

LINEAR PROGRAMMING — BASIC ASSUMPTIONS AND PROBLEM FORMULATION

OBJECTIVES

After studying the material in this chapter, you should be able to:

- *Understand the concept of a linear programming problem.*
- *Know the basic assumptions of linear programming.*
- *Formulate a linear programming problem mathematically.*
- *Identify the feasible region.*

3.1 INTRODUCTION

The central issue in many problems arising in business, industry and government is that various activities compete for limited resources. It is the responsibility of the decision maker to allocate the resources among the competing activities so as to achieve the best results. Linear programming problems deal with determining the best possible allocation of available resources among the competing activities so as to meet certain specifications. In particular, they may relate to situations where a number of resources, such as materials, machines, people, land etc., are available and are to be combined to yield several products. The objective or goal can whatever the user wants so long as it can be expressed in a linear manner. For example, the objective might be to maximize profits, to minimize costs or waste, or to minimize the number of employees. The constraints might reflect limited labour resources, limited raw materials, and limited demand for the products.

Linear programming is a mathematical technique for finding an optimum (maximum or minimum) value of a function, called objective function of several independent variables, these variables being subject to various restrictions (or constraints) expressed as equations or inequalities. The term 'linear' indicates that the function to be maximized or minimized is linear and that the corresponding constraints be represented by a system of linear inequalities or linear equations involving the variables.

A linear programming problem has the following essential parts :

(*i*) ***A set of non-negative restrictions.*** *This is to avoid getting negative solutions in which we are not interested. For example, negative price or negative output do not mean anything.*

(*ii*) ***A set of linear constraints*** *representing the limitations on the use of machinery, labour etc..*

(*iii*) ***An objective function*** *to be maximized or minimized subject to the above constraints.*

3.2 BASIC ASSUMPTIONS OF LINEAR PROGRAMMING

We now clearly state the assumptions that distinguish a linear programming problem from other mathematical problems. These assumptions are :

1. ***Proportionality.*** This assumption means that if the value of a variable is multiplied by a constant, then its contribution to the objective function and to each constraint is also multiplied by that constant. In actual situations, this is not always the case. For example, the law of diminishing returns, where a unit of activity contributes less than the previous unit, may be operating in a situation.

2. ***Additivity.*** This assumption means that the total value of the objective function is sum of the individual contributions and total impact on each constraint is the sum of the individual impacts of the decision variables. In other words, there is no "interaction" between the decision variables. To understand the concept of additivity better, consider the situation where a certain amount of water is to be mixed with a certain amount of alcohol. Since water and alcohol are partially miscible, the resulting amount will actually be much less than the sum of the amounts of the water and alcohol and thus the amount of the mixture is not additive.

3. ***Divisibility.*** Divisibility simply means that the decision variables can take any non-negative values, *i.e.*, fractional values of the decision variables are permitted. This is not always desirable. For example, it is impossible to produce one-third of an aeroplane. When it is necessary to have integer variables, a technique known as *integer programming* could be used.

4. ***Certainty.*** This assumption means that all parameters are known with certainty and do not change during the period being studied. Unfortunately, we will be solving problems about the future, not the past, and thus all data will be certain.

Remark. The proportionality and the additivity assumptions automatically imply that all the constraints in the problem are either linear equations or inequalities. They also imply that the objective function is linear.

3.3 APPLICATIONS OF LINEAR PROGRAMMING

Historically, the field of linear programming was developed by George B. Dantzig and his associates during World War II and was first used by the U.S. Air Force as an aid in decision making. Today it has application to a wide variety of problems. Some of these are as follows:

(*i*) ***Diet Problems.*** To determine the minimum requirement of nutrients subject to availability of foods and their prices.

(*ii*) ***Manufacturing Problems.*** To find the number of items of each type that should be manufactured so as to maximize the profit subject to production restrictions imposed by limitations on the use of machinery and labour.

(*iii*) ***Transportation Problems.*** To find the least costly way of transporting shipments from the warehouses to customers.

(*iv*) ***Blending Problems.*** To determine the optimum amount of several constituents to use in producing a set of products while determining the optimum quantity of each product to produce.

(*v*) ***Assembling Problems.*** To have the best combination of basic components to produce goods according to certain specifications.

(*vi*) ***Production Problems.*** To decide the production schedule to satisfy demand and minimize cost in face of fluctuating rates and storage expenses.

(*vii*) ***Job Assigning Problems.*** To assign job to workers for maximum effectiveness and optimum results subject to restrictions of wages and other costs.

(*viii*) ***Trim-Loss Problems.*** To determine the best way to obtain a variety of smaller rolls of paper from a standard width of roll that is kept in stock and, at the same time, minimize wastage.

3.4 GENERAL LINEAR PROGRAMMING PROBLEM

The mathematical statement of the general form of a linear programming problem (abbreviated *LPP*) may be written as follows :

Optimize (maximize or minimize) $Z = c_1 x_1 + c_2 x_2 + \ldots + c_n x_n$

subject to the constraints :

$$\left.\begin{array}{l} a_{11} x_1 + a_{12} x_2 + \ldots + a_{1n} x_n \; (\leq, =, \geq) b_1 \\ a_{21} x_1 + a_{22} x_2 + \ldots + a_{2n} x_n \; (\leq, =, \geq) b_2 \\ \quad \cdot \qquad\quad \cdot \qquad\qquad \cdot \qquad\quad \cdot \\ \quad \cdot \qquad\quad \cdot \qquad\qquad \cdot \qquad\quad \cdot \\ a_{m1} x_1 + a_{m2} x_2 + \ldots + a_{mn} x_n \; (\leq, =, \geq) b_m \end{array}\right\} \qquad \ldots(1)$$

and $$x_1, x_2, \ldots, x_n \geq 0 \qquad \ldots(2)$$

where

(*i*) $x_1, x_2, \ldots, x_n$ are the variables whose values we wish to determine and are called the *decision or structural variables,*

(*ii*) the linear function Z which is to be maximized or minimized is called the *objective function* of the general *LPP*,

(*iii*) the inequalities (1) are called the constraints of the general *LPP*,

(*iv*) the set of inequalities (2) is known as the set of *non-negative restrictions* of the general *LPP*,

(*v*) the constant c_j $(j = 1, 2, \ldots, n)$ represents the contribution (profit or cost) to the objective function of the *j*th variable,

(*vi*) the coefficients a_{ij} $(i = 1, 2, \ldots, m;\ j = 1, 2, \ldots, n)$ are referred to as the *technological* or *substitution coefficients,*

(*vii*) b_i $(i = 1, 2, \ldots, m)$ is the constant representing the requirement or availability of the *i*th constraint, and

(*viii*) the expression $(\leq, =, \geq)$ means that only one of the relationships in the set $(\leq, =, \geq)$ would hold for a particular constraint.

It is appropriate at this stage to give some definitions which pertain to general linear programming problems.

***Definition* 1.** A set of values of the decision variables which satisfy the constraints of general *LPP* is called a **solution** to the general *LPP.*

***Definition* 2.** Any solution to the general *LPP* which also satisfies the non-negative restrictions of the problem is called a **feasible solution.** The set of all feasible solutions constitutes what is called the **feasible region.**

***Definition* 3.** Any feasible solution which optimizes the objective function of a general *LPP* is called an **optimum solution** to the general *LPP.*

Notice that we spoke of *an* optimum solution. The reason being that a linear programming problem may have many optimum solutions. Moreover, the term "optimal solution" is also used for optimum solution

Remark. The feasible region for a linear programming problem is always a *polyhedral convex* set, *i.e.*, its boundary consists of line segments and has the property that a line segment that joins any two points in the set lies entirely in the set. See Figure 3.1 for some illustrations.

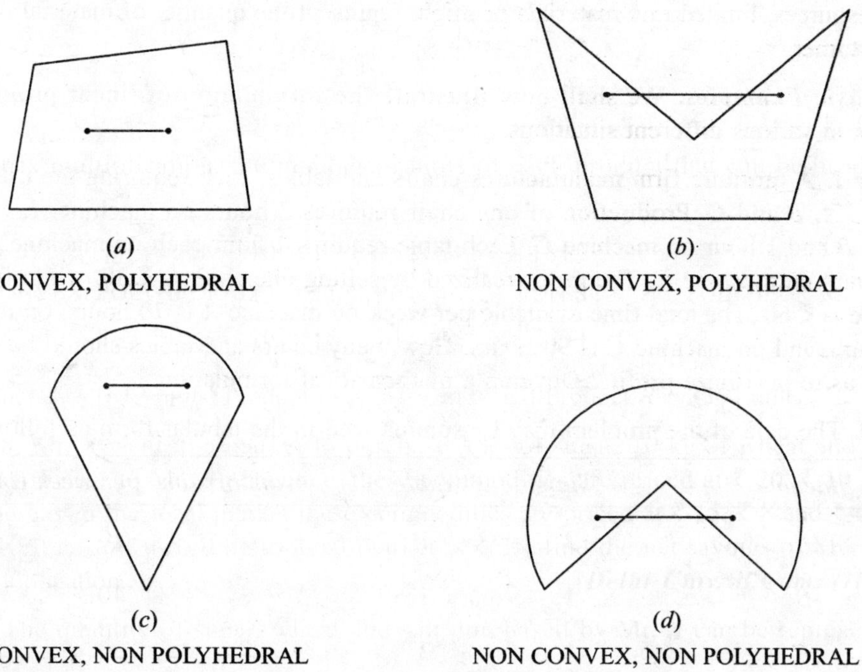

Fig. 3.1

3.5 FORMULATION OF A LINEAR PROGRAMMING PROBLEM

Problem formulation is the process of transforming the verbal description of a decision problem into a mathematical form that can be solved. Formulating a mathematical model of the problem is the first and probably the most important step in the solution of any real world problem.

Problem formulation is an art in itself, and there are no simple methods or rules available. Only with adequate practice can one learn to formulate problems correctly. In general, however, the following steps are involved.

Step* 1. *Identification of the variables. In any *LP* problem certain decisions need to be made. These decisions are represented by decision variables. These decision variables are those quantities whose values we wish to determine. In any event, the decision variables should be identified as a first step in the development of an *LP* model.

Step* 2. *Identification of the objective function. The objective function is a mathematical representation of the overall goal stated in terms of the decision variables. For an *LPP*, the objective function might take the form of maximizing profits or minimizing costs. Thus the second step in formulating an *LP* is to identify the type of optimization, i.e., maximization or minimization.

Step 3. Identification of the constraints. Finally, the third step is to identify the set of constraints, stated in terms of the decision variables, representing the conditions which must be satisfied in determining levels for the decision variables. The constraints might reflect limited labour resources, limited raw materials or might represent the quantity of material demanded by a customer.

Some Basic Examples. We shall now illustrate the formulation of linear programming problems in various different situations.

Example 1. A furniture firm manufactures chairs and tables, each requiring the use of three machines A, B and C. Production of one chair requires 2 hours on machine A, 1 hour on machine B and 1 hour on machine C. Each table requires 1 hour each on machines A and B and 3 hours on machine C. The profit realized by selling one chair is ₹ 30, while for a table the figure is ₹ 60. The total time available per week on machine A is 70 hours, on machine B is 40 hours, and on machine C is 90 hours. How many chairs and tables should be made per week so as to maximize profit ? Develop a mathematical formulation.

Solution. The data of the problem may be summarized in the tabular form as follows :

Machine	*Chair*	*Table*	*Available time per week (in hours)*
A	2	1	70
B	1	1	40
C	1	3	90
Profit per unit (in ₹)	30	60	

Let

x = the number of chairs to be produced per week

y = the number of tables to be produced per week

The problem imposes constraints upon the values of x and y, some of which are explicitly stated and some implied by the nature of the problem. For example, it makes no sense to assign negative values to x and y, so we must have

$$x \geq 0 \quad \text{and} \quad y \geq 0$$

We observe next that the total time (in hours) required on machine A for the production of x chairs and y tables is $2x + y$ and this should not exceed 70, the maximum time available on machine A. Accordingly, we obtain the following constraint :

$$2x + y \leq 70$$

Similarly, the total time required on machine B for the production of x chairs and y tables is $x + y$. Thus the limitation of 40 hours, at most on machine B is expressed by the constraint:

$$x + y \leq 40$$

Finally, machine C limitation takes the form $x + 3y \leq 90$.

Subject to these constraints, the value of the profit must be maximized, namely, $30x + 60y$. Let us designate the value of the profit by Z. Then we may restate our problem in the following mathematical form :

Maximize $Z = 30x + 60y$

subject to the constraints:

$$\begin{aligned} 2x + y &\le 70 \\ x + y &\le 40 \\ x + 3y &\le 90 \\ x, y &\ge 0 \end{aligned}$$

Example 2. A diet is to contain at least 400 units of carbohydrate, 500 units of fat, and 300 units of protein. Two foods are available : F_1, which costs ₹ 2 per unit, and F_2, which costs ₹ 4 per unit. A unit of food F_1 contains 10 units of carbohydrate, 20 units of fat, and 15 units of protein; a unit of food F_2 contains 25 units of carbohydrate, 10 units of fat, and 20 units of protein. Find the minimum cost for a diet that consists of a mixture of these two foods and also meets the minimum nutrition requirements. Formulate the problem as a linear programming problem.

Solution. The data of the problem may be summarized in the tabular form as follows:

Food	*Carbohydrate*	*Fat*	*Protein*	*Cost per unit*
F_1	10	20	15	₹ 2
F_2	25	10	20	₹ 4
Minimum requirement	400	500	300	

Let x = the number of units of food F_1 to be purchased

y = the number of units of food F_2 to be purchased

As in the preceding example, we must have

$$x \ge 0 \quad \text{and} \quad y \ge 0$$

The total amount of carbohydrate supplied from foods F_1 and F_2 is $10x + 25y$, and this is required to be at least as large as the minimum amount of carbohydrate required, namely, 400 units. Therefore, we must have

$$10x + 25y \ge 400$$

Similarly, the total amount of fat must amount to at least 500 units, and this requirement is expressed as follows :

$$20x + 10y \ge 500$$

A final constraint is imposed by the specification of at least 300 units of protein :

$$15x + 20y \ge 300$$

Subject to these constraints, our problem is to find the total cost (in ₹) of the mixture, namely, $2x + 4y$, as small as possible. Let us designate this total cost by Z. Then we may restate our problem in the following form :

Minimize $Z = 2x + 4y$

subject to the constraints:

$$10x + 25y \geq 400$$
$$20x + 10y \geq 500$$
$$15x + 20y \geq 300$$
$$x, y \geq 0$$

***Example* 3.** A company makes two kinds of leather belts, A and B. Belt A is a high quality belt, and B is of lower quality. The respective profits are ₹ 4 and ₹ 3 per belt. Each belt of type A requires twice as much time as required by a belt of type B, and if all belts were of type B, the company could make 1000 per day. The supply of leather is sufficient for only 800 belts per day (both A and B combined). Belt A requires a fancy buckle, and only 400 per day are available. There are only 700 buckles a day available for belt B.

What should be daily production of each type of belt ? Formulate the problem as a linear programming problem.

Solution. Let x denote the number of belts A and y denote the number of belts B produced per day. The quantity to be maximized is the profit Z (in rupees) given by

$$Z = 4x + 3y$$

Since the rate of production of belts of type B is 1000 per day, the total time taken to produce y belts of type B is $\frac{y}{1000}$. Also, since each belt of type A requires twice as much time as a belt of type B, the rate of production of belts of type A is 500 per day and consequently total time taken to produce x belts of type A is $\frac{x}{500}$. Thus we have the time constraint :

$$\frac{x}{500} + \frac{y}{1000} \leq 1 \quad \text{or} \quad 2x + y \leq 1000$$

The supply of leather imposes a constraint on the sum of x and y, which can not exceed 800. Thus,

$$x + y \leq 800$$

The constraint imposed by fancy buckles gives

$$x \leq 400$$

and the constraint imposed by buckles available for belt B gives

$$y \leq 700$$

Finally, we have non-negative constraints on the variables x and y :

$$x \geq 0 \quad \text{and} \quad y \geq 0$$

Thus the mathematical formulation of the problem is :

Maximize $Z = 4x + 3y$

subject to the constraints:

$$\begin{aligned} 2x + y &\le 1000 \\ x + y &\le 800 \\ x &\le 400 \\ y &\le 700 \\ x, y &\ge 0 \end{aligned}$$

***Example* 4.** A city hospital has the following minimal daily requirement for nurses :

Period	*Clock Time (24 hours day)*	*Minimal Number of Nurses Required*
1	6 a.m. – 10 a.m.	2
2	10 a.m. – 2 p.m.	7
3	2 p.m. – 6 p.m.	15
4	6 p.m. – 10 p.m.	8
5	10 p.m. – 2 a.m.	20
6	2 a.m. – 6 a.m.	6

Nurses report to the hospital at the beginning of each period and work for 8 consecutive hours. The hospital wants to determine the minimal number of nurses to be employed so that there will be sufficient number of nurses available for each period. Formulate this as a Linear Programming Problem by setting up appropriate constraints and objective function. DO NOT SOLVE. **[*Delhi Univ. B.Com. (H) 1982, 2006*]**

Solution. Let x_i $(i = 1, 2, \ldots, 6)$ be the number of nurses who report at the beginning of ith period. The objective is to minimize the total number of nurses employed in the hospital:

$$Z = x_1 + x_2 + \ldots + x_6$$

Since each period is of 4-hour duration and nurses report to the hospital at the beginning of each period and work for 8 consecutive hours, the x_1 nurses who reported at the beginning of first period shall still be on duty when second period starts. Thus during the second period there will be $x_1 + x_2$ nurses. But the minimal number of nurses required during the second period is specified to be 7. Thus we must have

$$x_1 + x_2 \ge 7$$

Similarly, the other constraints of the problem are :

$$\begin{aligned} x_2 + x_3 &\ge 15 \\ x_3 + x_4 &\ge 8 \end{aligned}$$

$$x_4 + x_5 \geq 20$$
$$x_5 + x_6 \geq 6$$
$$x_6 + x_1 \geq 2$$

Further, it makes no sense to assign negative values to $x_1, x_2, \ldots, x_6$, so we must have

$$x_1, x_2, \ldots, x_6 \geq 0.$$

***Example* 5.** The manager of an oil refinery must decide on the optimal mix of two possible blending processes of which the inputs and outputs per production run are as follows :

	Inputs (units)		*Outputs (units)*	
Process	*Crude A*	*Crude B*	*Gasoline X*	*Gasoline Y*
1	4	3	4	7
2	5	6	5	5

The maximum amounts available of crudes A and B are 225 units and 200 units respectively. Market demand shows that at least 150 units of gasoline X and 120 units of gasoline Y must be produced. The profits per production run from process 1 and process 2 are ₹ 200 and ₹ 300 respectively. Formulate the problem as a linear programming problem.

Solution. Let x_1 and x_2 denote the number of production runs of the two processes, respectively. Then the appropriate mathematical formulation of the problem is :

Maximize $\qquad Z = 200x_1 + 300x_2$

subject to the constraints:

$$4x_1 + 5x_2 \leq 225$$
$$3x_1 + 6x_2 \leq 200$$
$$4x_1 + 5x_2 \geq 150$$
$$7x_1 + 5x_2 \geq 120$$
$$x_1, x_2 \geq 0$$

***Example* 6.** A retired person wants to invest upto an amount of ₹ 30,000 in the fixed income securities. His broker recommends investing in two bonds, A and B. Bond A yields 7% return on the amount invested and bond B yields 10% return on the amount invested. After some consideration, he decides to invest at least ₹ 6,000 in bond A and at the most ₹ 12,000 in bond B. He also wants that the amount invested in bond A must be at least equal to the amount invested in bond B. What should his broker recommend if he wants to maximize his return on investment ? Formulate the problem mathematically.

[*Delhi Univ. B.Com. (H) 2022*]

Solution. Let the amount invested in bond A and bond B be ₹ x and ₹ y respectively. Then the appropriate mathematical formulation of the given problem is :

Maximize $\qquad Z = 0.7x + 0.10y$

subject to the constraints:

$$x + y \leq 30,000$$
$$x \geq 6,000$$
$$y \leq 12,000$$
$$x \geq y$$
$$x, y \geq 0$$

***Example* 7.** A person is interested in investing ₹ 50,00,000 in a mix of investments. The investment choices and expected rates of returns on each one of them are :

Investment	*Expected Rate of Return*
Mutual Fund A	0.12
Mutual Fund B	0.09
Money Market Fund	0.08
Government Bonds	0.085
Share Y	0.16
Share X	0.18

The investor wants at least 35 per cent of his investment in government bonds. Because of the higher perceived risk of the two shares, he has specified that the combined investment in these not to exceed ₹ 80,000. The investor also has specified that at least 20 per cent of the investment should be in the money market fund and that the amount of money invested in shares should not exceed the amount invested in mutual funds. His final investment condition is that the amount invested in mutual fund *A* should be no more than the amount invested in mutual fund *B*. The problem is to decide the amount of money to invest in each alternative so as to obtain the highest annual total return. Formulate the problem as a linear programming problem. **[*Delhi Univ. B.Com. (H) 1994*]**

Solution. Let

x_1 = the amount of money invested in Mutual Fund *A*

x_2 = the amount of money invested in Mutual Fund *B*

x_3 = the amount of money invested in Money Market Fund

x_4 = the amount of money invested in Government Bonds

x_5 = the amount of money invested in Share *Y*

x_6 = the amount of money invested in Share *X*

Then the appropriate mathematical formulation of the problem is as follows :

$$\text{Maximize} \quad Z = 0.12x_1 + 0.09x_2 + 0.08x_3 + 0.085x_4 + 0.16x_5 + 0.18x_6$$

subject to the constraints :

$$\begin{aligned} x_1 + x_2 + x_3 + x_4 + x_5 + x_6 &= 50,00,000 \\ x_4 &\geq 17,50,000 \\ x_5 + x_6 &\leq 80,000 \\ x_3 &\geq 10,00,000 \\ x_5 + x_6 &\leq x_1 + x_2 \\ x_1 &\leq x_2 \\ x_1, x_2, x_3, x_4, x_5, x_6 &\geq 0 \end{aligned}$$

***Example* 8.** A company manufacturing television sets and radios has four major departments: chassis, cabinet, assembly and final testing. Monthly capacities are as follows :

	Television Capacity		*Radio Capacity*
Chassis	1500	or	4500
Cabinet	1000	or	8000
Assembly	2000	or	4000
Testing	3000	or	9000

The contribution of television is ₹ 150 each and the contribution of radio is ₹ 25 each. Assuming that the company can sell any quantity of either product, determine the optimal combination of output.

Solution. Let x denote the number of television sets and y denote the number of radios manufactured per month. Then the quantity to be maximized is the total contribution Z given by

$$Z = 150x + 25y$$

Since the chassis department has a capacity to produce a maximum of 1500 *TV* chassis per month, therefore the time taken to produce x *TV* chassis is $\frac{x}{1500}$. Similarly, the time taken to produce y radio chassis is $\frac{y}{4500}$. Thus the total time taken in the chassis department is $\frac{x}{1500}$ + $\frac{y}{4500}$. This should not exceed the time limit of 1 month. Thus we have

$$\frac{x}{1500} + \frac{y}{4500} \leq 1 \quad \text{or} \quad 3x + y \leq 4500$$

Similarly, the constraints imposed by the other departments give

$$\frac{x}{1000} + \frac{y}{8000} \leq 1 \quad \text{or} \quad 8x + y \leq 8000$$

$$\frac{x}{2000} + \frac{y}{4000} \leq 1 \quad \text{or} \quad 2x + y \leq 4000$$

$$\frac{x}{3000}+\frac{y}{9000}\le 1 \quad \text{or} \quad 3x+y\le 9000$$

Finally, it makes no sense to assign negative values to x and y, so we must have

$$x \ge 0 \quad \text{and} \quad y \ge 0$$

Thus the mathematical formulation of the problem is :

Maximize $Z = 150x + 25y$

subject to the constraints:

$$3x + y \le 4500$$
$$8x + y \le 8000$$
$$2x + y \le 4000$$
$$3x + y \le 9000$$
$$x, y \ge 0$$

***Example* 9.** A firm buys casting of P and Q type of parts and sells them as finished products after machining, boring and polishing. The purchasing costs for casting are ₹ 3 and ₹ 4 each for parts P and Q respectively and selling costs are ₹ 8 and ₹ 10 respectively. The per hour capacity of machines used for machining, boring and polishing for the two products is given below :

Capacity (per hour)

	P	or	Q
Machine	30	or	50
Boring	30	or	45
Polishing	45	or	30

The running costs for machining, boring and polishing are ₹ 30, ₹ 22.5 and ₹ 22.5 per hour respectively. Formulate the linear programming problem to find out the product mix to maximize the profit. DO NOT SOLVE. **[*Delhi Univ. B.Com. (H) 2004*]**

Solution. Let the firm produce x_1 units of P and x_2 units of Q. Then

$$\text{Profit/unit of } P = 8-\left[3+\frac{30}{30}+\frac{22.5}{30}+\frac{22.5}{45}\right] = ₹2.75$$

$$\text{Profit/unit of } Q = 10-\left[4+\frac{30}{50}+\frac{22.5}{45}+\frac{22.5}{30}\right] = ₹4.15$$

Thus, the appropriate mathematical formulation of the problem is :

Maximize $Z = 2.75x_1 + 4.15x_2$

subject to the constraints:

$$\frac{x_1}{30} + \frac{x_2}{50} \le 1 \quad \text{or} \quad 5x_1 + 3x_2 \le 150$$

$$\frac{x_1}{30} + \frac{x_2}{45} \le 1 \quad \text{or} \quad 3x_1 + 2x_2 \le 90$$

$$\frac{x_1}{45} + \frac{x_2}{30} \le 1 \quad \text{or} \quad 2x_1 + 3x_2 \le 90$$

Finally, it makes no sense to assign negative values to x_1 and x_2, so we must have

$$x_1 \ge 0 \quad \text{and} \quad x_2 \ge 0$$

***Example* 10.** A media specialist plans to allocate advertising expenditure in three media where unit costs of a message are ₹ 1500, ₹ 1250 and ₹ 1000 respectively. The total advertising budget available for the year is ₹ 50,000. The first medium is a monthly magazine and it is desired to advertise not more than once in one issue. At least five advertisements should appear in the second medium and the number of advertisements in the third medium should strictly lie between 6 and 10. The effective audience for unit advertisement in the three media is given below :

Medium	1	2	3
Expected Effective Audience :	50,000	40,000	25,000

Formulate a linear programming problem to find the optimum allocation of advertisements in three media that would maximize the total effective audience

[*Delhi Univ. B.Com. (H) 1990*]

Solution. Let x_1, x_2 and x_3 be the number of advertisements in media 1, 2 and 3 respectively. Then the appropriate mathematical formulation of the given problem is :

Maximize $Z = 50{,}000x_1 + 40{,}000x_2 + 25{,}000x_3$

subject to the constraints:

$$1500x_1 + 1250x_2 + 1000x_3 \le 50{,}000$$

$$x_1 \le 12$$

$$x_2 \ge 5$$

$$6 \le x_3 \le 10$$

$$x_1, x_2, x_3 \ge 0$$

***Example* 11.** Shyam, an Agriculturist, has a farm with 125 acres. He produces radish, peas and potato. Whatever he raises is fully sold in the market. He gets ₹ 5 for radish per kg, ₹ 4 for peas per kg and ₹ 5 for potato per kg. The average yield is 1500 kg of radish per acre, 1800 of peas per acre and 1200 kg of potato per acre. To produce each 100 kg of radish and peas and to produce each 80 kg of potato a sum of ₹ 12.50 has to be used for manure. Labour required for each acre to raise the crop is 6 man days for radish and potato and each 5 days for peas. A total of 500 man days of labour at a rate of ₹ 40 per man day are available.

Formulate this as a linear programming model to maximize the Agriculturist's total profit.
[*Delhi Univ. B.Com. (H) 2001*]

Solution. The data of the problem may be summarized in the tabular form as follows :

	Radish	*Peas*	*Potato*
Revenue/acre	1500 × 5 = 7500	1800 × 4 = 7200	1200 × 5 = 6000
Cost of labour/acre	40 × 6 = 240	40 × 5 = 200	40 × 6 = 240
Cost of manure/acre	$\frac{12.50}{100} \times 1500 = 187.5$	$\frac{12.50}{100} \times 1800 = 225$	$\frac{12.50}{80} \times 1200 = 187.5$
Net profit/acre	7500 – 240 – 187.5 = 7072.5	7200 – 200 – 225 = 6775	6000 – 240 – 187.5 = 5572.5

Let x_1, x_2 and x_3 acres of farm be used to produce radish, peas and potato, respectively. Then the appropriate mathematical formulation of the problem is :

Maximize $$Z = 7072.5x_1 + 6775x_2 + 5572.5x_3$$
subject to the constraints:

$$x_1 + x_2 + x_3 \le 125$$
$$6x_1 + 5x_2 + 6x_3 \le 500$$
$$x_1, x_2, x_3 \ge 0$$

***Example* 12.** A pharmaceutical company has 100 kg of *A*, 180 kg of *B* and 120 kg of *C* available per month. They can use these materials to make three basic pharmaceutical products, *viz*., 5–10–5, 5–5–10 and 20–5–10, where the numbers in each represent the percentage by weight of *A*, *B* and *C* respectively in each of the products. The cost of these ingredients are given below :

Ingredient	*Cost per kg* (₹)
A	80
B	20
C	50
Inert ingredients	20

Selling prices of these products are ₹ 40.50, ₹ 43 and ₹ 45 per kg respectively. There is a capacity restriction of the company for the product 5–10–5 so as they cannot produce more than 30 kg per month. Determine how much of each of the products they should produce in order to maximize their monthly profit.

Solution. Let the products to be manufactured be denoted by P_1, P_2 and P_3. Then the given information may be summarrized in the tabular form as follows :

Products	*Ingredients*			
	A	*B*	*C*	*Inert*
P_1	5%	10%	5%	80%
P_2	5%	5%	10%	80%
P_3	20%	5%	10%	65%
Cost per kg (in ₹)	80	20	50	20

Cost/kg of P_1 = 5% of 80 + 10% of 20 + 5% of 50 + 80% of 20

= 4 + 2 + 2.50 + 16 = ₹ 24.50

Cost/kg of P_2 = 5% of 80 + 5% of 20 + 10% of 50 + 80% of 20

= 4 + 1 + 5 + 16 = ₹ 26

Cost/kg of P_3 = 20% of 80 + 5% of 20 + 10% of 50 + 65% of 20

= 16 + 1 + 5 + 13 = ₹ 35

Profit/kg of P_1 = 40.50 – 24.50 = ₹ 16

Profit/kg of P_2 = 43 – 26 = ₹ 17

Profit/kg of P_3 = 45 – 35 = ₹ 10

Let the company produce x_1 kg of P_1, x_2 kg of P_2 and x_3 kg of P_3 per month. Then the appropriate mathematical formulation of the problem is :

Maximize $Z = 16x_1 + 17x_2 + 10x_3$

subject to the constraints:

$$\frac{1}{20}x_1 + \frac{1}{20}x_2 + \frac{1}{5}x_3 \le 100 \qquad\qquad x_1 + x_2 + 4x_3 \le 2000$$

$$\frac{1}{10}x_1 + \frac{1}{20}x_2 + \frac{1}{20}x_3 \le 180 \quad \text{or} \quad 2x_1 + x_2 + x_3 \le 3600$$

$$\frac{1}{20}x_1 + \frac{1}{10}x_2 + \frac{1}{10}x_3 \le 120 \qquad\qquad x_1 + 2x_2 + 2x_3 \le 2400$$

$$x_1 \le 30 \qquad\qquad x_1 \le 30$$

$$x_1, x_2, x_3 \ge 0 \qquad\qquad x_1, x_2, x_3 \ge 0$$

***Example* 13.** A box manufacturer makes small and large boxes from a large piece of cardboard. The large boxes require 4 sq. ft. per box, while the small boxes require 3 sq. ft. per box. The manufacturer is required to make at least three large boxes and at least twice as many small boxes as large boxes. If 60 sq. ft. of cardboard is in stock, and if the profits on the small and large boxes are ₹ 2 and ₹ 3 per box, respectively, how many of each should be made in order to maximize the total profit ? Formulate the problem as a linear programming problem.

Solution. Let box manufacturer make x large boxes and y small boxes.

Then the appropriate mathematical formulation of the problem is :

Maximize (Total profit) $\quad Z = 3x + 2y$

subject to the constraints:

$$4x + 3y \leq 60$$
$$x \geq 3$$
$$y \geq 2x$$
$$x, y \geq 0$$

***Example* 14.** Alpha Heavy Engineering Company produces earthmovers and harvesters. Each product passes through two assembly departments A and B, which, respectively, have 300 hours and 320 hours of available time for the next month's production. Each earthmover requires 20 hours in department A and 40 in department B and each harvester requires 30 hours in department A and 20 in department B. The two products are tested in a third department. Each earthmover is given 60 hours of testing and each harvester 20, and, as per the agreement with the labour union, the total labour hours devoted to testing cannot fall below 270. The management has the operating policy of manufacturing at least one harvester for every two earthmovers produced. A major customer has placed an order for a minimum of 5 earthmovers and harvesters (in any combination, whatever) for next month, and, so, at least that many must be produced. Each earthmover gives a profit of ₹ 10,000 and each harvester ₹ 8000. Formulate the problem as a linear programming problem.

[Delhi Univ. B.Com. (H) 2002]

Solution. Let x denote the number of earthmovers and y denote the number of harvesters produced. Then the appropriate mathematical formulation of the problem is :

Maximize $\quad Z = 10{,}000x + 8000y$

subject to the constraints:

$$20x + 30y \leq 300$$
$$40x + 20y \leq 320$$
$$60x + 20y \geq 270$$
$$y \geq x/2 \quad \text{or} \quad x - 2y \leq 0$$
$$x + y \geq 5$$
$$x, y \geq 0$$

Example* 15.** A carpenter has started a workshop in which he manufactures handcarts. Each cart consists of a frame and two wheels. A frame uses 3 hours and a wheel 2 hours of labour of which 90 hours per week are available. The carpenter wants that he should manufacture at least 10 carts during a week. The cost of frame is ₹ 500 and that of a wheel is ₹ 200. Formulate this as a linear programming model. ***[Delhi Univ. B.Com. (H) 2004]

Solution. Let x be the number of frames (or handcarts) and y the number of wheels produced per week. Then the appropriate mathematical formulation of the problem is:

Minimize $Z = 500x + 200y$

subject to the constraints:

$$3x + 2y \leq 90$$
$$x \geq 10$$
$$y = 2x$$
$$x, y \geq 0$$

EXERCISE 3.1

1. Write a note on the business applications of linear programming.

[*Delhi Univ. B.Com. (H) 1989*]

2. A company sells two different products, *A* and *B*. The two products are produced in a common production process and are sold in two different markets. The production process has a total capacity of 45,000 man-hours. It takes 5 hours to produce a unit of *A* and 3 hours to produce a unit of *B*. The market has been surveyed and company officials feel that the maximum numbers of units of *A* that can be sold is 7000 and that of *B* is 10,000. If the profit is ₹ 60 per unit for the product *A* and ₹ 20 for the product *B*, how many units of each product should be sold to maximize profit ? Formulate the problem as a linear programming problem.

3. A firm manufactures two products, each of which must be processed through two departments, 1 and 2. The hourly requirements per unit for each product in each department, the weekly capacities in each department, selling price per unit, labour cost per unit, and raw material cost per unit are summarized as follows :

	Product A	*Product B*	*Weekly Capacity*
Department 1	3	2	120
Department 2	4	6	260
Selling price per unit	₹25	₹30	
Labour cost per unit	₹16	₹20	
Raw material cost per unit	₹4	₹4	

The problem is to determine the number of units to produce of each product so as to maximize total contribution to profit. Formulate this as a linear programming problem.

4. A rubber company is engaged in producing three types of tyres, *A*, *B* and *C*. To produce each type requires processing in two plants, Plant *I* and *II*. The capacities of the two plants, in number of tyres per day, are as follows :

Plant	*A*	*B*	*C*
I	50	100	100
II	60	60	200

The monthly demand for type A, B and C is 2500, 3000 and 7000, respectively. If Plant I costs ₹ 2500 per day, and Plant II costs ₹ 3500 per day to operate, how many days should each be run per month to minimize cost while meeting the demands ? Formulate the problem as a linear programming problem.

5. A diet is to contain at least 80 units of vitamin A and 100 units of vitamin B. Two foods are available, F_1 and F_2. Food F_1 costs ₹ 4 per unit and Food F_2 costs ₹ 5 per unit. A unit of food F_1 contains 3 units of vitamin A and 4 units of vitamin B; a unit of food F_2 contains 6 units of vitamin A and 3 units of vitamin B. Find the minimum cost for a diet that consists of a mixture of these two foods and also meets the minimal nutrition requirements.

6. An Airline agrees to charter planes for a group. The group needs at least 160 first class seats and at least 300 tourist class seats. The airline must use at least two of its model 314 planes which have 20 first class and 30 tourist class seats. The airline will also use some of its model 435 planes which have 20 first class seats and 60 tourist class seats. Each flight of a model 314 plane costs the company ₹ 1 lakh, and each flight of a model 535 plane costs ₹ 1.5 lakh. How many of each type plane should be used to minimize the flight cost ?

7. The meat department of a super market deals in three types of meat, say, A, B and C. The manager of the department finds that he has 240 pounds of A, 900 pounds of B and 450 pounds of C on a saturday morning. From past experience, he knows that he can sell two third of these quantities as straight cuts. The remaining meat will have to be ground into hamburger patties and picnic patties for which there is a long weekend demand. Each pound of hamburger patties contains 20% A and 60% B. Each pound of picnic patties contains 50% B and 30% C. The remainder of each product consists of an inexpensive non-meat filler which is available in unlimited quantities. How many pounds of each product should be made so that the maximum amount of meat is used ?

8. The daily Florist company is planning to make up floral arrangements for the upcoming festival. The company has available the following supply of flowers at the costs shown :

Type	*Number available*	*Cost per flower* (in ₹)
Red roses	800	0.20
Garnenias	456	0.25
Carnations	4000	0.15
White roses	920	0.20
Yellow roses	422	0.22

These flowers can be used in any of the four popular arrangements whose make up and selling prices are as follows :

Arrangement	*Requirements*	*Selling price* (in ₹)
Economy	4 red roses, 2 garnenias, 8 carnations	6
May time	8 white roses, 5 garnenias, 10 carnations, 4 yellow roses	8
Spring colour	9 red roses, 10 carnations, 9 white roses, 6 yellow roses	10
Deluxe rose	12 red roses, 12 white roses, 12 yellow roses	12

How many units of each arrangement should be made up in order to maximize profit, assuming all arrangements can be sold.

9. Rajan's mathematics teacher has given him three very long lists of problems with the instruction to submit no more than 100 of them (correctly solved) for credit. The problems in the first set are worth 5 points each, those in the second set are worth 4 points each, and those in the third set are worth 6 points each. Rajan knows from experience that he requires on the average 3 minutes to solve a 5-point problem, 2 minutes to solve a 4-point problem, and 4 minutes to solve a 6-point problem. Because he has other subjects to worry about, he can not afford to devote more than $3\frac{1}{2}$ hours altogether to this mathematics assignment. Moreover, the first two sets of problems involve numerical calculations and he knows that he cannot stand more than $2\frac{1}{2}$ hours of work on this type of problem. Under these circumstances, how many problems in each of these categories shall he do in order to get maximum possible credit for his efforts ?

10. A manufacturer can produce two products, *A* and *B*, during a given time period. Each of these products requires four different manufacturing operations : grinding, turning, assembly and testing. The manufacturing requirements in hours per unit of product are given below for *A* and *B*.

	A	*B*
Grinding	1	2
Turning	3	1
Assembly	6	3
Testing	5	4

The available capacities of these operation in hours for the given time period are : grinding 30; turning 60; assembly 200; testing 200. The contribution to profit is ₹ 2 for each unit of *A* and ₹ 3 for each unit of *B*. The firm can sell all that it produces at the prevailing market price. Determine the optimum amount of *A* and *B* to produce during the given time period.

11. A 24 hour supermarket has the following minimal requirement for cashiers :

Period :	1	2	3	4	5	6
Time of day : (*24 hour clock*)	3 – 7	7 – 11	11 – 15	15 – 19	19 – 23	23 – 03
Minimum No. Required :	7	20	14	20	10	5

Period 1 follows immediately after period 6. A cashier works eight consecutive hours, starting at the beginning of one of the six periods. To determine a daily employee work sheet which satisfies the requirements with the least number of personnel, formulate the problem as a linear programming problem. **[*Delhi Univ. B.Com. (H) 2000*]**

12. Hifashion Furnishers manufacture two models of dining table *P* and *Q*. The market standing of the firm is such that they are able to sell at profit margins of ₹ 300 and ₹ 400 respectively of as many numbers of P and Q as they make. The dining tables are manufactured in two stages, *viz.*, carpentry followed by paint shop whose capacities per month are as follows :

Carpentry :	70 of model P only or 120 of mode Q only or some intermediate mix
Paint Shop :	90 of model P only or 60 of model Q only or some intermediate mix

Within these manufacturing capacity constraints, find how many of each model should be manufactured per month so that the profit of the firm is maximized. Formulate the problem as a linear programming problem.

13. A publisher of a text book is planning on the type of binding to be used. The book may be cloth-bound or paperback. A cost analysis indicates that the cloth-bound book contributes ₹ 25 towards profitability while the contribution from the paperback edition is only ₹ 10. It takes 90 minutes for binding a cloth-bound book and 30 minutes for a paperback edition. The total time available for binding is 4000 hours. The market survey indicates that not more than 3000 copies of the cloth-bound would be sold, while the demand for paperback edition is unlimited. Formulate the problem as a linear programming problem to maximize the total profit.

14. A company produces two types of pens, say A and B. Pen A is superior in quality while pen B is of lower quality. Net profits on pen A and pen B are ₹ 5 and ₹ 3 respectively. Raw material required for pen A is twice as that of pen B. The supply of raw material is sufficient only for 1000 pens of B per day. Pen A requires a special nib and only 400 such nibs are available. For pen B, only 700 nibs are available in a day. Find the daily product mix so that the company can make maximum profit.

15. A scrap-metal dealer has received a bulk order from a customer for supply of 2000 kgs of scrap metal. The customer has specified that at least 1000 kgs of the order must be of high quality copper that can be melted easily and can be used to produce tubes. Further, the customer has specified that the order should not contain more than 10 per cent of the total order that is unfit for any commercial use.

The scrap-metal dealer purchases scrap from two different sources in unlimited quantities with following percentages (by weight) of high quality copper and unfit scrap :

	Source A	*Source B*
Copper	40%	75%
Unfit Scrap	7.5%	10%

The cost of metal purchased from source A and source B are ₹ 12.50 and ₹ 14.50 per kg respectively. Determine the optimum quantities of metal to be purchased from the two sources by the scrap metal dealer so as to minimize the total cost.

16. A farmer has a 100 acre farm. He can sell all the tomatoes, lettuce or radishes he can raise. The price he can obtain is ₹ 5 per kg for tomatoes, ₹ 2 per kg for lettuce and ₹ 3 per kg for radishes. The average yield per acre is 2000 kgs of tomatoes, 3000 kg of lettuce and 1000 kgs of radishes. Fertilizer is available at ₹ 2 per kg and the amount required per acre is 100 kgs each for tomatoes and lettuce and 50 kg for radishes. Labour required for sowing, cultivating and harvesting per acre is 5 man-days each for tomatoes and radishes and 6 man-days for lettuce. A total of 400 man-days of labour are available at ₹ 20 per man-day. Formulate this problem as a linear programming model to maximize the farmer's total profit.

17. The forest department of a state was contemplating introduction of a new concept known as "social-forestry". The various activities under the purview of social forestry are road side plantation, canal side tree plantation, tribal welfare activities, creation of village forests, creation of wood-based industries. An in-depth study of social cost and social benefits of each activity and the expected rate of return for each activity was carried out. The department has received a grant of ₹ 500 lakhs from the government and other world renowned organizations. The department has decided to spend not more than ₹ 100 lakhs for road side plantation, and at least ₹ 50 lakhs for canal side plantation, ₹ 75 lakhs for tribal welfare activities and ₹ 25 lakhs for wood-based industries. The rate of return for each activity is given as :

Activity	*Expected Rate of Return (percentage)*
Road side plantation	15
Canal side plantation	12
Tribal welfare activities	10
Creation of village forests	18
Creation of wood-based industries	14

Find how should the forest department allocate the available funds so as to optimize the rate of return. DO NOT SOLVE.

18. An aeroplane can carry a maximum of 200 passengers. A profit of ₹ 400 is made on each first class ticket and a profit of ₹ 300 is made on each economy class ticket. The airline reserves at least 20 tickets for first class seats. However, at least 4 times as many passengers prefer to travel by economy class as to the first class. Determine how many tickets of each type must be sold in order to maximize the total profit for the airline. Formulate the problem mathematically.

19. A firm is planning to advertize a special sale on radio and television during a particular week. A maximum budget of ₹ 1,60,000 is approved for this purpose. It is known that radio commercials cost ₹ 10,000 per slot with a minimum contract of 4 slots, which the firm intends to enter. Television commercials, on the other hand, cost ₹ 40,000 per slot. Because of heavy demand, at the most 2 TV slots are available in the week. A survey indicates that a radio slot attracts 10,000 customers, while a TV slot attracts 50,000 customers. How should the firm allocate its advertising to attract the largest number of customers ?

20. Floor cleaning detergent for domestic usage are formulated with three characteristics in mind : (*i*) cleaning capacity, (*ii*) disinfectant power, and (*iii*) skin irritation. Each of these characteristics is measured on a linear scale 0 -100. The market product must have a cleaning capacity of at least 60 units and disinfectant power of at least 60 units as measured on these scales. At the same time it is desired to minimize the skin irritation level. This final product is to be a blend of three basic detergents whose characteristics are detailed below :

Basic detergent	*Cleaning capacity*	*Irritation level (per kg)*	*Disinfectant power (per kg)*
A	90	70	30
B	65	50	85
C	45	10	70

Formulate the problem of finding the optimal blend of detergents *A*, *B*, *C* to form packs of 1 kg each as a linear programming problem. DO NOT SOLVE.

21. A manufacturer produces three models of trekking bicycles: Model A, Model B and Model C, whose selling prices per unit are ₹ 6000, ₹ 7,000 and ₹ 10000 respectively. The cost of the frame for Model A, Model B and Model C is ₹ 3000, ₹ 4,000 and ₹ 6,000 respectively. The capacity (bicycles per hour) of assembling, painting and packaging of each model is as follows :

Capacity per hour

Machine	Model A	Model B	Model C
Assembling	30	25	30
Painting	20	25	25
Packaging	50	50	25

The manufacturer possesses only one of each type of machine. Cost per hour to run each of the three machines is ₹ 1,500 for assembling, ₹ 2,000 for painting and ₹ 2,500 for packaging. Formulate LLP to maximize profit. Do not solve. ***[Delhi Univ. B.Com. (H) 2019]***

ANSWERS

2. Maximize $Z = 60x + 20y$
subject to : $5x + 3y \leq 45{,}000;\ x \leq 7000;\ y \leq 10{,}000;\ x, y \geq 0$

3. Maximize $Z = 5x + 6y$
subject to : $3x + 2y \leq 120;\ 4x + 6y \leq 260;\ x, y \geq 0$

4. Minimize $Z = 2500x + 3500y$
subject to : $50x + 60y \geq 2500;\ 100x + 60y \geq 3000;\ 100x + 200y \geq 7000;\ x, y \geq 0$

5. Minimize $Z = 4x + 5y$
subject to : $3x + 6y \geq 80;\ 4x + 3y \geq 100;\ x, y \geq 0$

6. Minimize $Z = x + 1.5y$
subject to : $20x + 20y \geq 160;\ 30x + 60y \geq 300;\ x \geq 2;\ x, y \geq 0$

7. Maximize $Z = 0.8x + 0.8y$
subject to : $0.2x \leq 80;\ 0.6x + 0.5y \leq 300;\ 0.3y \leq 150;\ x, y \geq 0$

8. Maximize $Z = 3.5x_1 + 2.77x_2 + 3.58x_3 + 4.56x_4$
subject to : $4x_1 + 9x_3 + 12x_4 \leq 800;\ 2x_1 + 5x_2 \leq 456;\ 8x_1 + 10x_2 + 10x_3 \leq 4000;$
$8x_2 + 9x_3 + 12x_4 \leq 920;\ 4x_2 + 6x_3 + 12x_4 \leq 422;\ x_1, x_2, x_3, x_4 \geq 0$

9. Maximize $Z = 5x_1 + 4x_2 + 6x_3$
subject to : $x_1 + x_2 + x_3 \le 100$; $3x_1 + 2x_2 + 4x_3 \le 210$; $3x_1 + 2x_2 \le 150$; $x_1, x_2, x_3 \ge 0$

10. Minimize $Z = 2x + 3y$
subject to : $x + 2y \le 30$; $3x + y \le 60$; $6x + 3y \le 200$; $5x + 4y \le 200$; $x, y \ge 0$

11. Minimize $Z = x_1 + x_2 + x_3 + x_4 + x_5 + x_6$
subject to : $x_1 + x_2 \ge 20$; $x_2 + x_3 \ge 14$; $x_3 + x_4 \ge 20$; $x_4 + x_5 \ge 10$; $x_5 + x_6 \ge 5$; $x_6 + x_1 \ge 7$;
$x_1, x_2, x_3, x_4, x_5, x_6 \ge 0$

12. Maximize $Z = 300x + 400y$
subject to : $12x + 7y \le 840$; $2x + 3y \le 180$; $x, y \ge 0$

13. Maximize $Z = 25x + 10y$
subject to : $90x + 30y \le 2,40,000$; $x \le 3000$; $x, y \ge 0$

14. Maximize $Z = 5x + 3y$
subject to : $2x + y \le 1000$; $x \le 400$; $y \le 700$; $x, y \ge 0$

15. Minimize $Z = 12.50x + 14.50y$
subject to : $x + y = 2000$; $8x + 15y \ge 20,000$; $3x + 4y \le 8000$; $x, y \ge 0$

16. Maximize $Z = 9700x_1 + 5680x_2 + 2800x_3$
subject to : $x_1 + x_2 + x_3 \le 100$; $5x_1 + 6x_2 + 5x_3 \le 400$; $x_1, x_2, x_3 \ge 0$

17. Maximize $Z = 0.15x_1 + 0.12x_2 + 0.10x_3 + 0.18x_4 + 0.14x_5$
subject to : $x_1 + x_2 + x_3 + x_4 + x_5 = 500$; $x_1 \le 100$; $x_2 \ge 50$; $x_3 \ge 75$; $x_5 \ge 25$
$x_1, x_2, x_3, x_4, x_5 \ge 0$

18. Maximize $Z = 400x + 300y$
subject to : $x + y \le 200$; $x \ge 20$; $y \ge 4x$; $x, y \ge 0$

19. Maximize $Z = 10,000x + 50,000y$
subject to : $10,000x + 40,000y \le 1,60,000$; $x \ge 4$; $y \le 2$; $x, y \ge 0$

20. Minimize $Z = 70x_1 + 50x_2 + 10x_3$
subject to : $90x_1 + 65x_2 + 45x_3 \ge 60$; $30x_1 + 85x_2 + 70x_3 \ge 60$; $x_1 + x_2 + x_3 = 1$; $x_1, x_2, x_3 \ge 0$

21. Maximize $Z = 2800x_1 + 2810x_2 + 3770x_3$
subject to : $5x_1 + 6x_2 + 5x_3 \le 150$; $5x_1 + 4x_2 + 4x_3 \le 100$; $x_1 + x_2 + x_3 \le 50$; $x_1, x_2, x_3 \ge 0$.

CHAPTER

4

LINEAR PROGRAMMING — THE SIMPLEX METHOD

OBJECTIVES

After studying the material in this chapter, you should be able to:

- *Understand the concept of basic feasible solution.*
- *Know principles of the simplex method.*
- *Solve the standard form of a maximization linear programming problem using simplex method.*
- *Use the artificial variables technique (big M-method) to solve maximization linear programming problems having mixed constraints.*
- *Solve the minimization case of a linear programming problem using simplex method.*
- *Recognize unbounded solutions, multiple optimum solutions, no optimum solution from the optimum simplex tableau.*

4.1 SOME USEFUL DEFINITIONS

We begin by introducing some definitions which are relevant for solving a linear programming problem using simplex method.

***Definition* 1. (*Slack Variable*)** A variable added to the left-hand side of a less than or equal to constraint to convert the constraint into an equality is called a **slack variable.** It makes up for the "slack" on the left side of the inequality so that we have equality. For example, consider the constraint

$$x + 2y \leq 30$$

To convert this into an equation, add a slack variable s to the left-hand side which results

$$x + 2y + s = 30$$

It must be nonnegative since it is the difference between 30 and a number that is less than or equal to 30. **In economic terminology slack variable represents the unused capacity.** If, for example, 5 units of x and 10 units of y are to be produced, then 25 units of the resource will be consumed and the slack variable should be set equal to 5 which represents the unused capacity.

***Definition* 2. (*Surplus Variable*)** A variable subtracted from the left-hand side of a greater than or equal to constraint to convert it into an equality is called a **surplus variable.** It subtracts the surplus of the left-hand side over the right side to convert the constraint into an equivalent equation. For example, consider the constraint

$$x + y \geq 9$$

To convert this into an equation, we subtract a surplus variable s from the left-hand side so that we have equality. Thus

$$x + y - s = 9$$

where s must be nonnegative.

***Definition* 3. (*Basic Solution*)** For a system of m simultaneous linear equations in n variables ($n > m$), a solution obtained by setting $(n - m)$ variables equal to zero and solving for the remaining m variables is called a **basic solution.** In addition, the solution for the m variables must be unique. The $(n - m)$ variables set equal to zero in any solution are called **nonbasic variables.** The other m variables whose values are obtained by solving the remaining system of equations are referred to as **basic variables.**

***Definition* 4. (*Basic Feasible Solution*)** A **basic feasible solution** (abbriviated *BFS*) to a linear programming problem is a basic solution for which the m variables solved for are all greater than or equal to zero. In other words, a basic solution which happens to be feasible is called a basic feasible solution.

To illustrate, let us consider the following example.

***Example* 1.** Identify all the basic and the basic feasible solutions of the following *LPP* :

Maximize $\qquad Z = 3x + 4y$

subject to the constraints:

$$2x + 2y \leq 80$$
$$2x + 4y \leq 120$$
$$x, y \geq 0$$

Solution. By introducing slack variables s_1, s_2, we can transform the constraint set into the equivalent set

$$2x + 2y + s_1 = 80$$
$$2x + 4y + s_2 = 120$$

This constraint set involves two equations and four variables. A basic solution to the given problem occurs when two of the four variables in the equivalent set of constraints are set equal to zero and the system is solved for the other two variables. The question is :

Which two variables should be set equal to zero ?

Let's enumerate the different possibilities. If s_1 and s_2 are set equal to 0, then the constraint equations become

$$2x + 2y = 80$$
$$2x + 4y = 120$$

Solving for x and y results in $x = 20$ and $y = 20$. If s_1 and x are set equal to zero, the system becomes

$$2y = 80$$
$$4y + s_2 = 120$$

TABLE 4.1

Solution	*Nonbasic Variables*	*Basic Variables*
1	s_1 and s_2	$x = 20,\quad y = 20$
2*	x and s_1	$y = 40,\quad s_2 = -40$
3	x and s_2	$y = 30,\quad s_1 = 20$
4	y and s_1	$x = 40,\quad s_2 = 40$
5*	y and s_2	$x = 60,\quad s_1 = -40$
6	x and y	$s_1 = 80,\quad s_2 = 120$

Solving for y and s_2 results in $y = 40$ and $s_2 = -40$. Proceeding in the same way, we develop Table 4.1, which summarizes all the solutions possibilities given that two of the four variables are assigned values of 0.

Solutions 2 and 5 have an asterisk next to them. They have received special attention because they are not feasible solutions. Each of these contains a variable which has a negative value. However, solutions 1, 3, 4 and 6 are basic feasible solutions to the linear programming problem.

Remark. If we look at Fig. 4.1 which is the graphical representation of the two constraints of the preceding example, we find that the values of x and y for solutions 1, 3, 4 and 6 in Table 4.1 correspond to the four corner points of the feasible region. Specifically, solution 1 corresponds to corner point B, solution 3 corresponds to corner point C, solution 4 corresponds to corner point A and solution 6 corresponds to corner point O. In fact, solutions 2 and 5, which are not feasible, correspond to points E and D respectively.

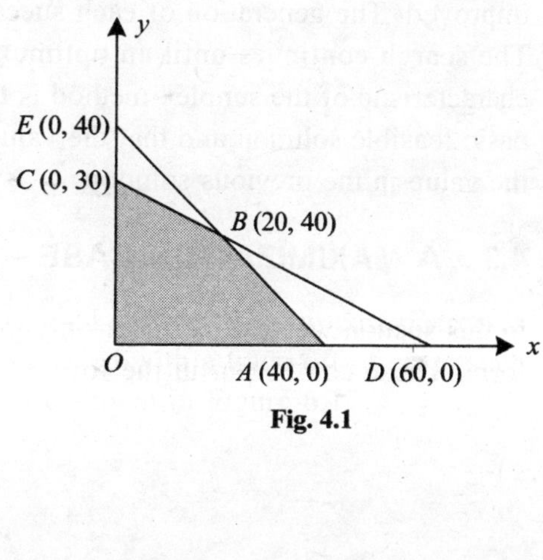

Fig. 4.1

Thus the six points O, A, B, C, D and E are basic points while the points O, A, B and C are basic feasible points and we clearly see the correspondence between corner points and the basic feasible points.

The following theorem gives us an algebraic description of corner points and reconfirms the above argument.

***Theorem* 1.** *The set of corner points of the feasible region corresponds to the set of basic feasible solutions. In other words, corner points are basic feasible solutions, and vice versa.*

Example 2. Find basic solutions for the following equations:

$$x + y + 2z = 6$$

$$3x + 2y + z = 10$$

Which of these solutions are basic feasible solution? **[*Delhi Univ. B.Com. (H) 2010*]**

Solution. When $x = 0$, $y + 2z = 6$ and $2y + z = 10 \Rightarrow y = 14/3,\ z = 2/3$

When $y = 0$, $x + 2z = 6$ and $3x + z = 10 \Rightarrow x = 14/5,\ z = 8/5$

When $z = 0$, $x + y = 6$ and $3x + 2y = 10 \Rightarrow x = -2,\ y = 8$

First and second basic solutions are basic feasible solutions.

4.2 PRINCIPLES OF THE SIMPLEX METHOD

The most popular non-graphical procedure for solving linear programming problems is called the *simplex method.* The simplex method is based on a **fundamental existence theorem** *which states that whenever there exists an optimum solution to a linear programming problem, there exists one which is also basic feasible solution.* Thus in the simplex method, the first step is always to obtain a basic feasible solution. This solution is then tested for optimality, *i.e.*, it searches to see whether the value of the objective function can be improved. The generation of each successive solution requires a system of linear equations. The search continues until an optimum solution, if it exists, is determined. An important characteristic of the simplex method is that it guarantees that each successive solution will be basic feasible solution and that the value of the objective function will be at least as good as the value in the previous solution.

4.3 A MAXIMIZATION CASE – ALL CONSTRAINTS OF THE TYPE ≤

In this section we shall consider only maximization linear programming problems in standard form. These can be put in the form :

Maximize $Z = c_1 x_1 + c_2 x_2 + \ldots + c_n x_n$
subject to the constraints:

$$\begin{aligned} a_{11} x_1 + a_{12} x_2 + \ldots + a_{1n} x_n &\le b_1 \\ a_{21} x_1 + a_{22} x_2 + \ldots + a_{2n} x_n &\le b_2 \\ \vdots \\ a_{m1} x_1 + a_{m2} x_2 + \ldots + a_{mn} x_n &\le b_m \end{aligned}$$

where $x_1, x_2, \ldots, x_n$ and $b_1, b_2, \ldots, b_m$ are nonnegative.

Thus a maximization problem is said to be in standard form provided the following conditions are satisfied :

1. *All the variables must be nonnegative.*
2. *Each constraint is written as a linear expression that is less than or equal to a non negative constant.*

The simplex method is now outlined for problems in the standard form. Other types of linear programming problems will be discussed in Sections 4.5 and 4.6.

Step 1. To Introduce Slack Variables. The first step in the simplex method is to introduce slack variables $s_1, s_2, \ldots, s_m$ (one for each constraint) to convert constraint inequalities to equations. Thus we can restate the problem in terms of equations :

Maximize $Z = c_1 x_1 + c_2 x_2 + \ldots + c_n x_n + 0s_1 + 0s_2 + \ldots + 0s_m$
subject to the constraints:

$$\begin{aligned} a_{11} x_1 + a_{12} x_2 + \ldots + a_{1n} x_n + s_1 &= b_1 \\ a_{21} x_1 + a_{22} x_2 + \ldots + a_{2n} x_n + s_2 &= b_2 \\ \vdots \\ a_{m1} x_1 + a_{m2} x_2 + \ldots + a_{mn} x_n + s_m &= b_m \end{aligned}$$

where $x_1, x_2, \ldots, x_n$ and $s_1, s_2, \ldots, s_m$ are nonnegative. Note that the slack variables have been assigned zero coefficients in the objective function. The reason is that these variables typically contribute nothing to the value of the objective function.

Step 2. To set up the Initial Basic Feasible Solution. The second step is to find the initial basic feasible solution. For this, we can take the m slack variables $s_1, s_2, \ldots, s_m$ as the basic variables. We can determine their values directly, because the remaining variables $x_1, x_2, \ldots, x_n$ are non-basic and therefore each has the value zero.

Thus $s_1 = b_1, s_2 = b_2, \ldots, s_m = b_m$.

The simplex method then proceeds to solve the above problem by designing and redesigning successively better basic feasible solutions until an optimal solution is obtained.

Step 3. To set up the Initial Simplex Tableau. For computational efficiency and simplicity, the simplex algorithm is usually performed in a table called a *simplex tableau.* The initial simplex tableau is shown in Tableau I.

The interpretation of the data in Tableau I is given after the table. Other simplex tableau will have similar interpretations.

Simplex Tabuleau I

$C_j \rightarrow$	*Basic*	*Solution*	c_1	c_2	...	c_n	0	0	...	0
$\downarrow$	*Variable*		x_1	x_2	...	x_n	s_1	s_2	...	s_m
0	s_1	b_1	a_{11}	a_{12}	...	a_{1n}	1	0	...	0
0	s_2	b_2	a_{21}	a_{22}	...	a_{2n}	0	1	...	0
.	.	.	.	.		.	.			.
.	.	.	.	.		.	.			.
0	s_m	b_m	a_{m1}	a_{m2}	...	a_{mn}	0	0	...	1
	Z_j	0	0	0	...	0	0	0	...	0
	$C_j - Z_j$		c_1	c_2	...	c_n	0	0	...	0

(*i*) In the first row labelled "C_j", we write the coefficients of the variables in the objective function. These values will remain the same in subsequent tableaux.

(*ii*) The second row shows the major column headings. These headings remain the same in the subsequent tableaux and apply to the values listed in the first *m* rows.

(*iii*) In the first column labelled "C_j", we write the coefficients (in the objective function) of the basic variables included in the specific program. Thus the coefficients of $s_1, s_2, ..., s_m$ which are included in the first program are written in the C_j column.

(*iv*) In the second column labelled "*Basic Variable*" (also called *Product Mix Column*) are listed the basic variables. Thus in the initial simplex tableau, these variables are the slack variables.

(*v*) In the column labelled "*Solution*" (also called *Quantity Column*), we write the solution values of the basic variables. Since the values of our initial basic feasible solution are represented by the right-hand sides of the constraints, these values are listed in the initial simplex tableau.

(*vi*) The *Body Matrix* (under nonbasic variables) in the initial simplex tableau consists of the coefficients of the decision variables in the constraints set. The entries in the body matrix can be interpreted as physical rates of substitution. In the initial tableau, for example, the number a_{ij} gives the rate of substitution between s_i and x_j. In other words, if we wish to produce a unit of x_j then a_{ij} units of s_i must be sacrificed.

(*vii*) The *identity matrix* (under $s_1, s_2, ..., s_m$) in the initial simplex tableau represents the coefficients of the slack variables in the constraints set. Each simplex tableau contains an identity matrix under basic variables.

(*viii*) To get an entry in the Z_j row under a column, we multiply the entries in that column by the corresponding entries of C_j column and add the products. The Z_j row entries will all be equal to zero in the initial simplex tableau. The Z_j entry under the "*Solution*" column gives the current value of the objective function. The other Z_j entries represent the decrease in the objective function that would result if one of the variables not included in the solution were brought into the solution.

(*ix*) The last row labelled "$C_j - Z_j$", called the *index row or net evaluation row,* is used to determine whether or not the current solution is optimal. The calculation of $C_j - Z_j$ row simply involves subtracting each Z_j value from the corresponding C_j value for that column, which is written at the top of that column. We observe that $C_j - Z_j$ values are meaningful for the nonbasic variables only. This is because for a basic variable $Z_j = 1 \times C_j = C_j$ so that $C_j - Z_j = C_j - C_j = 0$.

The numbers in the $C_j - Z_j$ row represent the net contribution to the objective function that results by introducing one unit of each of the respective column variables. A plus value indicates that a greater contribution can be made by bringing the variable for that column into the solution. A negative value indicates the amount by which contribution would decrease if one unit of the variable for that column were brought into the solution.

Step 4. Test the Solution for Optimality. Examine the entries of $C_j - Z_j$ row. If all the entries in $C_j - Z_j$ row are negative or zero, the optimal solution has been obtained, otherwise the presence of any positive entry indicates that the solution can further be improved.

Step 5. Revision of the Current Simplex Tableau. At each iteration, the simplex method moves from the current basic feasible solution to a better basic feasible solution. This involves replacing one current basic variable (called the *departing variable*) by a new nonbasic variable (called the *entrering variable*).

Determine the Entering Variable. The column with the largest positive entry in the $C_j - Z_j$ row is called the *key* (or *pivot*) column (indicated by ↑). This locates the variable to be introduced into the basis. Thus the nonbasic variable which will replace a basic variable is the one lying in the key column.

Determine the Departing Variable. The decision of which basic variable to replace is made by focusing upon the key column and the solution column. Divide each entry of the solution column by the corresponding positive entry in the key column. These quotients are written in the last column labelled "*Ratio*" of the simplex tableau to be replaced. The row that corresponds to the smallest nonnegative quotient is called the *key* (or *pivot*) row (indicated by ←). The departing variable is the corresponding variable in this row. The number that lies at the intersection of the key column and the key row of a given tableau is called the *key* (or *pivot*) number. We place a circle around this number.

Derivation of New Tableau. After identifying the entering and the departing variables, all that remains is to find the new basic feasible solution by constructing a new simplex tableau from the current one. This is accomplished by using elementary row operations so that the key number is reduced to 1 and other entries in the key column are transformed to 0. To change the key row so that the key number is 1, we can use the following formula :

$$\text{New key row} = \frac{\text{old key row}}{\text{key number}}$$

To change the non-key rows, we can use the formula :

New non-key row = old non-key row – (key column entry) × new key row,

where "*key column entry*" is the entry in this row that is in the key column.

Step 6. Test for Optimality. Again we apply the test for optimality to determine whether the new solution is optimal. If there are no positive $C_j - Z_j$ entries, we have an optimal solution and have finished the problem. Otherwise, go back to Step 5.

***Example* 3.** A firm makes two types of furnitures, chairs and tables. Profits are ₹ 20 per chair and ₹ 30 per table. Both products are processed on three machines M_1, M_2 and M_3. The time required for each product in hours and total time available in hours per week on each machine are as follows :

Machine	*Chair*	*Table*	*Available Time*
M_1	3	3	36
M_2	5	2	50
M_3	2	6	60

How should the manufacturer schedule his production in order to maximize profit ? (Use simplex method).

Solution. Step* 1. *Formulation of the problem. Let x and y be the number of chairs and tables to be produced per week respectively. Then the appropriate mathematical formulation of the given problem is :

Maximize $Z = 20x + 30y$

subject to the constraints:

$$3x + 3y \leq 36$$
$$5x + 2y \leq 50$$
$$2x + 6y \leq 60$$
$$x, y \geq 0$$

This problem is of standard form.

Step* 2. *Introduce slack variables. To convert the constraint inequalities into equalities, we introduce slack variables, s_1, s_2 and s_3. Thus we can restate the problem in terms of equations :

Maximize $Z = 20x + 30y + 0s_1 + 0s_2 + 0s_3$

subject to the constraints:

$$3x + 3y + s_1 = 36$$
$$5x + 2y + s_2 = 50$$
$$2x + 6y + s_3 = 60$$
$$x, y, s_1, s_2, s_3 \geq 0$$

Step* 3. *Set up the initial basic feasible solution. We shall first find an initial basic feasible solution and then determine whether the corresponding value of Z can be increased by a different basic feasible solution. By setting $x = 0$, $y = 0$, we get $s_1 = 36$, $s_2 = 50$, $s_3 = 60$ and $Z = 0$. Thus our initial basic feasible solution is :

$$x = 0, \quad y = 0, \quad s_1 = 36, \quad s_2 = 50, \quad s_3 = 60$$

at which $Z = 0$. This initial basic feasible solution is summarized in the initial simplex tableau.

Simplex Tableau I

$C_j \rightarrow$	Basic	Solution	20	30	0	0	0	Ratio
$\downarrow$	Variable		x	y	s_1	s_2	s_3	
0	s_1	36	3	3	1	0	0	36/3 = 12
0	s_2	50	5	2	0	1	0	50/2 = 25
0	s_3	60	2	(6)	0	0	1	$\leftarrow$ 60/6 = 10
	Z_j	0	0	0	0	0	0	
	$C_j - Z_j$		20	30 $\uparrow$	0	0	0	

Step* 4. *Test for optimality. The last row of the simplex tableau is used to determine whether or not the optimal solution is obtained. Since there is a positive entry in $C_j - Z_j$ row, the current solution is not optimal.

Step* 5. *Determine the key column and the key row. It is clear from the Tableau I that for each unit of x brought into the solution, the objective function is increased by ₹ 20 and for each unit of y brought into the solution, the objective function is increased by ₹ 30. Hence we get a *greater increase* in the value of Z if y, rather than x, enters the next program. Thus in this case, y is the entering variable and the column y is the key column (indicated by ↑). In terms of the simplex tableau, the key column corresponds to the largest positive entry in the $C_j - Z_j$ row.

To determine the variable that will depart from the current solution, we divide each entry in the solution column by the corresponding positive entry in the key column. The quotients 36/3, 50/2 and 60/6 are indicated in the column labelled "*Ratio*" of Tableau I. The row with the minimum nonnegative quotient is the key row (indicated by ←). This indicates that the variable s_3 is the departing variable. The number at the intersection of the key column and the key row is the key number. Thus the number, 6, is the key number. A circle is placed around this number.

Step* 6. *Derivation of Tableau II. Our goal is now to construct Tableau II and to see how much profit can be made by producing only tables and no chairs.

(*i*) *Transformation of the key row.* We first replace the key row with a row labelled y. Entries for the *y-row* are obtained by dividing each entry of key row by the key number, *i.e.*, 6. Thus the entries for the y-row in the next tableau are :

$$10 \quad \frac{1}{3} \quad 1 \quad 0 \quad 0 \quad \frac{1}{6}$$

(*ii*) *Transformation of non-key rows.* To replace non-key rows, we use the formula :

New non-key row = old non-key row – (key column entry) × new key row,

where "*key column entry*" is the entry in this row that is in the key column.

Based upon the above formula, the values in Tableau II for the new s_1 and s_2 rows are :

Row s_1:		*Row* s_2:	
	$36 - 3 \times 10 = 6$		$50 - 2 \times 10 = 30$
	$3 - 3 \times \frac{1}{3} = 2$		$5 - 2 \times \frac{1}{3} = \frac{13}{3}$
	$3 - 3 \times 1 = 0$		$2 - 2 \times 1 = 0$
	$1 - 3 \times 0 = 1$		$0 - 2 \times 0 = 0$
	$0 - 3 \times 0 = 0$		$1 - 2 \times 0 = 1$
	$0 - 3 \times \frac{1}{6} = -\frac{1}{2}$		$0 - 2 \times \frac{1}{6} = -\frac{1}{3}$

The above information is summarized in Tableau II.

Simplex Tableau II

$C_j \rightarrow$	*Basic*	*Solution*	20	30	0	0	0		*Ratio*
↓	*Variable*		x	y	s_1	s_2	s_3		
0	s_1	6	②	0	1	0	– 1/2	←	6/2 = 3
0	s_2	30	13/3	0	0	1	– 1/3		30/(13/3) = 90/13
30	y	10	1/3	1	0	0	1/6		10/(1/3) = 30
	Z_j	300	10	30	0	0	5		
	$C_j - Z_j$		10 ↑	0	0	0	– 5		

Note that a Z_j entry under a column in the above tableau is obtained by multiplying the entries in that column by the corresponding entries in the C_j column and adding the products. For example, the Z_j entry for x-column was obtained as follows :

$$0\,(2) + 0\,(13/3) + 30\,(1/3) = 10$$

The reader is reminded that the value of the objective function for this program is the Z_j entry in the "solution" column. Thus we get $Z = 300$, which is larger value than we had before (it was $Z = 0$). The entries for the $C_j - Z_j$ row in the above tableau were obtained by subtracting the Z_j row from the C_j row.

As the index row $C_j - Z_j$ of Tableau II shows, we still have a positive entry. Hence the solution contained in Tableau II is not an optimal solution. Thus we must revise this program and derive Tableau III. The procedure for deriving Tableau III is exactly the same as was followed in deriving Tableau II. Repeating the procedure, we get the revised program as given in Tableau III.

Simplex Tableau III

$C_j \rightarrow$	*Basic*	*Solution*	20	30	0	0	0
$\downarrow$	*Variable*		x	y	s_1	s_2	s_3
20	x	3	1	0	1/2	0	– 1/4
0	s_2	17	0	0	– 13/6	1	3/4
30	y	9	0	1	– 1/6	0	1/4
	Z_j	330	20	30	5	0	5/2
	$C_j - Z_j$		0	0	– 5	0	– 5/2

Since all the entries in $C_j - Z_j$ row of Tableau III are zero or negative, it follows that an optimal solution has been obtained, and the maximum value of Z is 330 and it occurs when $x = 3$ and $y = 9$ (and $s_2 = 17$, $s_1 = 0$ and $s_3 = 0$).

Opportunity Cost

In Testing for optimality, we want to know the impact of bringing a new variable into the solution. This is done in two steps. First, we find out what it will "cost" to bring in the new variable. Since the new variable displaces some of each of the current basic variables, the objective function will be reduced by whatever contribution the basic variable made. For example, if we look at the Simplex Tableau II of the preceding example, for x, bringing in one unit displaces 2 units of s_1, which has a unit contribution of 0. We will lose $2 \times 0 = 0$ from the objective function. Also, we'll lose $0 \times 13/3 = 0$ by displacing some of s_2 and lose $30 \times 1/3 = 10$ by displacing some of y. The total loss from the objective function is $0 + 0 + 10 = 10$. It is called the *opportunity cost* because it shows the penalty incurred if we pass the "opportunity" of leaving the solution as is and bring in a new variable.

Of course, adding one unit of a new variable to the solution will increase the objection function by an amount equal to the objective function coefficient. So to see whether the variable should be added, we compare its "marginal contribution" with its opportunity cost. The $C_j - Z_j$ row, then, shows the net impact one unit of a new variable would have on the objective function. This is why the rule for optimality calls for all $C_j - Z_j$ values to be zero or negative.

Shadow Price

The simplex method provides more than just the optimal solution. It provides additional information which can be useful for managerial decision making, specifically, the shadow price for each resource.

An existence of a slack variable in the final simplex tableau indicates that the resource has not been fully utilised. Non-existence of slack variables indicate full utilisation of the existing resources. Thus if the company plans to enlarge resources, it can look for only those resources which are fully utilised. To assess the value of additional resources, we can consider what difference it would make if we could provide an extra hour of each of the resources fully utilised. The increase in profit as a result of the additional unit of resource is seen to be the marginal value of the resource and is referred to as the shadow price for that resource.

The shadow price for each resource is shown in the $C_j - Z_j$ row of the final simplex tableau under the corresponding slack variable. For example, if we look at simplex tableau III of the preceding example, the slack variable s_1, corresponding to the first constraint, and it has $C_j - Z_j = -5$ in the final tableau. We say that *shadow price* for the first resource is + 5 (absolute value). This means that we could increase the objective function by 5 if we had an additional hour of that resource. Thus if the manager were to pay below ₹ 5 for the additional hour of first resource, then profits could be increased, and if the manager were to pay above this figure then profits would decrease. Hence the *maximum price* the manufacturer should pay for an additional hour of first resource is ₹ 5. Similarly, for the second resource, the shadow price is $s_2 = 0$. An additional hour of the second resource won't help, since this resource has not been fully utilised. For the third resource, the shadow price is $s_3 = 5/2$, which indicates that we could increase the objective function by 5/2 if we had an additional hour of this resource.

Example 4. Given the simplex tableau for a maximum problem of linear programming :

C_j	*Product Mix*	x_1	x_2	s_1	s_2	*Quantity*
4	x_2	1	1	1	0	6
0	s_2	1	0	– 1	1	2
	C_j	3	4	0	0	

(*i*) Test whether the above table gives optimum solution. If not, improve it and find the optimum solution.

(*ii*) If s_1 and s_2 are slack variables at machine I and machine II, what is the maximum price you will like to pay for 1 hour of each machine ?

[*Delhi Univ. B.Com. (H) 1997, 2006*]

Solution. Given simplex table can be represented as follows :

$C_j \rightarrow$	*Product*	*Quantity*	3	4	0	0
↓	*Mix*		x_1	x_2	s_1	s_2
4	x_2	6	1	1	1	0
0	s_2	2	1	0	– 1	1
	Z_j	24	4	4	4	0
	$C_j - Z_j$		– 1	0	– 4	0

(*i*) Looking at the $C_j - Z_j$ row, all entries are zero or negative; therefore the above table gives optimum solution. The optimum solution is : $x_1 = 0, x_2 = 6$ and max $Z = 24$.

(*ii*) The slack variable s_1 corresponds to the first constraint, and it has $C_j - Z_j = -4$ in the final table. Thus the shadow price for machine I is + 4. This means that we could increase the objective function by 4 if we had an additional unit of this machine; hence it is the maximum price one should pay for 1 hour of this machine. Similarly, for the second machine, the shadow price is $s_2 = 0$. An additional unit of the second machine won't help, since this machine has not been fully utilised.

***Example* 5.** Two materials *A* and *B* are required to construct tables and book cases. For one table 12 units of *A* and 16 units of *B* are needed while for a book case 16 units of *A* and 8 units of *B* are required. The profit on a book case is ₹ 25 and ₹ 20 on a table. 100 units of material *A* and 80 units of *B* are available. How many book cases and tables be produced to have maximum profit ?

Formulate this as a linear programming problem and solve by simplex method.

[Delhi Univ. B.Com. (H) 2014]

Solution. Let *x* denote the number of tables and *y* denote the number of book cases produced. Then the mathematical formulation of the problem is :

Maximize $Z = 20x + 25y$

subject to the constraints:

$$12x + 16y \le 100$$
$$16x + 8y \le 80$$
$$x, y \ge 0$$

Introducing slack variables s_1 and s_2, we can restate the problem in terms of equations :

Maximize $Z = 20x + 25y + 0s_1 + 0s_2$

subject to the constraints:

$$12x + 16y + s_1 = 100$$
$$16x + 8y + s_2 = 80$$
$$x, y, s_1, s_2 \ge 0$$

An initial basic feasible solution is : $x = 0, \quad y = 0, \quad s_1 = 100$ and $s_2 = 80$ at which $Z = 0$. The initial simplex tableau corresponding to this solution is Tableau I.

Simplex Tableau I

$C_j \rightarrow$	*Basic*	*Solution*	20	25	0	0	*Ratio*
$\downarrow$	*Variable*		x	y	s_1	s_2	
0	s_1	100	12	16	1	0	← 100/16 = 25/4
0	s_2	80	(16)	8	0	1	80/8 = 10
	Z_j	0	0	0	0	0	
	$C_j - Z_j$		20	25 ↑	0	0	

Since there is a positive $C_j - Z_j$ entry, we must continue. The largest positive entry, 25, occurs in the y-column. Thus y is the entering variable and y-column is the key column. To find the departing variable, we divide each entry of the "solution" column by the corresponding positive entry of the key column. The smallest nonnegative ratio is 25/4, so s_1 is the departing variable and s_1- row is the key row. The key number, 16, is circled. Replacing key-and non-key rows using the technique discussed in Section 4.4, we obtain second tableau.

Simplex Tableau II

$C_j \rightarrow$ ↓	*Basic Variable*	*Solution*	20 x	25 y	0 s_1	0 s_2	*Ratio*
25	y	25/4	3/4	1	1/16	0	25/3
0	s_2	30	(10)	0	– 1/2	1	← 3
	Z_j	625/4	75/4	25	25/16	0	
	$C_j - Z_j$		5/4 ↑	0	– 25/16	0	

Note that the value of the objective function has been improved from 0 to 625/4 However, since there is a positive entry in the $C_j - Z_j$ row, the current solution is not optimal and we must continue our process. The entering variable is now x. The smallest nonnegative ratio is 3. Thus s_2 is the departing variable and the key number, 10, is circled. Replacing key-and non-key rows using the technique discussed in Section 4.4, we have Tableau III.

Simplex Tableau III

$C_j \rightarrow$ ↓	*Basic Variable*	*Solution*	20 x	25 y	0 s_1	0 s_2
25	y	4	0	1	1/10	–3/40
20	x	3	1	0	– 1/20	1/10
	Z_j	160	20	25	3/2	1/8
	$C_j - Z_j$		0	0	– 3/2	– 1/8

Since all $C_j - Z_j$ entries in Tableau III are zero or negative, optimal solution has been obtained, and maximum value of Z is 160 which occurs when $x = 3$ and $y = 4$.

***Example* 6.** Following data are available for a firm which manufactures three items A, B and C :

Product	*Time Required in Hours*		*Profit*
	Assembly	*Finishing*	(in ₹)
A	10	2	80
B	4	5	60
C	5	4	30
Firm's Capacity	2000	1009	

Express the above data in the form of linear programming problem to maximize the profit from the production and solve it by simplex method. **[*Delhi Univ. B.Com. (H) 1983, 1993*]**

Solution. Let x_1, x_2 and x_3 denote the number of units of items *A*, *B* and *C* produced by the firm. Then the appropriate mathematical formulation of the problem is :

Maximize $Z = 80x_1 + 60x_2 + 30x_3$

subject to the constraints:

$$10x_1 + 4x_2 + 5x_3 \le 2000$$
$$2x_1 + 5x_2 + 4x_3 \le 1009$$
$$x_1, x_2, x_3 \ge 0$$

Introducing slack variables s_1 and s_2 so as to convert the constraint inequalities into equations, the given problem can be restated as follows :

Maximize $Z = 80x_1 + 60x_2 + 30x_3 + 0s_1 + 0s_2$

subject to the constraints:

$$10x_1 + 4x_2 + 5x_3 + s_1 = 2000$$
$$2x_1 + 5x_2 + 4x_3 + s_2 = 1009$$
$$x_1, x_2, x_3, s_1, s_2 \ge 0$$

The initial simplex tableau is Tableau I.

Simplex Tableau I

$C_j \rightarrow$ ↓	*Basic Variable*	*Solution*	80 x_1	60 x_2	30 x_3	0 s_1	0 s_2	*Ratio*
0	s_1	2000	(10)	4	5	1	0	← 200
0	s_2	1009	2	5	4	0	1	1009/2
	Z_j	0	0	0	0	0	0	
	$C_j - Z_j$		80 ↑	60	30	0	0	

A look at the $C_j - Z_j$ row reveals that x_1 is the entering variable. The smallest nonnegative quotient in the ratio column is 200, so s_1 is the departing variable and the key number, 10, is circled. Using elementary row operations, we get Tableau II.

Simplex Tableau II

$C_j \rightarrow$ ↓	*Basic Variable*	*Solution*	80 x_1	60 x_2	30 x_3	0 s_1	0 s_2	*Ratio*
80	x_1	200	1	2/5	1/2	1/10	0	500
0	s_2	609	0	(21/5)	3	– 1/5	1	←145
	Z_j	16,000	80	32	40	8	0	
	$C_j - Z_j$		0	28 ↑	– 10	– 8	0	

Note that the value of the objective function has been improved from 0 to 16,000. However, there is a positive entry in the $C_j - Z_j$ row, hence the current solution is not optimal. The entering variable is now x_2. The smallest nonnegative quotient in the ratio column is 145. Thus s_2 is the departing variable and 21/5 is the key number. Replacing key-and non-key rows using the technique discussed in Section 4.3, we have Tableau III.

Simplex Tableau III

$C_j \rightarrow$	*Basic*	*Solution*	80	60	30	0	0
$\downarrow$	*Variable*		x_1	x_2	x_3	s_1	s_2
80	x_1	142	1	0	3/14	5/42	– 2/21
60	x_2	145	0	1	5/7	– 1/21	5/21
	Z_j	20,060	80	60	60	20/3	20/3
	$C_j - Z_j$		0	0	– 30	– 20/3	– 20/3

Since all entries in $C_j - Z_j$ row of tableau III are zero or negative, optimum solution has been obtained, and the maximum value of Z is 20,060 which occurs when $x_1 = 142$, $x_2 = 145$ and $x_3 = 0$.

***Example* 7.** A firm uses three machines in the manufacture of three products. Each unit of product *A* requires 3 hours on machine I, 2 hours on machine II and one hour on machine III. Each unit of product *B* requires 4 hours on machine I, one hour on machine II and 3 hours on machine III, while each unit of product *C* requires 2 hours on each of the three machines. The contribution margin of the three product is ₹ 30, ₹ 40 and ₹ 35 per unit respectively. The machine hours available on the three machines are 90, 54 and 93 respectively. Formulate the above as a linear programming problem and solve for maximum profit, using simplex method.

[*Delhi Univ. B.Com. (H) 1995*]

Solution. (*i*) Let x_1, x_2 and x_3 denote the number of units produced of products *A*, *B* and *C* respectively. Then the appropriate mathematical formulation of the problem is :

Maximize $\quad Z = 30x_1 + 40x_2 + 35x_3$

subject to the constraints :

$$3x_1 + 4x_2 + 2x_3 \le 90$$
$$2x_1 + x_2 + 2x_3 \le 54$$
$$x_1 + 3x_2 + 2x_3 \le 93$$
$$x_1, x_2, x_3 \ge 0$$

(*ii*) Introducing 3 slack variables s_1, s_2 and s_3, the given problem can be restated as follows :

Maximize $\quad Z = 30x_1 + 40x_2 + 35x_3 + 0s_1 + 0s_2 + 0s_3$

subject to :

$$3x_1 + 4x_2 + 2x_3 + s_1 = 90$$
$$2x_1 + x_2 + 2x_3 + s_2 = 54$$
$$x_1 + 3x_2 + 2x_3 + s_3 = 93$$
$$x_1, x_2, x_3, s_1, s_2, s_3 \ge 0$$

An initial basic feasible solution is :

$$x_1 = x_2 = x_3 = 0, \quad s_1 = 90, \quad s_2 = 54 \quad \text{and} \quad s_3 = 93$$

Starting with this solution, we obtain a sequence of tableaux that leads to the final solution.

$C_j \rightarrow$ ↓	*Basic Variable*	*Solution*	30 x_1	40 x_2	35 x_3	0 s_1	0 s_2	0 s_3	*Ratio*
0	s_1	90	3	(4)	2	1	0	0	←45/2
0	s_2	54	2	1	2	0	1	0	54
0	s_3	93	1	3	2	0	0	1	31
	Z_j	0	0	0	0	0	0	0	
	$C_j - Z_j$		30	40 ↑	35	0	0	0	
40	x_2	45/2	3/4	1	1/2	1/4	0	0	45
0	s_2	63/2	5/4	0	(3/2)	– 1/4	1	0	←21
0	s_3	51/2	– 5/4	0	1/2	– 3/4	0	1	51
	Z_j	900	30	40	20	10	0	0	
	$C_j - Z_j$		0	0	15 ↑	– 10	0	0	
40	x_2	12	1/3	1	0	1/3	– 1/3	0	
35	x_3	21	5/6	0	1	– 1/6	2/3	0	
0	s_3	15	– 5/3	0	0	– 2/3	– 1/3	1	
	Z_j	1215	85/2	40	35	15/2	10	0	
	$C_j - Z_j$		– 25/2	0	0	– 15/2	– 10	0	

A look at the $C_j - Z_j$ row of last tableau shows that it is impossible to further increase the objective function. Hence the optimal solution is $x_1 = 0$, $x_2 = 12$ and $x_3 = 21$ with max $Z = 1215$.

***Example* 8.** A factory produces three different products *A*, *B* and *C*. The profits per unit of these products are ₹ 3, ₹ 4 and ₹ 6 respectively. The products are processed in three operations, *viz.*, *X*, *Y* and *Z* and the time (in hours) required in each operation for each unit is given below :

Operations	*Products* A	B	C
X	4	1	6
Y	5	3	1
Z	1	2	3

The factory has 3 machines for operation *X*, 2 machines for operation *Y* and only one machine for operation *Z*. The factory works 25 days in a month, at the rate of 16 hours a day in two shifts. The effective working of all the processes is only 80% due to power cuts or breakdown of machines.

(*i*) Formulate the problem mathematically.

(*ii*) Use simplex method to find how many units of each product should be produced monthly in order to maximize profit ? **[*Delhi Univ. B.Com. (H) 2019*]**

Solution. (*i*) Let x_1, x_2 and x_3 be the number of units produced per month of products A, B and C respectively. Then the appropriate mathematical formulation of the given problem is:

$$\begin{aligned} &\text{Maximize} && Z = 3x_1 + 4x_2 + 6x_3 \\ &\text{subject to the constraints:} && 4x_1 + x_2 + 6x_3 \le 25 \times 16 \times 0.8 \times 3 = 960 \\ &&& 5x_1 + 3x_2 + x_3 \le 25 \times 16 \times 0.8 \times 2 = 640 \\ &&& x_1 + 2x_2 + 3x_3 \le 25 \times 16 \times 0.8 \times 1 = 320 \\ &&& x_1, x_2, x_3 \ge 0 \end{aligned}$$

Introducing slack variables s_1, s_2 and s_3, the given problem can be restated as follows :

$$\begin{aligned} &\text{Maximize} && Z = 3x_1 + 4x_2 + 6x_3 + 0s_1 + 0s_2 + 0s_3 \\ &\text{subject to the constraints:} && 4x_1 + x_2 + 6x_3 + s_1 = 960 \\ &&& 5x_1 + 3x_2 + x_3 + s_2 = 640 \\ &&& x_1 + 2x_2 + 3x_3 + s_3 = 320 \\ &&& x_1, x_2, x_3, s_1, s_2, s_3 \ge 0 \end{aligned}$$

The complete set of simplex tableaux is shown below :

C_j	*Basic variable*	*Solution*	3 x_1	4 x_2	6 x_3	0 s_1	0 s_2	0 s_3	Ratio
0	s_1	960	4	1	6	1	0	0	160
0	s_2	640	5	3	1	0	1	0	640
0	s_3	320	1	2	(3)	0	0	1	← 320/3
	Z_j	0	0	0	0	0	0	0	
	$C_j - Z_j$		3	4	6 ↑	0	0	0	
0	s_1	320	2	– 3	0	1	0	– 2	160
0	s_2	1600/3	(14/3)	7/3	0	0	1	– 1/3	← 800/7
6	x_3	320/3	1/3	2/3	1	0	0	1/3	320
	Z_j	640	2	4	6	0	0	2	
	$C_j - Z_j$		1 ↑	0	0	0	0	– 2	

0	s_1	640/7	0	– 4	0	1	– 3/7	– 13/7
3	x_1	800/7	1	1/2	0	0	3/14	– 1/14
6	x_3	480/7	0	1/2	1	0	– 1/14	5/14
	Z_j	5280/7	3	9/2	6	0	3/14	27/14
	$C_j - Z_j$		0	– 1/2	0	0	– 3/14	– 27/14

Since all entries in $C_j - Z_j$ row are negative or zero optimum solution has been obtained, and the maximum value of Z is 5280/7 and it occurs when $x_1 = 800/7$, $x_2 = 0$ and $x_3 = 480/7$.

***Example* 9.** A firm buys casting of P and Q type of parts and sells them as finished products after machining, boring and polishing. The purchasing costs for casting are ₹ 3 and ₹ 4 each for parts P and Q respectively and selling costs are ₹ 8 and ₹ 10 respectively. The per hour capacity of machines used for machining, boring and polishing for the two products is given as follows :

Capacity (per hour)

	P		*Q*
Machine	30	or	50
Boring	30	or	45
Polishing	45	or	30

The running cost for machining, boring and polishing are ₹ 30, ₹ 22.5 and ₹ 22.5 per hour respectively. Formulate the linear programming problem to find out the product mix to maximize the profit and solve the same using simplex method.

[*Delhi Univ. B.Com. (H) 2004*]

Solution. Let the firm produce x_1 units of P and x_2 units of Q. Then

$$\text{Profit/unit of } P = 8 - \left[3 + \frac{30}{30} + \frac{22.5}{30} + \frac{22.5}{45}\right] = ₹\ 2.75$$

$$\text{Profit/unit of } Q = 10 - \left[4 + \frac{30}{50} + \frac{22.5}{45} + \frac{22.5}{30}\right] = ₹\ 4.15$$

Thus, the appropriate mathematical formulation of the problem is :

Maximize $Z = 2.75x_1 + 4.15x_2$

subject to :

$$\frac{x_1}{30} + \frac{x_2}{50} \le 1 \quad \text{or} \quad 5x_1 + 3x_2 \le 150$$

$$\frac{x_1}{30} + \frac{x_2}{45} \le 1 \quad \text{or} \quad 3x_1 + 2x_2 \le 90$$

$$\frac{x_1}{45} + \frac{x_2}{30} \le 1 \quad \text{or} \quad 2x_1 + 3x_2 \le 90$$

Introducing three slack variables s_1, s_2 and s_3, the given problem can be reformulated as follows:

$$\text{Maximize} \quad Z = 2.75x_1 + 4.15x_2 + 0s_1 + 0s_2 + 0s_3$$

subject to :

$$5x_1 + 3x_2 + s_1 = 150$$
$$3x_1 + 2x_2 + s_2 = 90$$
$$2x_1 + 3x_2 + s_3 = 90$$
$$x_1, x_2, s_1, s_2, s_3 \geq 0$$

An initial basic feasible solution is :

$$x_1 = 0, \quad x_2 = 0, \quad s_1 = 150, \quad s_2 = 90 \text{ and } s_3 = 90,$$

at which $Z = 0$. Starting with this solution, we obtain a sequence of simplex tableaux that leads to the final solution.

$C_j \rightarrow$ ↓	*Basic Variable*	*Solution*	2.75 x_1	4.15 x_2	0 s_1	0 s_2	0 s_3	*Ratio*
0	s_1	150	5	3	1	0	0	50
0	s_2	90	3	2	0	1	0	45
0	s_3	90	2	(3)	0	0	1	← 30
	Z_j	0	0	0	0	0	0	
	$C_j - Z_j$		2.75	4.15 ↑	0	0	0	
0	s_1	60	3	0	1	0	– 1	
0	s_2	30	5/3	0	0	1	– 2/3	
4.15	x_2	30	2/3	1	0	0	1/3	
	Z_j	124.5	83/30	4.15	0	0	83/60	
	$C_j - Z_j$		– 1/60	0	0	0	– 83/60	

Since all entries in $C_j - Z_j$ row are negative or zero, optimum solution has been obtained, and the maximum value of Z is ₹ 124.50 and it occurs when $x_1 = 0$ and $x_2 = 30$.

***Example* 10.** A firm assembles and sells two different types of outboard motors *A* and *B* using four resources. The production process can be described as follows :

Resources	*Capacity per Month*
I Motor unit shop	400 Type *A* or 250 Type *B* units or any linear combination of the two
II Type *A* gear and drive shop resource	175 Type *A* units
III Type *B* gear and drive shop resource	225 Type *B* units
IV Final assembly resource	200 Type *A* units or 350 Type *B* units or any linear combination of the two

Type A units bring in a profit of ₹ 90 each and Type B units ₹ 60 each. Formulate the above as a linear programming problem to maximize profit and solve the same by simplex method. Determine the shadow prices of the resources. **[*Delhi Univ. B.Com. (H) 2009*]**

Solution. Let x_1 and x_2 denote the number of outboard motors of Type A and Type B respectively. Then the appropriate mathematical formulation of the problem is :

$$\text{Maximize} \quad Z = 90x_1 + 60x_2$$

$$\text{subject to the constraints:} \quad \frac{x_1}{400} + \frac{x_2}{250} \leq 1$$

$$x_1 \leq 175$$

$$x_2 \leq 225$$

$$\frac{x_1}{200} + \frac{x_2}{350} \leq 1$$

$$x_1, x_2 \geq 0$$

Introducing slack variables s_1, s_2, s_3 and s_4, the given problem can be restated as follows :

$$\text{Maximize} \quad Z = 90x_1 + 60x_2 + 0s_1 + 0s_2 + 0s_3 + 0s_4$$

subject to the constraints:

$$\frac{x_1}{400} + \frac{x_2}{250} + s_1 = 1$$

$$x_1 + 0x_2 + s_2 = 175$$

$$0x_1 + x_2 + s_3 = 225$$

$$\frac{x_1}{200} + \frac{x_2}{350} + s_4 = 1$$

$$x_1, x_2, s_1, s_2, s_3, s_4 \geq 0$$

The complete set of simplex tableaux is shown as follows :

$C_j \rightarrow$ ↓	*Basic Variables*	*Solution*	90 x_1	60 x_2	0 s_1	0 s_2	0 s_3	0 s_4	*Ratio*
0	s_1	1	1/400	1/250	1	0	0	0	400
0	s_2	175	(1)	0	0	1	0	0	← 175
0	s_3	225	0	1	0	0	1	0	—
0	s_4	1	1/200	1/350	0	0	0	1	200
	Z_j	0	0	0	0	0	0	0	
	$C_j - Z_j$		90↑	60	0	0	0	0	

$C_j \rightarrow$ $\downarrow$	Basic Variables	Solution	90 x_1	60 x_2	0 s_1	0 s_2	0 s_3	0 s_4	Ratio
0	s_1	9/16	0	1/250	1	–1/400	0	0	1125/8
90	x_1	175	1	0	0	1	0	0	—
0	s_3	225	0	1	0	0	1	0	225
0	s_4	1/8	0	(1/350)	0	–1/200	0	1	← 175/4
	Z_j	15750	90	0	0	90	0	0	
	$C_j - Z_j$		0	60↑	0	–90	0	0	
0	s_1	31/80	0	0	1	(9/2000)	0	–7/5	← 775/9
90	x_1	175	1	0	0	1	0	0	175
0	s_3	725/4	0	0	0	7/4	1	–350	725/7
60	x_2	175/4	0	1	0	–7/4	0	350	—
	Z_j	18,375	90	60	0	–15	0	21,000	
	$C_j - Z_j$		0	0	0	15↑	0	– 21,000	
0	s_2	775/9	0	0	2000/9	1	0	–2800/9	← 775/9
90	x_1	800/9	1	0	– 2000/9	0	1	2800/9	
0	s_3	275/9	0	0	– 3500/9	0	1	1750/9	
60	x_2	1750/9	0	1	3500/9	0	0	–1750/9	
	Z_j	1,77,000/9	90	60	10,000/3	0	0	49,000/3	
	$C_j - Z_j$		0	0	– 10,000/3	0	0	– 49,000/3	

Since all $C_j - Z_j$ entries of the last table are zero or negative, optimum solution has been obtained and the maximum value of Z is 1,77,000/9 which occurs when $x_1 = 800/9$ and $x_2 = 1750/9$.

Shadow prices of the resources are : 10,000/3, 0, 0 and 49,000/3 respectively.

***Example* 11.** Use simplex method to solve the following *LLP* :

Maximize $Z = 4x + 5y$

subject to the constraints:

$$x + y \leq 20$$
$$x + 2y \leq 35$$
$$x - 3y \leq 12$$
$$x, y \geq 0$$

Solution. This problem is of standard form. By introducing slack variables s_1, s_2 and s_3, the problem can be restated in terms of equations :

Maximize $$Z = 4x + 5y + 0s_1 + 0s_2 + 0s_3$$
subject to the constraints:

$$x + y + s_1 = 20$$
$$x + 2y + s_2 = 35$$
$$x - 3y + s_3 = 12$$
$$x, y, s_1, s_2, s_3 \geq 0$$

An initial basic feasible solution is :

$$x = 0, \quad y = 0, \quad s_1 = 20, \quad s_2 = 35 \quad \text{and} \quad s_3 = 12$$

Starting with this solution, we obtain a sequence of tableaux that leads to the final solution.

$C_j \rightarrow$ ↓	*Basic Variable*	*Solution*	4 x	5 y	0 s_1	0 s_2	0 s_3	*Ratio*
0	s_1	20	1	1	1	0	0	20
0	s_2	35	1	(2)	0	1	0	← 35/2
0	s_3	12	1	– 3	0	0	1	–
	Z_j	0	0	0	0	0	0	
	$C_j - Z_j$		4	5 ↑	0	0	0	
0	s_1	5/2	(1/2)	0	1	– 1/2	0	←5
5	y	35/2	1/2	1	0	1/2	0	35
0	s_3	129/2	5/2	0	0	3/2	1	129/5
	Z_j	175/2	5/2	5	0	5/2	0	
	$C_j - Z_j$		3/2 ↑	0	0	– 5/2	0	
4	x	5	1	0	2	– 1	0	
5	y	15	0	1	– 1	1	0	
0	s_3	52	0	0	– 5	4	1	
	Z_j	95	4	5	3	1	0	
	$C_j - Z_j$		0	0	– 3	– 1	0	

A look at the $C_j - Z_j$ row of last tableau shows that it is impossible to further increase the objective function. Hence the optimal solution is $x = 5$, $y = 15$ with max $Z = 95$.

Remark. In Example 11, one may wonder why no quotient is considered in the s_3- row of the initial tableau. The basic feasible solution for this tableau is :

$$x = 0, \quad y = 0, \quad s_1 = 20, \quad s_2 = 35 \quad \text{and} \quad s_3 = 12$$

The entering variable is y. In the new basic feasible solution, the larger the increase in y, the larger the increase in Z. Now, by how much can we increase y ? Since x is still held at 0, from rows 1, 2 and 3 of the initial simplex tableau, it follows that

$$s_1 = 20 - y, \quad s_2 = 35 - 2y \quad \text{and} \quad s_3 = 12 + 3y$$

Since s_1, s_2, s_3 are nonnegative, we have $y \leq 20$, $y \leq 35/2$ and $y \geq -4$. Hence y can increase at most by 35/2. The condition $y \geq -4$ has no influence in determining the maximum increase in y. That is why the quotient $12/(-3) = -4$ is not considered in s_3- row.

Note. Although the simplex procedure that has been developed in this section applies to linear programming problems of standard form, other forms may be adapted to fit this form as is illustrated in the following example.

***Example* 12.** Maximize $Z = 2x_1 + x_2 - x_3$

subject to the constraints:

$$x_1 + x_2 \leq 1$$
$$x_1 - 2x_2 - x_3 \geq -2$$
$$x_1, x_2, x_3 \geq 0$$

Solution. This is a maximization problem in which the second constraint

$$x_1 - 2x_2 - x_3 \geq -2$$

has the inequality symbol $\geq$. Thus this is not the standard form of a maximization problem. However, multiplying both sides of this constraint by -1, we get

$$-x_1 + 2x_2 + x_3 \leq 2$$

which is in the standard form. Thus, although the given problem as stated is not in standard form, it can easily be modified to put in standard form. Now by introducing slack variables s_1 and s_2, the problem can be restated in form of equations :

Maximize $Z = 2x_1 + x_2 - x_3 + 0s_1 + 0s_2$

subject to the constraints:

$$x_1 + x_2 + s_1 = 1$$
$$-x_1 + 2x_2 + x_3 + s_2 = 2$$
$$x_1, x_2, x_3, s_1, s_2 \geq 0$$

An initial basic feasible solution is : $x_1 = 0,\ x_2 = 0,\ x_3 = 0,\ s_1 = 1$ and $s_2 = 2$

Starting with this solution, we obtain a sequence of tableaux that leads to the final solution.

Simplex Tableau I

$C_j \rightarrow$ ↓	*Basic Variable*	*Solution*	2 x_1	1 x_2	−1 x_3	0 s_1	0 s_2	*Ratio*
0	s_1	1	(1)	1	0	1	0	← 1
0	s_2	2	−1	2	1	0	1	–
	Z_j	0	0	0	0	0	0	
	$C_j - Z_j$		2 ↑	1	−1	0	0	

Simplex Tableau II

$C_j \rightarrow$	Basic	Solution	2	1	−1	0	0	Ratio
↓	Variable		x_1	x_2	x_3	s_1	s_2	
2	x_1	1	1	1	0	1	0	
0	s_2	3	0	3	1	1	1	
	Z_j	2	2	2	0	2	0	
	$C_j - Z_j$		0	−1	−1	−2	0	

Since all entries in $C_j - Z_j$ row are zero or negative, optimal solution has been obtained and the maximum value of Z is 2 which occurs when $x_1 = 1$, $x_2 = 0$ and $x_3 = 0$.

***Example* 13.** A firm makes three types of furnitures, desks, chairs and bookshelves. It has no difficulty in selling these items. However, limited availability of machine time, labour and floor space restrict production. Data on usage of resources, supplies and profits on items are given below :

	Desk	*Chair*	*Bookshelf*	*Supply*
Machine hrs/unit	8	4	5	1000 hrs
Labour hrs/unit	5	3	3	650 hrs
Floor space sq.ft/unit	9	6	9	1260 sq.ft
Contribution ₹/unit	270	144	225	

(*i*) Formulate the above as a linear programming problem.

(*ii*) Using x_1, x_2, x_3 as the respective decision variables, simplex algorithm was initiated for the above problem. At one stage, the following tableau was reached.

$C_j \rightarrow$	Basic	Solution	270	144	225	0	0	0
↓	Variables		x_1	x_2	x_3	s_1	s_2	s_3
270	x_1	125	1	1/2	5/8	1/8	0	0
0	s_2	25	0	1/2	– 1/8	– 5/8	1	0
0	s_3	135	0	3/2	27/8	– 9/8	0	1
	$C_j - Z_j$		0	9	225/4	– 135/4	0	0

Test whether the above table gives optimum solution. If not, improve it and find the optimum solution.

(*iii*) The Personnel Manager of the firm claims that by recruiting additional labour force, the profit can be increased. Assuming no other changes, is the claim valid ? Justify your answer.

Solution. (*i*) The appropriate mathematical formulation of the given problem is :

Maximize $\quad Z = 270x_1 + 144x_2 + 225x_3$: *Total profit*

subject to the constraints:

$$8x_1 + 4x_2 + 5x_3 \leq 1000 \quad \textit{Machine constraint}$$
$$5x_1 + 3x_2 + 3x_3 \leq 650 \quad \textit{Labour constraint}$$
$$9x_1 + 6x_2 + 9x_3 \leq 1260 \quad \textit{Space constraint}$$
$$x_1, x_2, x_3 \geq 0$$

(*ii*) Given table can be represented as follows :

$C_j \rightarrow$	*Basic*	*Solution*	270	144	225	0	0	0	*Ratio*
↓	*Variables*		x_1	x_2	x_3	s_1	s_2	s_3	
270	x_1	125	1	1/2	5/8	1/8	0	0	200
0	s_2	25	0	1/2	– 1/8	– 5/8	1	0	–
0	s_3	135	0	3/2	(27/8)	– 9/8	0	1	← 40
	Z_j	33,750	270	135	675/4	135/4	0	0	
	$C_j - Z_j$		0	9	225/4↑	– 135/4	0	0	

The above table does not give optimum solution because there are positive entries in the $C_j - Z_j$ row. We must, therefore, improve the solution. A look at the $C_j - Z_j$ row reveals that x_3 is the entering variable. The smallest nonnegative quotient in the ratio column is 40, so s_3 is the departing variable and the key number, 27/8, is circled. Replacing key-and non-key rows using the technique discussed in Section 4.3, we get the following table :

$C_j \rightarrow$	*Basic*	*Solution*	270	144	225	0	0	0
↓	*Variables*		x_1	x_2	x_3	s_1	s_2	s_3
270	x_1	100	1	2/9	0	1/3	0	– 5/27
0	s_2	30	0	5/9	0	– 2/3	1	1/27
225	x_3	40	0	4/9	1	– 1/3	0	8/27
	Z_j	36,000	270	160	225	15	0	50/3
	$C_j - Z_j$		0	– 16	0	– 15	0	– 50/3

Since all entries in $C_j - Z_j$ row of the above table are zero or negative, optimum solution has been obtained, and the maximum value of Z is 36,000 which occurs when $x_1 = 100$, $x_2 = 0$ and $x_3 = 40$.

(*iii*) The slack variable s_2 corresponds to the labour constraint. As seen from the solution column of the above table, $s_2 = 30$. This shows that labour force has not been fully utilised. Thus, addition of further labour force will not change the optimum solution. The personnel Manager's claim is, therefore, not valid.

***Example* 14.** A company makes 3 kinds of boats. All can be made profitably in this company, but company's monthly production is constrained by the limited amount of labour, wood and screws available each month. The director will choose the combination of boats that maximize his revenue as per the information given in the following table :

Input	*Type I Boat*	*Type II Boat*	*Type III Boat*	*Monthly Availability*
Lab. (Hrs.)	12	7	9	1260 Hrs.
Wood (Sq. feet)	22	18	16	19,008 Sq. feet
Screws (Kg.)	2	4	3	396 kgs.
Selli¬g Price (₹)	4000	2000	5000	

Formulate the above problem. Solve it by simplex method. From optimal table answer the following questions :

(*i*) How many boats of each type will be produced and what will be the resulting maximum revenue?

(*ii*) Which, if any, of the resource is not fully utilised. If so, how much of spare capacity is left?

(*iii*) How much wood will be used to make all the boats in the optimal solution?

[Delhi Univ. B.Com. (H) 2001, 2007]

Solution. (*i*) Let x_1, x_2 and x_3 denote the number of boats of Type I, Type II and Type III respectively. Then the appropriate mathematical formulation of the problem is :

Maximize $Z = 4000x_1 + 2000x_2 + 5000x_3$ *Total revenue*

subject to the constraints:

$12x_1 + 7x_2 + 9x_3 \le 1260$ *Labour constraint*

$22x_1 + 18x_2 + 16x_3 \le 19{,}008$ *Wood constraint*

$2x_1 + 4x_2 + 3x_3 \le 396$ *Screws constraint*

$x_1, x_2, x_3 \ge 0$

To solve the problem by simplex method, we introduce three slack variables s_1, s_2 and s_3 so that the given problem can be reformulated as follows :

Maximize $Z = 4000x_1 + 2000x_2 + 5000x_3 + 0s_1 + 0s_2 + 0s_3$

subject to the constraints:

$$12x_1 + 7x_2 + 9x_3 + s_1 = 1260$$
$$22x_1 + 18x_2 + 16x_3 + s_2 = 19{,}008$$
$$2x_1 + 4x_2 + 3x_3 + s_3 = 396$$
$$x_1, x_2, x_3, s_1, s_2, s_3 \ge 0$$

An initial basic feasible solution is :

$$x_1 = 0,\ x_2 = 0,\ x_3 = 0,\ s_1 = 1260,\ s_2 = 19{,}008 \text{ and } s_3 = 396$$

Starting with this solution, we obtain a sequence of tableaux that leads to the final solution.

$C_j \rightarrow$ ↓	*Basic Variable*	*Solution*	4000 x_1	2000 x_2	5000 x_3	0 s_1	0 s_2	0 s_3	*Ratio*
0	s_1	1260	12	7	9	1	0	0	140
0	s_2	19,008	22	18	16	0	1	0	1188
0	s_3	396	2	4	(3)	0	0	1	←132
	Z_j	0	0	0	0	0	0	0	
	$C_j - Z_j$		4000	2000	5000↑	0	0	0	
0	s_1	72	(6)	–5	0	1	0	–3	←12
0	s_2	16,896	34/3	–10/3	0	0	1	–16/3	25,344/17
5000	x_3	132	2/3	4/3	1	0	0	1/3	198
	Z_j	6,60,000	10,000/3	20,000/3	5000	0	0	5000/3	
	$C_j - Z_j$		2000/3↑	–14,000/3	0	0	0	–5000/3	
4000	x_1	12	1	–5/6	0	1/6	0	–1/2	
0	s_2	16,760	0	55/9	0	–17/9	1	1/3	
5000	x_3	124	0	17/9	1	–1/9	0	2/3	
	Z_j	6,68,000	4000	55,000/9	500	1000/9	0	4000/3	
	$C_j - Z_j$		0	–37,000/9	0	–1000/9	0	–4000/3	

Since all entries in $C_j - Z_j$ row are negative or zero, optimal solution has been obtained, and the maximum value of Z is 6,68,000 and it occurs when $x_1 = 12$, $x_2 = 0$ and $x_3 = 124$.

(*ii*) The presence of s_2 in the final simplex table indicates that the second resource, namely, wood has not been fully utilized. In fact, the spare capacity of wood left is 16,760 feet.

(*iii*) The quantity of wood used to make all the boats in the optimal solution is :

$$19{,}008 - 16{,}760 = 2248 \text{ feet.}$$

***Example* 15.** A firm manufacturing office furniture provides you the following information regarding resource consumption and availability and profit contribution :

Resources	*Usage Per Unit*			*Availability*
	Tables	*Chairs*	*Book Case*	
Timber (cu.ft.)	8	4	3	640
Assembly deptt. (man hours)	4	6	2	540
Finishing deptt. (man hours)	1	1	1	100
Profit contribution per unit (₹)	30	20	12	

The firm wants to determine its optimal product mix :

(i) Formulate the *LPP* and solve with the simplex method.

(ii) Find the optimal product mix and the total maximum profit contribution.

(iii) What are shadow prices of the resources?

[Delhi Univ. B.Com. (H) 2008, 2017]

Solution. (*i*) Let x_1, x_2 and x_3 be the number of tables, chairs and book cases manufactured, respectively. Then the appropriate mathematical formulation of the given problem is :

Maximize $Z = 30x_1 + 20x_2 + 12x_3$

subject to the constraints:

$$8x_1 + 4x_2 + 3x_3 \leq 640$$
$$4x_1 + 6x_2 + 2x_3 \leq 540$$
$$x_1 + x_2 + x_3 \leq 100$$
$$x_1, x_2, x_3 \geq 0$$

(*ii*) Introducing slack variables s_1, s_2 and s_3 so as to convert the constraint inequalities into equations, the given problem can be restated as follows :

Maximize $Z = 30x_1 + 20x_2 + 12x_3 + 0s_1 + 0s_2 + 0s_3$

subject to the constraints:

$$8x_1 + 4x_2 + 3x_3 + s_1 = 640$$
$$4x_1 + 6x_2 + 2x_3 + s_2 = 540$$
$$x_1 + x_2 + x_3 + s_3 = 100$$
$$x_1, x_2, x_3, s_1, s_2, s_3 \geq 0$$

An initial basic feasible solution is :

$$x_1 = 0,\ x_2 = 0,\ x_3 = 0,\ s_1 = 640,\ s_2 = 540,\ s_3 = 100,$$

at which $Z = 0$. Starting with this solution, we obtain a sequence of simplex tableaux that leads to the final solution.

$C_j \rightarrow$ ↓	*Basic variable*	*Solution*	30 x_1	20 x_2	12 x_3	0 s_1	0 s_2	0 s_3	*Ratio*
0	s_1	640	(8)	4	3	1	0	0	←80
0	s_2	540	4	6	2	0	1	0	135
0	s_3	100	1	1	1	0	0	1	100
	Z_j	0	0	0	0	0	0	0	
	$C_j - Z_j$		30↑	20	12	0	0	0	

$C_j \rightarrow$ $\downarrow$	Basic variable	Solution	30 x_1	20 x_2	12 x_3	0 s_1	0 s_2	0 s_3	Ratio
30	x_1	80	1	1/2	3/8	1/8	0	0	160
0	s_2	220	0	4	(1/2)	–1/2	1	0	55
0	s_3	20	0	1/2	5/8	–1/8	0	1	←40
	Z_j	2400	30	15	45/4	15/4	0	0	
	$C_j - Z_j$		0	5↑	3/4	–15/4	0	0	
30	x_1	60	1	0	–1/4	1/4	0	–1	
0	s_2	60	0	0	–9/2	1/2	1	–8	
20	x_2	40	0	1	5/4	–1/4	0	2	
	Z_j	2600	30	20	35/2	5/2	0	10	
	$C_j - Z_j$		0	0	–11/2	–5/2	0	–10	

Since all $C_j - Z_j$ entries of the last table are zero or negative, optimum solution has been obtained, and the maximum value of Z is 2600 which occurs when $x_1 = 60$, $x_2 = 40$ and $x_3 = 0$.

(*iii*) Shadow prices of the resources are 5/2, 0, and 10 respectively.

***Example* 16.** A Pharmaceutical company produces two popular drugs *A* and *B* which are sold at the rate of ₹ 9.60 and ₹ 7.80 respectively. The main ingredients are *X*, *Y* and *Z* and they are required in the following proportions :

Drugs	*X*	*Y*	*Z*
A	50%	30%	20%
B	30%	30%	40%

The total available quantities (gms) of different ingredients are 1600 in *X*, 1400 in *Y* and 1200 in *Z*. The costs (₹) of *X*, *Y* and *Z* per gm are ₹ 8, ₹ 6 and ₹ 4 respectively. Estimate the most profitable quantities of *A* and *B* to produce, using simplex method.

Solution. Cost of *A* per unit = 50% of 8 + 30% of 6 + 20% of 4

= 4 + 1.8 + 0.8 = ₹ 6.60

Cost of *B* per unit = 30% of 8 + 30% of 6 + 40% of 4

= 2.4 + 1.8 + 1.6 = ₹ 5.80

Profit per unit of *A* = 9.60 – 6.60 = ₹ 3

Profit per unit of *B* = 7.80 – 5.80 = ₹ 2

Let x denote the number of units of drug *A* and y denote the number of units of drug *B* produced. Then the appropriate mathematical formulation of the given problem is :

Maximize $Z = 3x + 2y$

subject to the constraints:

$$\begin{aligned} 0.5x + 0.3y &\le 1600 \\ 0.3x + 0.3y &\le 1400 \\ 0.2x + 0.4y &\le 1200 \\ x, y &\ge 0 \end{aligned} \quad \text{or} \quad \begin{aligned} 5x + 3y &\le 16,000 \\ 3x + 3y &\le 14,000 \\ 2x + 4y &\le 12,000 \\ x, y &\ge 0 \end{aligned}$$

Introducing slack variables s_1, s_2 and s_3 so as to convert the constraint inequalities into equations, the given problem can be restated as follows :

Maximize $Z = 3x + 2y + 0s_1 + 0s_2 + 0s_3$

subject to the constraints:

$$\begin{aligned} 5x + 3y + s_1 &= 16,000 \\ 3x + 3y + s_2 &= 14,000 \\ 2x + 4y + s_3 &= 12,000 \\ x, y, s_1, s_2, s_3 &\ge 0 \end{aligned}$$

An initial basic feasible solution is :

$$x = 0, \quad y = 0, \quad s_1 = 16,000, \quad s_2 = 14,000, \quad s_3 = 12,000$$

at which $Z = 0$. Starting with this solution, we obtain a sequence of simplex tableaux that leads to the final solution.

$C_j \rightarrow$	*Basic*	*Solution*	3	2	0	0	0	*Ratio*
$\downarrow$	*Variable*		x	y	s_1	s_2	s_3	
0	s_1	16,000	(5)	3	1	0	0	← 3200
0	s_2	14,000	3	3	0	1	0	14,000/3
0	s_3	12,000	2	4	0	0	1	6000
	Z_j	0	0	0	0	0	0	
	$C_j - Z_j$		3 ↑	2	0	0	0	
3	x	3200	1	3/5	1/5	0	0	16,000/3
0	s_2	4400	0	6/5	– 3/5	1	0	11,000/3
0	s_3	5600	0	(14/5)	– 2/5	0	1	← 2000
	Z_j	9600	3	9/5	3/5	0	0	
	$C_j - Z_j$		0	1/5 ↑	– 3/5	0	0	

$C_j \rightarrow$ ↓	Basic Variable	Solution	3 x	2 y	0 s_1	0 s_2	0 s_3	Ratio
3	x	2000	1	0	2/7	0	– 3/14	
0	s_2	2000	0	0	– 3/7	1	– 3/7	
2	y	2000	0	1	– 1/7	0	5/14	
	Z_j	10,000	3	2	4/7	0	1/14	
	$C_j - Z_j$		0	0	– 4/7	0	– 1/14	

Since all $C_j - Z_j$ entries of the last table are zero or negative, optimum solution has been obtained, and the maximum value of Z is 10,000 which occurs when $x = 2000$ and $y = 2000$.

4.4 TIE BREAKING IN THE SIMPLEX METHOD

It may be recalled that the simplex method is based on a set of rules whereby we proceed from one basic feasible solution to the next until an optimal solution, if it exists, is obtained. However, nothing was said as what to do, if the various choice rules of the simplex method do not lead to a clear-cut decision, either because of ties or other similar ambiguities. In this section such things are considered.

Tie for the Entering Variable. In a simplex tableau, suppose that two or more nonbasic variables are tied for having the largest positive $C_j - Z_j$ entry. In such a case, we choose any one of these to find the entering variable. The optimal solution will be reached eventually, regardless of the tied variable chosen.

Tie for the Departing Variable – Degeneracy. Now suppose that two or more quotients in the ratio column are tied for being the smallest nonnegative. We may choose any one of these quotients to find departing variable. When a tie for the smallest quotient exists, then along with the nonbasic variables, a basic feasible solution will have a basic variable that is 0. In this case we say that the basic feasible solution is *degenerate* or that the linear programming problem has a *degeneracy*. It is possible, in a degenerate situation, to obtain a sequence of tableaux, that correspond to basic feasible solutions which give the same value of the objective function. Moreover, we may eventually return to the first tableau in the sequence. This is called *cycling*. When cycling occurs, it is possible that we may never obtain the optimum value of the objective function. However, this phenomenon seldom occurs in practical problems. Let us illustrate.

***Example* 17.** Maximize $Z = 80x + 100y$

subject to the constraints:

$$20x + 30y \leq 3$$
$$60x + 40y \leq 4$$
$$x, y \geq 0$$

Solution. Introducing slack variables s_1 and s_2, we can restate the problem as follows:

Maximize $Z = 80x + 100y + 0s_1 + 0s_2$

subject to the constraints:

$$20x + 30y + s_1 = 3$$
$$60x + 40y + s_2 = 4$$
$$x, y, s_1, s_2 \geq 0$$

An initial basic feasible solution is : $x = 0$, $y = 0$, $s_1 = 3$ and $s_2 = 4$

The initial simplex tableau is shown below.

Simplex Tableau I

$C_j \rightarrow$	*Basic*	*Solution*	80	100	0	0	*Ratio*
$\downarrow$	*Variable*		x	y	s_1	s_2	
0	s_1	3	20	(30)	1	0	←1/10
0	s_2	4	60	40	0	1	1/10
	Z_j	0	0	0	0	0	
	$C_j - Z_j$		80	100 ↑	0	0	

The entering variable is y. Since there is a tie for the smallest quotient, we can choose either s_1 or s_2 as the departing variable. Let us choose s_1. The key number, 30, is circled. Replacing key-and non-key rows using the technique discussed in Section 4.3, we get Tableau II.

Simplex Tableau II

$C_j \rightarrow$	*Basic*	*Solution*	80	100	0	0	*Ratio*
$\downarrow$	*Variable*		x	y	s_1	s_2	
100	y	1/10	2/3	1	1/30	0	3/10
0	s_2	0	(100/3)	0	– 4/3	1	←0
	Z_j	10	200/3	100	10/3	0	
	$C_j - Z_j$		40/3 ↑	0	– 10/3	0	

Tableau II corresponds to a basic feasible solution, where a basic variable s_2 is zero. Thus the basic feasible solution is degenerate. Since there is a positive $C_j - Z_j$ entry, we continue. The entering variable is now x, the departing variable is s_2 and the key number, 100/3, is circled. Replacing key-and non-key rows, we get Tableau III.

Simplex Tableau III

$C_j \rightarrow$	*Basic*	*Solution*	80	100	0	0	*Ratio*
$\downarrow$	*Variable*		x	y	s_1	s_2	
100	y	1/10	0	1	3/50	– 1/50	
80	x	0	1	0	– 1/25	3/100	
	Z_j	10	80	100	14/5	2/5	
	$C_j - Z_j$		0	0	– 14/5	– 2/5	

Since all $C_j - Z_j$ entries are negative or zero, Z is maximum when $x = 0$ and $y = 1/10$. The maximum value of Z is 10. Note that this value is same as the value of Z corresponding to Tableau II. Thus, in degenerate problems, it is possible to arrive at the same value of Z at various stages of the simplex process.

Unbounded Solutions. A maximization linear programming is said to have an *unbounded solution* if it has an unbounded feasible region (that is it cannot be contained in any circle). In other words, a maximization linear programming problem has an unbounded solution if given any value of the objective function Z, a feasible solution with a greater value of Z can always be obtained. In simplex method, such a situation occurs when each entry in the key column is either negative or zero. To illustrate, consider the following example :

***Example* 18.** Maximize $Z = 5x_1 + 6x_2 + x_3$

subject to the constraints:

$$9x_1 + 3x_2 - 2x_3 \le 5$$
$$4x_1 + 2x_2 - x_3 \le 2$$
$$x_1 - 4x_2 + x_3 \le 3$$
$$x_1, x_2, x_3 \ge 0$$

Solution. Introducing stack variables s_1, s_2 and s_3, we can restate the problem as follows :

Maximize $Z = 5x_1 + 6x_2 + x_3 + 0s_1 + 0s_2 + 0s_3$

suLiect to the constraints:

$$9x_1 + 3x_2 - 2x_3 + s_1 = 5$$
$$4x_1 + 2x_2 - x_3 + s_2 = 2$$
$$x_1 - 4x_2 + x_3 + s_3 = 3$$
$$x_1, x_2, x_3, s_1, s_2, s_3 \ge 0$$

The complete set of simplex tableaux is shown as follows :

Simplex Tableau I

$C_j \rightarrow$	*Basic*	*Solution*	5	6	1	0	0	0	*Ratio*
$\downarrow$	*Variable*		x_1	x_2	x_3	s_1	s_2	s_3	
0	s_1	5	9	3	−2	1	0	0	5/3
0	s_2	2	4	(2)	−1	0	1	0	←1
0	s_3	3	1	−4	1	0	0	1	–
	Z_j	0	0	0	0	0	0	0	
	$C_j - Z_j$		5	6↑	1	0	0	0	

Simplex Tableau II

$C_j \rightarrow$	Basic	Solution	5	6	1	0	0	0	Ratio
↓	Variable		x_1	x_2	x_3	s_1	s_2	s_3	
0	s_1	2	3	0	– 1/2	1	– 3/2	0	–
6	x_2	1	2	1	– 1/2	0	1/2	0	–
0	s_3	7	9	0	– 1	0	2	1	–
	Z_j	6	12	6	– 3	0	3	0	
	$C_j - Z_j$		– 7	0	4↑	0	– 3	0	

Here the entering variable is x_3. Since each entry in the key column is negative, no quotient is possible. So the simplex method would stop with the message that Z is unbounded.

Multiple Optimal Solutions. In linear programming problems, situation may arise when there is the possibility of more than one optimal solution. In such a situation, we say that there are multiple solutions to the problem. We can determine the possibility of multiple optimal solutions from the final simplex tableau that gives an optimum solution. In the final simplex tableau, a zero $C_j - Z_j$ entry for a nonbasic variable indicates that the problem has more than one optimal basic feasible solution. The other optimal solutions are usually obtained by performing additional iterations of the simplex method, each time choosing a nonbasic variable with a zero $C_j - Z_j$ entry as the entering variable. To illustrate, consider the following example.

***Example* 19.** Maximize $Z = 9x + 6y$

subject to the constraints:

$$x + 2y \le 8$$
$$3x + 2y \le 12$$
$$x, y \ge 0$$

Solution. The initial simplex tableau is shown as follows :

Simplex Tableau I

$C_j \rightarrow$	Basic	Solution	9	6	0	0	Ratio
↓	Variable		x	y	s_1	s_2	
0	s_1	8	1	2	1	0	8
0	s_2	12	(3)	2	0	1	←4
	Z_j	0	0	0	0	0	
	$C_j - Z_j$		9↑	6	0	0	

Since there is a positive $C_j - Z_j$ entry, we continue. The entering variable is x, the departing variable is s_2 and the key number, 3, is circled. Replacing key-and non-key rows using the technique discussed in Section 4.3, we obtain second tableau.

Simplex Tableau II

$C_j\rightarrow$ ↓	Basic Variable	Solution	9 x	6 y	0 s_1	0 s_2	Ratio
0	s_1	4	0	(4/3)	1	– 1/3	←3
9	x	4	1	2/3	0	1/3	6
	Z_j	36	9	6	0	3	
	$C_j - Z_j$		0	0 ↑	0	– 3	

Since all entries in $C_j - Z_j$ row are negative or zero, an optimum solution has been obtained and the maximum value of Z is 36 which occurs when $x = 4$ and $y = 0$. However, because y is a nonbasic variable and the corresponding $C_j - Z_j$ entry is 0, we perform one more iteration to identify the other optimal basic feasible solution. Treating y as an entering variable, we obtain third tableau.

Simplex Tableau III

$C_j\rightarrow$ ↓	Basic Variable	Solution	9 x	6 y	0 s_1	0 s_2
6	y	3	0	1	3/4	– 1/4
9	x	2	1	0	– 1/2	1/2
	Z_j	36	9	6	0	3
	$C_j - Z_j$		0	0	0	– 3

The optimal solution here is $x = 2$, $y = 3$ and the maximum value of $Z = 36$, as before. Notice that the last tableau also has a nonbasic variable s_1 with a zero $C_j - Z_j$ entry. This is inevitable because the extra iteration does not change $C_j - Z_j$ row. Making s_1 an entering variable now would only lead back to the second tableau. Therefore there are only two basic feasible solutions that are optimal, all other optimal solutions are a weighted average of these two.

***Example* 20.** Given below is a table obtained after few iterations using simplex method, to solve a linear programming problem to maximize total contribution from product A and product B.

Basic Variables	x_1	x_2	s_1	s_2	s_3	Quantity
x_2	0	1	3/5	–2/5	0	300
x_1	1	0	– 2/5	3/5	0	300
s_3	0	0	– 1/5	– 1/5	1	400
C_j	8.5	10.5	0	0	0	

Give short answers to the following giving reasons :

(*i*) Is the above solution optimal ?

(*ii*) Is the above solution feasible ?

(*iii*) Does the problem have an alternative solution ? If so, find the other solution.

(*iv*) What are the shadow prices of the resources ? **[*Delhi Univ. B.Com. (H) 1998*]**

Solution. Given table can be represented as follows :

$C_j \rightarrow$ ↓	*Basic Variables*	*Quantity*	8.5 x_1	10.5 x_2	0 s_1	0 s_2	0 s_3
10.5	x_2	300	0	1	3/5	– 2/5	0
8.5	x_1	300	1	0	– 2/5	3/5	0
0	s_3	400	0	0	– 1/5	– 1/5	1
	Z_j	5700	8.5	10.5	2.9	0.9	0
	$C_j - Z_j$		0	0	– 2.9	– 0.9	0

(*i*) Yes, it is an optimal solution because all the entries in $C_j - Z_j$ row are negative or zero.

(*ii*) Yes, it is feasible because optimal solution is always feasible.

(*iii*) No, it does not have any alternative solution because none of the non-basic variables has a zero $C_j - Z_j$ entry in the final simplex table.

(*iv*) Shadow prices of the resources are 2.9, 0.9, and 0 respectively.

***Example* 21.** The simplex tableau for a maximization problem of *LP* is given below:

$C_j \rightarrow$	4	5	0	0	
	x_1	x_2	s_1	s_2	*Quantity*
	1	1	1	0	10
	1	0	–1	1	3

Answer the following questions, giving reasons in brief:

(*i*) Is this solution optimal?

(*ii*) Is it a case of multiple solutions? If yes, find the alternative solution.

(*iii*) Is this solution degenerate?

(*iv*) Is this solution feasible?

(*v*) If s_1 is slack in machine *A* (in hours/week) and s_2 is slack in machine *B* (in hours/week), which of these machines is being used to the full capacity when producing according to this solution?

(*vi*) A customer would like to have one unit of product x_1 and is willing to pay in excess of the normal price in order to get it. How much should the price be increased in order to ensure no reduction of profits?

(*vii*) How many units of the two products x_1 and x_2 are being produced according to this solution and what is the total profit?

(*viii*) Machine *A* (associated with slack s_1, in hours/week) has to be shut down for repairs for 2 hours next week. What will be the effect on profits?

(*ix*) How much would you be prepared to pay for another hour (per week) of capacity each on machine *A* and machine *B* ? **[*Delhi Univ. B.Com. (H) 2009*]**

Solution. Given simplex tableau can be represented as follows:

C_j → ↓	*Product Mix*	*Quantity*	4 x_1	5 x_2	0 s_1	0 s_2
5	x_2	10	1	1	1	0
0	s_2	3	1	0	–1	1
	Z_j	50	5	5	5	0
	$C_j - Z_j$		–1	0	–5	0

(*i*) Yes, it is an optimal solution because all the entries in $C_j - Z_j$ row are negative or zero.

(*ii*) No, it does not have multiple solutions because none of the non-basic variables has a zero $C_j - Z_j$ entry in the final simplex table.

(*iii*) No, because none of the basic variables x_2 or s_2 has a zero value.

(*iv*) Yes, it is feasible because optimal solution is always feasible.

(*v*) Machine *A* is being used to the full capacity because the value of slack variable s_1 is zero in the optimal simplex table.

(*vi*) The $C_j - Z_j$ row entry in the final simplex tableau under the non-basic variable x_1 is –1. This means the profit will be reduced by ₹1 if one unit of x_1 is added to the solution. Hence in order to ensure no reduction of profit, price of x_1 should be increased by ₹1.

(*vii*) The optimum solution is : $x_1 = 0, x_2 = 10$ and max $Z = 50$

(*viii*) From the last row, we find that the shadow price for Machine *A* is 5. The means that profit could be increased by ₹ 5 for every additional hour available on this machine. Thus if machine *A* is shut down next week for 2 hours, the profit would be decreased by ₹ 5 × 2 = ₹ 10.

(*ix*) The shadow price for machine *A* is 5. This means that we could increase the objective function by ₹ 5 if we had an additional hour on this machine. Hence the maximum price we would like to pay for another hour of capacity on this machine is ₹ 5. Similarly, for the second machine, the shadow price is $s_2 = 0$. An additional hour on the second machine won't help since this machine has not been fully utilised.

***Example* 22.** Given below is the simplex tableau for a maximization type of *LP* problem:

$C_j \rightarrow$	*Basic*	3	4	0	0	b_i
$\downarrow$	*Variable*	x_1	x_2	s_1	s_2	
4	x_2	1	1	1	0	6
0	s_2	1	0	–1	1	2

Answer the following questions with reasons:

(*i*) Does the tableau represent an optimal soultion?

(*ii*) Are there more than one optimal solution?

(*iii*) Is this solution degenerate?

(*iv*) Is this solution feasible?

(*v*) If s_1 is slack in machine *A* (in hours/week) and s_2 is slack in machine *B* (in hours/week), which of these machines is being used to the full capacity when producing according to this solution?

(*vi*) A customer would like to have one unit of product x_1 and is willing to pay in excess of the normal price in order to get it. How much should the price be increased in order to ensure no reduction in profit?

(*vii*) How many units of the two products x_1 and x_2 are being produced according to this solution and what is the total profit?

(*viii*) Machine *A* (associated with slack s_1, in hours/week) has to be shut down for repairs for 2 hours next week. How much will the reduction in profits be?

(*ix*) What is the maximum you would be prepared to pay for another hour (per week) of capacity each on machine *A* and machine *B* ?

(*x*) What are the shadow prices of the machine hours?

[*Delhi Univ. B.Com. (H) 2011*]

Solution. Given simplex tableau can be represented as follows:

$C_j \rightarrow$ $\downarrow$	*Basic Variable*	*Quantity*	3 x_1	4 x_2	0 s_1	0 s_2
4	x_2	6	1	1	1	0
0	s_2	2	1	0	–1	1
	Z_j	24	4	4	4	0
	$C_j - Z_j$		–1	0	–4	0

(*i*) Yes, it is an optimal solution because all the entries in $C_j - Z_j$ row are negative or zero.

(*ii*) No, it does not have more than one optimal solution because none of the non-basic variables has a zero $C_j - Z_j$ entry in the final simplex table.

(*iii*) No, because none of the basic variables has solution value equal to zero.

(*iv*) Yes, it is feasible because optimal solution is always feasible.

(*v*) Machine A is being used to the full capacity because value of slack variable s_1 is zero in the optimal simplex table.

(*vi*) The entry in the final $C_j - Z_j$ row under x_1 is –1. This means the profit will be reduced by ₹1 if one unit of x_1 is added to the solution. Hence in order to ensure no reduction of profit, price of x_1 should be increased by ₹1.

(*vii*) $x_1 = 0, x_2 = 6$ and total profit is 24.

(*viii*) Since marginal profitability of machine *A* is ₹ 4 per hour, a reduction of 2 hours capacity in a week would cause a reduction of $2 \times 4 =$ ₹ 8 in profit.

(*ix*) Machine *A* : ₹ 4 per hour; Machine *B* : nil.

(*x*) Shadow prices are: Machine *A* - ₹ 4 per hour; Machine *B* - nil.

***Example* 23.** A firm produces three products, *A*, *B*, *C* using three resources (material, machine hours and labour hours). The manager of the firm wants to find out the best production strategy. A student of B.Com (Hons.) who was familiar with linear programming technique offered to help him. The student formulated the problem and solved the problem by simplex method. He gave the following solution :

	x_1	x_2	x_3	s_1	s_2	s_3	*Quantity*
Contribution →	30	40	10	0	0	0	
	1/2	1	3/2	1/2	0	0	45
	3/2	0	−1/2	−1/2	1	0	15
	5/2	0	1/2	−1/2	0	1	35

Do you agree that this is the best production strategy? If not, improve the solution to get the best production strategy.

Also answer the following questions :

(*i*) Are all the three resources completely used? If not, how much of which resource is unused?

(*ii*) Can there be an alternative product mix which gives the same total contribution?

(*iii*) What are the marginal worth of the resources?

(*iv*) What happens if 15 machine hours are lost due to some mechanical problem.

[*Delhi Univ. B.Com. (H) 2010, 2017*]

Solution. Given table can be represented as follows :

$C_j \rightarrow$ ↓	*BV*	*Sol.*	30 x_1	40 x_2	10 x_3	0 s_1	0 s_2	0 s_3	*Ratio*
40	x_2	45	1/2	1	3/2	1/2	0	0	90
0	s_2	15	(3/2)	0	–1/2	–1/2	1	0	←10
0	s_3	35	5/2	0	1/2	–1/2	0	1	14
	Z_j	1800	20	40	60	20	0	0	
	$C_j - Z_j$		10↑	0	–50	–20	0	0	

Given solution is not optimal as there exists a positive entry 10 in $C_j - Z_j$ row under x_1 variable and so it can't be best production strategy and we proceed to get optimal solution as shown below.

$C_j \rightarrow$ ↓	*BV*	*Sol.*	30 x_1	40 x_2	10 x_3	0 s_1	0 s_2	0 s_3	*Ratio*
40	x_2	40	0	1	5/3	2/3	–1/3	0	
30	x_1	10	1	0	–1/3	–1/3	2/3	0	
0	s_3	10	0	0	4/3	1/3	–5/3	1	
	Z_j	1900	30	40	170/3	50/3	20/3	0	
	$C_j - Z_j$		0	0	–140/3	–50/3	–20/3	0	

This is optimal solution as all entries in $C_j - Z_j$ are zero or negative. Hence best production strategy is to produce 10 units of product *A* and 40 units of *B* to get maximum profit of ₹ 1900.

(*i*) No, $s_3 = 10$ implies that 10 labour hours are left unused.

(*ii*) No, it does not have any alternative solution because none of the non-basic variables has a zero $C_j - Z_j$ entry in the final simplex table.

(*iii*) Marginal worth of resources are : 50/3, 20/3, 0.

(*iv*) When 15 machine hours are lost, then production of *B* would increase by 5 units and that of *A* would decrease by 10 units and the total contribution decreases by ₹ 100.

***Example* 24.** A company produces 2 types of leather belts—Type *A* and Type *B*. Contribution per belt is ₹ 4 for Type *A* and ₹ 3 for Type *B*. The time requirements of one belt of Type *A* and Type *B* are in the ratio 2 : 3. Time available is sufficient to produce 500 belts of Type *A*. The leather is sufficient for only 400 belts. Belt *A* required a fancy buckle and only 200 fancy buckles are available.

(*i*) Formulate above as a linear programming problem.

(*ii*) Solve it by simplex method and comment on this optimum solution.

[*Delhi Univ. B.Com. (H) 1999*]

Solution. Let x denote the number of belts of Type A and y denote the number of belts of Type B produced per unit time. Then the appropriate mathematical formulation of the problem is :

Maximize $Z = 4x + 3y$

subject to the constraints:

$$2x + 3y \leq 1000$$
$$x + y \leq 400$$
$$x \leq 200$$
$$x, y \geq 0$$

Introducing slack variables s_1, s_2 and s_3 so as to convert the constraint inequalities into equations, the given problem can be restated as follows :

Maximize $Z = 4x + 3y + 0s_1 + 0s_2 + 0s_3$

subject to the constraints:

$$2x + 3y + s_1 = 1000$$
$$x + y + s_2 = 400$$
$$x + s_3 = 200$$
$$x, y, s_1, s_2, s_3 \geq 0$$

An initial basic feasible solution is :

$$x = 0, \quad y = 0, \quad s_1 = 1000, \quad s_2 = 400 \quad \text{and} \quad s_3 = 200,$$

at which $Z = 0$. The initial simplex tableau corresponding to this solution is :

Simplex Tableau I

$C_j \rightarrow$ $\downarrow$	*Basic* *Variable*	*Solution*	4 x	3 y	0 s_1	0 s_2	0 s_3	*Ratio*
0	s_1	1000	2	3	1	0	0	500
0	s_2	400	1	1	0	1	0	400
0	s_3	200	(1)	0	0	0	1	← 200
	Z_j	0	0	0	0	0	0	
	$C_j - Z_j$		4 ↑	3	0	0	0	

A look at the $C_j - Z_j$ row reveals that x is the entering variable. The smallest nonnegative quotient in the ratio column is 200, so s_3 is the departing variable and the key number, 1, is circled. Replacing key-and non-key rows using the technique discussed in Section 4.3, we get Tableau II.

Simplex Tableau II

$C_j \rightarrow$ ↓	Basic Variable	Solution	4 x	3 y	0 s_1	0 s_2	0 s_3	Ratio
0	s_1	600	0	(3)	1	0	– 2	←200
0	s_2	200	0	1	0	1	– 1	200
4	x	200	1	0	0	0	1	–
	Z_j	800	4	0	0	0	4	
	$C_j - Z_j$		0	3 ↑	0	0	– 4	

The entering variable is now y. Since there is a tie for the smallest quotient, we can choose either s_1 or s_2 as the departing variable. Let us choose s_1. The key number, 3, is circled. Replacing key-and non-key rows using the technique discussed in Section 4.3, we get Tableau III.

Simplex Tableau III

$C_j \rightarrow$ ↓	Basic Variable	Solution	4 x	3 y	0 s_1	0 s_2	0 s_3
3	y	200	0	1	1/3	0	– 2/3
0	s_2	0	0	0	– 1/3	1	– 1/3
4	x	200	1	0	0	0	1
	Z_j	1400	4	3	1	0	2
	$C_j - Z_j$		0	0	– 1	0	– 2

Since all $C_j - Z_j$ entries in Tableau III are zero or negative, optimal solution has been obtained, and the maximum value of Z is 1400 which occurs when $x = 200$ and $y = 200$. Note that optimal solution is *degenerate* because Tableau III corresponds to a basic feasible solution, where a basic variable s_2 is zero.

***Example* 25.** A firm produces three products A, B and C. Unit contributions of the products are ₹ 5, 10, and 8 respectively. Each unit of product A requires 3 kg of material, 4 machine hours and 2 labour hours; each unit of product B requires 5 kg of material, 4 machine hours and 4 labour hours and each unit of product C requires 2 kg of material, 4 machine hours and 5 labour hours. Everyday 60 kg of material, 72 machine hours and 100 labour hours are available. Use simplex method to find out the best production strategy. Also, investigate the effect on the solution of the following :

(*i*) An increase of 12 machine hours.

(*ii*) A decrease of 6 kg of material.

(*iii*) 3 units of product A are to be produced.

Solution. Let x_1, x_2, x_3 denote the number of units of product A, B and C, respectively, produced by the firm.

Then the appropriate mathematical formulation of the problem is :

Maximize $\quad Z = 5x_1 + 10x_2 + 8x_3 \quad$ *Total contribution*

subject to the constraints:

$$3x_1 + 5x_2 + 2x_3 \le 60 \quad \textit{Material constraint}$$
$$4x_1 + 4x_2 + 4x_3 \le 72 \quad \textit{Machine constraint}$$
$$2x_1 + 4x_2 + 5x_3 \le 100 \quad \textit{Labour constraint}$$
$$x_1, x_2, x_3 \ge 0$$

Introducing slack variables s_1, s_2 and s_3, we can restate the problem in terms of equations :

Maximize $\quad Z = 5x_1 + 10x_2 + 8x_3 + 0s_1 + 0s_2 + 0s_3$

subject to the constraints:

$$3x_1 + 5x_2 + 2x_3 + s_1 = 60$$
$$4x_1 + 4x_2 + 4x_3 + s_2 = 72$$
$$2x_1 + 4x_2 + 5x_3 + s_3 = 100$$
$$x_1, x_2, x_3, s_1, s_2, s_3 \ge 0$$

An initial basic feasible solution is :

$$x_1 = 0,\ x_2 = 0,\ x_3 = 0,\ s_1 = 60,\ s_2 = 72,\ s_3 = 100,$$

at which $Z = 0$. Starting with this solution, we obtain a sequence of simplex tableaux that leads to the final solution.

$C_j \rightarrow$	*Basic*	*Solution*	5	10	8	0	0	0	*Ratio*
$\downarrow$	*Variable*		x_1	x_2	x_3	s_1	s_2	s_3	
0	s_1	60	3	(5)	2	1	0	0	←12
0	s_2	72	4	4	4	0	1	0	18
0	s_3	100	2	4	5	0	0	1	20
	Z_j	0	0	0	0	0	0	0	
	$C_j - Z_j$		5	10 ↑	8	0	0	0	
10	x_2	12	3/5	1	2/5	1/5	0	0	30
0	s_2	24	8/5	0	(12/5)	– 4/5	1	0	←10
0	s_3	52	– 2/5	0	17/5	– 4/5	0	1	260/17
	Z_j	120	6	10	4	2	0	0	
	$C_j - Z_j$		– 1	0	4 ↑	– 2	0	0	

$C_j \rightarrow$ ↓	Basic Variable	Solution	5 x_1	10 x_2	8 x_3	0 s_1	0 s_2	0 s_3	Ratio
10	x_2	8	1/3	1	0	1/3	– 1/6	0	
8	x_3	10	2/3	0	1	– 1/3	5/12	0	
0	s_3	18	– 8/3	0	0	1/3	– 17/12	1	
	Z_j	160	26/3	10	8	2/3	5/3	0	
	$C_j - Z_j$		– 11/3	0	0	– 2/3	– 5/3	0	

A look at the $C_j - Z_j$ row of last tableau shows that it is impossible to further increase the objective function. Hence the optimal solution is $x_1 = 0,\ x_2 = 8,\ x_3 = 10$ with maximum value of Z as ₹ 160. Now we'll investigate the effect of the following :

(i) ***An increase of 12 machine hours.*** s_2- column in the final table shows that if one machine hour is increased, it would decrease the production of product B by 1/6 units but increase the production of product C by 5/12 units. This means $x_2 = 6,\ x_3 = 15$ when machine hours are increased by 12 hours. Also, the change in contribution for an increase of 1 machine hour is :

$$\text{change in contribution} = -10 \times \frac{1}{6} + 8 \times \frac{5}{12} = 5/3 > 0.$$

In other words, contribution will increase by ₹ 5/3 for every additional machine hour. Thus increase in contribution when 12 machine hours are increased = 5/3 × 12 = ₹ 20 and hence new contribution is ₹ 180.

(ii) ***A decrease of 6 kg of material.*** s_1- column in the final table shows that if 1 kg of material is decreased, it would decrease the production of product B by 1/3 units but increase the production of product C by 1/3 units. This means $x_2 = 6,\ x_3 = 12$ when 6 kg of material is decreased. Also, the change in contribution for a decrease of 1 kg of material is :

$$\text{change in contribution} = -\frac{1}{3} \times 10 + \frac{1}{3} \times 8 = -\frac{2}{3} < 0.$$

In other words, contribution will decrease by ₹ 2/3 for a decrease of each 1 kg of material. Thus decrease in contribution when 6 kg of material is decreased = 2/3 × 6 = ₹ 4 and hence new contribution is ₹ 156.

(iii) ***3 units of product A are to be produced.*** x_1- column in the final table shows that when one unit of product A is produced, it would reduce the production of product B by 1/3 units and reduce the production of product C by 2/3 units. This means $x_1 = 3,\ x_2 = 7,\ x_3 = 8$ when 3 units of product A are to be produced. Also, the change in contribution for producing one unit of product A is :

$$\text{change in contribution} = -10 \times \frac{1}{3} - 8 \times \frac{2}{3} = -\frac{26}{3} < 0.$$

Of course, adding one unit of product A will increase the objective function by ₹ 5. Thus the net effect is $5 - \frac{26}{3} = -\frac{11}{3}$. That is, for each unit of A brought in, the objective function will decrease by 11/3. Hence decrease in contribution when 3 units of product A are to produced $= \frac{11}{3} \times 3 =$ ₹ 11. Therefore, new contribution is ₹ 149.

EXERCISE 4.1

1. Solve the following linear programming problems by simplex method :

(*i*) Maximize $Z = x_1 + 2x_2$

subject to the constraints:
$$2x_1 + x_2 \le 8$$
$$2x_1 + 3x_2 \le 12$$
$$x_1, x_2 \ge 0$$

(*ii*) Maximize $Z = 4x_1 + 3x_2 + 6x_3$

subject to the constraints:
$$3x_1 + x_2 + 3x_3 \le 30$$
$$2x_1 + 2x_2 + 3x_3 \le 40$$
$$x_1, x_2, x_3 \ge 0$$

(*iii*) Maximize $Z = 2x_1 + 5x_2$

subject to the constraints:
$$x_1 + 4x_2 \le 24$$
$$3x_1 + x_2 \le 21$$
$$x_1 + x_2 \le 9$$
$$x_1, x_2 \ge 0$$

(*iv*) Maximize $Z = 2x_1 + x_2 + x_3$

subject to the constraints:
$$-2x_1 + x_2 - 2x_3 \le 4$$
$$x_1 - 2x_2 + x_3 \le 2$$
$$x_1, x_2, x_3 \ge 0$$

(*v*) Maximize $Z = 6x_1 + 4x_2$

subject to the constraints:
$$x_1 + 2x_2 \le 240$$
$$3x_1 + 2x_2 \le 300$$
$$x_1, x_2 \ge 0$$

(*vi*) Maximize $Z = x_1 + 4x_2 - x_3$

subject to the constraints:
$$-5x_1 + 6x_2 - 2x_3 \le 30$$
$$-x_1 + 3x_2 + 6x_3 \le 12$$
$$x_1, x_2, x_3 \ge 0$$

(*vii*) Maximize $Z = 3x_1 + 5x_2$

subject to the constraints:
$$x_1 \le 4$$
$$x_2 \le 6$$
$$3x_1 + 2x_2 \le 12$$
$$x_1, x_2 \ge 0$$

2. Find the basic solutions of the following system of equations :
$$x + 2y + 3z = 4$$
$$2x + 3y + 5z = 7$$

Also determine which of these basic solutions are

(*i*) basic feasible (*ii*) non-degenerate basic feasible.

3. In the context of linear programming, what do you mean by no optimal solution, multiple optimum solution and unbounded optimum solution ? How would you recognise these in the optimum simplex tableau ? **[*Delhi Univ. B.Com. (H) 2002*]**

4. While solving a linear programming problem by Simplex Method, how will you detect that:

(*i*) problem has unbounded solution;

(*ii*) problem has multiple optimal solutions;

(*iii*) problem has degenerate solution;

(*iv*) problem is infeasible. **[*Delhi Univ. B.Com. (H) 2010*]**

5. A toy manufacturer produces two types of dolls; a basic version-doll *A* and a deluxe version-doll *B*. Each doll of Type *B* takes twice as much time as required by a doll of Type *A*. The company has time to make a maximum of 2000 dolls of Type *A* per day. The supply of plastic is sufficient to produce 1500 dolls per day and each type requires equal amount of it. The deluxe version, *i.e.*, Type *B* requires a fancy dress of which there are 600 per day available. If the company makes a profit of ₹ 3 and ₹ 5 per doll, respectively, on doll *A* and *B*, how many of each should be produced per day in order to maximize profit ? Solve it by simplex method.

6. A firm has available 240, 370 and 180 kilos of wood, plastic and steel respectively. The firm produces two products *A* and *B*. Each unit of product *A* requires 1, 3 and 2 kilos of wood, plastic and steel respectively. The corresponding requirements for each unit of *B* are 3, 4 and 1 respectively. If *A* sells for ₹ 4 and *B* for ₹ 6, determine how many units of *A* and *B* should be produced to obtain the maximum gross income. Formulate the problem mathematically. Solve the problem by simplex method. **[*Delhi Univ. B.Com. (H) 1985*]**

7. A manufacturer produces two items X_1 and X_2. X_1 needs 2 hours on machine *A* and 2 hours on machine *B*. X_2 needs 3 hours on machine *A* and 1 hour on machine *B*. If machine *A* can run for a maximum of 12 hours per day and *B* for 8 hours per day and profits from X_1 and X_2 are ₹ 4 and ₹ 5 per item respectively, find by simplex method how many items per day be produced to have maximum profit. Give the interpretation for the values of 'Indicators' corresponding to slack variable in the final iteration.

8. Hens of two types are available. Type *A* costs ₹ 8/- per hen and Type *B*, ₹ 10/-per hen. Type *A* lays on an average 4 eggs per week and Type *B*, 5 eggs per week. Eggs sell at 50 paise each. The feeding cost of Type *A* hen is ₹ 1.50 per week and of Type *B* is ₹ 1.90 per week. Maximum of ₹ 180/- is to be spent on purchasing hens. How many of each type are to be bought to have maximum profit ? Assume that it is not possible to house more than 20 hens at a time.

(*i*) Formulate this as a linear programming problem.

(*ii*) Solve it by simplex method. **[*Delhi Univ. B.Com. (H) 1984*]**

9. The XYZ company manufactures two products *A* and *B*. These are processed on same machine.

A takes 10 minutes per item and *B* takes 2 minutes per item on machine. Machine can run for a maximum of 35 hours in a week. Product *A* requires 1 kg and product *B*, 0.5 kg of the raw material per item, the supply of which is 600 kg per week. Not more than 800 items of product *B* are required per week. If the product *A* costs ₹ 5 per item and can be sold for ₹ 10 and product *B* costs ₹ 5 per item and can be sold for ₹ 8 per item, determine by simplex method how many items of *A* and *B* per week be produced in order to maximize the profit.

[*Delhi Univ. B.Com. (H) 2005*]

10. A cabinet manufacturer produces wood cabinets for TV sets, stereo-systems and radios, each of which must be assembled, decorated and crated. Each TV cabinet requires 3 hours to assemble, 5 hours to decorate and 1/10 hours to crate and returns a profit of ₹ 10. Each stereo cabinet requires 10 hours to assemble, 8 hours to decorate and 3/5 hour to crate and returns a profit of ₹ 25. Each radio cabinet requires 1 hour to assemble, 1 hour to decorate and 1/10 hour to crate and returns a profit of ₹ 3. The manufacturer has the maximum of 30,000, 40,000 and 120 hours available for assembling, decorating and crating respectively.

(*i*) Formulate the above as a linear programming problem.

(*ii*) Use simplex method to find how many units of each product should be manufactured to maximize profit.

(*iii*) Does the problem have multiple solutions ?

11. A farmer has 1000 acres of land on which he can grow corn, wheat or soyabeans. Each acre of corn costs ₹ 100 for preparation, requires 7 man-days of work and yields a profit of ₹ 30. An acre of wheat costs ₹ 120 to prepare, requires 10 man-days of work and yields a profit of ₹ 40. An acre of soyabeans costs ₹ 70 to prepare, requires 8 man-days of work and yields a profit of ₹ 20. If the farmer has 1,00,000 for preparation and can count on 8000 man-days of work, how many acres should be allocated to each crop to maximize profit? Formulate this as a problem in linear programming, and solve it by simplex method. **[*Delhi Univ. B.Com. (H) 1987*]**

12. A company produces three products *P*, *Q* and *R* whose prices per unit are 3, 5 and 4 respectively. One unit of product *P* requires 2 units of m_1 and 3 units of m_2. One unit of product *Q* requires 2 units of m_2 and 5 units of m_3 and one unit of product *R* requires 3 units of m_1, 2 units of m_2 and 4 units of m_3. The company has 8 units of material m_1, 10 units of material m_2 and 15 units of material m_3 available to it.

(*i*) Formulate the problem mathematically.

(*ii*) Use simplex method to find how many units of each product should be produced to maximize revenue ? **[*Delhi Univ. B.Com. (H) 1985*]**

13. A manufacturing company contemplates to produce two additional products, called *A* and *B*, which can be marketed at prevailing prices in any reasonable quantities without difficulty. It is known that product *A* requires 10 and 5 man-hours per unit in the foundry and machine departments respectively and product *B* requires only 6 and 4. However, the profit margin of *A* is ₹ 30 per unit as compared to ₹ 20 per unit for *B*. In the week immediately ahead, it is estimated that there will be 1000 and 600 man-hours available in the foundry and the machine departments respectively. How much of *A* and *B* should be produced in order to utilize most profitably the excess capacities ? Use simplex method only.

14. A manufacturer produces bicycles and scooters, each of which must be processed through two machines *A* and *B*. Machine *A* has a maximum of 120 hours available and machine *B* has a maximum of 180 hours available. Manufacturing a bicycle requires 6 hours in machine *A* and 3 hours in machine *B*. Manufacturing a scooter requires 4 hours in machine *A* and 10 hours in machine *B*. If profits are ₹ 45 for a bicycle and ₹ 55 for a scooter, determine the number of bicycles and the number of scooters that should be manufactured in order to maximize the profit. Use simplex method.

15. A company can make three different products *A*, *B* and *C* by combining steel and rubber. Product *A* requires 2 units of steel and 3 units of rubber and can be sold at a profit of ₹ 40 per unit. Product *B* requires 3 units of steel and 3 units of rubber and can be sold at a profit of ₹ 45 per unit. Product *C* requires 1 unit of steel and 2 units of rubber and yields a ₹ 24 profit per unit. There are 100 units of steel and 120 units of rubber available per day. What should be the daily production of each of the three products in order to maximize profit ? Use simplex method.

16. A forging company producing ashtrays and serving platters can hire upto 20 workers. Three types of workers are available. Workers of Type I are capable of producing 5 ashtrays and 5 serving platters during one day and Type II workers make 10 serving platters per day. The other workers, Type III, produce 15 ashtrays during a day's production. The company can use no more than 160 ashtrays and 100 serving platters per day. The company makes a profit of ₹ 2 on ashtrays and ₹ 3 on serving platters. How many of each type of workers should management hire to maximize profit ? (*Use simplex method*).

17. A company produces three products P_1, P_2 and P_3 from two raw materials *A* and *B*, and labour *L*. One unit of product P_1 requires one unit of *A*, 3 units of *B* and 2 units of *L*. A unit of product P_2 requires 2 units of *A* and *B* each, and 3 units of *L*, while one unit of P_3 needs 2 units of *A*, 6 units of *B* and 4 units of *L*. The company has a daily availability of 8 units of *A*, 12 units of *B* and 12 units of *L*. It is further known that the unit contribution margin for the products is ₹ 3, 2 and 5 respectively for P_1, P_2 and P_3. Formulate this problem as a linear programming problem, and then solve it to determine the optimal product mix.

Is the solution obtained by you unique ? Identify an alternate optimal solution, if any. Also, obtain the shadow prices of the resources. **[*Delhi Univ. B.Com. (H) 1994, 2022*]**

18. The simplex table for a maximization problem of linear programming problem is given below :

$C_j \rightarrow$ $\downarrow$	*Product Mix*	4 x_1	5 x_2	0 s_1	0 s_2	*Quantity*
5	x_2	1	1	1	0	10
0	s_2	1	0	–1	1	3
	Z_j	5	5	5	0	50
	$C_j - Z_j$	–1	0	–5	0	

Answer the following questions, giving reasons in brief :

(*i*) Is the above solution optimal ?

(*ii*) Are there more than one optimal solutions ?

(*iii*) Is the solution degenerate ?

(*iv*) Is the solution feasible ?

(*v*) If s_1 is slack in machine A and s_2 is slack in machine B, which of these machines is being used to the full capacity when producing according to this solution ?

(*vi*) How many units of the two products x_1 and x_2 are being produced according to this solution and what is the total profit ?

(*vii*) What is the maximum price you will like to pay for one hour of each machine ?

19. Given below are the objective function, the constraints and the final simplex tableau for a linear programming product mix problem :

Maximize $Z = 2x_1 + 5x_2 + 8x_3$ *Objective function*

subject to the constraints:

$$6x_1 + 8x_2 + 4x_3 \le 96 \quad \textit{Department I}$$
$$2x_1 + x_2 + 2x_3 \le 40 \quad \textit{Department II}$$
$$5x_1 + 3x_2 + 2x_3 \le 60 \quad \textit{Department III}$$
$$x_1, x_2, x_3 \ge 0$$

Final Simplex Tableau

$C_j \rightarrow$	*Product*	2	5	8	0	0	0	*Quantity*
$\downarrow$	*Mix*	x_1	x_2	x_3	s_1	s_2	s_3	
5	x_2	1/3	1	0	1/6	– 1/3	0	8/3
8	x_3	5/6	0	1	– 1/12	2/3	0	56/3
0	s_3	7/3	0	0	– 1/3	– 1/3	1	44/3
	Z_j	25/3	5	8	1/6	11/3	0	$162\frac{2}{3}$
	$C_j - Z_j$	– 19/3	0	0	– 1/6	– 11/3	0	

(*i*) Write the optimum product mix and the profit contribution shown by the above solution.

(*ii*) Is this solution feasible ? Why ? Give brief reason.

(*iii*) Does the problem have any alternative optimum solution ? If so, find the other solution.

(*iv*) What are the shadow prices of three departments.

(*v*) If the company wishes to expand the production capacity, which of the three departments should be given priority.

(*vi*) If the company produces 6 units of x_1, how many units of x_2 and x_3 shall have to be reduced, if any ?

(*vii*) If a customer is prepared to pay higher price for product x_1, how much should the price be increased so that the company's profit remains unchanged ?

20. Given below is a table obtained after few iterations using simplex method, to solve a linear programming problem to maximize total contribution from product *A* and product *B*.

Basic Variables	x_1	x_2	s_1	s_2	s_3	*Solution*
x_2	0	1	1/3	0	– 2/3	200
s_2	0	0	– 1/3	1	– 1/3	0
x_1	1	0	0	0	1	200
C_j	4	3	0	0	0	

Give short answers to the following giving reasons :

(*i*) Is the above solution optimal ?

(*ii*) Does the problem have alternative solution ? If so, find the other solution.

(*iii*) Is the above solution degenerate ? Why ? Give brief reason.

(*iv*) How many units of the two products x_1 and x_2 are being produced according to this solution and what is the total profit ?

21. Given below is the simplex table for a maximization type of linear programming problem :

$C_j \rightarrow$ *Basic*	10	6	4	0	0	0	*Quantity*
Variable	x_1	x_2	x_3	s_1	s_2	s_3	
x_2	0	1	5/6	5/3	–1/6	0	200/3
x_1	1	0	1/6	–2/3	1/6	0	100/3
s_3	0	0	4	–2	0	1	100

Answer the following questions with reasons :

(*i*) Does the table above give an optimal solution?

(*ii*) Is this solution feasible?

(*iii*) Does the problem have alternative optimal solution?

(*iv*) Is this solution degenerate?

(v) Write the optimal product mix and the profit contribution shown by the above solution.

(*vi*) Indicate the shadow prices of the three resources.

(*vii*) If the company wishes to expand the production capacity, which of the three resources should be given priority?

(*viii*) If a customer is prepared to pay higher price for x_3, how much should the price be increased so that the company's profit remains unchanged?

(*ix*) Write down the objective function of above problem. **[*Delhi Univ. B.Com. (H) 2012*]**

22. Given below are the objective function, the constraints and the final simplex tableau for a linear programming product mix problem :

Max $Z = 15x_1 + 12x_2 + 11x_3$

subject to the constraints :

$3x_1 + 4x_2 + 5x_3 5 \leq 63$ (hours, Deptt A)

$4x_1 + x_2 + 4x_3 \leq 64$ (hours, Deptt B)

$x_1 + 4x_2 + 2x_3 \leq 40$ (hours, Deptt C)

$x_1, x_2, x_3 \geq 0$

Final Simplex Table

C_j	*Product*	15	12	11	0	0	0	*Quantity*
	Mix	x_1	x_2	x_3	S_1	S_2	S_3	
		0	1	8/13	4/13	–3/13	0	60/13
		1	0	11/13	–1/13	4/13	0	193/13
		0	0	–17/13	–15/13	8/13	1	87/13
	Z_j							
	$C_j - Z_j$							

(*i*) Complete the table and test whether the solution is optimal or not.

(*ii*) Write the optimal product mix and the profit contribution shown by the above solution.

(*iii*) Is this solution feasible? Give reasons.

(*iv*) Does the problem have any alternative solution? If so, show one such solution.

(*v*) Indicate the shadow prices of three departments.

(*vi*) If the company wishes to expand the production capacity, which of the three departments should be given priority?

(*vii*) If the company produces thirteen units of x_3, how many units of x_1 and x_2 shall have to be reduced, if any?

(*viii*) If a customer is prepared to pay higher prices for product x_3, how much should the price be increased so that the company's profit remains unchanged?

(*ix*) Indicate whether the solution given in the table is degenerate. If yes, which variable is degenerated?

[Delhi Univ. B.Com. (H) 2019]

23. Given below is the simplex tableau for a maximization type of linear programming problem:

C_j	*Basic*	30	40	35	0	0	0	*Quantity*
Variable	x_1	x_2	x_3	s_1	s_2	s_3		
40	x_2	1/3	1	0	1/3	– 1/3	0	12
35	x_3	5/6	0	1	– 1/6	2/3	0	21
0	s_3	– 5/3	0	0	– 2/3	– 1/3	1	15

Answer the following questions, giving reasons in brief:

(*i*) Does the table above give an optimal solution?

(*ii*) Is this solution feasible?

(*iii*) Is the solution degenerate?

(*iv*) Does the problem have alternative solutions? If so, show one such solution.

(*v*) If s_1 is the slack in Machine *A* (in hours/week), s_2 is the slack in Machine *B* (in hours/ week) and s_3 is the slack in machine *C* (in hours/week) which of these machine(s) is being used to the full capacity when producing according to this solution?

(*vi*) What are the shadow prices of the machines?

(*vii*) If a customer is prepared to pay a higher price for product x_1, how much should the price be increased so that the company's profit remains unchanged?

(*viii*) How many units of the three products x_1, x_2 and x_3 are being produced according to this solution and what is the total profit?

(*ix*) What is the maximum price you would be prepared to pay for another hour (per week) of capacity on machine *A*, *B* and *C* respectively?

(*x*) If the company produces 6 units of x_1, how many units of x_2 and x_3 shall have to be reduced, if any? **[*Delhi Univ. B.Com. (H) 2022*]**

ANSWERS

1. (*i*) $x_1 = 0,\ x_2 = 4$ and max $Z = 8$ (*ii*) $x_1 = 0,\ x_2 = 10,\ x_3 = 20/3$ and max $Z = 70$

(*iii*) $x_1 = 4,\ x_2 = 5$ and max $Z = 33$ (*iv*) Unbounded solution

(*v*) $x_1 = 100,\ x_2 = 0$ and max $Z = 600$ or $x_1 = 30,\ x_2 = 105$ and max $Z = 600$

(*vi*) Unbounded solution (*vii*) $x_1 = 0,\ x_2 = 6$ and max $Z = 30$

2. Basic solutions are :

(*a*) $y = -1,\ \ z = 2$ (with x as non-basic)

(*b*) $x = 1,\ \ z = 1$ (with y as non-basic)

(*c*) $x = 2,\ \ y = 1$ (with z as non-basic)

(*i*) (*b*) and (*c*) are basic feasible solutions

(*ii*) (*b*) and (*c*) are non-degenerate basic feasible solutions.

5. 1000 dolls of Type *A*, 500 dolls of Type *B* with max profit = ₹ 5500

6. Maximize $Z = 4x + 6y$, subject to : $x + 3y \leq 240,\ \ 3x + 4y \leq 370,\ \ 2x + y \leq 180, x, y \geq 0$; $x = 30,\ \ y = 70$ with max Z = ₹ 540

7. 3 units of X_1, 2 units of X_2 with max profit = ₹ 22

8. (*i*) Maximize $Z = 0.50x + 0.60y$, subject to : $8x + 10y \leq 180,\ \ x + y \leq 20,\ \ x, y \geq 0$

(*ii*) $x = 10,\ \ y = 10$ with max Z = ₹ 11

9. 50 items of *A*, 800 items of *B* with max profit = ₹ 2650

10. (*i*) Maximize $Z = 10x_1 + 25x_2 + 3x_3$, subject to : $3x_1 + 10x_2 + x_3 \leq 30{,}000,\ \ 5x_1 + 8x_2 + x_3 \leq 40{,}000$, $\frac{1}{10}x_1 + \frac{3}{5}x_2 + \frac{1}{10}x_3 \leq 120,\ \ x_1, x_2, x_3 \geq 0$

(*ii*) $x_1 = 1200,\ \ x_2 = 0,\ \ x_3 = 0$ with max $Z = 12{,}000$ (*iii*) No

11. Maximize $Z = 30x_1 + 40x_2 + 20x_3$, subject to : $x_1 + x_2 + x_3 \leq 1000,\ \ 100x_1 + 120x_2 + 70x_3 \leq 1{,}00{,}000$, $7x_1 + 10x_2 + 8x_3 \leq 8000,\ \ x_1, x_2, x_3 \geq 0;\ \ x_1 = 250,\ \ x_2 = 625,\ \ x_3 = 0$ with max $Z = 32{,}500$

12. (*i*) Maximize $Z = 3x_1 + 5x_2 + 4x_3$, subject to : $2x_1 + 3x_3 \leq 8,\ \ 3x_1 + 2x_2 + 2x_3 \leq 10,\ \ 5x_2 + 4x_3 \leq 15$, $x_1, x_2, x_3 \geq 0$, (*ii*) $x_1 = 4/3,\ \ x_2 = 3,\ \ x_3 = 0$ with max $Z = 19$

13. 40 units of *A*, 100 units of *B* with max profit = ₹ 3200

14. 10 bicycles, 15 scooters with max profit = ₹ 1275

15. 20 units of *A*, 20 units of *B*, 0 units of *C* with max profit = ₹ 1700

16. Maximize $Z = 25x_1 + 30x_2 + 30x_3$ subject to : $x_1 + x_2 + x_3 \leq 20,\ \ 5x_1 + 15x_3 \leq 160,\ \ 5x_1 + 10x_2 \leq 100$, $x_1, x_2, x_3 \geq 0;\ \ x_1 = 0,\ \ x_2 = 10,\ \ x_3 = 10$ with max $Z = 600$

17. Maximize $Z = 3x_1 + 2x_2 + 5x_3$ subject to : $x_1 + 2x_2 + 2x_3 \leq 8,\ \ 3x_1 + 2x_2 + 6x_3 \leq 12,\ \ 2x_1 + 3x_2 + 4x_3 \leq 12$, $x_1, x_2, x_3 \geq 0;\ \ x_1 = 4\ x_2 = 0, x_3 = 0$ and max $Z = 12$. The optimal solution is not unique. The alternate optimum solution is : $x_1 = 12/5, x_2 = 12/5,\ \ x_3 = 0$ and max $Z = 12$. Shadow prices are : 0, 1, and 0

18. (*i*) Yes, because all the entries in $C_j - Z_j$ row are zero or negative.

(*ii*) No, because none of the non-basic variables has a zero entry in the final $C_j - Z_j$ row.

(*iii*) No, because none of the basic variables takes a zero value.

(*iv*) Yes, because optimal solution is always feasible.

(*v*) The first machine has been used to the full capacity because $s_1 = 0$.

(*vi*) $x_1 = 0, x_2 = 10$, and max $Z = 50$.

(*vii*) The shadow price for the first machine is 5 and hence the maximum price we would like to pay for an additional hour of this machine is ₹ 5. The shadow price for the second machine is 0 which indicates that this machine has not been fully utilized and hence we would not like to pay anything for an additional hour of this machine.

19. (*i*) $x_1 = 0, x_2 = 8/3, x_3 = 56/3$ and max $Z = 162\frac{2}{3}$.

(*ii*) Yes, because each of the basic variables is nonnegative.

(*iii*) No, because none of the non-basic variables has a zero entry in the final $C_j - Z_j$ row.

(*iv*) Shadow prices are : 1/6, 11/3 and 0 respectively.

(*v*) Department II, because shadow price of this department is maximum.

(*vi*) If company wishes to produce 6 units of x_1, it has to reduce the production of x_2 by 2 units and production of x_3 by 5 units.

(*vii*) The $C_j - Z_j$ row entry in the final simplex tableau under the non-basic variable x_1 is $-19/3$. This means the profit will be reduced by ₹ 19/3 if one unit of x_1 is added to the solution. Hence in order to ensure that company's profit remains unchanged, price of x_1 should be increased by ₹ 19/3.

20. (*i*) Yes, because all the entries in $C_j - Z_j$ row are zero or negative.

(*ii*) No, because none of the non-basic variables has a zero entry in the final $C_j - Z_j$ row.

(*iii*) Yes, because the given table corresponds to a basic feasible solution, where a basic variable s_2 is zero.

(*iv*) $x_1 = 200, x_2 = 200$ and max $Z = 1400$

21. (*i*) Yes, it is an optimal solution because all the entries in $C_j - Z_j$ row are negative or zero.

(*ii*) Yes, it is feasible because optimal solution is always feasible.

(*iii*) No, it does not have alternative optimal solutions because none of the non-basic variables has a zero $C_j - Z_j$ entry in the final simplex table.

(*iv*) No, because none of the basic variables (x_1, x_2 or s_2) has a zero value.

(*v*) The optimum solution is : $x_1 = 100/3,\ x_2 = 200/3,\ x_3 = 0$ and max $Z = 2200/3$.

(*vi*) Shadow prices are : 10/3, 2/3 and 0.

(*vii*) Resource I, because shadow price of this resource is maximum.

(*viii*) The $C_j - Z_j$ row entry in the final simplex tableau under the non-basic variable x_3 is $-8/3$. This means the profit will be reduced by ₹ 8/3 if one unit of x_3 is added to the solution. Hence in order to ensure no reduction of profit, price of x_3 should be increased by ₹ 8/3.

(*ix*) The objective function is : Maximize $Z = 10x_1 + 6x_2 + 4x_3$.

22. *(i)* The completed final simplex table is shown below:.

C_j	Product Mix	Quantity	15 x_1	12 x_2	11 x_3	0 s_1	0 s_2	0 s_3
12	x_2	60/13	0	1	8/13	4/13	–3/13	0
15	x_1	193/13	1	0	11/13	–1/13	4/13	0
0	s_3	87/13	0	0	–17/13	–15/13	8/13	1
	Z_j	3615/13	15	12	261/13	33/13	24/13	0
	$C_j - Z_j$		0	0	–118/13	–33/13	–24/13	0

The above table represents the optimal solution because all the entries in $C_j - Z_j$ row are negative or zero.

(ii) The optimal solution is : $x_1 = 193/13$, $x_2 = 60/13$, and $x_3 = 0$, with max $Z = 3615/13$

(iii) Yes, it is feasible because optimal solution is always feasible.

(iv) No, it does not have an alternative solution because none of the non-basic variables has a zero $C_j - Z_j$ entry in the final simplex table.

(v) Shadow prices of three departments are: 33/13, 24/13, 0.

(vi) Department I, because shadow price of this department is maximum.

(vii) If the company wishes to produce 13 units of x_3, it has to reduce the production of x_1 by 11 units and production of x_2 by 8 units.

(viii) The $C_j - Z_j$ row entry in the final simplex table under the non-basic variable x_3 is –118/13. This means the profit will be reduced by ₹ 118/13 if one unit of x_3 is added to the solution. Hence in order to ensure that company's profit remains unchanged, price of x_3 should be increased by ₹ 118/13.

(ix) No, because none of the basic varibles takes a zero value.

23. The completed final simplex table is shown below:

$C_j \rightarrow$ ↓ Variable	Basic	Quantity	30 x_1	40 x_2	35 x_3	0 s_1	0 s_2	0 s_3
40	x_2	12	1/3	1	0	1/3	– 1/3	0
35	x_3	21	5/6	0	1	– 1/6	2/3	0
0	s_3	15	– 5/3	0	0	– 2/3	– 1/3	1
	Z_j	1215	31.875	40	35	7.5	10	0
	$C_j - Z_j$		– 1.875	0	0	– 7.5	– 10	0

(i) Yes, it is an optimal solution because all the entries in $C_j - Z_j$ now are negative or zero.

(ii) Yes, it is feasible because optimal solution is always feasible.

(*iii*) No, because none of the basic variables has a zero solution value.

(*iv*) No, it does not have any alternative solution because none of the non-basic variables has a zero $C_j - Z_j$ entry in the final simplex table.

(*v*) Machines *A* and *B* are being used to the full capacity because slack variables s_1 and s_2 have a value of zero in the optimal simplex table.

(*vi*) Shadow prices are :

Machine *A* – ₹ 7.5 per hour; Machine *B* – ₹ 10 per hour; Machine *C* – nil

(*vii*) The entry in the final $C_j - Z_j$ row under x_1 is – 1.875. This means the profit will be reduced by ₹ 1.875 if one unit of x_1 is added to this solution. Hence in order to ensure no reduction of profit, price of x_1 should is increased by ₹ 1.875.

(*viii*) $x_1 = 0, x_2 = 12, x_3 = 21$ and Total profit is 1215.

(*ix*) Machine *A* : ₹ 7.5 per hour; Machine *B* : ₹ 10 per hour; Machine *C* : nil

(*x*) A look at the x_1 – colomn in the final simplex table shows that when one unit of product x_1 is produced, the production of x_2 would be reduced by 1/3 units and production of x_3 would be reduced by 5/6 units. Thus, if the company produces 6 units of x_1, then the production of x_2 would be reduced by 2 units and production of x_3 would be reduced by 5 units.

4.5 ARTIFICIAL VARIABLES

Thus far we have presented the details of the simplex method for solving the standard form of a maximization linear programming problem. However, other forms may be adapted to fit this form. The only real problem that the other forms (=, $\geq$ or $b_i \leq 0$) introduce is in identifying an initial basic feasible solution. Before, this initial solution was found very conveniently by letting the slack variables be the initial basic variables, so that each one just equals the nonnegative right-hand side of its equation.

In this section we point out how to make adjustments required for other forms of the linear programming problem. You will see that all these adjustments can be made in the initialization step, so that the rest of the simplex method can then be applied just as you have already learned it. The standard approach that is used for all these cases is the **artificial variables technique.** This technique constructs a related linear programming problem called the *artificial problem* by introducing into each constraint that needs one a variable called an *artificial variable* just for the purpose of being the initial basic variable for that equation. The usual non-negativity restrictions are placed on these variables and the objective function also is modified to impose an exorbitant penalty on their having values larger than zero. The iterations of the simplex method then automatically force the artificial variables to disappear one at a time until they are all gone, after which the *original* problem is solved. This technique is best illustrated with the help of the following example.

***Example* 26.** Maximize $Z = 2x + y$

subject to the constraints:

$$x + y \leq 6 \quad ...(1)$$
$$-x + y \geq 4 \quad ...(2)$$
$$x, y \geq 0$$

Solution. Note that this is a maximization problem which cannot be written in standard form. To solve it, we convert each of the constraints (1) and (2) into equations. For constraint (1), we introduce a nonnegative slack variable s_1 so that we have

$$x + y + s_1 = 6$$

For constraint (2), we subtract a nonnegative surplus variable s_2 from $-x + y$ so that we have equality. Thus

$$-x + y - s_2 = 4$$

We can now restate the problem :

Maximize $Z = 2x + y$

subject to the constraints:

$$x + y + s_1 = 6 \quad ...(3)$$
$$-x + y - s_2 = 4 \quad ...(4)$$
$$x, y, s_1, s_2 \geq 0$$

Unfortunately, this problem does not have an obvious initial basic feasible solution because there is no longer a slack variable to use as the initial basic variable for Eq. (4). The artificial variable technique overcomes this difficulty by adding a nonnegative variable A to the left side of Eq. (4) in which a surplus variable is introduced. The variable A is called an *artificial variable.* Thus Eqs. (3) and (4) become

$$x + y + s_1 = 6 \quad ...(5)$$
$$-x + y - s_2 + A = 4 \quad ...(6)$$

An obvious solution to Eqs. (5) and (6) is found by setting x, y and s_2 equal to 0. This gives

$$x = y = s_2 = 0, \quad s_1 = 6 \text{ and } A = 4$$

Note that these values do not satisfy Eqs. (3) and (4). However, it is clear that any solution of Eqs. (5) and (6) for which $A = 0$ will give a solution to Eqs. (3) and (4) and vice versa. We can eventually force A to be 0 if we alter the original objective function. Therefore, if the original objective function $Z = 2x + y$ is changed to

$$Z^* = Z - MA = 2x + y - MA,$$

where M denotes some very large positive number. Then the maximum value of Z must occur when $A = 0$ (A cannot be negative). Thus the modified problem is :

$$\begin{aligned} \text{Maximize} \quad & Z^* = 2x + y - MA \\ \text{subject to the constraints:} \quad & x + y + s_1 = 6 \qquad \text{...(7)} \\ & -x + y - s_2 + A = 4 \qquad \text{...(8)} \\ & x, y, s_1, s_2, A \geq 0 \end{aligned}$$

The simplex procedure is now applied to this problem until we obtain a B.F.S. in which the artificial variable A equals 0. That solution will be an initial B.F.S. for the original problem and we then proceed as before. An initial basic feasible solution to the modified problem is

$$x = y = s_2 = 0, \; s_1 = 6 \text{ and } A = 4$$

In this B.F.S., the basic variables s_1 and A correspond to the nonstructural variables in Eqs. (7) and (8) that have positive coefficients. The corresponding value of Z^* is $Z^* = -4M$, which is "extremely" negative. The initial simplex tableau corresponding to this solution is shown below.

Simplex Tableau I

$C_j \rightarrow$ ↓	*Basic Variable*	*Solution*	2 x	1 y	0 s_1	0 s_2	$-M$ A	*Ratio*
0	s_1	6	1	1	1	0	0	6
$-M$	A	4	-1	(1)	0	-1	1	←4
	Z_j^*	$-4M$	M	$-M$	0	M	$-M$	
	$C_j - Z_j^*$		$2-M$	$1+M$↑	0	$-M$	0	

Since M is a large positive number, $1 + M$ is the largest positive entry. Thus the entering variable is y. From the ratio column, we choose A as the departing variable. The key number, 1, is circled. Replacing key- and non-key rows using technique of Section 4.3, we get Tableau II.

Simplex Tableau II

$C_j \rightarrow$ ↓	*Basic Variable*	*Solution*	2 x	1 y	0 s_1	0 s_2	$-M$ A	*Ratio*
0	s_1	2	(2)	0	1	1	-1	←1
1	y	4	-1	1	0	-1	1	–
	Z_j^*	4	-1	1	0	-1	1	
	$C_j - Z_j^*$		3↑	0	0	1	$-1-M$	

From Tableau II, we have the following B.F.S. : $s_1 = 2, \; y = 4, \; x = 0, \; s_2 = 0, \; A = 0$

Since $A = 0$, the values $s_1 = 2, y = 4, x = 0$ and $s_2 = 0$ form an initial B.F.S. for the *original* problem! The artificial variable has served its purpose. For succeeding tableaux, we shall delete the A-column (since we want to solve the original problem) and change Z^*'s to Z's (since $Z^* = Z$ for $A = 0$). From Tableaux II, the entering variable is x, the departing variable is s_1 and the key number, 2, is circled. Proceeding (omitting the A-column), we get Tableau III.

Simplex Tableau III

$C_j \rightarrow$ ↓	*Basic Variable*	*Solution*	2 x	1 y	0 s_1	0 s_2
2	x	1	1	0	1/2	1/2
1	y	5	0	1	1/2	– 1/2
	Z_j	7	2	1	3/2	1/2
	$C_j - Z_j$		0	0	– 3/2	– 1/2

Since all the entries in $C_j - Z_j$ row are negative or zero, the optimum solution has been obtained. The maximum value of Z is 7 and it occurs when $x = 1$ and $y = 5$.

Remark. The above example involved only one "≥" constraint. If a linear programming problem has more than one, each would be handled in just the same way. Thus each such constraint would be given an artificial variable to serve as its initial basic variable and the same coefficient $-M$ would be assigned to each of these variables in the objective function. The approach to other kinds of constraints requiring artificial variables is completely analogous. For example, if a problem involves an equality constraint of the form

$$a_1 x_1 + a_2 x_2 + ... + a_n x_n = b,$$

where $b \geq 0$, then an artificial variable A is inserted as follows :

$$a_1 x_1 + a_2 x_2 + ... + a_n x_n + A = b$$

***Example* 27.** Maximize $\quad Z = -3x + 2y$

subject to the constraints:

$$x - y \leq 4 \quad ...(1)$$

$$-x + y = 4 \quad ...(2)$$

$$x \geq 6 \quad ...(3)$$

$$x, y \geq 0$$

Solution. Since constraints (2) and (3) involve "=" and "≥", two artificial variables will occur. The equations for the artificial problem are

$$x - y + s_1 = 4$$

$$-x + y + A_1 = 4$$

$$x - s_2 + A_2 = 6$$

where s_1 and s_2 are, respectively, slack and surplus variables and A_1 and A_2 are artificial variables. The artificial objective function is

$$Z^* = Z - MA_1 - MA_2 = -3x + 2y - MA_1 - MA_2,$$

where M is a large positive number. The simplex method is applied to this artificial problem until we obtain a B.F.S. in which $A_1 = A_2 = 0$. We then delete the artificial variable columns, change the Z^*'s to Z's and continue by using the procedures of the previous section. An obvious initial B.F.S. to the artificial problem is :

$$x = y = s_2 = 0,\ s_1 = 4,\ A_1 = 4 \text{ and } A_2 = 6$$

The initial simplex tableau corresponding to this solution is shown.

Simplex Tableau I

$C_j \rightarrow$	*Basic*	*Solution*	-3	2	0	0	$-M$	$-M$	*Ratio*
$\downarrow$	*Variable*		x	y	s_1	s_2	A_1	A_2	
0	s_1	4	1	-1	1	0	0	0	–
$-M$	A_1	4	-1	(1)	0	0	1	0	$\leftarrow 4$
$-M$	A_2	6	1	0	0	-1	0	1	–
	Z_j^*	$-10M$	0	$-M$	0	M	$-M$	$-M$	
	$C_j - Z_j^*$		-3	$2+M\uparrow$	0	$-M$	0	0	

Since M is a large positive number, $2 + M$ is the largest positive entry. The entering variable is y, the departing variable is A_1 and the key number is circled. Replacing key- and non-key rows using technique discussed in Section 4.3, we get Tableau II.

Simplex Tableau II

$C_j \rightarrow$	*Basic*	*Solution*	-3	2	0	0	$-M$	$-M$	*Ratio*
$\downarrow$	*Variable*		x	y	s_1	s_2	A_1	A_2	
0	s_1	8	0	0	1	0	1	0	–
2	y	4	-1	1	0	0	1	0	–
$-M$	A_2	6	(1)	0	0	-1	0	1	$\leftarrow 6$
	Z_j^*	$8-6M$	$-2-M$	2	0	M	2	$-M$	
	$C_j - Z_j^*$		$-1+M\uparrow$	0	0	$-M$	$-2-M$	0	

Since there is a positive $C_j - Z_j^*$ entry, we continue. The entering variable is x, the departing variable is A_2 and the key number is circled. Replacing key- and non-key rows using technique discussed in Section 4.3, we obtain Tableau III.

Simplex Tableau III

$C_j \rightarrow$	Basic	Solution	-3	2	0	0	$-M$	$-M$
$\downarrow$	Variable		x	y	s_1	s_2	A_1	A_2
0	s_1	8	0	0	1	0	1	0
2	y	10	0	1	0	-1	1	1
-3	x	6	1	0	0	-1	0	1
	Z_j^*	2	-3	2	0	1	2	-1
	$C_j - Z_j^*$		0	0	0	-1	$-2-M$	$1-M$

The B.F.S. corresponding to Tableau III has both artificial variables equal to 0. Thus the A_1- and A_2- columns are no longer needed. However, a look at the $C_j - Z_j^*$ of last tableau shows that an optimum solution has been obtained. Since $Z^* = Z$ when $A_1 = A_2 = 0$, the maximum value of Z is 2 and it occurs when $x = 6$ and $y = 10$.

***Example* 28.** Solve the following linear programming problem by using simplex method :

Maximize $Z = 8x + 12y$

subject to the constraints:

$$x + y = 5$$
$$x \geq 2$$
$$y \leq 4$$
$$x, y \geq 0$$

[*Delhi Univ. B.Com. (H) 2003, 2014*]

Solution. Introducing a surplus variable s_1, a slack variable s_2 and two artificial variables A_1 and A_2, the problem can be restated as

Maximize $Z = 8x + 12y + 0s_1 + 0s_2 - MA_1 - MA_2$

subject to the constraints:

$$x + y + A_1 = 5$$
$$x - s_1 + A_2 = 2$$
$$y + s_2 = 4$$
$$x, y, s_1, s_2, A_1, A_2 \geq 0$$

where M is a large positive number. An initial basic feasible solution is :

$$x = y = s_1 = 0,\ A_1 = 5,\ A_2 = 2 \text{ and } s_2 = 4$$

Starting with this solution, we obtain a sequence of simplex tableaux that leads to the final solution.

C_j	Basic Variable	Solution	8 x	12 y	0 s_1	0 s_2	$-M$ A_1	$-M$ A_2	Ratio
$-M$	A_1	5	1	1	0	0	1	0	5
$-M$	A_2	2	(1)	0	-1	0	0	1	$\leftarrow 2$
0	s_2	4	0	1	0	1	0	0	–
	Z_j	$-7M$	$-2M$	$-M$	M	0	$-M$	$-M$	
	$C_j - Z_j$		$8+2M\uparrow$	$12+M$	$-M$	0	0	0	
$-M$	A_1	3	0	(1)	1	0	1	-1	$\leftarrow 3$
8	x	2	1	0	-1	0	0	1	–
0	s_2	4	0	1	0	1	0	0	4
	Z_j	$16-3M$	8	$-M$	$-8-M$	0	$-M$	$8+M$	
	$C_j - Z_j$		0	$12+M\uparrow$	$8+M$	0	0	$-8-2M$	
12	y	3	0	1	1	0	1	-1	
8	x	2	1	0	-1	0	0	1	
0	s_2	1	0	0	-1	1	-1	1	
	Z_j	52	8	12	4	0	12	-4	
	$C_j - Z_j$		0	0	-4	0	$-M-12$	$-M+4$	

A look at the $C_j - Z_j$ row of last tableau shows that it is impossible to further increase the objective function. Hence the optimum solution is : $x = 2,\ y = 3$ with max $Z = 52$.

***Example* 29.** Solve the following linear programming problem by using simplex method :

Maximize $\quad Z = x_1 + 3x_2 - 2x_3$

subject to the constraints:

$$-x_1 - 2x_2 - 2x_3 = -6 \quad ...(1)$$
$$-x_1 - x_2 + x_3 \le -2 \quad ...(2)$$
$$x_1, x_2, x_3 \ge 0$$

Solution. The constants on the right side of each of the constraints (1) and (2) are negative. Thus multiplying both sides of each constraint by – 1, we obtain

$$x_1 + 2x_2 + 2x_3 = 6 \quad ...(3)$$
$$x_1 + x_2 - x_3 \ge 2 \quad ...(4)$$

Introducing two artificial variables A_1 and A_2, the equations for the artificial problem are :

$$x_1 + 2x_2 + 2x_3 + A_1 = 6$$
$$x_1 + x_2 - x_3 - s + A_2 = 2,$$

where s is a surplus variable. The artificial objective function is :

$$Z^* = Z - MA_1 - MA_2 = x_1 + 3x_2 - 2x_3 - MA_1 - MA_2,$$

where M is a large positive number. The simplex tableaux appear below :

Simplex Tableau I

$C_j \rightarrow$	Basic	Solution	1	3	−2	0	−M	−M	Ratio
↓	Variable		x_1	x_2	x_3	s	A_1	A_2	
$-M$	A_1	6	1	2	2	0	1	0	3
$-M$	A_2	2	1	①	−1	−1	0	1	←2
	Z_j^*	$-8M$	$-2M$	$-3M$	$-M$	M	$-M$	$-M$	
	$C_j - Z_j^*$		$1+2M$	$3+3M$↑	$-2+M$	$-M$	0	0	

Since M is a large positive number, 3 + 3M is the largest positive entry. Thus the entring variable is x_2 . From the ratio coloumn we choose A_2 as the departing variable. The key number, 1, is circled. Proceeding, we get Tableau II.

Simplex Tableau II

$C_j \rightarrow$	Basic	Solution	1	3	−2	0	−M	−M	Ratio
↓	Variable		x_1	x_2	x_3	s	A_1	A_2	
$-M$	A_1	2	−1	0	④	2	1	−2	←1/2
3	x_2	2	1	1	−1	−1	0	1	–
	Z_j^*	$6-2M$	$3+M$	3	$-3-4M$	$-3-2M$	$-M$	$3+2M$	
	$C_j - Z_j^*$		$-2-M$	0	$1+4M$↑	$3+2M$	0	$-3-3M$	

The entring variable is x_3 . From the ratio coloumn we choose A_1 as the departing variable. The key number, 4, is circled. Proceeding, we get Tableau III.

Simplex Tableau III

$C_j \rightarrow$	Basic	Solution	1	3	−2	0	−M	−M	Ratio
↓	Variable		x_1	x_2	x_3	s	A_1	A_2	
−2	x_3	1/2	−1/4	0	1	(1/2)	1/4	−1/2	←1
3	x_2	5/2	3/4	1	0	−1/2	1/4	1/2	–
	Z_j^*	13/2	11/4	3	−2	−5/2	1/4	5/2	
	$C_j - Z_j^*$		−7/4	0	0	5/2↑	$(-1/4)-M$	$(-5/2)-M$	

The B.F.S. corresponding to Tableau III has both artificial variables equal to 0. We can thus delete A_1- and A_2- columns and change the Z^*'s to Z's. Continuing, we obtain simplex Tableau IV.

Simplex Tableau IV

$C_j \rightarrow$	Basic	Solution	1	3	−2	0
↓	*Variable*		x_1	x_2	x_3	s
0	s	1	−1/2	0	2	1
3	x_2	3	1/2	1	1	0
	Z_j	9	3/2	3	3	0
	$C_j - Z_j$		−1/2	0	−5	0

Since all the entries in $C_j - Z_j$ row are negative or zero, we have reached the final tableau. The maximum value of Z is 9 and it occurs when $x_1 = 0$, $x_2 = 3$ and $x_3 = 0$.

***Example* 30.** Maximize $Z = 2x_1 + 3x_2 + x_3$
subject to the constraints:

$$4x_1 + 3x_2 + x_3 = 6$$
$$x_1 + 2x_2 + 5x_3 = 4$$
$$x_1, x_2, x_3 \geq 0$$

[*Delhi Univ. B.Com. (H) 2005*]

Solution. Introducting artificial variables A_1 and A_2, the given problem can be restated as follows :

Maximize $Z = 2x_1 + 3x_2 + x_3 - MA_1 - MA_2$
subject to the constraints:

$$4x_1 + 3x_2 + x_3 + A_1 = 6$$
$$x_1 + 2x_2 + 5x_3 + A_2 = 4$$
$$x_1, x_2, x_3, A_1, A_2 \geq 0$$

where M is a large positive number. An initial basic feasible solution is :

$$x_1 = x_2 = x_3 = 0, \; A_1 = 6 \text{ and } A_2 = 4$$

Starting with this solution, we obtain a sequence of simplex tableaux that leads to the final solution.

C_j	Basic	Solution	2	3	1	−M	−M	*Ratio*
	Variable		x_1	x_2	x_3	A_1	A_2	
−M	A_1	6	4	3	1	1	0	6
−M	A_2	4	1	2	(5)	0	1	← 4/5
	Z_j	−10M	−5M	−5M	−6M	−M	−M	
	$C_j - Z_j$		2 + 5M	3 + 5M	1 + 6M ↑	0	0	

$-M$	A_1	26/5	(19/5)	13/5	0	1	–1/5	← 26/19
1	x_3	4/5	1/5	2/5	1	0	1/5	4
	Z_j	$\frac{4}{5}-\frac{26}{5}M$	$\frac{1}{5}-\frac{19}{5}M$	$\frac{2}{5}-\frac{13}{5}M$	1	$-M$	$\frac{1}{5}+\frac{1}{5}M$	
	C_j-Z_j		$\frac{9}{5}+\frac{19}{5}M\uparrow$	$\frac{13}{5}+\frac{13}{5}M$	0	0	$-\frac{1}{5}-\frac{6}{5}M$	
2	x_1	26/19	1	(13/19)	0	5/19	– 1/19	←2
1	x_3	10/19	0	5/19	1	–1/19	4/19	2
	Z_j	62/19	2	31/19	1	9/19	2/19	
	C_j-Z_j		0	26/19 ↑	0	$-\frac{9}{19}-M$	$-\frac{2}{19}-M$	
3	x_2	2	19/13	1	0	5/13	– 1/13	
1	x_3	0	– 5/13	0	1	– 2/13	3/13	
	Z_j	6	4	3	1	1	0	
	C_j-Z_j		–2	0	0	$-1-M$	$-M$	

Since M is a large positive number, all $C_j - Z_j$ entries in the last Table are zero or negative. Thus the optimum solution has been obtained and the maximum value of Z is 6 which occurs when $x_1 = 0$, $x_2 = 2$ and $x_3 = 0$.

Remark. It is possible for a linear programming problem involving artificial variables that the simplex procedure terminates and not all artificial variables are 0. It can be shown that in this situation the *feasible region of the original problem is empty* and hence there is *no optimum solution.* To illustrate, let us consider the following example :

***Example* 31.** Maximize $Z = 3x + 2y$
subject to the constraints:

$$-x + y \geq 3 \qquad \text{...(1)}$$
$$x + 2y \leq 2 \qquad \text{...(2)}$$
$$x, y \geq 0$$

Solution. Since constraint (1) contains the symbol "≥" and the constant on the right side is nonnegative, its equation will involve an artificial variable. The equations to consider are :

$$-x + y - s_1 + A_1 = 3$$
$$x + 2y + s_2 = 2,$$

where s_1, s_2 and A_1 are, respectively, surplus, slack and artificial variables. The artificial objective function is :

$$Z^* = Z - MA_1 = 3x + 2y - MA_1$$

The simplex tableaux appear below.

Simplex Tableau I

$C_j \rightarrow$	Basic	Solution	3	2	0	0	$-M$	Ratio
$\downarrow$	Variable		x	y	s_1	s_2	A_1	
$-M$	A_1	3	-1	1	-1	0	1	3
0	s_2	2	1	(2)	0	1	0	$\leftarrow 1$
	Z_j^*	$-3M$	M	$-M$	M	0	$-M$	
	$C_j - Z_j^*$		$3-M$	$2+M\uparrow$	$-M$	0	0	

Simplex Tableau II

$C_j \rightarrow$	Basic	Solution	3	2	0	0	$-M$
$\downarrow$	Variable		x	y	s_1	s_2	A_1
$-M$	A_1	2	$-3/2$	0	-1	$-1/2$	1
2	y	1	1/2	1	0	1/2	0
	Z_j^*	$2-2M$	$1+(3/2)M$	2	M	$1+(1/2)M$	$-M$
	$C_j - Z_j^*$		$2-(3/2)M$	0	$-M$	$-1-(1/2)M$	0

Since M is a large positive number, the entries in $C_j - Z_j^*$ row of simplex tableau II are negative or 0, so the simplex procedure terminates. The value of the artificial variable A_1 is 2. Therefore, as previously stated, the feasible region of the original problem is empty and hence no optimum solution exists.

***Example* 32.** Given the following initial simplex table :

$C_j \rightarrow$			30	20	0	0	$-M$	$-M$
$\downarrow$	*BV*	*Quantity*	x_1	x_2	s_1	s_2	A_1	A_2
0	s_1	8	1	1	1	0	0	0
$-M$	A_1	12	6	4	0	-1	1	0
$-M$	A_2	20	5	8	0	0	0	1
	Z_j	$-32M$	$-11M$	$-12M$	0	M	$-M$	$-M$
	$C_j - Z_j$		$30+11M$	$20+12M$	0	$-M$	0	0

Write down the original problem represented by the above table. Find the optimum solution.

[*Delhi Univ. B.Com. (H) 2010*]

Solution. The original problem is:

$$\begin{aligned} \text{Maximize} \quad & Z = 30x_1 + 20x_2 \\ \text{subject to the constraints:} \quad & x_1 + x_2 \le 8 \\ & 6x_1 + 4x_2 \ge 12 \\ & 5x_1 + 8x_2 = 20 \\ & x_1, x_2 \ge 0 \end{aligned}$$

The complete set of simplex tableaux that leads to the optimum solution is shown as follows:

C_j	*BV*	*Quantity*	30	20	0	0	$-M$	$-M$	*Ratio*
			x_1	x_2	s_1	s_2	A_1	A_2	
0	s_1	8	1	1	1	0	0	0	8
$-M$	A_1	12	6	4	0	-1	1	0	3
$-M$	A_2	20	5	(8)	0	0	0	1	← 5/2
	Z_j	$-32M$	$-11M$	$-12M$	0	M	$-M$	$-M$	
	$C_j - Z_j$		$30+11M$	$20+12M$ ↑	0	$-M$	0	0	
0	s_1	11/2	3/8	0	1	0	0	$-1/8$	88/6
$-M$	A_1	2	(7/2)	0	0	-1	1	$-1/2$	← 4/7
20	x_2	5/2	5/8	1	0	0	0	1/8	4
	Z_j	$50-2M$	$25/2-7M/2$	20	0	M	$-M$	$M/2+5/2$	
	$C_j - Z_j$		$35/2+7M/2$ ↑	0	0	$-M$	0	$-3M/2-5/2$	
0	s_1	37/7	0	0	1	3/28	$-3/28$	$-1/14$	148/3
30	x_1	4/7	1	0	0	$-2/7$	2/7	$-1/7$	–
20	x_2	15/7	0	1	0	(5/28)	$-5/28$	3/14	← 12
	Z_j	60	30	20	0	$-35/7$	35/7	0	
	$C_j - Z_j$		0	0	0	35/7 ↑	$-M-35/7$	$-M$	
0	s_1	4	0	$-3/5$	1	0	0	$-1/5$	
30	x_1	4	1	8/5	0	0	0	1/5	
0	s_2	12	0	28/5	0	1	-1	6/5	
	Z_j	120	30	48	0	0	0	6	
	$C_j - Z_j$		0	-28	0	0	$-M$	$-M-6$	

Since all entries in $C_j - Z_j$ row are negative or zero, optimal solution has been obtained, and the maximum value of Z is 120 and it occurs when $x_1 = 4$ and $x_2 = 0$.

EXERCISE 4.2

Use the simplex method to solve the following problems :

1. Maximize $Z = 3x + 4y$
subject to $x + 2y \le 8$
$x + 6y \ge 12$
$x, y \ge 0$

2. Maximize $Z = 20x + 15y$
subject to $x + y \ge 7$
$9x + 5y \le 45$
$2x + y \ge 8$
$x, y \ge 0$

3. Maximize $Z = 3x + 4y$
subject to $x + y \le 12$
$5x + 2y \ge 36$
$7x + 4y \ge 14$
$x, y \ge 0$

4. Maximize $Z = x + 2y$
subject to $x + y \le 12$
$x + y \ge 5$
$x \le 10$
$x, y \ge 0$

5. Maximize $Z = 3x_1 - 2x_2 + x_3$
subject to $x_1 + x_2 + x_3 \le 1$
$x_1 - x_2 + x_3 \ge 2$
$x_1 - x_2 - x_3 \le -6$
$x_1, x_2, x_3 \ge 0$

6. Maximize $Z = 3x_1 + 2x_2 - x_3$
subject to $x_1 + 3x_2 + x_3 \le 9$
$2x_1 + 3x_2 - x_3 \ge 2$
$3x_1 - 2x_2 + x_3 \ge 5$
$x_1, x_2, x_3 \ge 0$

7. Maximize $Z = 4x_1 + x_2 + 2x_3$
subject to $2x_1 + x_2 + 3x_3 \le 10$
$x_1 - x_2 + x_3 = 4$
$x_1, x_2, x_3 \ge 0$

8. Maximize $Z = 3x_1 + 2x_2 - x_3$
subject to $2x_1 - x_2 - x_3 \le 2$
$x_1 + 2x_2 + x_3 \ge 2$
$x_1 - 3x_2 - 2x_3 \le -5$
$x_1, x_2, x_3 \ge 0$

9. Maximize $Z = 2x_1 + x_2 - x_3$
subject to $x_1 + 2x_2 + x_3 \le 5$
$-x_1 + x_2 + x_3 \ge 1$
$x_1, x_2, x_3 \ge 0$

10. Maximize $Z = 3x_1 + 5x_2 + 2x_3$
subject to $-2x_1 + x_3 \ge 2$
$x_1 + 4x_2 + 2x_3 = 5$
$x_1, x_2, x_3 \ge 0$

11. Maximize $Z = 10x_1 + 15x_2 + 12x_3$
subject to $5x_1 + 3x_2 + x_3 \le 9$
$-5x_1 + 6x_2 + 15x_3 \le 15$
$2x_1 + x_2 + x_3 \ge 5$
$x_1, x_2, x_3 \ge 0$

12. Maximize $Z = x_1 + 3x_2 + 2x_3$
subject to $x_1 + x_2 + 4x_3 \ge 6$
$2x_1 + x_2 + 3x_3 \le 4$
$x_1, x_2, x_3 \ge 0$

13. An advertising agency wishes to reach two types of audiences. Customers with annual incomes greater than ₹ 40,000 (target audience *A*) and customers with annual incomes of less than ₹ 40,000 (target audience *B*). The total advertising budget is ₹ 2,00,000. One programme of TV advertising costs ₹ 50,000, one programme of radio advertising costs ₹ 20,000. For contract reasons, at least 3 programmes ought to be on TV and the number of radio programmes must not exceed 5. Survey indicates that a single TV programme reaches 4,50,000 customers in target audience *A* and 50,000 in target audience *B*. One radio programme reaches 20,000 in target audience *A* and 80,000 in target audience *B*. Determine the media mix to maximize the total reach.

14. XYZ Company during the festival season combines two items *A* and *B* to form gift packs. Each pack must weigh 5 kg and should contain at least 2 kg of *A* and not more than 4 kg of *B*. The net contribution to the company is ₹ 10 per kg of *A* and ₹ 12 per kg of *B*. The company wants to determine the optimum mix. Formulate this as a linear programming problem to maximize net contribution per pack and solve the same by using simplex method.

[*Delhi Univ. B.Com. (H) 1996*]

15. Given below is a table appearing in the course of the simplex solution of a maximization problem :

$C_j \rightarrow$	Basic	Solution	2	4	1	0	0	0	$-M$
$\downarrow$	Variable		x_1	x_2	x_3	s_1	s_2	s_3	A_1
2	x_1	4	1	2	1/2	0	0	7/8	0
0	s_2	12	0	0	–1	0	1	–1/2	0
0	s_1	12	0	6	0	1	0	1	–1
	Z_j	8	2	4	1	0	0	7/4	0
	$C_j - Z_j$		0	0	0	0	0	–7/4	$-M$

Give short answers to the following questions, briefly stating the reasons :

(i) Is the solution indicated above optimal ?

(ii) What is the optimal product-mix and the total maximum profit ?

(iii) Is the above solution degenerate ?

(iv) Does the problem have multiple optimal solutions ?

(v) Write alternative solution, if any.

(vi) What is the objective function of the problem ?

ANSWERS

1. $x = 6,\ y = 1$ and max $Z = 22$

2. $x = 0,\ y = 9$ and max $Z = 135$

3. $x = 3,\ y = 8$ and max $Z = 44$

4. $x = 0,\ y = 12$ and max $Z = 24$

5. No optimum solution

6. $x_1 = 9,\ x_2 = 0,\ x_3 = 0$ and max $Z = 27$

7. $x_1 = 14/3,\ x_2 = 2/3,\ x_3 = 0$ and max $Z = 58/3$

8. Unbounded solution

9. $x_1 = 1,\ x_2 = 2,\ x_3 = 0$ and max $Z = 4$

10. $x_1 = 1/5,\ x_2 = 0,\ x_3 = 12/5$ and max $Z = 27/5$

11. No optimum solution

12. No optimum solution

13. $x = 4,\ y = 0$ and max $Z = 20{,}00{,}000$

14. $x = 2,\ y = 3$ and max $Z = 56$.

15. *(i)* Yes, it is an optimal solution because all the entries in $C_j - Z_j$ row are negative or zero.

(ii) The optimal product mix is : $x_1 = 4, x_2 = 0$ and $x_3 = 0$. The total maximum profit is : $Z = 8$.

(iii) No, the above solution is not degenerate because none of the basic variables has a zero value.

(iv) Yes, it does have multiple optimum solutions because non-basic variables, namely x_2 and x_3, have zero $C_j - Z_j$ entries in the final simplex table.

(v) Alternate solutions are :

(a) $x_1 = 0,\ x_2 = 2,\ x_3 = 0$ with max $Z = 8$

(b) $x_1 = 0,\ x_2 = 0,\ x_3 = 8$ with max $Z = 8$

(vi) Objective function of the problem is : Maximize $Z = 2x_1 + 4x_2 + x_3$.

4.6 THE SIMPLEX METHOD—A MINIMIZATION CASE

So far in this chapter, we have used the simplex method to maximum linear programming problems, in which the optimum solution yields the largest possible value for the objective function. In this section we will illustrate the simplex method for solving a minimization problem. Just as we did in the maximization case, we now outline the steps used in the simplex method for solving a minimization problem.

Steps of the Simplex Method (Minimization Case)

***Step* 1.** *Convert the inequality constraints into equality constraints by introducing slack, surplus, and artificial variables as needed. Each of the slack/surplus variables will have a coefficient of zero in the objective function, whereas each of the artificial variables will have a coefficient of + M in the objective function, M being a very large number. This is just to keep them out of the final solution.*

***Step* 2.** *Set up the initial table. The initial basic feasible variables are the slack or artificial variables associated with each constraint (all decision variables are zero).*

***Step* 3.** *Test the solution for optimality by calculating the values of $C_j - Z_j$ row. The solution is optimal if all $C_j - Z_j$ values are nonnegative.*

***Step* 4.** *If the solution is not optimal, then identify the variable to enter the basis. Select the variable with the largest negative $C_j - Z_j$ value as the entering variable. The departing variable is determined in exactly the same manner as when maximizing.*

***Step* 5.** *Revise the solution to develop a new table in exactly the same fashion as discussed in the maximization case.*

***Step* 6.** *Go to step 3 and repeat the procedure until all $C_j - Z_j$ values are nonnegative.*

The following example will serve to demonstrate the simplex method when minimizing.

***Example* 33.** Use simplex method to solve the following *LPP.*

Minimize $\quad Z = 8x + 12y$

subject to the constraints: $\quad 2x + 2y \geq 1$

$$x + 3y \geq 2$$

$$x,\ y \geq 0$$

Solution. Introducing two surplus variables s_1, s_2 and two artificial variables A_1, A_2, the problem can be restated as :

Minimize $Z = 8x + 12y + 0s_1 + 0s_2 + MA_1 + MA_2$

subject to the constraints: $2x + 2y - s_1 + A_1 = 1$

$$x + 3y - s_2 + A_2 = 2$$

$$x, y, s_1, s_2, A_1, A_2 \geq 0$$

where M is a large positive number. Note that the artificial variables have $+ M$ as objective function coefficients to ensure that they are driven out of the final solution.

An initial basic feasible solution is : $x = 0,\ y = 0,\ s_1 = 0,\ s_2 = 0,\ A_1 = 1$ and $A_2 = 2$

The initial simplex table for this problem is shown in simplex Tableau I.

Simplex Tableau I

$C_j \rightarrow$	Basic	Solution	8	12	0	0	M	M	Ratio
$\downarrow$	Variable		x	y	s_1	s_2	A_1	A_2	
M	A_1	1	2	(2)	-1	0	1	0	$\leftarrow 1/2$
M	A_2	2	1	3	0	-1	0	1	2/3
	Z_j	$3M$	$3M$	$5M$	$-M$	$-M$	M	M	
	$C_j - Z_j$		$8 - 3M$	$12 - 5M \uparrow$	M	M	0	0	

The simplex criterion for optimality shows that the solution is not optimal, since both x and y have negative $C_j - Z_j$ values. The largest negative, y, is chosen for the entering variable and the quotients to the right show A_1 is outgoing. (Continue the rule of smallest non-negative value as designating the outgoing variable). After revision, a second solution is generated in Simplex Tableau II.

Simplex Tableau II

$C_j \rightarrow$	Basic	Solution	8	12	0	0	M	M	Ratio
$\downarrow$	Variable		x	y	s_1	s_2	A_1	A_2	
12	y	1/2	1	1	$-1/2$	0	1/2	0	–
M	A_2	1/2	-2	0	(3/2)	-1	$-3/2$	1	$\leftarrow 1/3$
	Z_j	$6 + M/2$	$12 - 2M$	12	$-6 + 3M/2$	$-M$	$6 - 3M/2$	M	
	$C_j - Z_j$		$-4 + 2M$	0	$6 - 3M/2 \uparrow$	M	$-6 + 5M/2$	0	

Variable y has replaced A_1. Again, this solution is not optimal, since s_1 has negative $C_j - Z_j$ value. As shown, s_1 will be the incoming variable and A_2 the outgoing variable. The third solution is shown in Simplex Tableau III.

Simplex Tableau III

$C_j \rightarrow$	*Basic*	*Solution*	8	12	0	0	M	M
↓	*Variable*		x	y	s_1	s_2	A_1	A_2
12	y	2/3	1/3	1	0	– 1/3	0	1/3
0	s_1	1/3	– 4/3	0	1	– 2/3	– 1	2/3
	Z_j	8	4	12	0	– 4	0	4
	$C_j - Z_j$		4	0	0	4	M	$M-4$

The *BFS* corresponding to Tableau III has both artificial variables equal to 0. Thus the A_1- and A_2-columns are no longer needed. However, the entries in $C_j - Z_j$ row are nonnegative and hence an optimum solution has been reached. The solution is $x = 0$, $y = 2/3$ with min $Z = 8$.

***Example* 34.** The standard weight of a special purpose brick is 5 kg and it contains two basic ingredients B_1 and B_2. B_1 costs ₹ 5 per kg and B_2 costs ₹ 8 per kg. Strength considerations dictate that the brick contains not more than 4 kg of B_1, and minimum of 2 kg of B_2. Since the demand for the product is likely to be related to the price of the brick, find the minimum cost of brick satisfying the above conditions.

[*Delhi Univ. B.Com. (H) 1997, 2007 (SOL)*]

Solution. Let x kg of ingredient B_1 and y kg of ingredient B_2 be used in the brick. Then the appropriate mathematical formulation of the problem is :

Minimize $\quad Z = 5x + 8y$

subject to the constraints:

$$x + y = 5$$
$$x \leq 4$$
$$y \geq 2$$
$$x, y \geq 0$$

To solve it, we convert each of the constraints into equations. Thus introducing slack, surplus and artificial variables, the problem can be restated as :

Minimize $\quad Z = 5x + 8y + 0s_1 + 0s_2 + MA_1 + MA_2$

subject to the constraints:

$$x + y + A_1 = 5$$
$$x + s_1 = 4$$
$$y - s_2 + A_2 = 2$$
$$x, y, s_1, s_2, A_1, A_2 \geq 0$$

where s_1 and s_2 are, respectively, slack and surplus variables and A_1 and A_2 are artificial variables. An initial basic feasible solution to the problem is :

$$x = 0, \; y = 0, \; s_2 = 0, \; A_1 = 5, \; s_1 = 4 \text{ and } A_2 = 2$$

The initial simplex tableau corresponding to this solution is shown below :

Simplex Tableau I

$C_j \rightarrow$	Basic	Solution	5	8	0	0	M	M	Ratio
$\downarrow$	Variable		x	y	s_1	s_2	A_1	A_2	
M	A_1	5	1	1	0	0	1	0	5
0	s_1	4	1	0	1	0	0	0	–
M	A_2	2	0	(1)	0	-1	0	1	$\leftarrow 2$
	Z_j	$7M$	M	$2M$	0	$-M$	M	M	
	$C_j - Z_j$		$5-M$	$8-2M\uparrow$	0	M	0	0	

A look at the $C_j - Z_j$ row reveals that y is the entering variable. The smallest nonnegative quotient in the ratio column is 2, so A_2 is the departing variable and the key number, 1, is circled. Replacing key- and non-key rows using the technique of Section 4.4, we get Tableau II.

Simplex Tableau II

$C_j \rightarrow$	Basic	Solution	5	8	0	0	M	M	Ratio
$\downarrow$	Variable		x	y	s_1	s_2	A_1	A_2	
M	A_1	3	(1)	0	0	1	1	-1	$\leftarrow 3$
0	s_1	4	1	0	1	0	0	0	4
8	y	2	0	1	0	-1	0	1	–
	Z_j	$16+3M$	M	8	0	$-8+M$	M	$8-M$	
	$C_j - Z_j$		$5-M\uparrow$	0	0	$8-M$	0	$-8+2M$	

Since there are negative values in $C_j - Z_j$ row, this solution is not optimal and we continue. Proceeding, we obtain Tableau III.

Simplex Tableau III

$C_j \rightarrow$	Basic	Solution	5	8	0	0	M	M
$\downarrow$	Variable		x	y	s_1	s_2	A_1	A_2
5	x	3	1	0	0	1	1	-1
0	s_1	1	0	0	1	-1	-1	1
8	y	2	0	1	0	-1	0	1
	Z_j	31	5	8	0	-3	5	3
	$C_j - Z_j$		0	0	0	3	$M-5$	$M-3$

Since M is a large positive number, all values in $C_j - Z_j$ row are now zero or positive. Thus the optimal solution is obtained. The solution is $x = 3, y = 2$ with minimum $Z = 31$.

***Example* 35.** Given the following initial simplex table of minimization problem:

BV	X_1	X_2	S_1	S_2	A_1	A_2	*Quantity*
A_1	3	1	-1	0	1	0	27
A_2	1	1	0	0	0	1	21
S_2	1	2	0	1	0	0	40
$C_j \rightarrow$	8	4	0	0	M	M	

(*i*) Write the corresponding linear programming problem.

(*ii*) Solve it by simplex method.

(*iii*) Is it a case of multiple solutions? If yes, also find the alternative solution.

[*Delhi Univ. B.Com. (H) 2018*]

Solution. (*i*) The original problem is:

Minimize $Z = 8x_1 + 4x_2$

Subject to

$$3x_1 + x_2 \geq 27$$
$$x_1 + x_2 = 21$$
$$x_1 + 2x_2 \leq 40$$
$$x_1, x_2 \geq 0$$

(*ii*) The complete set of simplex tableaux that leads to the optimum solution is shown as follows:

C_j	*BV*	*Quantity*	8 x_1	4 x_2	0 s_1	0 s_2	M A_1	M A_2	Ratio
M	A_1	27	(3)	1	–1	0	1	0	9
M	A_2	21	1	1	0	0	0	1	21
O	s_2	40	1	2	0	1	0	0	40
	Z_j	$48M$	$4M$	$2M$	$-M$	0	M	M	
	$C_j - Z_j$		$8 - 4M\uparrow$	$4 - 2M$	M	0	0	0	
8	x_1	9	1	1/3	–1/3	0	1/3	0	27
M	A_2	12	0	(2/3)	1/3	0	–1/3	1	18
O	s_2	31	0	5/3	1/3	1	–1/3	0	93/5

	Z_j	$72 + 12M$	8	$\frac{8}{3}+\frac{2}{3}M$	$-\frac{8}{3}+\frac{1}{3}M$	0	$\frac{8}{3}-\frac{1}{3}M$	M
	$C_j - Z_j$		0	$-\frac{4}{3}-\frac{2}{3}M \uparrow$	$\frac{8}{3}-\frac{1}{3}M$	0	$-\frac{8}{3}+\frac{4}{3}M$	0
8	x_1	3	1	0	–1/2	0	1/2	–1/2
4	x_2	18	0	1	–1/2	0	–1/2	3/2
0	s_2	14/5	1	0	–2/5	–1/5	2/5	0
	Z_j	96	8	4	–2	0	2	2
	$C_j -$	Z_j	0	0	2	0	M–2	M–2

Since all entries in $C_j - Z_j$ row are zero or positive, optimal solution has been obtained, and the minimum value of Z is 96 and it occurs when $x_1 = 3$ and $x_2 = 18$.

(*iii*) No, it does not have multiple solutions because none of the non-basic variables has a zero $C_j - Z_j$ entry in the final simplex table.

EXERCISE 4.3

Use the simplex method to solve the following problems :

1. Minimize $Z = 3x + 6y$

subject to $-x + y \leq 6$

$x + y \geq 10$

$x, y \geq 0$

2. Minimize $Z = 4x + 2y$

subject to $x + 2y \geq 4$

$x + 4y \geq 6$

$x, y \geq 0$

3. Minimize $Z = 6x + 4y$

subject to $-x + y \leq 1$

$x + y \geq 3$

$x, y \geq 0$

4. Minimize $Z = 5x_1 + 4x_2 + 3x_3$

subject to $x_1 + x_2 + x_3 \geq 100$

$2x_1 + x_2 \geq 50$

$x_1, x_2, x_3 \geq 0$

5. Minimize $Z = x_1 - x_2 - 3x_3$

subject to $x_1 + 2x_2 + x_3 = 4$

$x_2 + x_3 = 1$

$x_1 + x_2 \leq 6$

$x_1, x_2, x_3 \geq 0$

6. Minimize $Z = x_1 + 8x_2 + 5x_3$

subject to $x_1 + x_2 + x_3 \geq 8$

$-x_1 + 2x_2 + x_3 \geq 2$

$x_1, x_2, x_3 \geq 0$

7. Minimize $Z = x_1 + x_2 + 2x_3$

subject to $-x_1 - x_2 + x_3 \leq 1$

$x_1 - x_2 + x_3 \geq 2$

$x_1, x_2, x_3 \geq 0$

8. Minimize $Z = 4x_1 + 4x_2 + 6x_3$

subject to $x_1 - x_2 - x_3 \leq 3$

$x_1 - x_2 + x_3 \geq 3$

$x_1, x_2, x_3 \geq 0$

9. Minimize $Z = x_1 + 2x_2 + x_3$

subject to $x_1 - x_2 - x_3 \leq -1$

$6x_1 + 3x_2 + 2x_3 = 12$

$x_1, x_2, x_3 \geq 0$

10. Minimize $Z = 2x_1 + 3x_2 + x_3$

subject to $x_1 + 4x_2 + 2x_3 \geq 8$

$3x_1 + 2x_2 \geq 6$

$x_1, x_2, x_3 \geq 0.$

11. The final product of a firm has a requirement that it must weigh exactly 150 kg. The two raw materials used in the manufacture of this product are *A* with a cost of ₹ 2 per unit and *B* with a cost of ₹ 8 per unit. Each unit of *A* weighs 5 kg and each unit of *B* weighs 10 kg. At least 14 units of *B* and not more than 20 unit of *A* must be used. How much of each type of raw material should be used for each unit of the final product if cost is to be minimized ? Use simplex method.

12. An animal feed company must produce 200 kg of a mixture consisting of ingredient X_1 and X_2 daily. X_1 costs ₹ 3 per kg and X_2 ₹ 8 per kg. No more than 80 kgs of X_1 can be used, and at least 60 kgs of X_2 must be used. Find how much of each ingredient should be used if the company wants to minimize cost. Use simplex method.

13. A diet for a sick person must contain at least 4000 units of vitamins, 50 units of minerals and 1400 units of calories. Two foods *A* and *B* are available at a cost of ₹ 4 and ₹ 3 per unit respectively. If one unit of *A* contains 200 units of vitamins, 1 unit of mineral and 40 units of calories and one unit of *B* contains 100 units of vitamins, 2 units of minerals and 40 units of calories, find by simplex method, what combination of food be used to have least cost ?

14. A carpenter has started a workshop in which he manufactures handcarts. Each cart consists of a frame and two wheels. *A* frame uses 3 hours and a wheel 2 hours of labour of which 90 hours per week are available. The carpenter wants that he should manufacture at least 10 carts during a week. The cost of frame is ₹ 500 and that of a wheel is ₹ 200. Formulate this as a linear programming model and using simplex method, determine an optimal production plan.

[*Delhi Univ. B.Com. (H) 2004*]

15. XYZ Chemical Corporation must produce exactly 1000kg of a special mixture of chemical *A* and chemical *B* for a customer. Chemical *A* costs ₹ 50 per kg and chemical *B* costs ₹ 60 per kg. No more than 300 kg of chemical *A* can be used and atleast 150 kg of chemical *B* must be used. Using simplex method determine the least cost blend of the two ingredients. **[*Delhi Univ. B.Com. (H) 2012*]**

ANSWERS

1. $x = 10,\ y = 0$ and min $Z = 30$

2. $x = 0,\ y = 2$ and min $Z = 4$

3. $x = 1,\ y = 2$ and min $Z = 14$

4. $x_1 = 0,\ x_2 = 50,\ x_3 = 50$ and min $Z = 350$

5. $x_1 = 3,\ x_2 = 0,\ x_3 = 1$ and min $Z = 0$

6. $x_1 = 3,\ x_2 = 0,\ x_3 = 5$ and min $Z = 28$

7. $x_1 = 2,\ x_2 = 0,\ x_3 = 0$ and min $Z = 2$

8. $x_1 = 3,\ x_2 = 0,\ x_3 = 0$ and min $Z = 12$

9. $x_1 = 5/4,\ x_2 = 0,\ x_3 = 9/4$ and min $Z = 7/2$

10. $x_1 = 4/5,\ x_2 = 9/5,\ x_3 = 0$ and min $Z = 7$

11. $x = 2,\ y = 14$ and min $Z = 116$

12. $x_1 = 80,\ x_2 = 120$ and min $Z = 1200$

13. $x = 5,\ y = 30$ and min $Z = 110$

14. 10 frames and 20 wheels with min cost ₹ 9000

15. 300 kg of chemical *A* and 700 kg of chemical *B* with min cost ₹ 57,000.

CHAPTER

5

DIFFERENTIATION

OBJECTIVES

After studying the material in this chapter, you should be able to:

- *Understand the meaning of the derivative of a function.*
- *Differentiate a function using the rules for differentiation.*
- *Use logarithmic differentiation to find derivatives of certain algebraic expressions.*
- *Find derivatives of functions defined parametrically.*
- *Determine higher-order derivatives of a function.*

5.1 THE DERIVATIVE

Let $y = f(x)$ be a function defined in an interval (a, b). Let $x = c$ be any point of this interval so that $f(c)$ is the corresponding value of the function. Let $c + h$ be any other point of this interval which lies to the right or left of c according as h is positive or negative. The corresponding value of the function is $f(c + h)$. Then $f(c + h) - f(c)$ is the change in the dependent variable y corresponding to the change h in the independent variable x. Consider the ratio

$$\frac{f(c+h) - f(c)}{h}$$

of these two changes which is a function of h and is not defined for $h = 0$, c being a fixed point.

***Definition*.** A function $y = f(x)$ is said to be **differentiable** at $x = c$ if

$$\lim_{h \to 0} \frac{f(c + h) - f(c)}{h}$$

exists and the limit is called the **derivative** of the function $f(x)$ at $x = c$. This limit is denoted by the symbol $f'(c)$. Other notations for $f'(c)$ are $\left(\dfrac{dy}{dx}\right)_{x=c}$ and $y_1(c)$.

***Note*.** It may be noted that the limit must be the same whether h tends to zero through positive or negative values.

A function $y = f(x)$ is said to be *differentiable on an interval* (a, b) if $f(x)$ is differentiable at each point of this interval.

***Example* 1.** Show that the function $f(x) = x^2$ is differentiable at $x = 2$ and obtain its derivative at $x = 2$.

***Solution*.** We have

$$f(2) = (2)^2 = 4$$

and

$$f(2+h) = (2+h)^2 = 4 + h^2 + 4h$$

$$\therefore \quad \frac{f(2+h) - f(2)}{h} = \frac{4 + h^2 + 4h - 4}{h} = h + 4$$

which approaches 4 as $h \to 0$.

Hence $f(x)$ is differentiable at $x = 2$ and its derivative at $x = 2$ is $f'(2) = 4$.

***Example* 2.** Show that the function $f(x) = |x|$ is not differentiable at $x = 0$.

***Solution*.** We have

$$f(0) = |0| = 0$$

and

$$f(0+h) = |h|$$

$$\frac{f(0+h) - f(0)}{h} = \frac{|h|}{h} = \begin{cases} 1, & \text{if } h > 0 \\ -1, & \text{if } h < 0 \end{cases}$$

$$\therefore \quad \frac{f(0+h) - f(0)}{h} = \frac{|h|}{h} = \begin{cases} 1, & \text{if } h > 0 \\ -1, & \text{if } h < 0 \end{cases}$$

Hence

$$\lim_{h \to 0^+} \frac{f(0+h) - f(0)}{h} = 1$$

and

$$\lim_{h \to 0^-} \frac{f(0+h) - f(0)}{h} = -1$$

Thus $\displaystyle\lim_{h \to 0} \frac{f(0+h) - f(0)}{h}$ does not exist and, therefore, f is not differentiable at $x = 0$.

Remark. We have defined the derivative of a function $f(x)$ at $x = c$, where c is a fixed constant. Of course, $f'(c)$ is also a constant. However, if instead of c, we use any number x, then the definition of the derivative has the form :

$$\lim_{h \to 0} \frac{f(x+h) - f(x)}{h}$$

In this case, the derivative depends on x. In other words, the derivative is a new function denoted by $f'(x)$. The process of finding the derivative is called *differentiation.*

Besides $f'(x)$, other notations for the derivative of $y = f(x)$ are :

$$\frac{dy}{dx}, \quad y', \quad y_1, \quad \frac{d}{dx}[f(x)] \quad \text{and} \quad D_x\,[f(x)].$$

Note **.** $\frac{dy}{dx}$ is a single symbol denoting derivative and is not to be regarded as the quotient of dy by dx which have not so far been defined as individual symbols.

Example **3.** Find the derivative of $f(x) = x^2$.

Solution **.** By definition,

$$\begin{aligned} f'(x) &= \lim_{h \to 0} \frac{f(x+h) - f(x)}{h} \\ &= \lim_{h \to 0} \frac{(x+h)^2 - x^2}{h} \\ &= \lim_{h \to 0} \frac{x^2 + h^2 + 2xh - x^2}{h} \\ &= \lim_{h \to 0} (h + 2x) \\ &= 2x \end{aligned}$$

Thus $2x$ is the derivative of the function x^2.

5.2 RULES FOR DIFFERENTIATION

It is apparent that to differentiate a function by the direct use of the definition of a derivative often involves a long and tedious process. However, there are rules or formulas that allow us to evaluate derivatives quickly without explicitly going through the limit process. We shall look at some rules in this section.

***Rule* 1.** *If c is a constant, then* $\frac{d}{dx}(c) = 0$. *That is, the derivative of a constant function is zero.*

For example, if $y = 5$, then $\frac{dy}{dx} = \frac{d\,5}{dx} = 0$.

***Rule* 2.** *If n is any real number, then* $\frac{d}{dx}(x^n) = nx^{n-1}$. *That is, the derivative of a constant power of x is the exponent times x raised to a power one less than the given power.*

***Example* 4.** Differentiate each of the following :

(*a*) $x^{3/2}$ (*b*) x (*c*) $\frac{1}{\sqrt{x}}$ (*d*) $\frac{1}{x}$.

***Solution* .** (*a*) By rule 2, $\frac{d}{dx}\left(x^{3/2}\right) = \frac{3}{2}x^{(3/2)-1} = \frac{3}{2}x^{1/2} = \frac{3}{2}\sqrt{x}$.

(*b*) $\frac{d}{dx}(x) = \frac{d}{dx}(x^1) = 1.x^{1-1} = 1.x^0 = 1$. Thus the derivative of x with respect to itself is 1.

(*c*) $\frac{d}{dx}\left(\frac{1}{\sqrt{x}}\right) = \frac{d}{dx}\left(x^{-1/2}\right) = -\frac{1}{2}x^{(-1/2)-1} = -\frac{1}{2}x^{-3/2}$.

(*d*) $\frac{d}{dx}\left(\frac{1}{x}\right) = \frac{d}{dx}(x^{-1}) = -1x^{-1-1} = -x^{-2} = -\frac{1}{x^2}$.

***Rule* 3.** *If f is a differentiable function and c is a constant, then* $\frac{d}{dx}[cf(x)] = c\frac{d}{dx}[f(x)]$.

That is, the derivative of a constant times a function is the constant times the derivative of the function.

***Example* 5.** Differentiate the following functions :

(*a*) $5x^2$ (*b*) $\frac{7}{3}x^{3/2}$ (*c*) $\frac{1{,}000}{\sqrt[3]{x}}$

***Solution* .** (*a*) $\frac{d}{dx}(5x^2) = 5\frac{d}{dx}(x^2) = 5\,(2x) = 10x$

(*b*) $\frac{d}{dx}\left(\frac{7}{3}x^{3/2}\right) = \frac{7}{3}\frac{d}{dx}(x^{3/2}) = \frac{7}{3}\left(\frac{3}{2}x^{1/2}\right) = \frac{7}{2}\sqrt{x}$.

(*c*) $\frac{d}{dx}\left(\frac{1{,}000}{\sqrt[3]{x}}\right) = 1{,}000\frac{d}{dx}\left(\frac{1}{\sqrt[3]{x}}\right) = 1{,}000\frac{d}{dx}(x^{-1/3}) = 1{,}000\left(-\frac{1}{3}x^{-4/3}\right) = -\frac{1{,}000}{3}x^{-4/3}$

***Rule* 4.** *If f and g are differentiable functions, then*

$$\frac{d}{dx}[f(x) \pm g(x)] = \frac{d}{dx}[f(x)] \pm \frac{d}{dx}[g(x)].$$

That is, the derivative of the sum or difference of two functions is the sum or difference of their derivatives.

Note : Rule 4 can be extended to the derivative of a finite number of sums and differences of functions.

***Example* 6.** Differentiate the following functions :

(*a*) $x^5 + x^2$ (*b*) $\frac{x^3+1}{x}$ (*c*) $2x^4 - 3x^3 + 5x^2 + x + 4$

***Solution* .** (*a*) $\frac{d}{dx}(x^5 + x^2) = \frac{d}{dx}(x^5) + \frac{d}{dx}(x^2) = 5x^4 + 2x$

(*b*) $\frac{d}{dx}\left(\frac{x^3+1}{x}\right) = \frac{d}{dx}\left(x^2 + \frac{1}{x}\right) = \frac{d}{dx}(x^2) + \frac{d}{dx}\left(\frac{1}{x}\right) = 2x + \frac{d}{dx}(x^{-1}) = 2x - x^{-2} = 2x - \frac{1}{x^2}$

(*c*) $\frac{d}{dx}(2x^4 - 3x^3 + 5x^2 + x + 4) = \frac{d}{dx}(2x^4) - \frac{d}{dx}(3x^3) + \frac{d}{dx}(5x^2) + \frac{d}{dx}(x) + \frac{d}{dx}(4)$

$$= 2\frac{dx^4}{dx} - 3\frac{dx^3}{dx} + 5\frac{dx^2}{dx} + \frac{dx}{dx} + 0$$

$$= 2(4x^3) - 3(3x^2) + 5(2x) + 1$$

$$= 8x^3 - 9x^2 + 10x + 1.$$

EXERCISE 5.1

1. Find the derivative of each of the following functions :

(*i*) $f(x) = 6$ (*ii*) $f(x) = 3x^5$ (*iii*) $f(x) = x^7$ (*iv*) $f(x) = 7x - 4$

(*v*) $f(x) = \sqrt{3}\, x^{3/4}$ (*vi*) $f(x) = \frac{5x+8}{16}$ (*vii*) $f(x) = \log 5$ (*viii*) $f(x) = e^2$

2. Differentiate the following functions :

(*i*) $f(x) = 3x^3 - 7x^2 + 7$ (*ii*) $f(x) = 3x^2 + 4x - 8$

(*iii*) $f(x) = 5(6 - x^4)$ (*iv*) $f(x) = 5x^3 - 6x^2 + 9x - e^3$

3. Differentiate with respect to x the following functions :

(*i*) $3x^6 - x^4 + 2x^{-2}$ (*ii*) $x^m + a^m$ (*iii*) $(2x + 3)(x + 1)$

ANSWERS

1. *(i)* 0 *(ii)* $15x^4$ *(iii)* $7x^6$ *(iv)* 7 *(v)* $\frac{3\sqrt{3}}{4}x^{-1/4}$ *(vi)* $\frac{5}{16}$ *(vii)* 0 *(viii)* 0

2.(*i*) $9x^2 - 14x$ (*ii*) $6x + 4$ (*iii*) $-20x^3$ (*iv*) $15x^2 - 12x + 9$

3.(*i*) $18x^5 - 4x^3 - 4x^{-3}$ (*ii*) mx^{m-1} (*iii*) $4x + 5$

5.3 PRODUCT AND QUOTIENT RULES

So far we have seen how to compute derivatives of sums, differences and constant multiples of functions. In this section, we find the derivatives of products and quotients. Since the derivative of a sum of functions is the sum of their derivatives, one might think that the derivative of a product of two functions is the product of their derivatives. However, this is not the case, as the next rule shows.

***Rule* 5. (*Product Rule*).** *If f and g are differentiable functions, then*

$$\frac{d}{dx}[f(x)\,g(x)] = f(x)\frac{d}{dx}[g(x)] + g(x)\frac{d}{dx}[f(x)]$$

That is, the derivative of the product of two functions is the first function times the derivative of the second, plus the second function times the derivative of the first.

***Example* 7.** If $y = (4x^3 - 5x + 7)(3x^4 - 2x^3 + 2)$, find $\frac{dy}{dx}$.

***Solution* .** Applying the product rule, we obtain

$$\frac{dy}{dx} = (4x^3 - 5x + 7)\frac{d}{dx}(3x^4 - 2x^3 + 2) + (3x^4 - 2x^3 + 2)\frac{d}{dx}(4x^3 - 5x + 7)$$

$$= (4x^3 - 5x + 7)(12x^3 - 6x^2) + (3x^4 - 2x^3 + 2)(12x^2 - 5)$$

$$= 84x^6 - 48x^5 - 75x^4 + 120x^3 - 18x^2 - 10. \quad \textit{(On simplification)}$$

***Rule* 6. (*Quotient Rule*).** *If f and g are differentiable functions and g* $(x) \neq 0$, then

$$\frac{d}{dx}\left[\frac{f(x)}{g(x)}\right] = \frac{g(x)f'(x) - f(x)g'(x)}{[g(x)]^2}$$

That is, the derivative of the quotient of two functions is the denominator times the derivative of the numerator, minus the numerator times the derivative of the denominator, all divided by the square of the denominator.

***Example* 8.** If $y = \frac{2x+1}{x-1}$, find $\frac{dy}{dx}$.

***Solution*.** Using the quotient rule, we obtain

$$\frac{dy}{dx} = \frac{(x-1)\dfrac{d}{dx}(2x+1) - (2x+1)\dfrac{d}{dx}(x-1)}{(x-1)^2}$$

$$= \frac{(x-1)(2)-(2x+1)(1)}{(x-1)^2} = -\frac{3}{(x-1)^2}$$

***Example* 9.** If $y = \dfrac{x^3 - x^2 + 1}{x^2 + 1}$, find $\dfrac{dy}{dx}$.

***Solution*.** Applying the quotient rule, we obtain

$$\frac{dy}{dx} = \frac{(x^2+1)\dfrac{d}{dx}(x^3 - x^2 + 1) - (x^3 - x^2 + 1)\dfrac{d}{dx}(x^2+1)}{(x^2+1)^2}$$

$$= \frac{(x^2+1)(3x^2 - 2x) - (x^3 - x^2 + 1)(2x)}{(x^2+1)^2}$$

$$= \frac{x^4 + 3x^2 - 4x}{(x^2+1)^2} = \frac{x(x^3 + 3x - 4)}{(x^2+1)^2}.$$

5.4 THE CHAIN RULE

The chain rule is one of the most important rules for finding derivatives. Before we state it, let us suppose that y is a function of u :

$$y = f(u)$$

and u in turn is a function of x :

$$u = g(x).$$

Then by substituting $g(x)$ for u in the first function, we can consider y as a function of x :

$$y = f(g(x)).$$

Functions of this type are called *composite functions.* To find the derivative of $f(g(x))$, we could, of course, explicitly set up $f(g(x))$ and then differentiate it. Very often, the algebra we would have to go through to find $f(g(x))$ explicitly is very cumbersome and we want to avoid it. Fortunately, the chain rule will allow us to differentiate $f(g(x))$ without explicitly knowing $f(g(x))$.

Rule 7. (Chain Rule). If $y = f(u)$ *is a differentiable function of* u *and* $u = g(x)$ *is a differentiable function of* x, *then* $y = f(g(x))$ *is a differentiable function of* x *and*

$$\frac{dy}{dx} = \frac{dy}{du} \cdot \frac{du}{dx}$$

The following examples illustrate applications of the chain rule to find derivatives.

***Example* 10.** If $y = u^2$ and $u = 2x^3 + 5x + 1$, find $\frac{dy}{dx}$.

***Solution*.** By the chain rule,

$$\frac{dy}{dx} = \frac{dy}{du} \cdot \frac{du}{dx}$$

$$= \frac{d}{du}(u^2) \cdot \frac{d}{dx}(2x^3 + 5x + 1)$$

$$= 2u(6x^2 + 5)$$

$$= 2(2x^3 + 5x + 1)(6x^2 + 5).$$

***Example* 11.** If $y = \sqrt{u}$ and $u = 5 + 7x + x^3$, find $\frac{dy}{dx}$.

***Solution*.** By the chain rule,

$$\frac{dy}{dx} = \frac{dy}{du} \cdot \frac{du}{dx}$$

$$= \frac{d}{du}(\sqrt{u}) \cdot \frac{d}{dx}(5 + 7x + x^3)$$

$$= \frac{1}{2}u^{-1/2} \cdot (7 + 3x^2)$$

$$= \frac{1}{2\sqrt{u}} \cdot (7 + 3x^2) = \frac{(7 + 3x^2)}{2\sqrt{5 + 7x + x^3}}.$$

***Example* 12.** If $y = (2x^3 - 5x^2 + 7x + 1)^{25}$, find $\frac{dy}{dx}$.

***Solution*.** The function y can be written as $y = u^{25}$, where $u = 2x^3 - 5x^2 + 7x + 1$. Thus by the chain rule,

$$\frac{dy}{dx} = \frac{dy}{du} \cdot \frac{du}{dx}$$

$$= (25u^{24})(6x^2 - 10x + 7)$$

$$= 25(2x^3 - 5x^2 + 7x + 1)^{24}(6x^2 - 10x + 7).$$

***Example* 13.** Differentiate $\left(\dfrac{2x+5}{5x-7}\right)^4$ with respect to x.

***Solution* .** Let $y = \left(\dfrac{2x+5}{5x-7}\right)^4$. Then y can be written as $y = u^4$, where $u = \dfrac{2x+5}{5x-7}$. Thus

by the chain rule, $\dfrac{dy}{dx} = \dfrac{dy}{du} \cdot \dfrac{du}{dx}$

$$= (4u^3)\left[\frac{(5x-7)(2)-(2x+5)(5)}{(5x-7)^2}\right]$$

$$= 4\left(\frac{2x+5}{5x-7}\right)^3\left[\frac{-39}{(5x-7)^2}\right] = -156\,\frac{(2x+5)^3}{(5x-7)^5}.$$

***Remark* .** The chain rule is particularly helpful in finding the derivative of quantities raised to powers, as the derivative of $[f(x)]^n$. Let $y = u^n$ and $u = f(x)$. Applying the chain rule, we obtain

$$\frac{dy}{dx} = \frac{dy}{du} \cdot \frac{du}{dx} = nu^{n-1}\, f'(x)$$

Therefore, $$\boldsymbol{\frac{d}{dx}[f(x)]^n = n[f(x)]^{n-1} . f'(x)}$$

***Example* 14.** If $y = (2x^5 - 5x^2 + 7)^9$, find $\dfrac{dy}{dx}$.

***Solution* .** We have $\dfrac{dy}{dx} = 9(2x^5 - 5x^2 + 7)^8 \cdot \dfrac{d}{dx}(2x^5 - 5x^2 + 7)$

$$= 9(2x^5 - 5x^2 + 7)^8\,(10x^4 - 10x)$$

$$= 90x\,(2x^5 - 5x^2 + 7)^8 (x^3 - 1).$$

***Note* :** In actual practice the u substitution is not made.

Derivatives of Logarithmic and Exponential Functions

The following result (proof omitted) helps us in determining derivatives of logarithmic and exponential functions.

Result (*a*) If $f(x) = \log_a x$, where a and x are positive, then

$$f'(x) = \frac{1}{x}\log_a e \quad \textit{i.e.,} \quad \frac{d}{dx}[\log_a x] = \frac{1}{x}\log_a e$$

In particular, $$\frac{d}{dx}[\log x] = \frac{1}{x}$$

(*b*) If $f(x) = a^x$, where $a > 0$, then

$$f'(x) = a^x \log_e a \quad i.e., \quad \frac{d}{dx}(a^x) = a^x \log_e a$$

In particular, $$\frac{d}{dx}(e^x) = e^x.$$

Remark . The chain rule is also useful in finding the derivatives of $y = \log u$ and $y = e^u$ when u is an appropriate function of x. For example, if $y = \log u$, then by the chain rule,

$$\frac{dy}{dx} = \frac{dy}{du} \cdot \frac{du}{dx} = \frac{1}{u}\frac{du}{dx}$$

Thus $$\frac{d}{dx}(\log u) = \frac{1}{u}\frac{du}{dx}.$$

Similarly, $$\frac{d}{dx}(e^u) = e^u \frac{du}{dx}.$$

***Example* 15.** If $y = \log(5x^2 + 7)$, find $\frac{dy}{dx}$.

Solution . This function is of the form y = $\log u$, where $u = 5x^2 + 7$. Thus using the remark, we obtain

$$\frac{dy}{dx} = \frac{1}{5x^2+7}\frac{d}{dx}(5x^2 + 7) = \frac{1}{5x^2+7}(10x) = \frac{10x}{5x^2+7}.$$

***Example* 16.** If $y = e^{2x^2 + 3x + 7}$, find $\frac{dy}{dx}$.

Solution . Applying the remark, we have

$$\frac{dy}{dx} = e^{2x^2+3x+7}\frac{d}{dx}(2x^2 + 3x + 7) = e^{2x^2+3x+7}(4x + 3).$$

***Example* 17.** If $y = x^2 \log\left(3x + 7\right)$, find $\frac{dy}{dx}$.

Solution . Using the product rule, we obtain

$$\frac{dy}{dx} = x^2\frac{d}{dx}[\log(3x + 7)] + \log(3x + 7)\frac{d}{dx}(x^2)$$

$$= x^2\left(\frac{1}{3x+7}\right)(3) + [\log(3x + 7)]\ (2x)$$

$$= \frac{3x^2}{3x+7} + 2x\log(3x + 7).$$

EXERCISE 5.2

1. Find the derivatives of the following functions :

(*i*) $(x^2 - 3x + 7)(2x^2 + 5x + 9)$ (*ii*) $x^2 \log x$ (*iii*) $x^3 e^{2x}$

2. Differentiate with respect to x each of the following functions :

(*i*) $\dfrac{x-1}{x+1}$ (*ii*) $\dfrac{\log x}{x^2}$ (*iii*) $\dfrac{x}{\sqrt{1+x^2}}$ (*iv*) $\dfrac{2x^3+1}{3x^2+1}$

3. Find the derivatives of the following functions :

(*i*) $(5x+4)^5$ (*ii*) $\log(\log x)$ (*iii*) $\log(3x^2 + 2x - 5)$ (*iv*) $\log(xe^x)$

4. Differentiate with respect to x each of the following functions :

(*i*) 4^{3x+1} (*ii*) $(e^{3x} + 1)^4$ (*iii*) $e^{1+\sqrt{x}}$

ANSWERS

1. (*i*) $8x^3 - 3x^2 + 16x + 8$ (*ii*) $x(1 + 2\log x)$ (*iii*) $(2x+3)x^2 e^{2x}$

2. (*i*) $\dfrac{2}{(x+1)^2}$ (*ii*) $\dfrac{1 - 2\log x}{x^3}$ (*iii*) $\dfrac{1}{(1+x^2)^{3/2}}$ (*iv*) $\dfrac{6x(x^3+x-1)}{(3x^2+1)^2}$

3. (*i*) $25(5x+4)^4$ (*ii*) $\dfrac{1}{x \log x}$ (*iii*) $\dfrac{6x+2}{3x^2+2x-5}$ (*iv*) $\dfrac{x+1}{x}$

4. (*i*) $(3\log 4)\, 4^{3x+1}$ (*ii*) $12e^{3x}(e^{3x}+1)^3$ (*iii*) $\dfrac{e^{1+\sqrt{x}}}{2\sqrt{x}}$

5.5 IMPLICIT DIFFERENTIATION

Until now we have considered exclusively functions of x that were explicitly (directly) given in terms of x. However, sometimes we deal with relationships between x and y in which it may be very complicated or even impossible to find an explicit expression for y and yet we need to find dy/dx. For example, it would be difficult to solve $x^3 + y^3 - xy = 0$ for y.

An equation of the form $f(x, y) = 0$, such as $x^3 + y^3 - xy = 0$, is said to express y *implicitly* as a function of x. The word "implicitly" is used since y is not given explicitly as a function of x. It is assumed or implied that the equation $f(x, y) = 0$ defines y as at least one differentiable function of x. Hence to find dy/dx, we differentiate each term with respect to x treating y as a function of x and then separating dy/dx. This method of finding dy/dx is called **implicit differentiation.** The method is best illustrated with the help of following examples.

***Example* 18.** Find $\frac{dy}{dx}$ when $x^3 + y^3 = xy$.

***Solution* .** By treating y as a function of x and differentiating both sides with respect to x, we get

$$\frac{d}{dx}(x^3) + \frac{d}{dx}(y^3) = \frac{d}{dx}(xy)$$

$$\Rightarrow \qquad 3x^2 + 3y^2\frac{dy}{dx} = x\frac{dy}{dx} + y$$

$$\Rightarrow \qquad (3y^2 - x)\frac{dy}{dx} = y - 3x^2 \qquad \textit{(Regrouping)}$$

$$\Rightarrow \qquad \frac{dy}{dx} = \frac{y - 3x^2}{3y^2 - x}.$$

***Example* 19.** If $x\sqrt{1+y} + y\sqrt{1+x} = 0$, prove that $\frac{dy}{dx} = -\frac{1}{(1+x)^2}$.

***Solution* .** We have $\quad x\sqrt{1+y} + y\sqrt{1+x} = 0$

$$\Rightarrow \qquad x\sqrt{1+y} = -y\sqrt{1+x}$$

$$\Rightarrow \qquad x^2(1+y) = y^2(1+x) \qquad \textit{(Squaring)}$$

$$\Rightarrow \qquad x^2 + x^2y = y^2 + y^2x$$

$$\Rightarrow \qquad x^2 - y^2 = y^2x - x^2y \qquad \textit{(Regrouping)}$$

i.e., $\quad (x-y)(x+y) = -xy(x-y)$

or $\quad (x+y) = -xy$, assuming $y \neq x$

Solving this for y gives

$$y = -\frac{x}{1+x}$$

Differentiating both sides with respect to x, we get

$$\frac{dy}{dx} = -\left[\frac{(1+x).1 - x.1}{(1+x)^2}\right] = -\left[\frac{1+x-x}{(1+x)^2}\right] = -\frac{1}{(1+x)^2}.$$

5.6 DERIVATIVE OF THE INVERSE OF A FUNCTION

Let $y = f(x)$ be a function of x and suppose that we can solve this equation for x in terms of y so that we may write x as a function of y, say

$$x = g(y)$$

Then $g(y)$ is called the *inverse* of $f(x)$. The following result describes the relationship between $\frac{dy}{dx}$ and $\frac{dx}{dy}$.

Let $y = f(x)$ be a function which admits of an inverse function $x = g(y)$. Let $y = f(x)$ be differentiable at x such that $f'(x) \neq 0$. Then $x = g(y)$ is differentiable at the corresponding value of y and $\frac{dx}{dy} = \frac{1}{dy/dx}$.

We will illustrate this result with the help of following example.

***Example* 20.** Find $\frac{dx}{dy}$ if (*i*) $y = 5x - 3$ (*ii*) $y = \sqrt{x+2}$ (*iii*) $y = e^{1-x}$

Solution **.** (*i*) We have $y = 5x - 3$

$$\therefore \quad \frac{dy}{dx} = 5 \Rightarrow \frac{dx}{dy} = \frac{1}{dy/dx} = \frac{1}{5}$$

(*ii*) We have $y = \sqrt{x+2}$

$$\therefore \quad \frac{dy}{dx} = \frac{1}{2\sqrt{x+2}} \Rightarrow \frac{dx}{dy} = \frac{1}{dy/dx} = 2\sqrt{x+2}$$

(*iii*) We have $y = e^{(1-x)}$

$$\therefore \quad \frac{dy}{dx} = e^{(1-x)}(-1) = -e^{(1-x)} = -y \Rightarrow \frac{dx}{dy} = \frac{1}{dy/dx} = -\frac{1}{y}.$$

5.7 DERIVATIVE OF A FUNCTION DEFINED PARAMETRICALLY

A function $y = \varphi(x)$ is said to be defined **parametrically** if both x and y are given in terms of a new variable, say t. The variable t is called the **parameter.**

Let $x = f(t)$ and $y = g(t)$ be two differentiable functions of 't'. We assume that $x = f(t)$ possesses an inverse function $t = h(x)$. Thus we obtain $y = g(h(x))$ and as such y can be considered as a composite function. Hence by the chain rule

$$\frac{dy}{dx} = \frac{dy}{dt} \cdot \frac{dt}{dx} = \frac{dy/dt}{dx/dt}, \text{ provided } \frac{dx}{dt} \neq 0$$

***Example* 21.** Find $\dfrac{dy}{dx}$ if

(*i*) $x = \dfrac{1-t}{1+t}$ and $y = 2t^3 + 4^t$ (*ii*) $x = a \log t$ and $y = bt^2$

***Solution* .** (*i*) We have $x = \dfrac{1-t}{1+t}$ and $y = 2t^3 + 4^t$

$$\therefore \quad \frac{dx}{dt} = \frac{(1+t)(-1)-(1-t)(1)}{(1+t)^2} \quad \text{and} \quad \frac{dy}{dt} = 6t^2 + 4^t \log 4$$

$$\text{i.e.,} \quad \frac{dx}{dt} = \frac{-2}{(1+t)^2} \quad \text{and} \quad \frac{dy}{dt} = 6t^2 + 4^t \log 4$$

$$\text{Thus} \quad \frac{dy}{dx} = \frac{dy/dt}{dx/dt} = \frac{(6t^2 + 4^t \log 4)}{-2/(1+t)^2} = \frac{(6t^2 + 4^t \log 4)(1+t)^2}{-2}$$

(*ii*) We have $x = a \log t$ and $y = bt^2$

$$\therefore \quad \frac{dx}{dt} = \frac{a}{t} \quad \text{and} \quad \frac{dy}{dt} = 2bt$$

$$\text{Thus} \quad \frac{dy}{dx} = \frac{dy/dt}{dx/dt} = \frac{2bt^2}{a}.$$

5.8 LOGARITHMIC DIFFERENTIATION

The method of logarithmic differentiation is used to differentiate functions of the form $y = f(x)^{g(x)}$. With this method, we first take the natural logarithm of both sides of $y = f(x)^{g(x)}$ to obtain $\log y = \log [f(x)^{g(x)}]$. After simplifying $\log [f(x)^{g(x)}]$ by using properties of logarithms, we differentiate both sides with respect to x and then solve for *dy*/*dx*. This method can also be used to differentiate functions which are product of several functions. The following examples illustrate the method of logarithmic differentiation.

***Example* 22.** If $y = x^x$, find $\dfrac{dy}{dx}$.

***Solution* .** The given function has the form $f(x)^{g(x)}$. Taking the natural logarithm of both sides, we get

$$\log y = \log (x^x) = x \log x$$

Differentiating both sides with respect to x gives

$$\frac{1}{y}\frac{dy}{dx} = x\left(\frac{1}{x}\right) + (\log x)\,(1) = 1 + \log x$$

Multiplying both sides by y and then substituting x^x for y, we obtain

$$\frac{dy}{dx} = y\,[1 + \log x] = x^x\,(1 + \log x).$$

***Example* 23.** If $x^y = e^{x-y}$, show that $\frac{dy}{dx} = \frac{\log x}{(1+\log x)^2}$.

***Solution* .** Taking the natural logarithm of both sides gives

$$\log\,(x^y) = \log\,(e^{x-y})$$

or, $\quad y \log x = (x - y) \log e = x - y \qquad$ (*since* $\log e = 1$)

$\therefore \quad y \log x + y = x$

or, $\quad y\,(1 + \log x) = x$

or, $\quad y = \dfrac{x}{1 + \log x}$

$$\therefore \quad \frac{dy}{dx} = \frac{(1 + \log x)\,1 - x\left(\frac{1}{x}\right)}{(1 + \log x)^2} = \frac{1 + \log x - 1}{(1 + \log x)^2} = \frac{\log x}{(1 + \log x)^2}.$$

***Example* 24.** If $x^y + y^x = a^b$, find $\frac{dy}{dx}$.

***Solution* .** Here we can not take the natural logarithm of both sides because

$$\log\,(m + n) \neq \log m + \log n$$

However, if we let $u = x^y$ and $v = y^x$

then

$$u + v = a^b$$

$$\therefore \quad \frac{du}{dx} + \frac{dv}{dx} = 0 \qquad ...(1)$$

Now $u = x^y$. Using logarithmic differentiation, we obtain

$$\log u = \log x^y = y \log x$$

Differentiating both sides with respect to x gives

$$\frac{1}{u}\frac{du}{dx} = y\left(\frac{1}{x}\right) + (\log x)\frac{dy}{dx}$$

$$\Rightarrow \quad \frac{du}{dx} = u\left(\frac{y}{x} + \log x\,\frac{dy}{dx}\right)$$

i.e., $$\frac{du}{dx} = x^y\left(\frac{y}{x} + \log x\,\frac{dy}{dx}\right)$$

or, $$\frac{du}{dx} = yx^{y-1} + x^y \log x\,\frac{dy}{dx} \qquad ...(2)$$

Again $v = y^x$. Using logarithmic differentiation, we get

$$\log v = \log(y^x) = x \log y$$

Differentiating both sides with respect to x gives

$$\frac{1}{v}\frac{dv}{dx} = x\left(\frac{1}{y}\frac{dy}{dx}\right) + (\log y).1 = \frac{x}{y}\frac{dy}{dx} + \log y$$

$$\Rightarrow \quad \frac{dv}{dx} = v\left[\frac{x}{y}\frac{dy}{dx} + \log y\right]$$

i.e., $$\frac{dv}{dx} = y^x\left[\frac{x}{y}\frac{dy}{dx} + \log y\right]$$

or, $$\frac{dv}{dx} = y^{x-1}\,x\,\frac{dy}{dx} + y^x \log y \qquad ...(3)$$

Substituting the values of $\frac{du}{dx}$ and $\frac{dv}{dx}$ from (2) and (3), respectively, in (1), we obtain

$$yx^{y-1} + x^y \log x\,\frac{dy}{dx} + xy^{x-1}\frac{dy}{dx} + y^x \log y = 0$$

or, $$(x^y \log x + xy^{x-1})\frac{dy}{dx} = -(x^{y-1}y + y^x \log y)$$

$$\Rightarrow \quad \frac{dy}{dx} = -\left(\frac{x^{y-1}y + y^x \log y}{x^y \log x + xy^{x-1}}\right).$$

5.9 SUCCESSIVE DIFFERENTIATION

Let $y = f(x)$ be a differentiable function of x. Then its derivative, $f'(x)$, is called the **first derivative** of f. The first derivative, in general, is itself a function of x which can further be differentiated with respect to x. The resulting derivative, if it exists, is called the **second derivative** of f. It is denoted by $f''(x)$ (read "f double prime of x"). Similarly, if the second derivative is differentiable, its derivative is called the **third derivative**, written $f'''(x)$. Continuing in this way, we get *higher-order derivatives.* The process of finding higher-order derivatives of a function is called **successive differentiation.**

Some other notations for higher-order derivatives are :

First derivative : $y', \; y_1, \; \dfrac{dy}{dx}, \; \dfrac{d}{dx}[f(x)]$

Second derivative : $y'', \; y_2, \; \dfrac{d^2y}{dx^2}, \; \dfrac{d^2}{dx^2}[f(x)]$

Third derivative : $y''', \; y_3, \; \dfrac{d^3y}{dx^3}, \; \dfrac{d^3}{dx^3}[f(x)]$

$\vdots \qquad \vdots \qquad \vdots \qquad \vdots$

nth derivative : $y^{(n)}, \; y_n, \; \dfrac{d^ny}{dx^n}, \; \dfrac{d^n}{dx^n}[f(x)]$

***Remark* 1.** To avoid clumsy notation, primes are not used beyond the third derivative.

***Remark* 2.** The nth derivative of a function $y = f(x)$ evaluated at a point $x = c$ is denoted by

$$f^n(c), \; y^{(n)}(c) \; \text{or} \; y_n(c) \; \text{or} \; \left(\frac{d^nf}{dx^n}\right)_{x=c}$$

***Example* 25.** If $y = 2x^3 + 5x^2 - 7x + 4$, find $\dfrac{d^2y}{dx^2}$.

***Solution* .** Differentiating y with respect to x gives $\dfrac{dy}{dx} = 6x^2 + 10x - 7$

Differentiating $\dfrac{dy}{dx}$ gives $\dfrac{d^2y}{dx^2} = 12x + 10$.

***Example* 26.** If $y = \left(\dfrac{1}{x}\right)^x$, show that $\left(\dfrac{d^2y}{dx^2}\right)_{x=1} = 0$.

Solution . We have $y = \left(\frac{1}{x}\right)^x$. Taking the natural logarithm of both sides gives

$$\log y = x \log\left(\frac{1}{x}\right) = -x \log x$$

Differentiating both sides with respect to x gives

$$\frac{1}{y}\frac{dy}{dx} = -\left[x\left(\frac{1}{x}\right) + (\log x).1\right] = -(1 + \log x)$$

Solving for $\frac{dy}{dx}$, we obtain $\qquad \frac{dy}{dx} = -y(1 + \log x) \qquad$ ...(1)

Differentiating both sides of this equation with respect to x, we have

$$\frac{d^2y}{dx^2} = -\left[y\left(\frac{1}{x}\right) + (1 + \log x)\frac{dy}{dx}\right]$$

$$\frac{d^2y}{dx^2} = -\left[\frac{y}{x} - (1 + \log x)^2\, y\right] \qquad \text{[From (1)]}$$

$$\frac{d^2y}{dx^2} = -y\left[\frac{1}{x} - (1 + \log x)^2\right] = -\left(\frac{1}{x}\right)^x\left[\frac{1}{x} - (1 + \log x)^2\right]$$

$$\therefore \quad \left(\frac{d^2y}{dx^2}\right)_{x=1} = -\left(\frac{1}{1}\right)^1\left[\frac{1}{1} - (1 + \log 1)^2\right] = -[1 - 1] = 0.$$

Example 27. If $x = at^2$, $y = 2at$, *find* $\frac{d^2y}{dx^2}$.

Solution . We have $\qquad \frac{dx}{dt} = 2at \quad and \quad \frac{dy}{dt} = 2a$

$$\therefore \quad \frac{dy}{dx} = \frac{dy/dt}{dx/dt} = \frac{2a}{2at} = \frac{1}{t}$$

$$\therefore \quad \frac{d^2y}{dx^2} = \frac{d}{dt}\left(\frac{1}{t}\right).\frac{dt}{dx} = -\frac{1}{t^2}.\frac{1}{2at} = -\frac{1}{2at^3}.$$

EXERCISE 5.3

1. (*i*) If $x = at^2$ and $y = 2at$, find $\frac{dy}{dx}$.

(*ii*) $x = \frac{a(1 - t^2)}{1 + t^2}$ and $y = \frac{2bt}{1 + t^2}$, find $\frac{dy}{dx}$.

(*iii*) If $x = \dfrac{3at}{1+t^3}$ and $y = \dfrac{3\,at^2}{1+t^3}$, find $\dfrac{dy}{dx}$.

(*iv*) If $x = t\,(\log t)$ and $y = \dfrac{\log t}{t}$, find $\dfrac{dy}{dx}$ when $t = 1$.

2. Differentiate with respect to x each of the following functions :

(*i*) $x^5 + y^5 = 5ax^2\,y^2$ (*ii*) $x^{2/3} + y^{2/3} = a^{2/3}$

(*iii*) $ax^2 + 2\,h\,xy + by^2 + 2\,g\,x + 2\,f\,y + c = 0$ (*iv*) $3x^3 - 5x^2y + 2xy^2 + 4y^3 = 0$

3. Find $dx\,/\,dy$ from each of the following functions :

(*i*) $y = 2x^3 + 7x^2 + 4$ (*ii*) $y = \sqrt{(625 - x^2)}$ (*iii*) $y = ae^{-bx}$ (*iv*) $y = \dfrac{5}{x+2}$

4. (*i*) If $x^y \,.\, y^x = k$, where k is a constant, find $\dfrac{dy}{dx}$. ***[Delhi Univ. B.Com. (H) 1989]***

(*ii*) If $x^m\ y^n = (x+y)^{m+n}$, prove that $\dfrac{dy}{dx} = \dfrac{y}{x}$.

5. (*i*) If $x^y = y^x$, show that $\dfrac{dy}{dx} = \dfrac{y\,(y - x\log y)}{x\,(x - y\log x)}$.

(*ii*) If $e^{x+y} = x^y$, show that $\dfrac{dy}{dx} = \dfrac{y\,(x-1)}{x\,(1-y)}$.

6. (*i*) If $y = \dfrac{\log x}{x}$, show that $y_2 = \dfrac{2\log x - 3}{x^3}$.

(*ii*) If $y = ax^3 + bx^2 + cx + d$, show that $y_4 = 0$.

(*iii*) If $y = e^x + \log x$, prove that $y_3 = e^x + \dfrac{2}{x^3}$.

7. (*i*) If $x = \log t$ and $y = \dfrac{1}{t}$, prove that $\dfrac{d^2y}{dx^2} + \dfrac{dy}{dx} = 0$.

(*ii*) If $x = \dfrac{t^2}{1+t}$ and $y = \dfrac{t}{1+t}$ find y_2.

(*iii*) If $x = \dfrac{2at}{1+t^2}$ and $y = \dfrac{b\,(1-t^2)}{1+t^2}$, find y_2.

ANSWERS

1. (*i*) $\frac{1}{t}$ (*ii*) $-\frac{b}{2a}\left(\frac{1-t^2}{t}\right)$ (*iii*) $\frac{t\left(2-t^2\right)}{t-2t^3}$ (*iv*) 1

2. (*i*) $\frac{x\left(2ay^2-x^3\right)}{y\left(y^3-2ax^2\right)}$ (*ii*) $-\frac{y^{1/3}}{x^{1/3}}$ (*iii*) $-\left(\frac{ax+hy+g}{hx+by+f}\right)$ (*iv*) $\frac{10xy-9x^2-2y^2}{4xy-5x^2+12y^2}$

3. (*i*) $\frac{1}{6x^2+14x}$ (*ii*) $-\frac{\sqrt{625-x^2}}{x}$ (*iii*) $-\frac{e^{bx}}{ab}$ (*iv*) $\frac{-(x+2)^2}{5}$

4. (*i*) $-\frac{y\,(y+x\log y)}{x\,(x+y\log x)}$.

7. (*ii*) $\frac{-2(t+1)^3}{t^2\,(t+2)^3}$ (*iii*) $-\frac{b}{a^2}\left(\frac{1+t^2}{1-t^2}\right)^2$

CHAPTER

6

APPLICATIONS OF DIFFERENTIATION

OBJECTIVES

After studying the material in this chapter, you should be able to:

- *Interpret the derivative as an instantaneous rate of change.*
- *Find the intervals on which a given function is increasing or decreasing.*
- *Understand the concepts of convexity and concavity.*
- *Find maximum and minimum values of a given function.*
- *Understand the concept of marginal analysis.*
- *Evaluate price elasticity of demand.*
- *Determine the price and quantity for which total revenue is maximum.*
- *Find the level of output for which average cost is minimum.*
- *Solve profit maximization problems.*
- *Use max-min theory to determine the economic order quantity (EOQ).*

6.1 INTRODUCTION

Derivatives have a wide range of applications to a very large number of disciplines such as Engineering, Business Economics, Physics, Social Sciences etc. One of the primary interpretations of the derivative is the rate of change concept which, in turn, has many applications in science and engineering as well as in social sciences. In this chapter we will see how this rate of change concept can be of aid in understanding the marginal concept in economics.

We will also illustrate the use of the derivative to obtain information about the shape of the graph of a function. For example, we show how the sign of the derivative is used to settle questions about intervals over which the graph of a function is rising or falling. The principal application of this is in locating high and low points on graphs and, in turn, these points are used to determine the maximum and minimum values attained by the function. For example, if we are dealing with a profit function in terms of price, we can determine the maximum profit and the price at which the profit is maximum.

6.2 THE DERIVATIVE AS AN INSTANTANEOUS RATE OF CHANGE

In this section we give the interpretation of the derivative as a rate of change which, in turn, has applications to science, business and economics.

Let us consider a function $y = f(x)$. We use the symbol Δx to denote the change in the variable x. For example, if x changes from 2 to 4, then the change in x is $\Delta x = 4 - 2 = 2$. Δx may be positive or negative. Similarly, we use the symbol Δy to denote the change in the variable y. Usually, Δy depends on the values of x and Δx. According to the equation, at any point x, the corresponding value of y is given by $y = f(x)$. Suppose that there is a change, Δx, in x, so that the new value of x is $x + \Delta x$. Consequently, the new value of y is $f(x + \Delta x)$. Thus the change, Δy, in y is given by

$$\Delta y = f(x + \Delta x) - f(x)$$

The ratio $\dfrac{\Delta y}{\Delta x}$ is called the *average rate of change* of y with respect to x over the interval from x to $x + \Delta x$.

If Δx were to become smaller and smaller, the average rate of change over the interval from x to $(x + \Delta x)$, would be close to what we might call the *instantaneous rate of change* of y *with respect to* x. We define the limit of the average rate of change of the function $y = f(x)$ as $\Delta x \to 0$ to be the **instantaneous** or **point** or simply **the rate of change** of y with respect to x. Thus the instantaneous rate of change of a function $y = f(x)$ with respect to x is defined to be

$$\lim_{\Delta x \to 0} \frac{\Delta y}{\Delta x} = \lim_{\Delta x \to 0} \frac{f(x + \Delta x) - f(x)}{\Delta x}$$

The last limit is just the definition of the derivative of f with respect to x. Thus the **instantaneous rate of change** of y with respect to $x = \dfrac{dy}{dx} = \lim\limits_{\Delta x \to 0} \dfrac{\Delta y}{\Delta x}$.

Example 1. Find the rate of change of $f(x) = 2x^2 - 12x + 45$ with respect to x. Evaluate the rate when $x = 4$.

Solution. The rate of change of y with respect to x is : $\dfrac{dy}{dx} = 4x - 12$

In particular, when $x = 4$, $\dfrac{dy}{dx} = 4(4) - 12 = 4$.

6.3 INCREASING AND DECREASING FUNCTIONS

Increasing Function. A function f is said to be **increasing** (abbreviated ↑) on the interval (a, b) if $f(x)$ increases as x increases on the interval (a, b), that is, if the graph of f rises from left to right (See Figure 6.1).

Thus we see that f is increasing on (a, b) if $f(x_1) < f(x_2)$ whenever $x_1 < x_2$ with x_1 and x_2 in (a, b).

Note. **The term upward sloping** is also sometimes used for an increasing function.

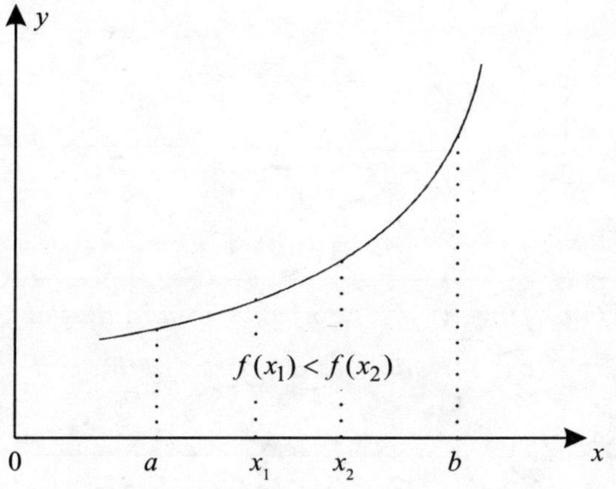

Fig. 6.1 : *f is increasing on (a, b)*

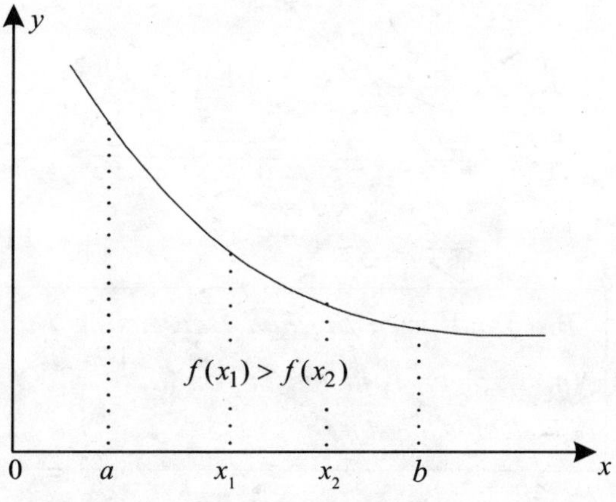

Fig. 6.2 : *f is decreasing on (a, b)*

Decreasing Function. A function f is said to be **decreasing** (abbreviated ↓) on the interval (a, b) if $f(x)$ decreases as x increases on the interval (a, b), that is, if the graph of f is falling from left to right (See Figure 6.2).

Thus f is decreasing on (a, b) if we have $f(x_1) > f(x_2)$ whenever $x_1 < x_2$ with x_1 and x_2 in (a, b).

Note. **The term downward sloping** is also sometimes used for a decreasing function.

If we examine the graph of an increasing function (see Figure 6.3 (a)), we see that the slope of the tangent line at any point on the curve is positive, whereas if we examine the graph of a decreasing function (see Figure 6.3 (b)), we see that the slope of the tangent line at any point on the curve is negative.

These observations together with the fact the derivative gives the slope of the tangent line lead to the following test which allows us to use the derivative to determine when a function is increasing or decreasing.

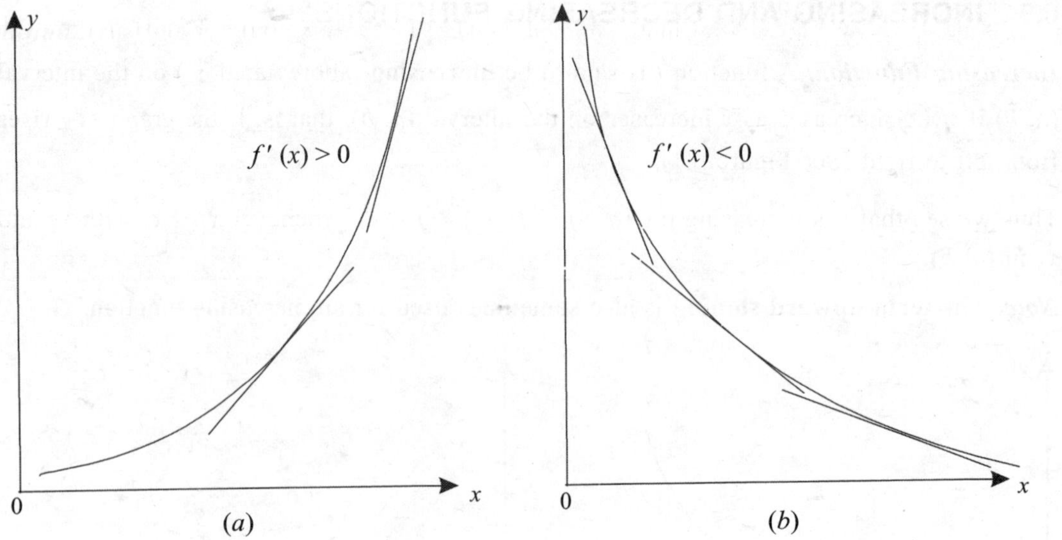

Fig.6.3

> ***Test for Increasing and Decreasing Functions.*** *If* $f'(x) > 0$ *for all x in an interval* (a, b)*, then f is increasing on* (a, b) *and if* $f'(x) < 0$ *for all x in an interval* (a, b)*, then f is decreasing on* (a, b)*.*

Let us consider few examples.

Example 2. Find the intervals on which the function $f(x) = 2x^3 - 9x^2 + 12x$ is increasing or decreasing.

Solution. Differentiating the function $f(x) = 2x^3 - 9x^2 + 12x$ gives

$$f'(x) = 6x^2 - 18x + 12 = 6(x-1)(x-2)$$

We must determine when $f'(x)$ is positive and when $f'(x)$ is negative. We can determine the sign of $f'(x)$ by considering the intervals determined by the roots of $6(x-1)(x-2)$, namely 1 and 2.

If $x < 1$, then $f'(x) = 6(-)(-) > 0$ so $f \uparrow$;

If $1 < x < 2$, then $f'(x) = 6(+)(-) < 0$ and so $f \downarrow$;

If $x > 2$, then $f'(x) = 6(+)(+) > 0$ and so $f \uparrow$.

Thus f is increasing on $(-\infty, 1)$ and $(2, \infty)$ and is decreasing on $(1, 2)$.

Example 3. Find the intervals on which the function $f(x) = 6 + 12x + 3x^2 - 2x^3$ is increasing or decreasing.

Solution. We have $f(x) = 6 + 12x + 3x^2 - 2x^3$

$$\Rightarrow \qquad f'(x) = 12 + 6x - 6x^2 = -6(x^2 - x - 2) = -6(x+1)(x-2)$$

Setting $f'(x) = 0$ gives $x = -1,\ 2$. These two values lead us to consider three intervals as shown below.

If $x < -1$, then $f'(x) = (-)(-)(-) = (-)$, so f is decreasing on $(-\infty, -1)$;
If $-1 < x < 2$, then $f'(x) = (-)(+)(-) = (+)$, so f is increasing on $(-1, 2)$;
If $x > 2$, then $f'(x) = (-)(+)(+) = (-)$, so f is decreasing on $(2, \infty)$;

Example 4. Determine the intervals on which the function $f(x) = x^4 - 8x^3 + 22x^2 - 24x + 21$ is decreasing or increasing.

Solution. We have $\qquad f(x) = x^4 - 8x^3 + 22x^2 - 24x + 21$

$$\therefore \qquad f'(x) = 4x^3 - 24x^2 + 44x - 24 = 4(x-1)(x-2)(x-3)$$

Setting $f'(x) = 0$ gives $x = 1,\ 2,\ 3$. These three values lead us to consider four intervals as shown below.

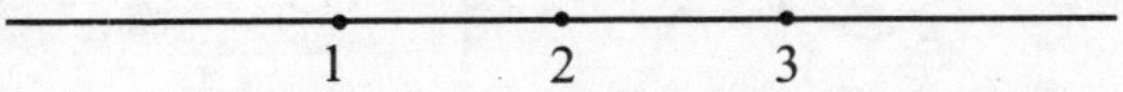

If $x < 1$, then $f'(x) = 4(-)(-)(-) = (-)$, so f is decreasing on $(-\infty, 1)$;
If $1 < x < 2$, then $f'(x) = 4(+)(-)(-) = (+)$, so f is increasing on $(1, 2)$;
If $2 < x < 3$, then $f'(x) = 4(+)(+)(-) = (-)$, so f is decreasing on $(2, 3)$;
If $x > 3$, then $f'(x) = 4(+)(+)(+) = (+)$, so f is increasing on $(3, \infty)$.

6.4 CONCAVITY AND CONVEXITY

In the preceding section we have seen that the first derivative can be used to determine when a function is increasing and when it is decreasing. However, a function can be increasing (or, respectively, decreasing) and be concave up (See Figure 6.4) or it can be increasing (or, respectively, decreasing) and be concave down (See Figure 6.5), but the first derivative cannot distinguish between these possibilities. As we shall see the second derivative can.

Geometrically, a function is said to be **concave up** (or **convex down)** on an interval if its curve bends upward (Figure 6.4) and **concave down** (or **convex up)** on an interval if its curve bends downward (Figure 6.5).

In Figure 6.4 note that each curve lies *above* the tangent lines whereas in Figure 7.5, each curve lies *below* the tangent lines.

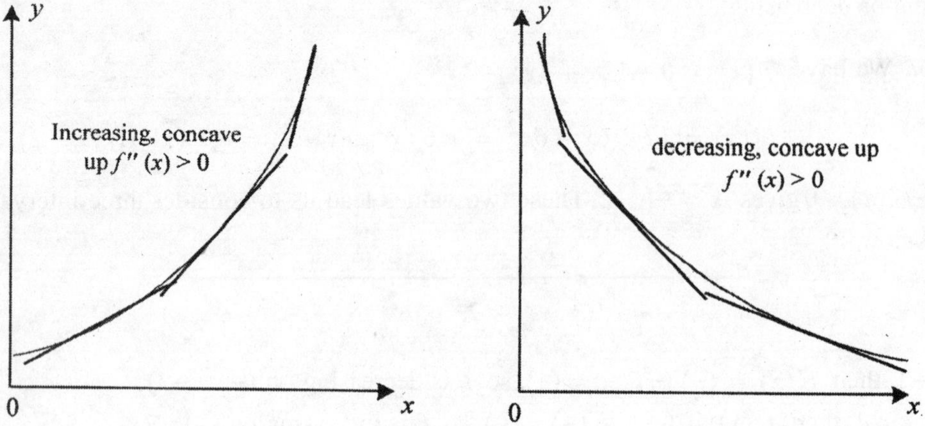

Fig. 6.4

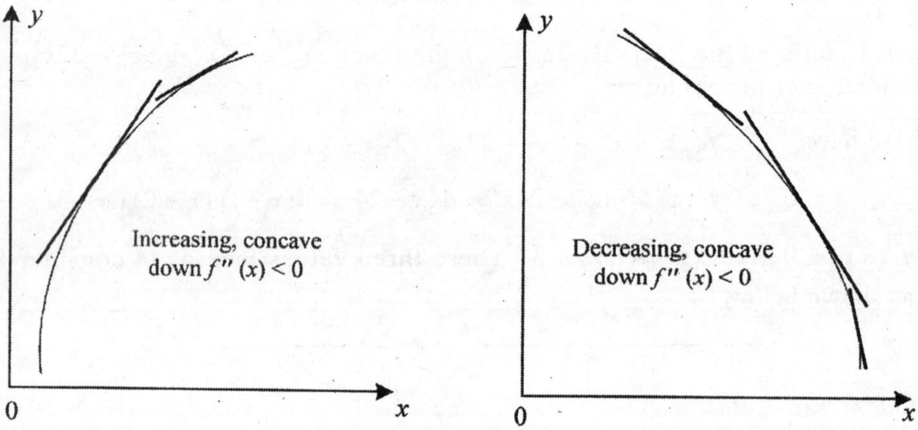

Fig. 6.5

We now state a rule which determines when a function is concave up and when it is concave down.

> ***Test for Concavity.*** *If* $f''(x) > 0$ *for all x in an interval I, then f is concave up on I; and if* $f''(x) < 0$ *for all x in I, then f is concave down on I.*

Fig. 6.6

Inflection Point. A point on the graph of $y = f(x)$ at which the curve changes from concave upward to concave downward or from concave downward to concave upward is called an **inflection point** (See Figure 6.6).

We describe below two different methods for determining the points of inflection.

Test for Inflection Points

FIRST METHOD

***Step* 1.** *Find $f''(x)$.*

***Step* 2.** *Set $f''(x)$ equal to zero and solve for x. These values of x are the candidates for inflection points. Let $x = x_0$ be one such value.*

***Step* 3.** *If $f''(x)$ changes sign as x passes through x_0, then f has a point of inflection at $x = x_0$. However, if $f''(x)$ does not change sign as x passes through x_0, then x_0 is not a point of inflection of f. Repeat this step for other values of x obtained in step 2.*

SECOND METHOD

***Step* 1.** *Find $f''(x)$.*

***Step* 2.** *Solve $f''(x) = 0$ for x. Let $x = x_0$ be one of its solutions.*

***Step* 3.** *Find $f'''(x_0)$. If $f'''(x_0) \neq 0$, then x_0 is a point of inflection of f. Repeat this step for other solutions obtained in Step 2.*

Remark. If $f'''(x_0) = 0$, the above test fails to give any information and we follow the general rule described as follows:

Let $f''(x_0) = f'''(x_0) \ldots = f^{(n-1)}(x_0) = 0$, but $f^{(n)}(x_0) \neq 0$. If n is odd, f has a point of inflection at x_0. If n is even, f is concave up or below according as $f^{(n)}(x) > 0$ or < 0.

***Example* 5.** Examine the function $f(x) = x^3 - 3x^2 + 3x - 3$ for concavity and points of inflection.

Solution. We have $\quad f'(x) = 3x^2 - 6x + 3$ and $f''(x) = 6x - 6 = 6(x - 1)$

For f to be concave up, we must have

$$f''(x) > 0 \quad \text{i.e.,} \quad 6(x-1) > 0 \quad \Rightarrow \quad x > 1$$

For f to be concave down, we must have

$$f''(x) < 0 \quad \text{i.e.,} \quad x < 1$$

To find points of inflection, we set $f''(x) = 0 \quad \Rightarrow \quad x = 1$

Also $\qquad f'''(x) = 6$

which is never zero, so $x = 1$ is a point of inflection of f.

Example 6. The cost of producing x units is given as $C(x) = .001x^3 - .3x^2 + 30x + 42$.

Determine where the cost function is concave up and where it is concave down. Also find the inflection point. **[*Delhi Univ. B. Com. (H) 1988*]**

Solution. We are given $C(x) = .001x^3 - .3x^2 + 30x + 42$

$\therefore$ $C'(x) = .003x^2 - .6x + 30$

and $C''(x) = .006x - .6$

For the given cost function to be concave up, we must have

$$C''(x) > 0 \quad i.e., \quad .006x - .6 > 0 \quad \Rightarrow \quad x > 100$$

For the given cost function to be concave down, we must have

$$C''(x) < 0 \quad i.e., \quad .006x - .6 < 0 \quad \Rightarrow \quad x < 100$$

Thus C is concave up when $x > 100$ and is concave down when $x < 100$. To find the inflection point, we solve

$$C''(x) = 0$$

for x. This gives $x = 100$. Since $C'''(x) = .006 \neq 0$, $x = 100$ is a point of inflection.

Example 7. An indifference map is defined by the relation $(x + h)\sqrt{y + k} = a$, where h and k are fixed positive numbers and a is a positive parameter. By expressing y as a function of x and by finding derivatives, show that each indifference curve is downward sloping and convex from below. **[*Delhi Univ. B. Com. (H) 2009*]**

Solution. We have $(x + h)\sqrt{y + k} = a$

$$\Rightarrow \quad \sqrt{y+k} = \frac{a}{x+h} \qquad i.e., \qquad y + k = \frac{a^2}{(x+h)^2} \qquad or \qquad y = \frac{a^2}{(x+h)^2} - k$$

Now
$$\frac{dy}{dx} = -\frac{2a^2}{(x+h)^3} < 0$$

Thus each indifference curve is downward sloping. Also

$$\frac{d^2y}{dx^2} = \frac{6a^2}{(x+h)^4} > 0,$$

so each indifference curve is convex from below.

Example 8. Examine the nature of the cost function $TC = a(x - 5)^3 + b$, where x is quantity produced and a and b are positive constants. Give a rough sketch of the function.

[*Delhi Univ. B. Com. (H) 1990*]

Solution. Differentiating the given cost function with respect to x, we obtain

$$\frac{d\,TC}{dx} = 3a\,(x-5)^2 > 0 \qquad (\text{since } a > 0)$$

Differentiating again with respect to x gives

$$\frac{d^2\,TC}{dx^2} = 6a\,(x-5)$$

which is positive or negative according as $x > 5$ or < 5. Thus the given cost function is upward sloping for all values of x but concave up or down according as $x > 5$ or < 5. Clearly, the function has a point of inflection at $x = 5$. A rough sketch of the function is shown in Figure 6.7.

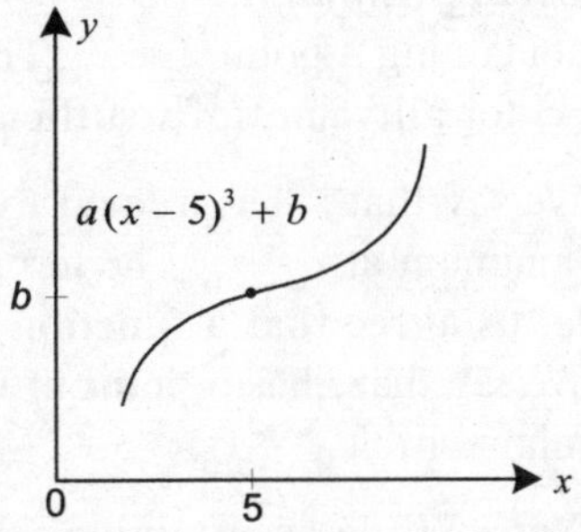

Fig. 6.7

Example 9. Show that the demand curve $p = \dfrac{a}{x-b}$ is downward sloping and convex from below whereas the demand curve $p = \sqrt{x-b}$ is upward sloping and concave from below (a and b are positive constants).

Solution. Differentiating the demand curve $p = \dfrac{a}{x-b}$ with respect to x, we obtain

$$\frac{dp}{dx} = -\frac{a}{(x-b)^2} < 0 \qquad (\text{since } a > 0)$$

Thus the demand curve $p = \dfrac{a}{x-b}$ is downward sloping. Also,

$$\frac{d^2p}{dx^2} = \frac{2a}{(x-b)^3} > 0$$

Therefore, it is convex from below.

For the demand curve $p = \sqrt{x-b}$, we have $\dfrac{dp}{dx} = \dfrac{1}{2\sqrt{x-b}} > 0$

and
$$\frac{d^2p}{dx^2} = -\frac{1}{4}\,\frac{1}{(x-b)^{3/2}} < 0$$

Thus the demand curve $p = \sqrt{x-b}$ is upward sloping and concave from below.

6.5 LOCAL MAXIMA AND LOCAL MINIMA

In this section we shall be concerned with the most important application of the derivative in determining the "high points" and "low points" on a curve technically known as *maximum* and *minimum* values.

Local Maximum. Let $y = f(x)$ be a real-valued function that is defined in some interval containing a point $x = x_0$. Then $f(x)$ is said to have a **local maximum** at $x = x_0$ if $f(x_0) \geq f(x)$ for all values of x sufficiently close to x_0. The **local maximum value** is $f(x_0)$.

Local Minimum. Let $y = f(x)$ be a real-valued function that is defined in some interval containing a point $x = x_0$. Then $f(x)$ is said to have a **local minimum** at $x = x_0$ if $f(x_0) \leq f(x)$ for all values of x sufficiently close to x_0. The **local minimum value** is $f(x_0)$.

We say that f has a **local extremum** at $x = x_0$ if f has either a local maximum or a local minimum at $x = x_0$. The term **relative** is also sometimes used instead of local. **Furthermore, let us agree that a function cannot have a local extremum at an end point of its domain.** We say that f has a **point of inflection** at $x = x_0$ if f has neither a local maximum nor a local minimum at $x = x_0$.

Note. While ascertaining whether or not a point x_0 is a local extremum, we compare the function value at x_0 to those of neighbouring points, so that the values of the function outside the neighbourhood do not come into picture at all. Thus a function defined in a given domain can have several local maximum and minimum values. For example, in Figure 6.8, the function f has a local maximum at x_1, x_3 and x_5, and a local minimum at x_2 and x_4.

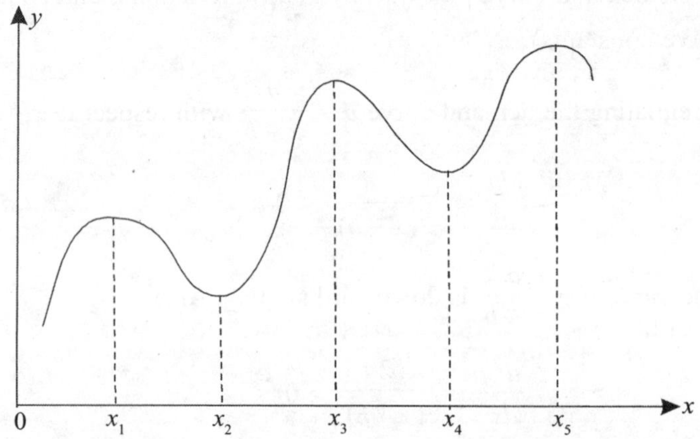

Fig. 6.8

> ***A Necessary Condition for Local Extrema.*** *Let $y = f(x)$ be a function that is differentiable on an open interval containing a point $x = x_0$. A necessary condition for f to have a local maximum or a local minimum at $x = x_0$ is that $f'(x_0) = 0$.*

Geometrically interpreted, the necessary condition states that if f has a local extremum at a point $x = x_0$ and that there is a tangent line to the graph of f at $(x_0, f(x_0))$, then the tangent line at $(x_0, f(x_0))$ must be horizontal. See, for example, the function graphed in Figure 6.9.

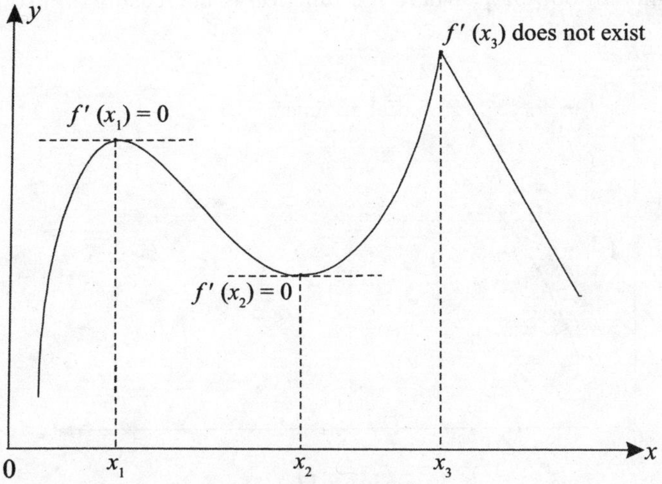

Fig. 6.9

Remark. The necessary condition stated above gives us a means for generating candidates for values of x where the differentiable function f may attain either a local maximum or a local minimum. Such candidates are called **stationary points** of f.

Stationary Point. A point x_0 in the domain of f at which $f'(x_0) = 0$ is called a **stationary point** of f.

Note. It may be noted that for a differentiable function, a local maximum or a local minimum point is always a stationary point but a stationary point may neither be a local maximum nor a local minimum. For example, the function $f(x) = x^3$ has as its derivative $f'(x) = 3x^2$. Consequently, $f'(0) = 0$. Thus $x = 0$ is a stationary point of the given function. However, it is neither a local maximum nor a local minimum (See Figure 6.10).

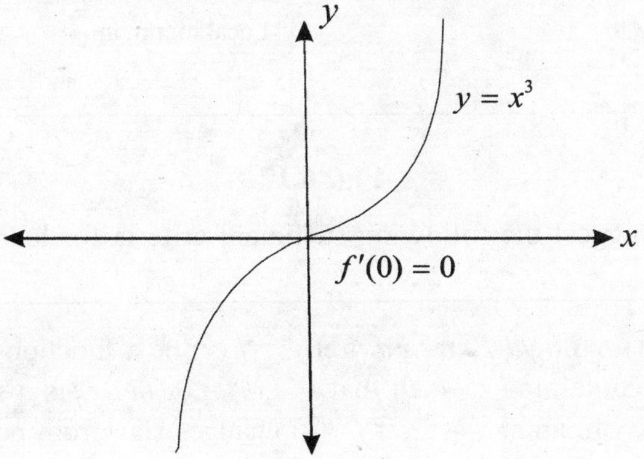

Fig. 6.10

We next obtain sufficient conditions for the differentiable function to have a local extremum at a point. Let us investigate Figure 6.11 which has a local maximum at $x = x_0$. It is clear that there exists a left handed neighbourhood of x_0 where the function is increasing and so $f'(x) > 0$, and a right handed neighbourhood of x_0 where the function is decreasing and so $f'(x) < 0$.

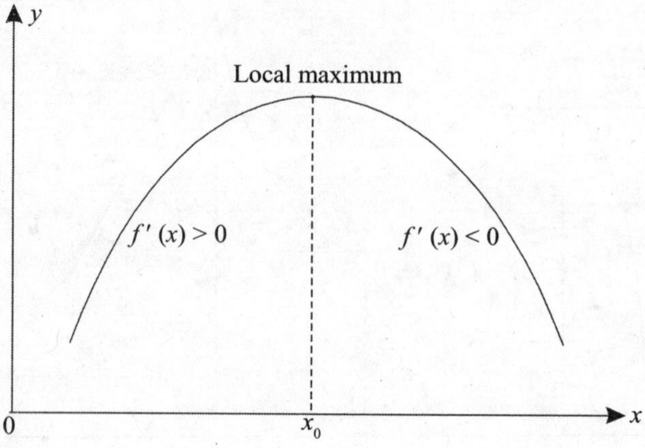

Fig. 6.11

In a similar fashion, let us investigate Figure 6.12 which has a local minimum at $x = x_0$. To the left of x_0, there is a neighbourhood where the function is decreasing and so $f'(x) < 0$; to the right of x_0, there is a neighbourhood where the function is increasing and so $f'(x)>0$.

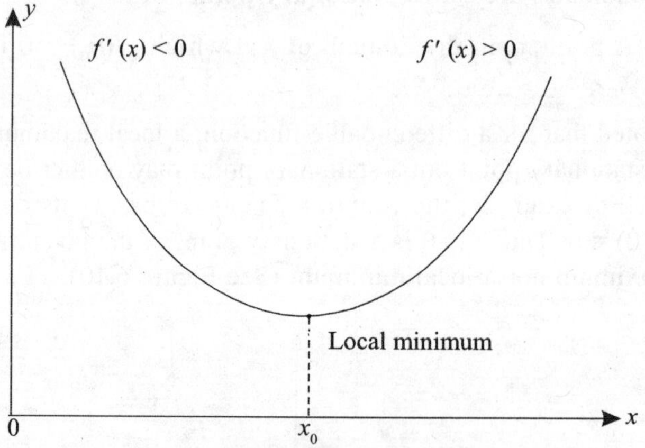

Fig. 6.12

These investigations suggest the following sufficient criteria for local maximum and local minimum.

> ***Sufficient Condition for Local Extrema.*** Let $y = f(x)$ be a function that is differentiable on an open interval containing x_0 such that $f'(x_0) = 0$, *i.e.* x_0 is a stationary point of f. Then f has a local maximum at $x = x_0$ if $f'(x)$ changes sign from positive to negative as x increases through x_0 *and* f has a local minimum at $x = x_0$ if $f'(x)$ changes sign from negative to positive as x increases through x_0.

Combining necessary and sufficient conditions, we have the *first-derivative test* for the local extrema of the differentiable function $y = f(x)$:

First-Derivative Test for Local Extrema

Step 1. *Find* $f'(x)$.

Step 2. *Find all stationary points of the function* $y = f(x)$, *i.e., find all values of x where* $f'(x) = 0$.

Step 3. *For each stationary point* x_0, *determine the sign of the first derivative slightly to the left but not beyond the next stationary point and slightly to the right but not beyond the next stationary point.*

Step 4. *There is a local maximum at* $x = x_0$ *iff* $f'(x)$ *changes sign from positive to negative, and a local minimum if* $f'(x)$ *changes sign from negative to positive. If* $f'(x)$ *does not change sign, there is no local extremum at* $x = x_0$. *Such a point is called a point of inflection.*

The following examples illustrate this test.

Example 10. Find all the points of local maxima and minima of the function:

$$f(x) = x^3 - 3x^2 - 9x + 15$$

Solution. We have $$f'(x) = 3x^2 - 6x - 9 = 3(x+1)(x-3)$$

$\therefore$ $$f'(x) = 0 \text{ when } x = -1 \text{ or } x = 3$$

Thus $x = -1$ and $x = 3$ are stationary points.

If $x < -1$, then $f'(x) = 3(-)(-) > 0$;

If $-1 < x < 3$, then $f'(x) = 3(+)(-) < 0$; and

If $x > 3$, then $f'(x) = 3(+)(+) > 0$.

Since $f'(x)$ changes sign from + to – as x increases through – 1, therefore $x = -1$ is a point of local maximum. Similarly, $f'(x)$ changes sign from – to + as x increases through 3, therefore $x = 3$ is a point of local minimum.

Example 11. Find all the points of local maxima and local minima as well as the corresponding local maximum and minimum values of the function:

$$f(x) = x^4 - 8x^3 + 22x^2 - 24x + 1.$$

Solution. We have $$f'(x) = 4x^3 - 24x^2 + 44x - 24 = 4(x-1)(x-2)(x-3)$$

Thus $f'(x) = 0$ when $x = 1$, $x = 2$ or $x = 3$. These values of x are the candidates for local maximum and local minimum. To find the nature of each point, we apply first-derivative test:

If $x < 1$, then $f'(x) = 4(-)(-)(-) = (-)$;

If $1 < x < 2$, then $f'(x) = 4(+)(-)(-) = (+)$;

If $2 < x < 3$, then $f'(x) = 4(+)(+)(-) = (-)$; and

If $x > 3$, then $f'(x) = 4(+)(+)(+) = (+)$;

Nature of x = 1. Since $f'(x)$ changes sign from – to + as x increases through 1, therefore $x = 1$ is a point of local minimum. The corresponding minimum value is $f(1) = 1 - 8 + 22 - 24 + 1 = -8$.

Nature of x = 2. Since $f'(x)$ changes sign from + to – as x increases through 2, therefore $x = 2$ is a point of local maximum. The corresponding maximum value is $f(2) = 16 - 64 + 88 - 48 + 1 = -7$.

Nature of x = 3. Since $f'(x)$ changes sign from – to + as x increases through 3, therefore $x = 3$ is a point of local minimum. The corresponding minimum value is $f(3) = 81 - 216 + 198 - 72 + 1 = -8$.

Second-Derivative Test for Local Extrema

In the preceding section we discussed the so called first derivative test for determining local extrema of a function. However, the same thing can sometimes be done very conveniently by the use of second derivative. We outline the procedure as follows :

Step 1. *Find* $f'(x)$.

Step 2. *Set* $f'(x)$ *equal to zero and solve for x to get stationary points.*

Step 3. *Find* $f''(x)$.

Step 4. *For each stationary point* x_0, *determine* $f''(x_0)$.

Step 5. *If* $f''(x_0) < 0$, *then f has a local maximum at* $x = x_0$ *and if* $f''(x_0) > 0$, *then f has a local minimum at* $x = x_0$.

Note. The second derivative test fails if $f''(x_0) = 0$ at a stationary point x_0. Under these conditions there may be a local maximum, a local minimum, or neither at $x = x_0$. In such cases, we either use the first derivative test or employ the following general test which makes use of higher order derivatives.

General Criteria for Local Extrema

Let $f'(x_0) = f''(x_0) = \ldots = f^{(n-1)}(x_0) = 0$, but $f^{(n)}(x_0) \neq 0$. Then

1. *If n is odd, f has neither a local maximum nor a local minimum at* $x = x_0$.

2. *If n is even, f has a local maximum if* $f^{(n)}(x_0) < 0$, *and has a local minimum if* $f^{(n)}(x_0) > 0$.

This method is illustrated in the following examples.

Example 12. Use the second derivative test to find the local maxima and minima of $f(x) = 2x^3 - 15x^2 + 36x + 18$.

Solution. Differentiating the given function with respect to *x*, we obtain

$$f'(x) = 6x^2 - 30x + 36 = 6(x-2)(x-3)$$

Now $\qquad f'(x) = 0$ when $x = 2$ and $x = 3$

Also $\qquad f''(x) = 12x - 30$

$\therefore \qquad f''(2) = 12(2) - 30 = -6 < 0,$

hence there is a local maximum at $x = 2$.

Further $\qquad f''(3) = 12(3) - 30 = 6 > 0,$

hence there is a local minimum at $x = 3$.

Example 13. Find all the points of local maxima and minima of the function :

$$f(x) = x^5 - 5x^4 + 5x^3 - 1$$

Solution. We have

$$f'(x) = 5x^4 - 20x^3 + 15x^2 = 5x^2(x^2 - 4x + 3) = 5x^2(x-1)(x-3)$$

Solving $f'(x) = 0$ for *x* gives $x = 0$, $x = 1$ and $x = 3$. Further

$$f''(x) = 20x^3 - 60x^2 + 30x$$

If $x = 1$, $f''(x) = -10 < 0$, so there is a local maximum at $x = 1$.

If $x = 3$, $f''(x) = 90 > 0$, so there is a local minimum at $x = 3$

If $x = 0$, $f''(x) = 0$ and the second-derivative test fails. We now turn either to the first-derivative test or to general criteria for local extrema to analyse what is happening at 0. Let us apply the general criteria. We have

$$f'''(x) = 60x^2 - 120x + 30 \quad \Rightarrow \quad f'''(0) = 30$$

Thus $\qquad f'(0) = f''(0) = 0$, but $f'''(0) \neq 0$

Since $n = 3$ is odd, therefore *f* has neither a maximum nor a minimum value at $x = 0$.

Example 14. Find the local maxima and minima of the function:

$$f(x) = x^4 - 8x^3 + 22x^2 - 24x + 1$$

Solution. We have $f'(x) = 4x^3 - 24x^2 + 44x - 24 = 4(x-1)(x-2)(x-3)$

Thus $\qquad f'(x) = 0$ when $x = 1$, $x = 2$ and $x = 3$.

Further $\qquad f''(x) = 12x^2 - 48x + 44$

If $x = 1$, $f''(x) = 8 > 0$, hence f has a local minimum at $x = 1$. The corresponding minimum value is $f(1) = -8$.

If $x = 2$, $f''(x) = -4 < 0$, hence f has a local maximum at $x = 2$. The corresponding maximum value is $f(2) = -7$.

If $x = 3$, $f''(x) = 8 > 0$, hence f has a local minimum at $x = 3$. The corresponding minimum value is $f(3) = -8$.

Example 15. Find all the points of local maxima and minima of the function $f(x) = (1-x)^2 e^x$. Also find the corresponding maximum and minimum values.

Solution. We have

$$f(x) = (1-x)^2 e^x = (1-2x+x^2)\, e^x$$

$$\therefore \quad f'(x) = (1-2x+x^2)\, e^x + e^x(-2+2x) = e^x(1-2x+x^2-2+2x) = e^x(x^2-1)$$

i.e., $\quad f'(x) = e^x(x^2-1)$

For local maxima and minima, $f'(x) = 0$. That is,

$$e^x(x^2-1) = 0 \Rightarrow x = \pm 1 \qquad (\because e^x \text{ is never } 0)$$

Also $\qquad f''(x) = e^x\, 2x + (x^2-1)\, e^x = e^x(x^2+2x-1)$

If $x = 1$, $f''(x) = e(1+2-1) = 2e > 0$ and hence there is a local minimum at $x = 1$. The corresponding minimum value is $f(1) = 0$.

If $x = -1$, $f''(x) = e^{-1}(1-2-1) = -\dfrac{2}{e} < 0$ and hence there is a local maximum at $x = -1$. The corresponding maximum value is $f(-1) = 4/e$.

Example 16. An efficiency study at a certain factory indicates that workers who arrive on the job at 9.00 a.m. assemble $f(x) = -x^3 + 6x^2 + 15x$ computers in x hours. The study indicates further that after one-hour lunch break, the workers assemble $g(x) = -\dfrac{1}{3}x^3 + x^2 + 23x$ computers in x hours. Determine the time between 9.00 a.m. and 6.00 p.m. at which one-hour break should be scheduled so that workers assemble the maximum number of computers. **[*Delhi Univ. B. Com. (H) 2010*]**

Solution. Let lunch break be scheduled x hours after 9 a.m.

Number of computers assembled before lunch break $= -x^3 + 6x^2 + 15x$

Number of hours after lunch break $= 8 - x$

Number of computers assembled after lunch break $= -\frac{1}{3}(8-x)^3 + (8-x)^2 + 23(8-x)$

$\therefore$ Total number of computers assembled is :

$$z = -x^3 + 6x^2 + 15x - \frac{1}{3}(8-x)^3 + (8-x)^2 + 23(8-x)$$

Our problem is to find x so that z is maximum. Differentiating z w.r.t. x, obtain

$$\frac{dz}{dx} = -3x^2 + 12x + 15 + (8-x)^2 - 2(8-x) - 23$$

$$= -3x^2 + 12x + 15 + 64 + x^2 - 16x - 16 + 2x - 23$$

$$= -2x^2 - 2x + 40$$

For z to be minimum, $\frac{dz}{dx} = 0$. That is,

$$-2x^2 - 2x + 40 = 0 \quad \Rightarrow \quad x = -5 \text{ or } x = 4 \quad \Rightarrow \quad x = 4 \quad \textit{(rejecting -ve value)}$$

Also, $\frac{d^2z}{dx^2} = -4x - 2$, which is negative for $x = 4$.

Hence z is maximum when $x = 4$, *i.e.*, when lunch-break is scheduled at 1 p.m.

EXERCISE 6.1

1. Find the intervals on which the following functions are increasing or decreasing:

 (*i*) $f(x) = x^3 - 6x^2 - 36x + 2$ (*ii*) $f(x) = 2x^3 - 15x^2 + 36x + 1$

 (*iii*) $f(x) = 8 + 36x + 3x^2 - 2x^3$ (*iv*) $f(x) = -2x^3 - 9x^2 - 12x + 1$

2. Find the intervals on which the following functions are increasing or decreasing:

 (*i*) $f(x) = x^4 - 8x^3 + 22x^2 - 24x + 1$ (*ii*) $f(x) = x^4 - \frac{x^3}{3}$

 (*iii*) $f(x) = (x+1)^3 (x-3)^3$ (*iv*) $f(x) = x^2 e^x$

3. Show that the demand curve $p = \frac{50}{x+3}$ is decreasing and concave up.

4. Show that the demand curve $p = \dfrac{a}{x+b} - c$ is decreasing and concave up, where a and b are positive constants and c is any constant.

5. Show that the demand curve $p = (a - bx)^2$, where a and b are positive constants, and x is less than a/b, is downward sloping and convex from below.

6. The supply function of a product is $x = a\sqrt{p-b}$, where x is quantity supplied, p (which is greater than b) is price and a and b are positive constants. Show, by using calculus, that the supply curve is increasing and concave down. **[Delhi Univ. B. Com. (H) 2012]**

7. Discuss the concavity of the following cost function of a firm :

$$C(x) = 6x - 0.03x^2 + 0.00005x^3$$ **[Delhi Univ. B. Com. (H) 1995]**

8. Find all points of local maxima or minima, if any, of the following functions. Find also the corresponding local maximum or local minimum values, as the case may be :

(*i*) $f(x) = x^3 - 6x^2 + 9x + 15$ (*ii*) $f(x) = 2x^3 + 3x^2 - 36x + 10$

(*iii*) $f(x) = 2x^3 - 9x^2 + 12x + 25$ (*iv*) $f(x) = (x - 1)(x + 2)^2$

9. Examine for maximum and minimum values the following functions :

(*i*) $f(x) = 3x^4 - 4x^3 + 1$ (*ii*) $f(x) = x^4 + 2x^3 - 3x^2 - 4x + 4$

10. Investigate the maximum and minimum values of the function :

$$f(x) = 12x^5 - 45x^4 + 40x^3 + 6.$$

ANSWERS

1. (*i*) Increasing on $(-\infty, -2)$ and $(6, \infty)$; Decreasing on $(-2, 6)$.

(*ii*) Increasing on $(-\infty, 2)$ and $(3, \infty)$; Decreasing on $(2, 3)$.

(*iii*) Increasing on $(-2, 3)$; Decreasing on $(-\infty, -2)$ and $(3, \infty)$.

(*iv*) Increasing on $(-2, -1)$; Decreasing on $(-\infty, -2)$ and $(-1, \infty)$.

2. (*i*) Increasing on $(1, 2)$ and $(3, \infty)$; Decreasing on $(-\infty, 1)$ and $(2, 3)$.

(*ii*) Increasing on $(1/4, \infty)$; Decreasing on $(-\infty, 1/4)$.

(*iii*) Increasing on $(1, \infty)$; Decreasing on $(-\infty, 1)$.

(*iv*) Increasing on $(-\infty, -2)$ and $(0, \infty)$; Decreasing on $(-2, 0)$.

7. Concave up when $x > 200$ and concave down when $x < 200$.

8. (*i*) Local maximum at $x = 1$, value $= 19$; local minimum at $x = 3$, value $= 15$.

(*ii*) Local maximum at $x = -3$, value $= 91$; local minimum at $x = 2$, value $= -34$.

(*iii*) Local maximum at $x = 1$, value $= 30$; local minimum at $x = 2$, value $= 29$.

(*iv*) Local maximum at $x = -2$, value $= 0$; local minimum at $x = 0$, value $= -4$.

9. (*i*) Local maximum at $x = 1$, value $= 0$; point of inflection at $x = 0$.

(*ii*) Local maximum at $x = -\frac{1}{2}$, value $= \frac{81}{16}$; local minimum at $x = 2$, value $= 0$; and local minimum at $x = 1$, value $= 0$.

10. Local maximum at $x = 1$, value $= 13$; local minimum at $x = 2$, value $= -10$; point of inflection at $x = 0$.

6.6 AVERAGE COST AND MARGINAL COST

In economics the variation of one quantity with respect to another quantity may be described by either an *average* concept or a *marginal* concept. The average concept expresses the variation of one quantity over a specified range of values of a second quantity, whereas the marginal concept is the instantaneous rate of change of one quantity with respect to another quantity.

For example, if $C = f(x)$ gives the total cost of producing and marketing x units of a product, then the **average cost,** AC, is defined to be

$$AC = \frac{C}{x}$$

and the **marginal cost,** MC, is defined to be the rate of change of C with respect to x. Thus

$$MC = \frac{dC}{dx}$$

We interpret marginal cost as the approximate cost of one additional unit of output.

Example 17. If a manufacturer's total cost function is $C = 0.1x^2 + 3$, find

(*a*) the average cost function,

(*b*) the marginal cost function, and

(*c*) the marginal cost when 4 units are produced. Interpret the result.

Solution . (*a*) The average cost function, AC, is $AC = \frac{C}{x} = \frac{(0.1x^2 + 3)}{x} = 0.1x + \frac{3}{x}$.

(*b*) The marginal cost function, MC, is obtained by differentiating C with respect to x. Thus

$$MC = \frac{dC}{dx} = 0.2x$$

(*c*) The marginal cost when 4 units are produced is $(MC)_{x=4} = (0.2)\,4 = 0.80$.

This means that, if production is increased by 1 unit from 4 units to 5 units, then the cost of additional unit is approximately 0.80. It may be noted that the actual cost of producing one more unit beyond 4 units is $C(5) - C(4) = 5.5 - 4.6 = 0.90$.

Example 18. The cost of producing x tons of steel is given by $C(x) = \frac{1}{10}x^2 + 5x + 1000$.

(*i*) Find the marginal cost function.

(*ii*) Find marginal cost when $x = 1000$ tons.

(*iii*) Interpret $C'(1000)$ and compare with the actual cost of 1001th ton.

[Delhi Univ. B. Com. (H) 2007]

Solution. Given $$C(x) = \frac{1}{10}x^2 + 5x + 1000$$

(*i*) $$MC = C'(x) = \frac{1}{5}x + 5$$

(*ii*) $$(MC)_{x=1000} = \frac{1}{5} \times 1000 + 5 = 205$$

(*iii*) *Interpretation of C' (1000)* : This means that if production is increased by 1 ton from 1000 tons to 1001 tons, then the cost of additional ton is approximately 205.

The actual cost of producing 1001th ton is : $C(1001) - C(1000) = 206.10$

Example 19. The total cost $C(x)$ of a firm is: $C(x) = 1500 + 30x + x^2$, where x is the output. Determine:

(*i*) the average cost,

(*ii*) the marginal cost,

(*iii*) the marginal cost when 20 units are produced, and

(*iv*) the actual cost of producing twenty first unit.

Solution. (*i*) The average cost is given by

$$AC = \frac{C}{x} = \frac{1500 + 30x + x^2}{x} = \frac{1500}{x} + 30 + x$$

(*ii*) The marginal cost, *MC*, is obtained by differentiating *C* with respect to *x*. Thus

$$MC = \frac{dC}{dx} = 30 + 2x$$

(*iii*) The marginal cost when 20 units are produced is:

$$(MC)_{x=20} = 30 + 2(20) = 70$$

(*iv*) The actual cost of producing twenty first unit is:

$$C(21) - C(20) = [1500 + 30(21) + (21)^2] - [1500 + 30(20) + (20)^2]$$
$$= 2571 - 2500 = 71.$$

Example 20. The total cost $C(x)$ of a firm is : $C(x) = 0.005x^3 - 0.02x^2 - 30x + 5000$, where x is the output. Determine :

(i) The average cost.

(ii) The marginal average cost, *i.e.*, the rate of change of average cost with respect to x.

(iii) The marginal cost.

(iv) The rate of change of MC with respect to x.

(v) The value of x for which $MVC = AVC$, where MVC and AVC denote, respectively, the marginal variable cost and the average variable cost.

Solution. (*i*) The average cost is : $AC = \frac{C}{x} = \frac{0.005x^3 - 0.02x^2 - 30x + 5000}{x}$

$$= 0.005x^2 - 0.02x - 30 + \frac{5000}{x}$$

(*ii*) The marginal average cost, MAC, is given by

$$MAC = \frac{dAC}{dx} = 0.01x - 0.02 - \frac{5000}{x^2}$$

(*iii*) The marginal cost, MC, is obtained by differentiating C with respect to x. Thus

$$MC = 0.015x^2 - 0.04x - 30$$

(*iv*) The rate of change of MC with respect to x is simply derivative of MC with respect to x :

$$\frac{dMC}{dx} = 0.03x - 0.04$$

(*v*) We have $VC = 0.005x^3 - 0.02x^2 - 30x$

$\therefore$ $MVC = \frac{dVC}{dx} = 0.015x^2 - 0.04x - 30$

and $AVC = \frac{VC}{x} = 0.005x^2 - 0.02x - 30$

Thus the equation $MVC = AVC \Rightarrow 0.015x^2 - 0.04x - 30 = 0.005x^2 - 0.02x - 30$

$\Rightarrow 0.01x^2 - 0.02x = 0$ *i.e.*, $x(0.01x - 0.02) = 0 \Rightarrow x = 0$ or $x = 2.$

Example 21. If the total cost function is given by $C = a + bx + cx^2$, where x is the quantity of output, show that

$$\frac{d}{dx}(AC) = \frac{1}{x}(MC - AC)$$

where MC and AC are marginal and average cost. **[*Delhi Univ. B. Com. (H) 1983*]**

Solution. The average cost function is:

$$AC = \frac{C}{x} = \frac{a}{x} + b + cx$$

Also $$MC = \frac{dC}{dx} = \frac{d}{dx}(a + bx + cx^2) = b + 2cx$$

$$\therefore \quad \frac{1}{x}(MC - AC) = \frac{1}{x}\left[(b + 2\,cx) - \left(\frac{a}{x} + b + cx\right)\right] = \frac{1}{x}\left[cx - \frac{a}{x}\right] = c - \frac{a}{x^2} \qquad ...(1)$$

Now $$\frac{d}{dx}(AC) = \frac{d}{dx}\left(\frac{a}{x} + b + cx\right) = -\frac{a}{x^2} + c \qquad ...(2)$$

From (1) and (2), we conclude that $\dfrac{d}{dx}(AC) = \dfrac{1}{x}[MC - AC]$.

Example 22. The rate of change of total cost (C) of a commodity per unit change of output x is called the marginal cost of the commodity. If there exists a relation between C and x in the form : $C = 3x\left(\dfrac{x + 7}{x + 5}\right) + 5$. Prove that the marginal cost falls continuously as the output increases. **[*Delhi Univ. B. Com. (H) 1982*]**

Solution. We have $$C = 3x\left(\frac{x + 7}{x + 5}\right) + 5 = 3\left(\frac{x^2 + 7x}{x + 5}\right) + 5$$

Differentiating C with respect to x, we obtain

$$\text{Marginal Cost} = \frac{dC}{dx} = 3\left[\frac{(x + 5)(2x + 7) - (x^2 + 7x)}{(x + 5)^2}\right]$$

$$= 3\left[\frac{x^2 + 10x + 35}{(x + 5)^2}\right]$$

$$= 3\left[\frac{(x^2 + 10x + 25) + 10}{(x + 5)^2}\right]$$

$$= 3\left[\frac{(x + 5)^2 + 10}{(x + 5)^2}\right]$$

$$= 3\left[1 + \frac{10}{(x + 5)^2}\right]$$

i.e., Marginal Cost $= MC = 3\left[1+\dfrac{10}{(x+5)^2}\right]$

$$\therefore \quad \frac{d}{dx}(MC) = 3\left[\frac{-20}{(x+5)^3}\right] = \frac{-60}{(x+5)^3} < 0$$

Hence the marginal cost falls continuously as the output increases.

Example 23. The average cost function (AC) for a commodity is given by $AC = x + 5 + \dfrac{36}{x}$, where x is the output. Find the output for which AC is increasing and the output for which AC is decreasing with increasing output. Also find the total cost, C, and the marginal cost, MC, as functions of x.

Solution. Differentiating both sides of the equation

$$AC = x + 5 + \frac{36}{x}$$

with respect to x, we obtain $\dfrac{d}{dx}(AC) = 1 - \dfrac{36}{x^2}$

For AC to be increasing, we have

$$\frac{d}{dx}(AC) > 0 \ \ i.e., \ 1 - \frac{36}{x^2} > 0 \quad \Rightarrow \quad x^2 > 36 \ \Rightarrow \ x > 6,$$

since x cannot be negative.

Similarly, for AC to be decreasing, we have

$$\frac{d}{dx}(AC) < 0 \ \ i.e., \ 1 - \frac{36}{x^2} < 0 \quad \Rightarrow \quad x^2 < 36 \ \text{ or } \ 0 < x < 6$$

To find the total cost function C, we have

$$AC = \frac{C}{x} \quad \Rightarrow \quad C = x\,(AC) = x\left[x + 5 + \frac{36}{x}\right] = x^2 + 5x + 36$$

Differentiating C, we have the marginal cost function : $MC = \dfrac{dC}{dx} = 2x + 5.$

Relation Between Average and Marginal Cost Curves

Although the cost functions may assume many different shapes under different circumstances, yet, usually under natural economic limitations, we assume average and marginal cost curves to have U-shapes as shown in Figure 6.13. The relation between them is assumed to be as follows :

1. *When AC is decreasing, MC < AC.*
2. *When AC is minimum, AC = MC.*
3. *When AC is increasing, MC > AC.*

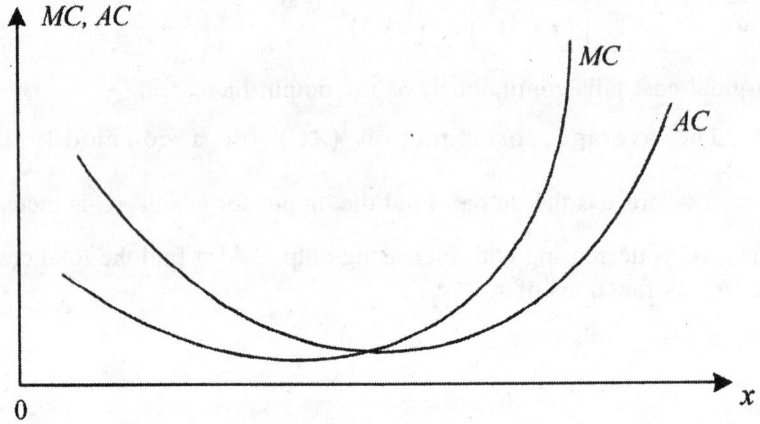

Fig. 6.13

We will make use of calculus to prove these relationships. Assume that $C = C(x)$ is the total cost function, where x is the output. Then

$$\text{Average Cost} = AC = \frac{C}{x} \quad \text{and} \quad \text{Marginal Cost} = MC = \frac{dC}{dx}$$

Differentiating the average cost function AC, we have

$$\frac{d}{dx}(AC) = \frac{d}{dx}\left(\frac{C}{x}\right) = \frac{x\dfrac{dC}{dx} - C}{x^2} = \frac{1}{x}\left[\frac{dC}{dx} - \frac{C}{x}\right] = \frac{1}{x}[MC - AC]$$

Thus $$\frac{d}{dx}(AC) = \frac{1}{x}[MC - AC]$$

1. *When the average cost curve is sloping downward,* $\frac{d}{dx}(AC) < 0$, and this implies

$$\frac{1}{x}[MC - AC] < 0, \quad i.e., \quad MC < AC$$

This means when AC curve is decreasing, MC curve is below the AC curve (Figure 6.13).

2. *When the average cost curve reaches at the lowest point,* $\frac{d}{dx}(AC) = 0$, and this means

$$\frac{1}{x}[MC - AC] = 0, \quad i.e., \quad MC = AC$$

Thus $MC = AC$ at the lowest point of AC curve (Figure 6.13).

3. *When the average cost curve is increasing,* $\frac{d}{dx}(AC) > 0$, and this means

$$\frac{1}{x}[MC - AC] > 0, \quad i.e., \quad MC > AC$$

Thus when AC curve is rising, MC curve is above AC curve (Figure 6.13).

6.7 AVERAGE REVENUE AND MARGINAL REVENUE

Recall that the total revenue, R, from the sale of x units of a product is given by $R = px$, where p is the price per unit. The **average revenue,** AR, is defined to be

$$AR = \frac{R}{x}$$

Since $R = px$, it follows that $AR = \frac{px}{x} = p$. Thus average revenue is same as the price per unit.

The **marginal revenue** is defined to be the rate of change of total revenue with respect to quantity demanded. Hence marginal revenue is merely the derivative of R with respect to x :

$$MR = \textit{Marginal revenue} = \frac{dR}{dx}$$

Marginal revenue indicates the rate at which revenue changes with respect to units sold. It is interpreted as the approximate revenue received from selling one additional unit of output.

We now give a brief marginal analysis of a firm operating under the two extreme conditions of *pure competition and monopoly.*

Case* 1. *Pure Competition

The following basic assumptions are in effect :

(*a*) *There are many firms producing homogeneous products.*

(*b*) *Price is determined by the market.*

(*c*) *Price is not changed by the firm's output. That is, competition is perfect in the sense that it can sell any amount of output it wishes at the going market price, which cannot be affected by the individual producer whose share in the market is very small.*

In this case, $R = px$ and p is not dependent on x. Therefore, $MR = p$.

Case* 2. *Monopoly

The following basic assumptions are in effect :

(*a*) *There is only one seller and there are no close substitutes for the product of the monopolist, i.e., the seller.*

(*b*) *Demand depends on price and vice versa.*

(*c*) *The firm must lower its price to sell more items.*

In this case, $R = px$ but p depends on x. Therefore, $MR = p + x\frac{dp}{dx}$ by the product rule.

Example 24. Find the total revenue and marginal revenue for a firm operating under pure competition with a current price of ₹ 2 per unit.

Solution . If x units are sold, the total revenue is given by $R = 2x$. The marginal revenue is:

$$MR = \frac{dR}{dx} = 2,$$

which is constant. Thus the marginal revenue is 2 regardless of the number of units sold.

Example 25. The demand function of a product is given by $p = 200 - 5x$, where p is the price per unit and x is the quantity demanded. Find MR.

Solution. The demand function is : $p = 200 - 5x$

$$\Rightarrow \qquad R = px = (200 - 5x)\, x = 200x - 5x^2$$

Hence $$MR = \frac{dR}{dx} = 200 - 10x$$

Example 26. Find the marginal revenue function for a monopolist confronted with demand $x = 10 - 2p$.

Solution. Demand function is $x = 10 - 2p$ which gives $p = 5 - \frac{x}{2}$.

$$\therefore \qquad R = px = \left(5 - \frac{x}{2}\right) x = 5x - \frac{x^2}{2}$$

Hence $$MR = \frac{dR}{dx} = 5 - x.$$

Example 27. The total revenue received from the sale of x units of a product is given by $R(x) = 20x - \frac{x^2}{2}$. Find:

(*i*) The average revenue, (*ii*) The marginal revenue,

(*iii*) The marginal revenue when $x = 5$ and interpret it,

(*iv*) The actual revenue from the sale of sixth unit.

Solution. (*i*) The average revenue, AR, is :

$$AR = \frac{R}{x} = \frac{1}{x}\left(20x - \frac{x^2}{2}\right) = 20 - \frac{x}{2}$$

(*ii*) The marginal revenue, MR, is obtained by differentiating R with respect to x. Thus

$$MR = \frac{dR}{dx} = 20 - x$$

(*iii*) The marginal revenue when $x = 5$ is : $(MR)_{x=5} = 20 - 5 = 15$.

Interpretation. This means that revenue is increased by ₹ 15 if an additional unit beyond 5 is sold.

(*iv*) The actual revenue from the sale of 6th unit is :

$$R(6) - R(5) = \left[20(6) - \frac{(6)^2}{2}\right] - \left[20(5) - \frac{(5)^2}{2}\right] = 102 - 87.5 = ₹14.50.$$

Example 28. Suppose that the demand function for a product is given by $p = a - bx$, where a and b are positive constants. Find:

(*i*) The total revenue function.

(*ii*) The average revenue function.

(*iii*) The marginal revenue function.

Solution. (*i*) If R denotes the total revenue function, then

$$R = px = (a - bx)\,x = ax - bx^2$$

(*ii*) The average revenue, AR, is given by

$$AR = \frac{R}{x} = \frac{ax - bx^2}{x} = a - bx$$

(*iii*) The marginal revenue, MR, is given by

$$MR = \frac{dR}{dx} = \frac{d}{dx}(ax - bx^2) = a - 2bx.$$

Example 29. A firm knows that the demand function for one of its products is linear. It also knows that it can sell 1000 units when the price is ₹ 4 per unit, and it can sell 1500 units when the price is ₹ 2 a unit. Determine (*a*) the demand function (*b*) the total revenue function (*c*) the average revenue function (*d*) the marginal revenue function.

Solution. (*a*) Since the demand function is assumed to be linear, let it be given by

$$p = a + bx,$$

where p is the price per unit and x is the quantity demanded at this price. Since $x = 1000$ when $p = 4$, and $x = 1500$ when $p = 2$, we have

$$4 = a + 1000\,b \quad \text{and} \quad 2 = a + 1500\,b$$

Solving these equations simultaneously for a and b, we obtain $a = 8$ and $b = -\dfrac{1}{250}$.

Hence the demand function is : $p = 8 - \dfrac{x}{250}$.

(*b*) The total revenue function is: $R = px = \left(8 - \frac{x}{250}\right)x = 8x - \frac{x^2}{250}$

(*c*) The average revenue function is: $AR = \frac{R}{x} = 8 - \frac{x}{250}$

(*d*) The marginal revenue function is: $MR = \frac{dR}{dx} = 8 - \frac{x}{125}$.

Example 30. A manufacturer surveys market conditions and finds that x, the number of units of his commodity the consumer will purchase if the commodity is offered at price p (per unit), is given by $x = \frac{1}{3}p^2 - 2p + 3$. Find the price at which total revenue R is (*a*) increasing, (*b*) decreasing.

Solution. The revenue R is given by: $R = px = p\left(\frac{1}{3}p^2 - 2p + 3\right) = \frac{1}{3}p^3 - 2p^2 + 3p$

Differentiating R with respect to p gives

$$R'(p) = p^2 - 4p + 3 = (p - 1)(p - 3)$$

Setting $R'(p) = 0$ gives $p = 1$ or $p = 3$. These two values lead us to consider three intervals as shown below:

If $p < 1$, then $R'(p) = (-)(-) = (+)$, so R is increasing if $p < 1$.
If $1 < p < 3$, then $R'(p) = (+)(-) = (-)$, so R is decreasing if $1 < p < 3$.
If $p > 3$, then $R'(p) = (+)(+) = (+)$, so R is increasing if $p > 3$.

Example 31. It is given that a demand curve is convex from below, *i.e.*, $\frac{d^2p}{dx^2} > 0$ at all points, where p is price per unit and x is quantity demanded. Show that the marginal revenue curve is also convex from below if either $\frac{d^3p}{dx^3}$ is positive or $\frac{d^3p}{dx^3}$ is negative but numerically less than $\left(\frac{3}{x}\right)\frac{d^2p}{dx^2}$. Does a similar property hold for demand curves that are concave from below? **[*Delhi Univ. B.A. (Eco. Hons.) 1984*]**

Solution. Assume that the demand function is $p = f(x)$. Then the total revenue is:

$$R = px$$

and the marginal revenue is:

$$MR = \frac{dR}{dx} = p + \frac{dp}{dx}$$

Differentiating MR with respect to x gives

$$\frac{d}{dx}(MR) = \frac{dp}{dx} + x\frac{d^2p}{dx^2} + \frac{dp}{dx}.1 = 2\frac{dp}{dx} + x\frac{d^2p}{dx^2}$$

Differentiating this equation with respect to x, we obtain

$$\frac{d^2}{dx^2}(MR) = 2\frac{d^2p}{dx^2} + x\frac{d^3p}{dx^3} + \frac{d^2p}{dx^2}.1 = 3\frac{d^2p}{dx^2} + x\frac{d^3p}{dx^3}$$

Thus $$\frac{d^2}{dx^2}(MR) = 3\frac{d^2p}{dx^2} + x\frac{d^3p}{dx^3}$$

Now the marginal revenue curve will be convex from below if

$$\frac{d^2}{dx^2}(MR) > 0 \quad \textit{i.e., if} \quad 3\frac{d^2p}{dx^2} + x\frac{d^3p}{dx^3} > 0 \qquad \text{...(1)}$$

But the demand function is given to be convex from below, therefore $\frac{d^2p}{dx^2} > 0$. Thus inequality (1) holds if

$$\text{either} \quad \frac{d^3p}{dx^3} > 0 \quad \text{or} \quad \frac{d^3p}{dx^3} < 0 \quad \text{but} \quad -\frac{d^3p}{dx^3} < \frac{3}{x}\frac{d^2p}{dx^2}$$

In other words, MR is convex from below if either $\frac{d^3p}{dx^3} > 0$ or $\frac{d^3p}{dx^3} < 0$ but numerically less than $\frac{3}{x}\frac{d^2p}{dx^2}$.

It can similarly be shown that if a demand curve is concave from below $\left(\frac{d^2p}{dx^2} < 0\right)$ at all points, then the marginal revenue curve is also concave from below if either $\frac{d^3p}{dx^3} < 0$ or $\frac{d^3p}{dx^3} > 0$ but less than $\left(-\frac{3}{x}\right)\frac{d^2p}{dx^2}$.

6.8 MARGINAL REVENUE PRODUCT

Suppose a manufacturer hires t employees who produce a total of x units of a product per day. We can think of x as a function of t. If R is the total revenue the manufacturer receives for selling these units, then R can also be considered as a function of t. Thus we can look at $\frac{dR}{dt}$, the rate of change of revenue with respect to the number of employees. This derivative is called the **marginal revenue product.** It is approximately the change in revenue that results when a manufacturer hires an extra employee. Since

$$R = px,$$

where p is the price per unit, therefore

$$\frac{dR}{dt} = p\frac{dx}{dt} + x\frac{dp}{dt}$$

$$= p\frac{dx}{dt} + x\frac{dp}{dx}\frac{dx}{dt} \qquad \textit{(Chain rule)}$$

$$= \frac{dx}{dt}\left[p + x\frac{dp}{dx}\right]$$

Hence marginal revenue product $= \dfrac{dR}{dt} = \dfrac{dx}{dt}\left[p + x\dfrac{dp}{dx}\right].$

Example 32. A manufacturer determines that t employees will produce a total of x units of a product per day, where $x = 5t$. If demand equation for the product is $p = -0.5x + 40$, determine the marginal revenue product when $t = 2$. Interpret your result.

[*Delhi Univ. B. Com. (H) 2009*]

Solution. If R is the total revenue, then

$$R = px = (-0.5x + 40)x = -0.5x^2 + 40x$$

$$= -0.5\,(5t)^2 + 40\,(5t)$$

$$= -12.5t^2 + 200\,t$$

$\therefore$ Marginal revenue product $= \dfrac{dR}{dt} = -25t + 200$

When $t = 2$, $\dfrac{dR}{dt} = (-25)(2) + 200 = 150$. This means if third employee is hired, the extra revenue generated is approximately 150.

6.9 MARGINAL PROPENSITY TO CONSUME

If $C = f(I)$ is a consumption function, where I is national income and C is national consumption, then **marginal propensity to consume** is defined as the rate of change of consumption with respect to income. Thus

$$\textbf{\textit{Marginal propensity to consume}} = MPC = \frac{dC}{dI}$$

If S denotes the savings, *i.e.*, $S = I - C$, then the **marginal propensity to save** is defined as the rate of change of S with respect to I. Thus

$$\textbf{\textit{Marginal propensity to save}} = MPS = \frac{dS}{dI} = 1 - \frac{dC}{dI}$$

It indicates how fast savings change with respect to income.

Example 33. If the consumption function is given by $C = 6 + \frac{3I}{4} - \frac{\sqrt{I}}{3}$, determine the marginal propensity to consume and the marginal propensity to save when $I = 25$.

Solution. Marginal propensity to consume $= \frac{dC}{dI} = \frac{3}{4} - \frac{1}{6}\frac{1}{\sqrt{I}}$

When $I = 25$, $\quad \frac{dC}{dI} = \frac{3}{4} - \frac{1}{6}\frac{1}{\sqrt{25}} = \frac{43}{60} \approx 0.716.$

The marginal propensity to save when $I = 25$ is : $1 - 0.716 = 0.284$.

EXERCISE 6.2

1. The total cost $C(x)$ of a firm is : $C(x) = 2x^2 + 4x + 3$, where x is the output. Determine :
   (*i*) the average cost, (*ii*) the marginal cost,
   (*iii*) the marginal cost when 5 units are produced and interpret it,
   (*iv*) the actual cost of producing sixth unit.
2. The total cost $C(x)$ of a firm is: $C(x) = 0.0005x^3 - 0.7x^2 - 30x + 3000$, where x is the output. Determine :
   (*i*) the average cost (AC), (*ii*) the marginal cost (MC), and
   (*iii*) the value of x for which $MVC = AVC$, where VC denotes the variable cost, MVC denotes marginal variable cost and AVC denotes average variable cost.
3. The total cost function of a firm is : $C(x) = 0.0005\,x^3 - 0.03x^2 + 6x + 2000$, where x is the output. Determine :
   (*i*) the average cost,
   (*ii*) the marginal average cost, the rate of change of average cost with respect to x,
   (*iii*) the marginal cost,
   (*iv*) the rate of change of MC with respect to x,

(*v*) the value of x for which $MVC = AVC$, where MVC and AVC denote, respectively, the marginal variable cost and the average variable cost.

4. If the total cost function for a commodity is given by $C(x) = \frac{1}{3}x^3 + 3x^2 - 7x + 16$, where x is the quantity of output, show that $\frac{d}{dx}(AC) = \frac{1}{x}(MC - AC)$.

5. The rate of change of total cost (C) of a commodity per unit change of output (x) is called the marginal cost of the commodity. If there exists a relation between C and x in the form

$$C = 2x\left(\frac{x+4}{x+1}\right) + 6,$$

show that the marginal cost falls continuously as the output increases.

6. By examining the signs of certain derivatives, show that each of the total cost functions:

$$C = \sqrt{(ax+b)} + c \quad \text{and} \quad C = ax\left(\frac{x+b}{x+c}\right) + d, \ (b > c)$$

gives average and marginal cost curves which fall continuously with increasing outputs.

7. Given, the total cost function for x units of a commodity, $C(x) = \frac{1}{3}x^3 + 3x^2 - 7x + 16$. Find :

(*i*) The marginal cost. (*ii*) The average cost.

(*ii*) The marginal average cost, the rate of change of average cost with respect to x.

(*iv*) Show that the marginal average cost $= \dfrac{x(MC) - C(x)}{x^2}$.

8. The total cost function is $C = 2x^3 - 5x^2 + 7x$, find the marginal average cost function (MAC). What type of curve is this MAC ? Find the points where it cuts the axes.

9. Given, the total cost function for x units of a commodity, $C(x) = ax^3 + bx^2 - cx + d$, where $a > 0$, $b < 0$, $c > 0$. Show that average variable cost and marginal cost curves intersect at minimum average variable cost.

10. The total revenue received from the sale of x units of a product is given by $R(x) = 200 + \frac{x^2}{5}$. Find :

(*i*) The average revenue. (*ii*) The marginal revenue.

(*iii*) The marginal revenue, when $x = 25$.

(*iv*) The actual revenue from the sale of twenty sixth unit.

11. The demand function for a product is given by $p = 8000 - 20x - x^2$. Find the marginal revenue function.

12. A firm knows that the demand function for one of its products is linear. It also knows that it can sell 1400 units when the price is ₹ 4 per unit, and it can sell 1800 units at a price of ₹ 2 per unit. Find *(i)* the demand function, *(ii)* the total revenue function, and *(iii)* the marginal revenue function.

13. Find the marginal revenue function for a monopolist confronted with demand $x = (600 - p)/8$.

14. A monopolist's demand function is $p = 300 - 5x$. Find (*i*) the marginal revenue function, (*ii*) the relationship between the slope of *MR* and *AR*, and (*iii*) at what price is the marginal revenue zero ?

15. A firm's demand function is $p = 10e^{-x/400}$, where p is price per unit and x is quantity demanded. Find at what level of output is the marginal revenue zero?

16. Show that the demand curve $p = a/(x + b)$ is downward sloping and convex from below (a and b are positive constants). Does the same property hold of marginal revenue curve?

17. A manufacturer determines that t employees will produce a total of x units of a product per day, where $x = 2t$. If the demand equation for the product is $p = -0.5x + 20$, determine the marginal revenue product when $t = 5$. Interpret your result.

18. A manufacturer determines that t employees will produce a total of x units of a product per day, where $x = 10\,t^2/\sqrt{t^2 + 19}$. If the demand equation for the product is $p = 900/(x + 9)$, determine the marginal revenue product when $t = 9$.

19. A manufacturer determines that t employees will produce a total of x units of a product per day, where $x = (200\,t - t^2)/20$. It the demand equation for the product is $p = -0.1x + 70$, determine the marginal revenue product when $t = 40$.

20. If $p = k/x$, where k is a constant, is the demand equation for a manufacturer's product, and $x = f(t)$ defines a function that gives the total number of units produced per day by t employees, show that the marginal revenue product is always zero.

21. If the consumption function is given by $C = 2 + 2\sqrt{I}$, determine the marginal propensity to consume and the marginal propensity to save when $I = 9$.

22. If $C = 7 + 0.6\,I - 0.25\sqrt{I}$ is a consumption function, find the marginal propensity to consume and the marginal propensity to save when $I = 16$.

23. If the consumption function is given by $C = 8 + 9\,I^{3/2}$, determine the marginal consumption function.

ANSWERS

1. (*i*) $AC = 2x + 4 + \dfrac{3}{x}$ (*ii*) $MC = 4x + 4$

(*iii*) 24. This means that if production is increased by 1 unit from 5 units to 6 units, then the change in cost is approximately ₹ 24.

(*iv*) ₹ 26.

2. (*i*) $AC = 0.0005x^2 - 0.7x - 30 + \dfrac{3000}{x}$ (*ii*) $MC = 0.0015x^2 - 1.4x - 30$ (*iii*) 700

3. (*i*) $AC = 0.0005x^2 - 0.03x + 6 + \dfrac{2000}{x}$ (*ii*) $MAC = 0.001x - 0.03 - \dfrac{2000}{x^2}$

(*iii*) $MC = 0.0015x^2 - 0.06x + 6$ (*iv*) $0.003x - 0.06$

(*v*) $x = 0$ or $x = 30$

7. (*i*) $MC = x^2 + 6x - 7$ (*ii*) $AC = \dfrac{1}{3}x^2 + 3x - 7 + \dfrac{16}{x}$ (*iii*) $MAC = \dfrac{2}{3}x + 3 - \dfrac{16}{x^2}$

8. $MAC = 4x - 5$, a straight line, cuts x-axis at (5/4, 0) and y-axis at (0, – 5)

10. (*i*) $AR = \dfrac{200}{x} + \dfrac{x}{5}$ (*ii*) $MR = \dfrac{2}{5}x$ (*iii*) 10 (*iv*) 10.2

11. $MR = 8{,}000 - 40x - 3x^2$ **12.** (*i*) $p = 11 - \dfrac{x}{200}$ (*ii*) $R = 11x - \dfrac{x^2}{200}$ (*iii*) $MR = 11 - \dfrac{x}{100}$

13. $MR = 600 - 16x$

14. (*i*) $MR = 300 - 10x$ (*ii*) slope of MR is twice the slope of AR (*iii*) $p = 150$ **15.** 400 units.

16. Yes

17. 20; *Interpretation :* If 6th employee is hired, the extra revenue generated is approximately 20

18. 10.71 **19.** 36 **21.** 1/3 and 2/3 **22.** 0.569 and 0.431

23. $MPC = \dfrac{27}{2}\sqrt{I}$

6.10 PRICE ELASTICITY OF DEMAND

We know that there exists an inverse relationship between price and quantity demanded. At lower prices, more is bought. But how much more? A great deal or just a little. Similarly at higher prices, less is bought. Much less or only a little less. If the price of petrol went up by 50 per cent, would people buy 50 per cent less petrol or would they buy 5 per cent less, or what? To answer questions like these is to make use of the concept of *price elasticity of demand.*

The Meaning of Elasticity. In economics, elasticity always has the same meaning : *It is the ratio of the percentage change in the dependent variable to a given percentage change in the independent variable.* For example, let us consider the function $y = f(x)$. Let the independent variable, x, change from x to $x + h$. Then the dependent variable, y, changes from $f(x)$ to $f(x + h)$. Thus the percentage change in the dependent variable

$$= \frac{f(x + h) - f(x)}{f(x)} \times 100$$

and the percentage change in the independent variable

$$= \frac{(x+h)-x}{x} \times 100 = \frac{h}{x} \times 100$$

The ratio of these percentage changes is

$$\frac{\dfrac{f(x+h)-f(x)}{f(x)} \times 100}{\dfrac{h}{x} \times 100} = \frac{x}{f(x)} \frac{f(x+h)-f(x)}{h} \quad \text{...(1)}$$

If f is differentiable, then as $h \to 0$, the limit of $\dfrac{f(x+h)-f(x)}{h}$ is $f'(x) = \dfrac{dy}{dx}$. Thus the limit of (1) is

$$\frac{x}{f(x)} \frac{dy}{dx} = \frac{x}{y} \frac{dy}{dx}$$

which is called **point elasticity of the function** $y = f(x)$ at the point x. It is denoted by the Greek letter η (eta). Thus

$$\eta = \frac{x}{y} \frac{dy}{dx}.$$

Example 34. A study of transport economics uses the relation $x = 0.4\, k^{1.06}$, where k is the expenditure on building roads and x is a measure of traffic volume. Find the elasticity of x with respect to k. An increase of 3% in traffic volume corresponds to how much increase in expenditure on roads?

[*Delhi Univ. B. Com. (H) 2010*]

Solution. The elasticity of x with respect to k is :

$$\eta_{x,k} = \frac{k}{x} \frac{dx}{dk} = \frac{k}{0.4k^{1.06}} \times (0.4)(1.06) k^{0.06} = 1.06$$

But $$n_{x,k} = \frac{\text{\% change in traffic volume}}{\text{\% change in expenditure}}$$

$$\Rightarrow \quad 1.06 = \frac{3}{\text{\% change in expenditure}}$$

$$\Rightarrow \quad \text{\% change in expenditure} = 3/1.06 = 2.83\%.$$

Example 35. If the elasticity of a function $f(x)$ of x is defined to be $\eta = \frac{x}{f(x)} \frac{d}{dx}[f(x)]$. Show that the elasticities of $x f(x)$ and $f(x)/x$ are, respectively, $(\eta + 1)$ and $(\eta - 1)$.

Solution. By definition, the elasticity of the function $x f(x)$ is :

$$\frac{x}{x f(x)} \cdot \frac{d}{dx}[x f(x)] = \frac{1}{f(x)}\left\{x \frac{d}{dx}[f(x)] + f(x).1\right\}$$

$$= \frac{x}{f(x)} \frac{d}{dx}[f(x)] + 1 = \eta + 1$$

Similarly, the elasticity of the function $f(x)/x$ is :

$$\frac{x}{f(x)/x} \cdot \frac{d}{dx}[f(x)/x] = \frac{x^2}{f(x)}\left[\frac{x \frac{d}{dx}[f(x)] - f(x).1}{x^2}\right]$$

$$= \frac{x}{f(x)} \frac{d}{dx}[f(x)] - 1 = \eta - 1.$$

Price Elasticity of Demand. *The price elasticity of demand is defined as the ratio of percentage change in quantity demanded to a percentage change in price :*

$$\frac{\textit{percentage change in quantity demanded}}{\textit{percentage change in price}}$$

The price elasticity of demand is a numerical measure of the degree to which quantity demanded responds to a change in price, other determinants of demand being kept constant.

For example, if the price of salt fell by 10% and the price of cigarettes fell by 10%, the % change (increase) in quantity demanded due to equal changes in price would be different for salt and cigarettes. Thus salt and cigarettes are said to have a different price elasticity of demand. Thus, for the same percentage decrease in price, the percentage increase in the quantity for different goods is different. Price elasticity of demand provides us with a numerical measure of this difference.

From the definition, if $x = f(p)$ is the demand function for a product, its price elasticity of demand is given by

$$\frac{\dfrac{f(p+h) - f(p)}{f(p)} \times 100}{\dfrac{(p+h) - p}{p} \times 100} = \frac{p}{f(p)} \frac{f(p+h) - f(p)}{h} \qquad \text{...(1)}$$

If f is differentiable, then as $h \to 0$, the limit of $\dfrac{f(p+h)-f(p)}{h}$ is $f'(p) = \dfrac{dx}{dp}$. Thus the limit of (1) is

$$\frac{p}{f(p)}\frac{dx}{dp} = \frac{p}{x}\frac{dx}{dp}$$

which is called **point elasticity of demand at price *p* (or quantity *x*). It is denoted by η_d or simply by η.** Other notations for point elasticity of demand are e_d and e.

Since there exists an inverse relationship between quantity demanded and price for most goods, $\dfrac{dx}{dp}$ is always likely to be negative and hence η_d will always likely to be negative. Thus we place a minus sign in the formula to make η_d positive:

$$\eta_d = -\frac{p}{x}\frac{dx}{dp}$$

Some economists disregard the minus sign. However, we shall not adopt this practice. The reason is that we want to equate "more elastic" with "more responsive". For example, suppose two commodities *X* and *Y* have elasticities of 5 and 2 (calculated after multiplying by (– 1) in accordance with the formula). The demand for commodity *X* is more responsive to price changes than is the demand for commodity *Y*, and *X* has a greater elasticity since 5 is greater than 2. Hence "more elastic" is equated with "more responsive". However, if we did not multiply by (– 1), the two elasticities would be – 5 and – 2. Since – 2 is greater than – 5, we would be likely to say that *Y* has a greater elasticity than *X* (in fact, it is the other way around). Hence without multiplying by (– 1) we would not be able to substitute "more elastic" for "more responsive".

There are five categories of elasticity on the basis of the absolute (or numerical) value of the elasticity of demand :

1. When $1 < |\eta_d| < \infty$, demand is said to be *elastic*. In this case, for a given percentage change in price, there is a greater percentage change in quantity demanded.

2. When $|\eta_d| = 1$, demand is said to have *unit elasticity*. In this case, for a given percentage change in price, there is an equal percentage change in quantity demanded.

3. When $0 < |\eta_d| < 1$, demand is said to be *inelastic*. In this case, for a given percentage change in price, there is a smaller percentage change in quantity demanded.

4. When $|\eta_d| = 0$, demand is said to be *perfectly inelastic*. In this case, a change in price is accompanied by no change at all in the quantity demanded.

5. When $|\eta_d| = \infty$, demand is said to be *perfectly elastic.* In this case, a price reduction caused the quantity demanded to increase infinitely.

Note. In general, remember, "luxuries" are elastic whereas "necessities" like food, etc., are inelastic.

Example 36. If $x = 25 - 3p - p^2$ be a demand function, find the point elasticity of demand at $p = 3$.

Solution. We have $\dfrac{dx}{dp} = -3 - 2p = -(3 + 2p)$

$$\therefore \qquad \eta_d = -\frac{p}{x}\frac{dx}{dp} = -\frac{p}{(25 - 3p - p^2)}[-(3 + 2p)] = \frac{3p + 2p^2}{25 - 3p - p^2}$$

When $p = 3$,
$$\eta_d = \frac{3(3) + 2(3)^2}{25 - 3(3) - (3)^2} = \frac{27}{7}.$$

Example 37. Determine the point elasticity of demand of the demand function $p = \dfrac{k}{x}$ (where $k > 0$).

Solution . We have $p = \dfrac{k}{x}$. Differentiating with respect to x gives $\dfrac{dp}{dx} = -\dfrac{k}{x^2}$.

$$\therefore \qquad \frac{dx}{dp} = \frac{1}{dp/dx} = -\frac{x^2}{k}$$

Thus
$$\eta_d = -\frac{p}{x}\frac{dx}{dp} = \left(-\frac{k}{x^2}\right)\left(-\frac{x^2}{k}\right) = 1$$

Hence the demand has unit elasticity for all $x > 0$.

Example 38. If the demand curve is of the form $p = ae^{-kx}$, where p is the price and x is the demand, prove that the elasticity of demand is $\dfrac{1}{kx}$. Hence deduce the elasticity of demand for the curve $p = 10e^{-x/2}$.

Solution. We have $p = ae^{-kx}$ Differentiating with respect to x gives $\dfrac{dp}{dx} = -ake^{-kx}$.

$$\therefore \qquad \frac{dx}{dp} = \frac{1}{dp/dx} = -\frac{1}{ake^{-kx}}$$

$$\Rightarrow \qquad \eta_d = -\frac{p}{x}\frac{dx}{dp} = -\left(\frac{ae^{-kx}}{x}\right)\left(-\frac{1}{ake^{-kx}}\right) = \frac{1}{kx}.$$

Hence the elasticity of demand of the demand curve $p = ae^{-kx}$ is $\frac{1}{kx}$.

To deduce the elasticity of demand for the curve $p = 10e^{-x/2}$, we compare the function $p = 10e^{-x/2}$ with the function $p = ae^{-kx}$ and find that $a = 10$ and $k = \frac{1}{2}$. Consequently, the elasticity of demand for the curve $p = 10e^{-x/2}$ is given by $\frac{2}{x}$.

Example 39. For the demand curve $aQ + bP - K = 0$, where a, b and K are positive constants. Determine the point elasticity of demand when marginal revenue is zero.

[*Delhi Univ. B. Com. (H) 1994*]

Solution. The demand function is given by

$$aQ + bP - K = 0 \Rightarrow P = \frac{K - aQ}{b}$$

Let R denote the revenue function. Then

$$R = PQ = \frac{K - aQ}{b}\,.\,Q = \frac{KQ - aQ^2}{b}$$

$$\Rightarrow \qquad MR = \frac{dR}{dQ} = \frac{K - 2aQ}{b}$$

Setting $\qquad MR = 0$ gives $Q = \frac{K}{2a}$.

Further, differentiating the demand function $P = \frac{K - aQ}{b}$ w.r.t. Q, we obtain

$$\frac{dP}{dQ} = -\frac{a}{b} \Rightarrow \frac{dQ}{dP} = -\frac{b}{a}$$

Thus $$\eta_Q = -\frac{P}{Q}\frac{dQ}{dP} = \frac{K - aQ}{bQ}\,.\,\frac{b}{a} = \frac{K - aQ}{aQ}$$

$$\therefore \qquad (\eta_Q)_{MR=0} = (\eta_Q)_{Q=K/2a} = 1$$

Hence demand has unit elasticity when $MR = 0$.

Example 40. If the demand function for a product is given by $p = 4 - 5x^2$, for what value of x does the demand have unit elasticity ?

Solution. Differentiating the demand equation $p = 4 - 5x^2$ with respect to x, we obtain

$$\frac{dp}{dx} = -10x \quad \therefore \quad \frac{dx}{dp} = -\frac{1}{10x}$$

$$\Rightarrow \qquad \eta_d = -\frac{p}{x}\frac{dx}{dp} = -\left(\frac{4-5x^2}{x}\right)\left(-\frac{1}{10x}\right) = \frac{4-5x^2}{10x^2}$$

Demand has unit elasticity if $\eta_d = 1$. That is,

$$\text{if } \frac{4-5x^2}{10x^2} = 1 \quad \text{or,} \quad 4 - 5x^2 = 10x^2 \quad \textit{i.e.,} \quad 15x^2 = 4 \Rightarrow x = \frac{2}{\sqrt{15}},$$

rejecting negative sign, since the quantity x can never be negative.

Example 41. For a linear demand function, show, using calculus, that price elasticity of demand increases with the increase in price and decreases with the increase in quantity.

[*Delhi Univ. B. Com. (H) 2004*]

Solution. Let the linear demand function be $x = a - bp$, where a and b are positive constants.

$$\Rightarrow \qquad \frac{dx}{dp} = -b$$

and hence $\quad \eta_d = -\frac{p}{x}\frac{dx}{dp} = -\frac{p}{a-bp} \times (-b) = \frac{bp}{a-bp}$ (*in terms of p*) ....(1)

Also, $\quad \eta_d = -\frac{p}{x}\frac{dx}{dp} = = -\frac{a-x}{bx} \times (-b) = \frac{a-x}{x}$ (*in terms of x*) ....(2)

Differentiating Eq. (1) w.r.t. p, we obtain

$$\frac{d}{dp}(\eta_d) = \frac{(a-bp)\,b - bp\,(-b)}{(a-bp)^2} = \frac{ab}{(a-bp)^2} > 0 \qquad (\because a, b > 0)$$

$\therefore \qquad \eta_d$ increases with increase in p

Similarly, differentiating Eq. (2) w.r.t. x, we obtain

$$\frac{d}{dx}(\eta_d) = \frac{x(-1) - (a-x)}{x^2} = -\frac{a}{x^2} < 0 \qquad (\because a > 0)$$

$\therefore \qquad \eta_d$ decreases with increase in x.

Example 42. The demand for a commodity is given by $x = 48 - 3p^2$. Find the point elasticity of demand when $p = 3$. If this price of 3 is decreased by 4%, determine the relative increase in demand and hence an approximation to the elasticity of demand.

Solution. We have $x = 48 - 3p^2$. Differentiating with respect to p gives $\frac{dx}{dp} = -6p$.

$$\therefore \quad \eta_d = -\frac{p}{x}\frac{dx}{dp} = -\left(\frac{p}{48-3p^2}\right)(-6p) = \frac{6p^2}{48-3p^2}$$

When $p = 3$, $\quad \eta_d = \frac{6(3)^2}{48-3(3)^2} = \frac{54}{21} = \frac{18}{7} = 2.57$

Now $\quad$ old price = 3 $\quad \Rightarrow \quad$ old demand = $48 - 3(3)^2 = 21$

Also decrease in price = 4% = $\frac{4}{100} \times 3 = 0.12$

$\therefore \quad$ new price = $3 - 0.12 = 2.88$ $\quad \Rightarrow \quad$ new demand = $48 - 3(2.88)^2 = 23.1168$

$$\therefore \quad \% \text{ increase in demand} = \frac{\text{new demand} - \text{old demand}}{\text{old demand}} \times 100$$

$$= \frac{23.1168 - 21}{21} \times 100 = \frac{2.1168}{21} \times 100 = 10.08$$

$$\therefore \quad \eta_d = \frac{\text{percentage change in quantity demanded}}{\text{percentage change in price}} = \frac{10.08}{4} = 2.52.$$

Example 43. The demand law is : $p = \alpha e^{-\beta x}, (\alpha, \beta > 0)$. Find the elasticity of demand. Evaluate the elasticity when $\alpha = 10, \beta = \frac{1}{2}$ and $x = 5$. Find the percentage fall in price necessary to increase the demand by 5%. **[*Delhi Univ. B. Com. (H) 2008, 2013*]**

Solution. The demand law is : $\quad p = \alpha e^{-\beta x} \quad (\alpha, \beta > 0)$

$$\Rightarrow \quad \frac{dp}{dx} = -\alpha\beta e^{-\beta x} \quad \Rightarrow \quad \frac{dx}{dp} = -\frac{1}{\alpha\beta e^{-\beta x}}$$

$$\therefore \quad \eta_d = -\frac{p}{x}\frac{dx}{dp} = -\frac{\alpha e^{-\beta x}}{x} \times -\frac{1}{\alpha\beta e^{-\beta x}} = \frac{1}{\beta x}$$

When $\alpha = 10$, $\beta = \frac{1}{2}$ and $x = 5$, $\eta_d = \frac{1}{\frac{1}{2} \times 5} = \frac{2}{5} = 0.4$

To find the percentage fall in price necessary to increase the demand by 5%, we have

$$\eta_d = \frac{\% \text{ change in demand}}{\% \text{ change in price}}$$

$$\Rightarrow \quad 0.4 = \frac{5}{\% \text{ change in price}} \quad \Rightarrow \quad \% \text{ change in price} = \frac{5}{0.4} = 12.5\ \%.$$

The Relationship between Marginal Revenue, Average Revenue and Point Elasticity of Demand. An important relationship between marginal revenue, *MR*, average revenue, *AR*, and point elasticity of demand, η_d is given by the formula :

$$MR = AR\left(1 - \frac{1}{\eta_d}\right)$$

The proof of the said relationship is given as follows : Assume that the demand function is

$$p = f(x)$$

The total revenue is: $R = px$

Therefore, the average revenue is: $AR = \frac{R}{x} = p,$

and the marginal revenue is: $MR = \frac{dR}{dx} = \frac{d}{dx}(px) = p + x\frac{dp}{dx}$

$$= p\left[1 + \frac{x}{p}\frac{dp}{dx}\right]$$

$$= p\left[1 - \frac{1}{\eta_d}\right] \qquad \left(\text{since } \eta_d = -\frac{p}{x}\frac{dx}{dp}\right)$$

$$= AR\left(1 - \frac{1}{\eta_d}\right).$$

$\because$

Remark. Simplifying for η_d, the above relationship can also be expressed as

$$\eta_d = \frac{AR}{AR - MR}.$$

Corollary. *The total revenue function is increasing on an interval where demand is elastic, decreasing on an interval where demand is inelastic, and attains its maximum value where demand is unitary.*

Proof. If demand is elastic, $\eta_d > 1 \quad \Rightarrow \quad \frac{1}{\eta_d} < 1 \quad \Rightarrow \quad 1 - \frac{1}{\eta_d} > 0$

$\therefore$ $$MR = AR\left(1 - \frac{1}{\eta_d}\right) > 0 \qquad (\because AR = p > 0)$$

i.e., $$\frac{dR}{dx} > 0 \Rightarrow R \uparrow$$

Thus the total revenue function R is increasing on an interval where demand is elastic. Similarly, it can be shown that R is decreasing on an interval where demand is inelastic. Finally, if demand is unitary,

$$\eta_d = 1 \quad \Rightarrow \quad MR = AR\left(1 - \frac{1}{\eta_d}\right) = 0 \quad \Rightarrow \quad \frac{dR}{dx} = 0$$

$\Rightarrow$ R attains its maximum value at a point where demand is unitary.

Example 44. (*i*) Verify that $\eta_d = \dfrac{AR}{AR - MR}$ for the linear demand law $p = a + bx$.

[*Delhi Univ. B. Com. (H) 1984*]

(*ii*) What is the marginal revenue for a demand curve which has infinite elasticity ?

[*Delhi Univ. B. Com. (H) 1986*]

Solution. (*i*) Differentiating the demand law $p = a + bx$ with respect to x gives

$$\frac{dp}{dx} = b \qquad \therefore \qquad \frac{dx}{dp} = \frac{1}{dp/dx} = \frac{1}{b}$$

Hence
$$\eta_d = -\frac{p}{x}\frac{dx}{dp} = -\frac{(a+bx)}{x}.\frac{1}{b} = -\frac{(a+bx)}{bx}$$

Also
$$R = px = (a+bx)\,x = ax + bx^2$$

$$\therefore \quad AR = \frac{R}{x} = p = a + bx \quad \text{and} \quad MR = \frac{dR}{dx} = a + 2bx$$

Thus
$$\frac{AR}{AR - MR} = \frac{(a + bx)}{(a + bx) - (a + 2bx)} = \frac{(a + bx)}{-bx} = \eta_d$$

(*ii*) We know that $MR = AR\left(1 - \dfrac{1}{\eta_d}\right)$. If η_d is infinite, $\dfrac{1}{\eta_d} = 0$, and hence $MR = AR$.

Example 45. If demand equation for a certain commodity is $q = 500 - 2p$ $(0 \le p \le 250)$, then

(*i*) determine where the demand is elastic, inelastic and of unit elasticity with respect to price.

(*ii*) use the result of Part (*i*) to determine the intervals of increase and decrease of the revenue function and the price at which revenue is maximized. **[*Delhi Univ. B. Com. (H) 2001*]**

Solution. We have $\dfrac{dq}{dp} = -2 \quad \therefore \quad \eta_q = -\dfrac{p}{q}\dfrac{dq}{dp} = -\dfrac{p}{500 - 2p} \times -2 = \dfrac{2p}{500 - 2p}$

(*i*) For demand to be elastic, $\eta_q > 1$. That is,

$$\frac{2p}{500-2p} > 1 \quad \Rightarrow \quad 2p > 500 - 2p \quad \Rightarrow \quad 4p > 500 \quad \text{or} \quad p > 125$$

Thus demand is elastic on the interval $125 < p \leq 250$.

For demand to be inelastic, $\eta_q < 1$. Th\at is,

$$\frac{2p}{500-2p} < 1 \quad \Rightarrow \quad 2p < 500 - 2p \quad \Rightarrow \quad 4p < 500 \quad \text{or} \quad p < 125$$

Thus demand is inelastic on the interval $0 \leq p < 125$.

For demand to have unit elasticity, $\eta_q = 1$. That is,

$$\frac{2p}{500-2p} = 1 \quad \Rightarrow \quad 2p = 500 - 2p \quad \Rightarrow \quad p = 125$$

Thus demand is of unit elasticity when $p = 125$.

(*ii*) The revenue function will be *increasing* on an interval where the demand is elastic *i.e.,* on the interval $125 < p \leq 250$. The revenue function will be *decreasing* on an interval where the demand is inelastic, *i.e.,* on the interval $0 \leq p < 125$. At the price $p = 125$ of unit elasticity, the revenue function has a local maximum.

Example 46. The manager of a bookstore determines that when a certain new paperback novel is priced at p rupees per copy, the daily demand will be $x = 300 - p^2$ copies. Calculate the elasticity of demand at a price where total revenue is maximum.

[*Delhi Univ. B. Com. (H) 2022*]

Solution. The demand function is : $x = 300 - p^2$

$\Rightarrow$ The total revenue, R, is given by

$$R = px = 300\,p - p^3$$

$$\Rightarrow \quad \frac{dR}{dp} = 300 - 3p^2$$

For R to be maximum, $\dfrac{dR}{dp} = 0$ *i.e.,* $300 - 3p^2 = 0 \Rightarrow p = 10$

Also, $\dfrac{d^2R}{dp^2} = -6p < 0$ when $p = 10$

Thus, R is maximum when $p = 10$.

To evaluate elasticity of demand, we differentiate the demand function $x = 300 - p^2$ w.r.t. p to obtain

$$\frac{dx}{dp} = -2p$$

$$\therefore \quad \eta_d = -\frac{p}{x}\frac{dx}{dp} = -\frac{p}{300 - p^2}(-2p) = \frac{2p^2}{300 - p^2}$$

Evaluating η_d at $p = 10$, we obtain $\eta_d = 1$.

Income Elasticity of Demand. Just as price determines price elasticity, so does income, another determinant of demand, determine income elasticity.

Income elasticity of demand is defined as the ratio of percentage change in demand to a given percentage change in income. It is a numerical measure of the degree to which quantity demanded responds to a change in income, other determinants of the demand being kept constant.

Just as we obtained the formula for price elasticity of demand, the formula for income elasticity of demand, η_I, is given by

$$\eta_I = \frac{I}{x}\frac{dx}{dI}$$

The income elasticity of demand may be positive or negative depending on whether the commodity is normal or inferior.

A normal good is one where a percentage increase in income causes a percentage increase in quantity demanded and vice versa. Thus for normal commodities (*e.g.*, clothing, cigarettes) income and quantity demanded vary in direct proportion, so $\frac{dx}{dI}$ is positive and hence income elasticity of demand is positive.

An inferior commodity is one where a percentage increase in income causes a percentage decrease in quantity demanded and vice versa. Thus for inferior commodities (e.g., cheap whisky, artificial jewellery, etc.) income and quantity demanded vary in an inverse proportion, so $\frac{dx}{dI}$ is negative and hence income elasticity of demand is negative.

Example 47. The demand, x, as a function of income, I, is given by $x = -75 + \frac{1}{2}I$. Determine the income elasticity of demand at $I = 200$.

Solution. We have $\quad \frac{dx}{dI} = \frac{1}{2} \quad \therefore \quad \eta_I = \frac{I}{x}\frac{dx}{dI} = \frac{I}{\left(-75 + \frac{1}{2}I\right)} \cdot \frac{1}{2}$

When $I = 200$, $\eta_I = \frac{200}{(-75 + 100)} \cdot \frac{1}{2} = 4$.

Example 48. The demand x as a function of income y is given by $30x = 10 + 2y$. Obtain the expression for the income elasticity of demand and its value when $y = 250$.

[Delhi Univ. B.Com. (H) 2005]

Solution. The demand x as a function of income y is given by

$$30x = 10 + 2y \quad or \quad x = \frac{1}{3} + \frac{1}{15}y \quad \Rightarrow \quad \frac{dx}{dy} = \frac{1}{15}$$

Thus, if η_y denotes the income elasticity of demand, then

$$\eta_y = \frac{y}{x}\frac{dx}{dy} = \frac{y}{\frac{1}{3}+\frac{1}{15}y}\cdot\frac{1}{15} = \frac{y}{y+5} \quad \Rightarrow \quad (\eta_y)_{y=250} = \frac{250}{250+5} = \frac{250}{255} = \frac{50}{51}.$$

Price Elasticity of Supply. *Price elasticity of supply is just as important as price elasticity of demand and has the same meaning : It is the ratio of the percentage change in the supply of a commodity to a given percentage change in price.* It is a measure of the degree of responsiveness of quantity supplied to changes in the price of the commodity, other determinant of supply being kept fixed.

Let η_s be the symbol for elasticity of supply. Then the formula for η_s retains the same form as that of η_d. Thus

$$\eta_s = \frac{p}{x}\frac{dx}{dp}$$

Since, in general, both price and quantity supplied go up and down together, η_s has a positive sign.

Example 49. The supply function for a commodity is $x = 2p^2 + 5$. Find the elasticity of supply when $p = 3$. **[Delhi Univ. B. Com. (H) 2005]**

Solution. Differentiating the supply function $x = 2p^2 + 5$ w.r.t. p, we obtain $\frac{dx}{dp} = 4p$.

$$\therefore \qquad \eta_s = \frac{p}{x}\frac{dx}{dp} = \frac{p}{2p^2+5}4p = \frac{4p^2}{2p^2+5}$$

$$\Rightarrow \qquad (\eta_s)_{p=3} = \frac{4(3)^2}{2(3)^2+5} = \frac{36}{23}.$$

Example 50. If η_1 and η_2 be the price elasticities of supply laws: $p = e^x$ and $p = e^x/x$, respectively, show that $\eta_1\,\eta_2 = \eta_2 - \eta_1$. **[Delhi Univ. B. Com. (H) 1993]**

Solution. To find η_1, we differentiate both sides of the equation $p = e^x$ to obtain

$$\frac{dp}{dx} = e^x \quad \therefore \quad \frac{dx}{dp} = \frac{1}{e^x} \quad \Rightarrow \quad \eta_1 = \frac{p}{x}\frac{dx}{dp} = \frac{e^x}{x}\cdot\frac{1}{e^x} = \frac{1}{x}$$

Similarly, to find η_2, we differentiate both sides of the equation $p = e^x/x$ to obtain

$$\frac{dp}{dx} = \frac{xe^x - e^x \cdot 1}{x^2} = \frac{e^x (x-1)}{x^2} \quad \therefore \quad \frac{dx}{dp} = \frac{x^2}{e^x (x-1)}$$

$$\Rightarrow \qquad \eta_2 = \frac{p}{x}\frac{dx}{dp} = \frac{e^x}{x^2} \cdot \frac{x^2}{e^x (x-1)} = \frac{1}{x-1}$$

Now, $$\eta_1 \eta_2 = \left(\frac{1}{x}\right)\left(\frac{1}{x-1}\right) = \frac{1}{x(x-1)} \qquad \text{...(1)}$$

and $$\eta_2 - \eta_1 = \frac{1}{x-1} - \frac{1}{x} = \frac{x-(x-1)}{x(x-1)} = \frac{1}{x(x-1)} \qquad \text{...(2)}$$

From Equations (1) and (2), we thus conclude that $\eta_1 \eta_2 = \eta_2 - \eta_1$.

Example 51. The supply of a certain good is given by $x = a\sqrt{p-b}$, where x is quantity supplied, p (which is greater than b) is price and a and b are positive constants. Find the expression for the elasticity of supply as a function of price and, by using calculus, show that the elasticity decreases as price increases and becomes unity at the price equal to $2b$.

[Delhi Univ. B. Com. (H) 2009, 2014 (SOL)]

Solution. We have $$x = a\sqrt{p-b}$$

$$\therefore \qquad \frac{dx}{dp} = a \cdot \frac{1}{2}(p-b)^{-1/2} \cdot 1 = \frac{a}{2\sqrt{p-b}}$$

$$\Rightarrow \qquad \eta_s = \frac{p}{x}\frac{dx}{dp} = \frac{p}{a\sqrt{p-b}} \cdot \frac{a}{2\sqrt{p-b}} = \frac{p}{2(p-b)}$$

Differentiating η_s with respect to p, we obtain

$$\frac{d}{dp}(\eta_s) = \frac{1}{2}\left[\frac{(p-b)(1) - p \cdot 1}{(p-b)^2}\right] = -\frac{b}{2(p-b)^2},$$

which is negative, since $b > 0$. Hence η_s is a decreasing function of p. In other words, elasticity decreases as price increases. Also, when $p = 2b$, $\eta_s = \dfrac{2b}{2(2b-b)} = 1$.

Example 52. Find the elasticities of demand and supply at equilibrium point for the demand function $p = \sqrt{100 - x^2}$ and supply function $x = 2p - 10$, where p is price and x is quantity.

[Delhi Univ. B. Com. (H) 2018]

Solution. We are given : $p = \sqrt{100 - x^2}$: *demand function*

and $x = 2p - 10$ *or* $p = \dfrac{x+10}{2}$: *supply function*

At the equilibrium point, demand = supply

$$\Rightarrow \quad \sqrt{100 - x^2} = \frac{x+10}{2} \quad or \quad 4(100 - x^2) = (x+10)^2$$

$$\Rightarrow \quad x^2 + 4x - 60 = 0 \quad i.e., \quad (x-6)(x+10) = 0 \quad \Rightarrow \quad x = 6 \text{ or } x = -10$$

We discard $x = -10$, since x represents quantity. Thus $x = 6$. Substituting $x = 6$ into the demand function or the supply function, we obtain $p = 8$. Thus the equilibrium point is (6, 8).

Computation of η_d **:** Differentiating the demand function $p = \sqrt{100 - x^2}$ *w.r.t.* x, we obtain

$$\frac{dp}{dx} = -\frac{x}{\sqrt{100 - x^2}} \quad \Rightarrow \quad \frac{dx}{dp} = -\frac{\sqrt{100 - x^2}}{x}$$

$$\therefore \quad \eta_d = -\frac{p}{x}\frac{dx}{dp} = -\frac{\sqrt{100 - x^2}}{x} \times -\frac{\sqrt{100 - x^2}}{x} = \frac{100 - x^2}{x^2}$$

When $x = 6$, $$\eta_d = \frac{100 - 36}{36} = \frac{64}{36} = \frac{16}{9}$$

Computation of η_s **:** Differentiating the supply function $x = 2p - 10$ *w.r.t.* p, we obtain

$$\frac{dx}{dp} = 2 \quad \therefore \quad \eta_s = \frac{p}{x}\frac{dx}{dp} = \frac{p}{2p - 10} \times 2 = \frac{p}{p - 5}$$

When $p = 8$, $$\eta_s = \frac{8}{3}$$

Elasticity of Cost. *The elasticity of cost is defined as the ratio of percentage change in total cost to a given percentage change in output.* Thus, if $C = f(x)$ is the total cost of a firm producing x unit of a commodity, then the elasticity of cost, denoted by η_C is given by

$$\eta_C = \frac{x}{C}\frac{dC}{dx}$$

Similarly, elasticity of average cost is

$$\eta_{AC} = \frac{x}{AC}\frac{d}{dx}(AC)$$

$$= \frac{x}{C/x}\frac{d}{dx}(C/x) = \frac{x^2}{C}\left[\frac{x\frac{dC}{dx} - C.1}{x^2}\right] = \frac{x}{C}\frac{dC}{dx} - 1 = \eta_C - 1$$

That is, $\eta_{AC} = \eta_C - 1$.

Example 53. Find the elasticity of total cost and of average cost of the function $C = 2x^2 + 4x + 3$.

Solution. Elasticity of total cost is:

$$\eta_C = \frac{x}{C}\frac{dC}{dx} = \frac{x}{C}(4x+4) = \frac{x(4x+4)}{2x^2+4x+3}$$

Elasticity of average cost is:

$$\eta_{AC} = \frac{x}{AC}\frac{d}{dx}(AC) = \frac{x}{\left(2x+4+\frac{3}{x}\right)}\frac{d}{dx}\left(2x+4+\frac{3}{x}\right)$$

$$= \frac{x^2}{2x^2+4x+3}\left(2-\frac{3}{x^2}\right) = \frac{2x^2-3}{2x^2+4x+3}.$$

Example 54. Show that for the total cost function $C = \sqrt{ax+b}$, where a and b are positive constants, the elasticity of total cost increases but remains less than unity as x increases.

[*Delhi Univ. B.A. (Eco. Hons.) 1984*]

Solution. Elasticity of total cost function is :

$$\eta_C = \frac{x}{C}\frac{dC}{dx} = \frac{x}{\sqrt{ax+b}}\frac{d}{dx}(\sqrt{ax+b})$$

$$= \frac{x}{\sqrt{ax+b}}\left(\frac{1}{2}\frac{1}{\sqrt{ax+b}}.a\right) = \frac{ax}{2(ax+b)}$$

Differentiating η_C with respect to x, we obtain

$$\frac{d}{dx}(\eta_C) = \frac{a}{2}\left[\frac{(ax+b).1 - x.a}{(ax+b)^2}\right] = \frac{ab}{2(ax+b)^2},$$

which is positive, since a and b are both positive. Hence η_C increases with an increase in the output x. However,

$$\eta_C = \frac{ax}{2(ax+b)} = \frac{1}{2}\left[\frac{(ax+b)-b}{(ax+b)}\right] = \frac{1}{2}\left[1 - \frac{b}{(ax+b)}\right] < 1.$$

Example 55. If a firm produces an output x at a total cost of $C = ax^2 + bx$, $(a, b > 0)$ find an expression for η_C, the elasticity of total cost. Show that η_C is always greater than unity and increases as x increases. **[*Delhi Univ. B. Com. (H) 2007, 2014*]**

Solution. Given $C = ax^2 + bx$: *Cost function*

$$\therefore \quad \text{Elasticity of cost} = \eta_C = \frac{x}{C}\frac{dC}{dx} = \frac{x}{ax^2 + bx}(2ax + b)$$

$$= \frac{2ax + b}{ax + b} = 1 + \frac{ax}{ax + b} > 1 \qquad (\because a, b > 0)$$

Also,
$$\frac{d\eta_C}{dx} = \frac{(ax + b)\,a - ax \cdot a}{(ax + b)^2} = \frac{ab}{(ax + b)^2},$$

which is positive, since a and b are positive. Hence η_C increases as x increases.

Example 56. A manufacturer's total cost is $C(x) = \frac{1}{6}x^3 + 642x + 400$. Use the concept of elasticity to estimate the amount by which the manufacturer should decrease the production to reduce the total cost by 3%, when the current level of production is 4 units.

Solution.
$$\eta_C = \frac{x}{C}\frac{dC}{dx} = \frac{x}{(x^3/6) + 642x + 40}\left(\frac{1}{2}x^2 + 642\right)$$

When $x = 4$,
$$\eta_C = \frac{4}{10.67 + 2568 + 400} \times 650 = \frac{2600}{2978.67} = 0.8278$$

But
$$\eta_C = \frac{\%\text{ change in cost}}{\%\text{ change in output}} \Rightarrow 0.8728 = \frac{3}{\%\text{ change in output}}$$

$$\Rightarrow \qquad \%\text{ change in output} = \frac{3}{0.8728} = 3.4372\%.$$

EXERCISE 6.3

1. The demand function for a commodity is $p = 10 + 5x - 2x^2$. Find the elasticity of demand when $x = 2$.
2. If the demand law is $x = 20/(p + 1)$, find the elasticity of demand at the point $p = 3$. What type of curve is the demand function ?
3. Find the elasticity of demand in terms of quantity if the demand law is given by $p = 1/(1 + x^2)$.
4. Suppose quantity x and price p are related by $xp^m = a$, where a and m are positive constants. Show that $\eta_d = m$. What m will make such demand laws inelastic ? elastic ? unitary ?

5. Suppose a demand function is given by $p = a - bx$, where a and b are positive constants. Find the point elasticity of demand in terms of price p. When is demand inelastic ? elastic ?

6. Calculate the price elasticity of demand of the demand function $x = 15p^{-0.75}$.

[Delhi Univ. B.A. (Eco. Hons) 1987]

7. If η_1 and η_2 are the price elasticities of demand for the demand laws : $x = 1 - p$ and $x = p - p^2$, show that $\eta_2 = \eta_1 - 1$.

8. (*a*) If the demand function is $p = 4 - 5x$, for what value of x does the demand have unit elasticity?

(*b*) If the demand function is $p = a - bx^2$, where a and b are positive constants, find for what value of x does the demand have unit elasticity ?

9. The demand function for a product is $x = 500 - 40p + p^2$, where p is the price per unit and x is the number of units demanded. Find the point elasticity of demand when $p = 15$. If this price of 15 is increased by 2%, what is the approximate percentage change in demand ? Hence find an approximation to the elasticity of demand.

10. Using calculus, show that $MR = AR\left(1 - \dfrac{1}{\eta_d}\right)$ and verify this for the demand function

$$x = \frac{a-p}{b}, \quad b \neq 0.$$

[Delhi Univ. B. Com. (H) 1997]

11. Verify the relationship $MR = AR\left(1 - \dfrac{1}{\eta_d}\right)$ for the demand function $p = 25e^{-x/400}$.

12. (*a*) If $AR = 20$ and elasticity of demand with respect to price is 2, find MR.

(*b*) Verify the relationship $MR = p\left(1 - \dfrac{1}{\eta_d}\right)$ for the demand function $p = (12 - x)^{1/2}$, $0 \leq x \leq 12$.

13. The demand function for a certain commodity is $5q = 400 - 2p$ $(0 \leq p \leq 200)$.

(*a*) Determine where the demand is elastic, inelastic and of unit elasticity with respect to price.

(*b*) Use the result of part (*a*) to determine the intervals of increase and decrease of the revenue function and the price at which revenue is maximized.

14. The demand, x, as a function of income, I, is given by $x = 5 + \dfrac{1}{10} I$. Determine the income elasticity of demand at $I = 200$.

15. The supply curve for a commodity is $p = \sqrt{9 + x}$. Find the elasticity of supply at price $p = 4$.

16. Find the elasticities of demand and supply at the equilibrium price for the demand function $p = 16 - x^2$ and supply function $p = 2x^2 + 4$, where p is price and x is quantity.

17. Suppose a manufacturer of shoes will place on the market 50 (thousand pairs) when the price is ₹ 350 per pair and 35 when the price is 300. Find the supply function, assuming that price p and quantity x are linearly related. Hence obtain the elasticity of supply.

18. Find the elasticity of total cost and of average cost of the function $C = ax^2 + bx + c$.

19. (*a*) Derive a relationship between the elasticity of cost and elasticity of average cost.

(*b*) A firm has for x items produced the total cost C given by the equation $C = 100 + 0.015x^2$. Use the concept of elasticity to estimate the amount by which the firm should decrease the production to reduce the total cost by 3%, when the current level of production is 100 units.

20. Using calculus, show that the elasticity of average cost is equal to the elasticity of total cost minus one. Using this relationship, find the elasticity of average cost for the total cost function $C(x) = 2000\, e^{1.5x}$ and verify. **[Delhi Univ. B.Com. (H) 2004]**

21. (*a*) Derive, using calculus, a relation between the elasticity of total product and elasticity of average product.

(*b*) A study of transport economics uses the relation $x = 0.4\, k^{1.06}$, where k is the expenditure on building roads and x is a measure of traffic volume. Find the elasticity of x with respect to k. An increase of 3% in traffic volume corresponds to how much increase in expenditure on roads? **[Delhi Univ. B. Com. (H) 2010]**

ANSWERS

1. 2 **2.** 3/4, rectangular hyperbola **3.** $\dfrac{1+x^2}{2x^2}$

4. Inelastic if $m < 1$, elastic if $m > 1$ and unitary if $m = 1$

5. $\dfrac{p}{a-p}$; elastic if $p > \dfrac{a}{2}$, inelastic if $p < \dfrac{a}{2}$

6. 0.75 **8.** (*a*) $\dfrac{2}{5}$ (*b*) $\sqrt{\dfrac{a}{3b}}$ **9.** 1.2; 2.328; 1.164

12. (*a*) $MR = 10$

13. (*a*) elastic on $100 < p \le 200$; inelastic on $0 \le p < 100$; and unitary at $p = 100$

(*b*) increasing on $100 < p \le 200$; decreasing on $0 \le p < 100$; and maximum at $p = 100$

14. 4/5 **15.** 32/7 **16.** $\eta_d = \dfrac{3}{2}, \eta_s = \dfrac{3}{4}$

17. $x = \dfrac{3p}{10} - 55$, $\eta_S = \dfrac{3p}{3p - 550}$ **18.** $\eta_C = \dfrac{x(2ax+b)}{ax^2+bx+c}$, $\eta_{AC} = \dfrac{ax^2 - c}{ax^2+bx+c}$

19. (*a*) $\eta_{AC} = \eta_C - 1$ (*b*) 2.5%

20. $\eta_{AC} = 1.5x - 1$ **21.** (*a*) $\eta_{AQ} = \eta_Q - 1$ (*b*) 2.83%

6.11 APPLIED MAX-MIN PROBLEMS

This section is concerned with the applications of maxima and minima to applied business problems. For example, a firm producing a product might be interested to find the number of units produced in order to maximize profit or minimize average cost. Similarly, a firm may be interested to find the level of output at which the total revenue is maximum.

Maximization of Total Revenue. If a firm's demand function is given, it is possible to determine the level of output at which the total revenue is maximum. Thus if $p = f(x)$ be the demand function for a firm's product, where p is the price per unit and x is the number of units produced and sold, then the total revenue, R, of the firm given by

$$R = (\text{price per unit})\ (\text{number of units sold}) = p.x = f(x)\,x,$$

is a function of x. Hence R is maximum when

$$\frac{dR}{dx} = 0 \quad \text{and} \quad \frac{d^2R}{dx^2} < 0$$

But $\frac{dR}{dx}$ is marginal revenue MR. Thus to maximize total revenue, it is necessary that $MR = 0$.
For this to indeed correspond to a maximum, it is required that

$$\frac{d^2R}{dx^2} = \frac{d}{dx}\left(\frac{dR}{dx}\right) = \frac{d}{dx}(MR) < 0.$$

Example 57. Given is the revenue function $R = 10{,}000 - (x - 20)^2$ where R is the revenue and x is the quantity sold. Find the revenue maximizing level of output and the total revenue at this level of output.

Solution. The revenue function is given by

$$R = 10{,}000 - (x - 20)^2$$

To maximize R, we differentiate R w. r. t. x :

$$\frac{dR}{dx} = -2(x - 20)$$

For R to be maximum, $\frac{dR}{dx} = 0 \Rightarrow x - 20 = 0$ or $x = 20$

Further, $\frac{d^2R}{dx^2} = -2 < 0$. Hence R is maximum when $x = 20$. Substituting $x = 20$ in the revenue function, we get $R\,(x = 20) = 10{,}000$.

Example 58. The demand function for a particular commodity is $p = 15e^{-x/3}$ for $0 \le x \le 8$, where p is the price per unit and x is the number of units demanded. Determine the price and quantity for which total revenue is maximum. **[*Delhi Univ. B. Com. (H) 1982*]**

Solution . Let R be the total revenue, then

$$R = \text{(price per unit) (number of units sold)} = px = 15xe^{-x/3}$$

$$\Rightarrow \quad \frac{dR}{dx} = 15\left[e^{-x/3}.1 + xe^{-x/3}\left(-\frac{1}{3}\right)\right] = 15e^{-x/3}\left(1-\frac{x}{3}\right)$$

For R to be maximum, $\frac{dR}{dx} = 0$, *i.e.*, $15e^{-x/3}\left(1-\frac{x}{3}\right) = 0 \Rightarrow x = 3$ (*since* $e^x > 0$)

Further, $\frac{d^2R}{dx^2} = \left[15e^{-x/3}\left(-\frac{1}{3}\right) + \left(1-\frac{x}{3}\right)15e^{-x/3}\left(-\frac{1}{3}\right)\right] = -5e^{-x/3}\left(2-\frac{x}{3}\right)$

When $x = 3$, $\frac{d^2R}{dx^2} = -5e^{-1} < 0$, therefore R is maximum when $x = 3$. Substituting this value of x in the demand function $p = 15e^{-x/3}$, we get price $p = 15e^{-1} = 15/e$.

Example 59. A shoe manufacturer produces both the high grade and the low grade shoes in the units of X hundreds of pairs and Y hundreds of pairs respectively. The plant is capable of producing the shoes according to the relationship $Y = \frac{42-5X}{10-X}$, $X < 10$. If the high grade shoes are sold at a price twice the low grade shoes, determine the number of pairs of both which should be produced to maximize total revenue. ***[Delhi Univ. B. Com. (H) 1998]***

Solution. Let p denote the price per pair of low grade shoes. Then the price per pair of high grade shoes would be $2p$. If R denotes the total revenue, then

$$R = 2pX + pY = 2pX + p\,\frac{42-5X}{10-X}$$

$$\Rightarrow \quad \frac{dR}{dX} = 2p + p\left[\frac{(10-X)(-5)-(42-5X)(-1)}{(10-X)^2}\right]$$

$$= 2p - \frac{8p}{(10-X)^2} = p\left[2 - \frac{8}{(10-X)^2}\right]$$

For R to be maximum, $\frac{dR}{dX} = 0$ *i.e.*, $p\left[2 - \frac{8}{(10-X)^2}\right] = 0$

$$\Rightarrow \quad (10-X)^2 = 4 \Rightarrow 10 - X = \pm 2 \Rightarrow X = 8 \text{ or } X = 12$$

The value $X = 12$ can be rejected since $X < 10$. Further $\frac{d^2R}{dX^2} = \frac{-16p}{(10-X)^3}$, which is negative at $X = 8$. Hence R is maximum when $X = 8$. Substituting $X = 8$ in $Y = \dfrac{42 - 5X}{10 - X}$, we obtain $Y = 1$. Thus the manufacturer should produce 800 pairs of high grade and 100 pairs of low grade shoes to maximize his total revenue.

Example 60. A company charges ₹ 6,000 for a television set on orders of 50 or less sets. The charge is reduced on every set by ₹ 75 per set for each set ordered in excess of 50. Find the largest size order the company should allow so as to receive a maximum revenue.

Solution. Let x be the number of sets ordered in *excess* of 50. Then the total number of television sets ordered is $50 + x$. Since for each set ordered in excess of 50, the charge is reduced by ₹ 75 per set, therefore for x sets ordered in excess of 50, the charge is reduced by $75x$ per set. Thus the new charge per set is $6000 - 75x$. Hence if R denotes the revenue of the company, then

$$\begin{aligned} R &= \text{(order size) (charge per set)} \\ &= (50 + x)(6000 - 75x) \\ &= 3{,}00{,}000 + 2250x - 75x^2 \end{aligned}$$

$$\Rightarrow \qquad \frac{dR}{dx} = 2250 - 150x$$

For R to be maximum, $\dfrac{dR}{dx} = 0$. That is, $2250 - 150x = 0 \quad \Rightarrow \quad x = 15$

Also, $\dfrac{d^2R}{dx^2} = -150$, which is always negative. Thus the maximum revenue is obtained when $(50 + 15) = 65$ sets are ordered.

Example 61. A tour operator charges ₹ 136 per passenger for 100 passengers with a discount of ₹ 4 for each 10 passengers in excess of 100. Determine the number of passengers that will maximize the amount of money the tour operator receives.

Solution. Let $10x$ be the number of passengers in *excess* of 100. Then the total number of passengers is $100 + 10x$. Since for each 10 passengers in excess of 100, there is a discount of ₹ 4 per passenger, therefore for $10x$ passengers in excess of 100, there is a discount of ₹ $4x$ per passenger. Hence the new charge per passenger is $136 - 4x$. Thus, if R denotes the amount the tour operator receives, then

$$\begin{aligned} R &= \text{(number of passengers) (charge per passenger)} \\ &= (100 + 10x)(136 - 4x) = 13{,}600 + 960x - 40x^2 \end{aligned}$$

$$\Rightarrow \quad \frac{dR}{dx} = 960 - 80x$$

For R to be maximum, $\frac{dR}{dx} = 0$. That is, $960 - 80x = 0 \quad \Rightarrow \quad x = 12$

Also, $\frac{d^2R}{dx^2} = -80$, which is always negative. Thus, R is maximum when $x = 12$. In other words, the tour operator will receive the maximum amount of money when $100 + 10(12) = 220$ passengers are travelling.

Example 61. A TV company has 1000 subscribers who are each paying ₹ 100 per month. The company proposes to increase the monthly subscription and it is believed that for every increase of ₹ 1, five subscribers will discontinue the service. Find what increase will yield maximum revenue and what will this revenue be ?

Solution . Let x be the increase in the monthly subscription (in ₹). Then the new rate will be $100 + x$. Since for each increase of ₹ 1, 5 subscribers will discontinue the service. Thus, the total number of subscribers left is $1000 - 5x$. Let R denote the revenue of the company, then

$$R = \text{(rate) (number of subscribers)}$$

$$= (100 + x)(1000 - 5x) = 100000 + 500x - 5x^2 \qquad \text{...(1)}$$

$$\Rightarrow \quad \frac{dR}{dx} = 500 - 10x$$

Setting $\frac{dR}{dx} = 0$, we get $x = 50$. Also $\frac{d^2R}{dx^2} = -10 < 0$

We, therefore, conclude that R is maximum when the monthly subscription is increased by ₹ 50 and the maximum revenue is obtained by substituting $x = 50$ into expression (1). Thus

$$\text{maximum revenue} = 1{,}00{,}000 + 500\,(50) - 5\,(50)^2 = ₹\ 1{,}12{,}500$$

Example 63. Show that, for maximum total revenue, elasticity of demand is 1.

Solution. The total revenue, R, is given by

$$R = px$$

where p is the price per unit and x is the number of units produced and sold. Differentiating R with respect to x gives

$$\frac{dR}{dx} = p + x\frac{dp}{dx}$$

For R to be maximum, it is necessary that $\frac{dR}{dx} = 0$. That is,

$$p + x\frac{dp}{dx} = 0 \quad \Rightarrow \quad -\frac{p}{x}\frac{dp}{dx} = 1 \quad \text{or} \quad \eta_d = 1.$$

Example 64. A firm's demand function is: $x = 400 \log \left(\frac{10}{p}\right)$. Find the price and quantity where total revenue is maximum. Also, find price elasticity of demand at that price.

[*Delhi Univ. B. Com. (H) 1999*]

Solution. Solving $x = 400 \log \left(\frac{10}{p}\right)$ for p, we obtain $p = 10e^{-x/400}$ and hence the total revenue, R, is given by : $R = px = 10xe^{-x/400}$

Differentiating R w.r.t. x, we obtain

$$\frac{dR}{dx} = 10\left[xe^{-x/400}\left(-\frac{1}{400}\right) + e^{-x/400}.1\right] = 10e^{-x/400}\left(-\frac{x}{400}+1\right)$$

For R to be maximum, $\frac{dR}{dx} = 0$ *i.e.,* $-\frac{x}{400} + 1 = 0 \quad (\because 10e^{-x/400} \neq 0) \Rightarrow x = 400$

Also
$$\frac{d^2R}{dx^2} = 10\left[e^{-x/400}\left(-\frac{1}{400}\right) + \left(-\frac{x}{400}+1\right)e^{-x/400}\left(-\frac{1}{400}\right)\right]$$

$$= -\frac{1}{40}e^{-x/400}\left[2 - \frac{x}{400}\right]$$

At $x = 400$, $\frac{d^2R}{dx^2} = -\frac{1}{40}e^{-1} < 0$. Therefore R is maximum when $x = 400$. Substituting $x = 400$ in $p = 10e^{-x/400}$, we obtain price $p = 10e^{-1}$.

To evaluate price elasticity of demand, we differentiate the demand function

$$x = 400 \log\left(\frac{10}{p}\right) = 400 (\log 10 - \log p)$$

with respect to p to get $\frac{dx}{dp} = -\frac{400}{p}$.

$$\therefore \qquad \eta_d = -\frac{p}{x}\frac{dx}{dp} = -\frac{p}{400 \log\left(\frac{10}{p}\right)}\left(-\frac{400}{p}\right) = \frac{1}{\log\left(\frac{10}{p}\right)} = \frac{400}{x}$$

Evaluating η_d at $p = 10e^{-1}$ or at $x = 400$, we obtain $\eta_d = 1$.

Minimization of Average Cost. If $C = f(x)$ gives the total cost of producing and marketing x units of a product, then the average cost, AC, is given by $AC = C/x$. An important problem in economics is to find the level of output for which the average cost is minimum. By using the techniques of maxima and minima, the average cost is minimum if

$$\frac{d}{dx} AC = 0 \quad \text{and} \quad \frac{d^2}{dx^2}(AC) > 0$$

Example 65. The manufacturing cost of an item consists of ₹ 1,000/- as overheads, material cost ₹ 2/- per item and the labour cost $x^2/90$ for x items produced. Find how many items be produced to have the average cost as minimum. **[*Delhi Univ. B. Com. (H) 2011*]**

Solution. The total cost C is given by : $C = 1000 + 2x + \dfrac{x^2}{90}$

and the average cost is given by : $AC = \dfrac{C}{x} = \dfrac{1000}{x} + 2 + \dfrac{x}{90}$

We wish to minimize AC. For AC to be minimum, $\dfrac{d}{dx}(AC) = 0$. That is,

$$-\frac{1000}{x^2} + \frac{1}{90} = 0 \quad \Rightarrow \quad x^2 = 90{,}000 \quad \text{or} \quad x = 300$$

Also $\qquad \dfrac{d^2}{dx^2}(AC) = \dfrac{2000}{x^3},$

which is positive when $x = 300$. Hence AC is minimum when $x = 300$.

Example 66. A company produces x units of output at a total cost of $\frac{1}{3}x^3 - 18x^2 + 160x$.

Find :

(*i*) the output at which marginal cost is minimum,

(*ii*) the output at which average cost is minimum, and

(*iii*) the output at which average cost is equal to marginal cost.

Solution. The total cost function is : $C = \frac{1}{3}x^3 - 18x^2 + 160x$

(*i*) The marginal cost, MC, is given by : $MC = \dfrac{dC}{dx} = x^2 - 36x + 160$

$$\Rightarrow \qquad \frac{dMC}{dx} = 2x - 36$$

For MC to be minimum, $\frac{dMC}{dx} = 0$. That is, $2x - 36 = 0 \Rightarrow x = 18$

Further, $$\frac{d^2MC}{dx^2} = 2 > 0$$

Thus MC is minimum when $x = 18$.

(*ii*) The average cost, AC, is given by : $AC = \frac{C}{x} = \frac{1}{3}x^2 - 18x + 160$

$$\Rightarrow \quad \frac{dAC}{dx} = \frac{2}{3}x - 18$$

For AC to be minimum, $\frac{dAC}{dx} = 0$. That is, $\frac{2}{3}x - 18 = 0 \Rightarrow x = 27$

Also, $$\frac{d^2AC}{dx^2} = \frac{2}{3} > 0$$

Thus AC is minimum when $x = 27$.

(*iii*) We wish to find the output x so that $AC = MC$

i.e., $$\frac{1}{3}x^2 - 18x + 160 = x^2 - 36x + 160$$

$$\Rightarrow \quad \frac{2}{3}x^2 - 18x = 0 \quad \text{or} \quad x\left(\frac{2}{3}x - 18\right) = 0 \quad \Rightarrow \quad x = 0 \quad \text{or} \quad x = 27.$$

Thus, the output at which average cost is equal to marginal cost is 27 units ($x = 0$ gives a trivial solution).

Example 67. The cost function of a manufacturing company is given by $C = 0.1x^3 - 3x^2 + 60\,x$. Find the value of x at which slope of average cost is zero. Also verify that $AC = MC$ at this value of x.

Solution. The cost function is given by: $C = 0.1x^3 - 3x^2 + 60x$

The average cost, AC, is given by: $AC = \frac{C}{x} = 0.1x^2 - 3x + 60$ . . . (1)

$$\therefore \quad \text{Slope of average cost} = \frac{dAC}{dx} = 0.2x - 3$$

Setting $\frac{dAC}{dx} = 0$, *i.e.,* $0.2x - 3 = 0 \Rightarrow x = 15$

Also, the marginal cost function is : $MC = \frac{dC}{dx} = 0.3x^2 - 6x + 60$ . . . (2)

Substituting $x = 15$ in (1) and (2) respectively, we obtain

$$AC = 0.1\,(225) - 3\,(15) + 60 = 37.5 \quad \text{and} \quad MC = 0.3\,(225) - 6\,(15) + 60 = 37.5$$

Thus $AC = MC$ at the value of x for which slope of average cost is zero.

Example 68. Given the total cost function as: $C = ax^2 + bx + c$, where $a > 0$, $b \geq 0$, $c \geq 0$. Show that average and marginal cost are equal at minimum average cost.

Solution. The average cost function is : $AC = \frac{C}{x} = ax + b + \frac{c}{x}$, ...(1)

and the marginal cost function is : $MC = \frac{dC}{dx} = 2ax + b$ ... (2)

We must show that $MC = AC$ at the level of output for which AC is minimum. Now

$$\frac{d}{dx}(AC) = a - \frac{c}{x^2}$$

For AC to be minimum, $\frac{d}{dx}(AC) = 0$. That is, $a - \frac{c}{x^2} = 0 \Rightarrow x = \sqrt{\frac{c}{a}}$ (*since* $x > 0$)

Also $$\frac{d^2}{dx^2}(AC) = \frac{2c}{x^3},$$

which is positive when $x = \sqrt{\frac{c}{a}}$. Thus AC is minimum when $x = \sqrt{\frac{c}{a}}$. Substituting $x = \sqrt{\frac{c}{a}}$ into expressions (1) and (2), we obtain

$$AC = a\sqrt{\frac{c}{a}} + b + \frac{c}{\sqrt{\frac{c}{a}}} = 2\sqrt{ac} + b \quad \text{and} \quad MC = 2a\sqrt{\frac{c}{a}} + b = 2\sqrt{ac} + b$$

Hence $MC = AC$ at the level of output for which AC is minimum.

Example 69. When travelling x km per hour a truck burns diesel oil at the rate of $\frac{1}{300}\left(\frac{900}{x} + x\right)$ litre per km. If diesel costs ₹ 4 per litre and the crew driver is paid ₹ 15 per hour, find the steady speed that will minimize the total cost of a trip of 500 km.

[*Delhi Univ. B. Com. (H) 2007, 2012 (Modified)*]

Solution. Time taken for the trip = $(500/x)$ hours

If C is the total cost of the trip (in ₹) then

$$C = \frac{500}{x}\times 15+\frac{1}{300}\left(\frac{900}{x}+x\right)\times 500\times 4 = \frac{13,500}{x}+\frac{20}{3}x$$

$$\Rightarrow \qquad \frac{dC}{dx} = -\frac{13,500}{x^2}+\frac{20}{3}$$

Setting $\frac{dC}{dx} = 0$, we get $20x^2 - 40,500 = 0 \quad \Rightarrow \quad x = 45$ (*rejecting –ve value*)

Also, $\frac{d^2C}{dx^2} = \frac{27,000}{x^3}$, which is positive at $x = 45$. Thus C is minimum when $x = 45$ km/hr.

Example 70. The cost of fuel consumed per hour in running a train is proportional to the square of its speed (in kms per hour), and it costs ₹ 3200 per hour at a speed of 40 kms per hour. What is the most economical speed, if the fixed charges are ₹ 12,800 per hour ?

[*Delhi Univ. B. Com. (H) 2008*]

Solution. Let v km/hour be the speed and c be the cost of fuel per hour. Then

$$c \propto v^2 \quad \Rightarrow \quad c = k\,v^2, \text{ } k \text{ being a constant}$$

It is given that $c = 3200$ when $v = 40$

$$\therefore \quad 3200 = k(40)^2 \quad \Rightarrow \quad k = 2 \quad \therefore \quad c = 2v^2$$

Other costs = ₹ 12,800/hour

∴ Total running cost per hour = $12,800 + 2v^2$

If the total distance covered be S kms with the speed v km/hour, the number of hours consumed in the journey is S/v. Hence the total cost of the journey is :

$$TC = \frac{S}{v}\left[12,800 + 2v^2\right] = \frac{12,800S}{v} + 2Sv$$

For TC to be minimum, $\frac{dTC}{dv} = 0$. That is, $\frac{-12,800S}{v^2} + 2S = 0 \quad \Rightarrow \quad v = 80$

Also, $\frac{d^2TC}{dv^2} = \frac{25,600S}{v^3} > 0$ when $v = 80$

Thus TC is minimum when $v = 80$ and hence the most economical speed is 80 km/hr.

Example 71. A machine initially costs ₹ 6400 with no scrap value. The cost of operation is ₹ 500 in the first year and increases by ₹ 800 in each successive year. Determine *(i)* the number of years it be operated for minimizing total cost per year, and *(ii)* corresponding cost per year.

Solution. *(i)* Let machine be operated for x years. Then

$$\begin{aligned}
\text{Operating cost in the first year} &= 500\\
\text{Operating cost in the second year} &= 500 + 800\\
\text{Operating cost in the third year} &= 500 + 2(800)\\
\dots\quad \dots &\quad \dots\quad \dots\\
\dots\quad \dots &\quad \dots\quad \dots\\
\text{Operating cost in the } x\text{th year} &= 500 + (x-1)\,800
\end{aligned}$$

$$\begin{aligned}
\therefore\quad \text{Total operating cost in } x \text{ years} &= 500x + [1 + 2 + \dots + (x-1)]800\\
&= 500x + \frac{(x-1)x}{2} \times 800\\
&= 500x + 400x^2 - 400x = 400x^2 + 100x
\end{aligned}$$

If C denotes the total cost, then

$$C = \text{Initial cost} + \text{Operating Cost} = 6400 + 400x^2 + 100x$$

$$\therefore\quad AC \text{ (total cost per year)} = \frac{C}{x} = \frac{6400}{x} + 400x + 100 \qquad \dots (1)$$

$$\Rightarrow\quad \frac{d}{dx}(AC) = -\frac{6400}{x^2} + 400 \quad \text{and} \quad \frac{d^2}{dx}(AC) = \frac{12800}{x^3}$$

For AC to be minimum, $\frac{d}{dx}(AC) = 0$. This implies

$$-\frac{6400}{x^2} + 400 = 0 \quad \Rightarrow \quad x^2 = 16 \quad \Rightarrow \quad x = 4 \quad \textit{(rejecting -ve value)}$$

Also, at $x = 4$, $\quad \frac{d^2}{dx^2}(AC) = 200 > 0$

Hence total cost per year is minimum when machine is operated for 4 years.

(ii) Substituting $x = 4$ into expression (1), we obtain the corresponding cost per year:

$$AC = \frac{6400}{4} + 1600 + 100 = ₹3300.$$

Maximization of Total Profit. We shall now develop an important principle in economics by using calculus. Recall that if $R(x)$ and $C(x)$ denote, respectively, the total revenue received and the total cost incurred in the production of x units of a commodity, then the profit $P(x)$ is given by

$$P(x) = R(x) - C(x).$$

Let us consider the most profitable output. We know that profit is maximum when

$$\frac{dP}{dx} = 0 \quad and \quad \frac{d^2P}{dx^2} < 0$$

But $$\frac{dP}{dx} = \frac{d}{dx}(R - C) = \frac{dR}{dx} - \frac{dC}{dx}$$

Thus by setting $\dfrac{dP}{dx} = 0$, we have $\dfrac{dR}{dx} = \dfrac{dC}{dx}$.

That is, at the level of maximum profit, the slope of the tangent to the total revenue curve must be equal to the slope of the tangent to the total cost curve (See Figure 6.14).

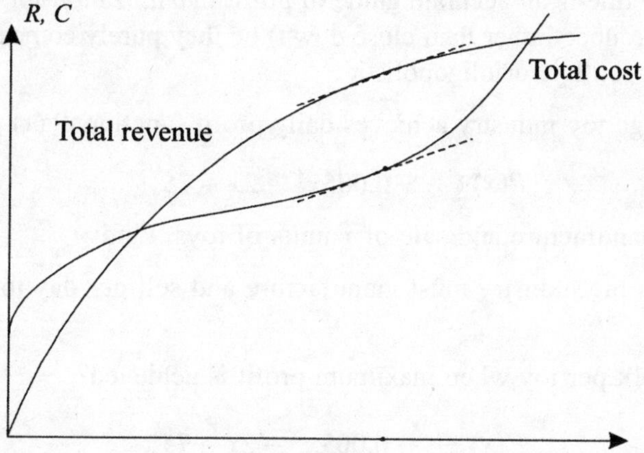

Fig. 6.14

But $\dfrac{dR}{dx}$ is marginal revenue and $\dfrac{dC}{dx}$ is marginal cost. Thus to maximize profit, it is necessary that $MR = MC$. For this to indeed correspond to a maximum, it is necessary that $\dfrac{d^2P}{dx^2} < 0$. But

$$\frac{d^2P}{dx^2} = \frac{d}{dx}\left(\frac{dP}{dx}\right) = \frac{d}{dx}(MR - MC) = \frac{dMR}{dx} - \frac{dMC}{dx}$$

Thus $$\frac{dMR}{dx} < \frac{dMC}{dx}$$

This means the slope of the marginal revenue curve must be less than the slope of the marginal cost curve at the point of maximum profit. In other words, at the point of maximum profit, the marginal cost curve must cut the marginal revenue curve from below (See Figure 6.15).

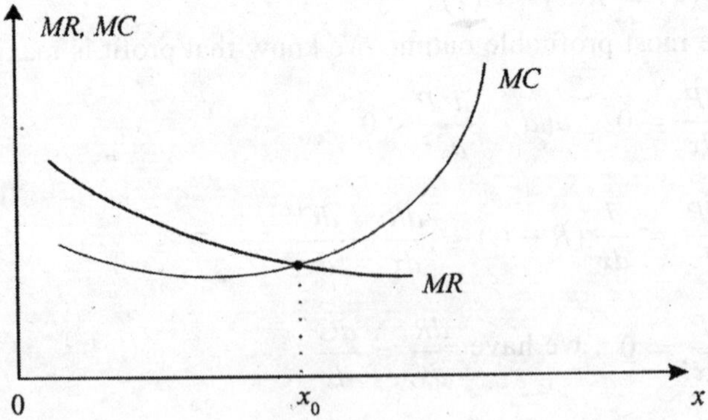

Fig. 6.15

Note. The $MR = MC$ rule is an accurate guide to profit maximization for all firms (assuming the firm chooses to produce rather than close down) be they purely competitive monopolistic, monopolistically competitive, or oligopolistic.

Example 72. A cottage toy industry achieves daily profit function $P(x)$ given by

$$P(x) = -0.005x^2 + 2x - 75$$

which is based on manufacture and sale of x units of toys. Find:

(i) How many toys the industry must manufacture and sell per day to achieve maximum profit?

(ii) What is the profit per toy when maximum profit is achieved?

Solution. We have
$$P(x) = -0.005x^2 + 2x - 75 \quad \ldots(1)$$

$$\therefore \quad P'(x) = -0.01x + 2$$

To maximize profit, we set $P'(x) = 0$, and solve for x. Thus

$$-0.01x + 2 = 0 \Rightarrow x = 200$$

Also, $P''(x) = -0.01$ (always negative)

Thus, the profit is maximum when 200 toys are manufactured and sold daily. Substituting $x = 200$ in (1), the maximum profit is:

$$P(200) = -0.005 \times (200)^2 + 2 \times 200 - 75 = ₹\ 125$$

$$\therefore \quad \text{Profit per toy} = \frac{125}{200} = ₹\ 0.625 \text{ or } 62.5 \text{ paise.}$$

Example 73. A manufacturer can sell x items per day at a price p rupees each, where $p = 125 - 5/3x$. The cost of production for x items is $500 + 13x + 0.2x^2$. Find how many items per day should he manufacture for maximum profit, assuming all items produced are sold. What is the maximum profit?

Solution. We are given $p = 125 - \frac{5}{3}x$: *demand function*

and $C = 500 + 13x + \frac{x^2}{5}$: *cost function*

The revenue, R, is given by $R = px = \left(125 - \frac{5}{3}x\right)x = 125x - \frac{5}{3}x^2$

The profit, P, is given by $P = R - C = \left(125x - \frac{5}{3}x^2\right) - \left(500 + 13x + \frac{x^2}{5}\right)$

$$= -\frac{28}{15}x^2 + 112x - 500 \qquad \ldots(1)$$

First order condition for maximum profit gives

$$\frac{dP}{dx} = 0 \quad i.e., \quad -\frac{56}{15}x + 112 = 0 \quad \Rightarrow \quad x = 30$$

Second order condition for maximum profit requires

$$\frac{d^2P}{dx^2} < 0 \qquad i.e., \qquad -\frac{56}{15} < 0,$$

which is always satisfied. Hence maximum profit is obtained when the production is 30 items per day and maximum profit is: max $P =$ ₹ 1180, which is obtained by substituting $x = 30$ in Eq. (1)

Example 74. A company has for x items produced the total cost C and total revenue R given by equations $C = 100 + 0.015x^2$ and $R = 3x$. Find how many items be produced to maximize the profit. What is this profit ? ***[Delhi Univ. B. Com. (H) 1982]***

Solution. Let P denote the profit function. Then

$$P = R - C = 3x - (100 + 0.015x^2) = 3x - 100 - 0.015x^2 \qquad \ldots(1)$$

$$\therefore \quad \frac{dP}{dx} = 3 - 0.030x$$

To maximize profit, we set $\frac{dP}{dx} = 0$ and solve for x. Thus $3 - 0.030x = 0 \Rightarrow x = 100$

Also $\frac{d^2P}{dx^2} = -0.030$, which is always negative. Thus profit P is maximum when $x = 100$ and the maximum profit is obtained by substituting $x = 100$ in expression (1):

$$\therefore \quad \text{Maximum } P = 3(100) - 100 - 0.015(100)^2 = 50.$$

Alternative Method : We have $R = 3x$ and $C = 100 + 0.015x^2$

$\therefore \quad MR = 3 \quad \text{and} \quad MC = 0.030x$

For profit maximization, $MR = MC \Rightarrow 3 = 0.030x$ *or* $x = 100$, as before

For this to indeed correspond to maximum, it is necessary that $\dfrac{dMR}{dx} < \dfrac{dMC}{dx}$ *i.e.*, $0 < 0.030$, which is true. Hence maximum profit is obtained when $x = 100$. To find this profit, we have $P = R - C = 3x - 100 - 0.015x^2$

$\therefore$ When $x = 100$; $P = 3(100) - 100 - 0.015(100)^2 = 50$.

Example 75. If the total cost of a firm is: $C(x) = \dfrac{1}{3}x^3 - 5x^2 + 30x + 10$, where C is the total cost and x is the output, and price under pure competition is given as 6, find for what value of x the profit will be maximized? Examine both first and second order conditions.

Solution. Since price is fixed at 6, therefore the revenue, R, from the sale of x units is given by $R = 6x$. Hence the profit function is given by

$$P = R - C = 6x - \left(\frac{1}{3}x^3 - 5x^2 + 30x + 10\right) = -24x + 5x^2 - \frac{1}{3}x^3 - 10$$

$$\therefore \quad \frac{dP}{dx} = -24 + 10x - x^2$$

For maximum profit, $\dfrac{dP}{dx} = 0$, That is, $-24 + 10x - x^2 = 0$

$$\Rightarrow \quad x^2 - 10x + 24 = 0 \quad \Rightarrow \quad (x-6)(x-4) = 0 \quad \Rightarrow \quad x = 6 \text{ or } x = 4$$

Also $$\frac{d^2P}{dx^2} = 10 - 2x \begin{cases} < 0, \text{ if } x = 6 \\ > 0, \text{ if } x = 4 \end{cases}$$

Thus maximum profit is obtained when $x = 6$.

Remark. Monopoly Problems in Economic Theory

When a firm is assumed to act as a monopolist with the objective of maximizing total profit, two alternative points of view can be taken. *Either:* the firm fixes its output and leaves the price to be determined by the demand conditions. *Or:* the firm fixes its price and the demand conditions determine the appropriate output. The analysis of the problem is different in the two cases but the results obtained are effectively identical. This is best illustrated in the next example.

Example 76. A monopolist firm has the following total cost and demand functions :

$$C(x) = ax^2 + bx + c \ (a, b, c > 0) \quad \text{and} \quad p(x) = \beta - \alpha x \ (\alpha, \beta > 0)$$

Show that the firm's optimum output level are equal when it fixes (*a*) output, (*b*) price.

[*Delhi Univ. B. Com. (H) 1985*]

Solution. (*a*) ***When the firm fixes the output***

We are given : $C = ax^2 + bx + c$ and $p = \beta - \alpha x$

The revenue from output x is : $R = px = (\beta - \alpha x)\, x = \beta x - \alpha x^2$,

and the profit function is : $P = R - C = (\beta x - \alpha x^2) - (ax^2 + bx + c)$

$$= -(\alpha + a)\, x^2 + (\beta - b)\, x - c$$

The output x fixed by the firm for maximum profit must satisfy the following conditions:

$$\frac{dP}{dx} = 0 \quad \text{and} \quad \frac{d^2P}{dx^2} < 0$$

The first condition $\frac{dP}{dx} = 0$ gives $-2\,(\alpha + a)\, x + (\beta - b) = 0 \quad \Rightarrow \quad x = \frac{\beta - b}{2\,(\alpha + a)}$.

and $$\frac{d^2P}{dx^2} = -2\,(\alpha + a) < 0$$

Thus the second condition $\frac{d^2P}{dx^2} < 0$ is also satisfied.

Hence the profit maximizing level of output is $\frac{\beta - b}{2\,(\alpha + a)}$.

(*b*) ***When the firm fixes the price***

Taking the second view, the firm fixes the price, and its output to meet the demand must be $x = \frac{\beta - p}{\alpha}$. Thus, in this case

$$R = xp = \frac{(\beta - p)}{\alpha}\, p = \frac{(\beta p - p^2)}{\alpha}$$

and $$C = ax^2 + bx + c = a\left(\frac{\beta - p}{\alpha}\right)^2 + b\left(\frac{\beta - p}{\alpha}\right) + c$$

Hence the profit at price p is : $P = R - C$

$$= \frac{(\beta p - p^2)}{\alpha} - a\left(\frac{\beta - p}{\alpha}\right)^2 - b\left(\frac{\beta - p}{\alpha}\right) - c$$

and this is maximum when $$\frac{dP}{dp} = 0 \quad \text{and} \quad \frac{d^2P}{dp^2} < 0$$

Now $$\frac{dP}{dp} = \left(\frac{\beta - 2p}{\alpha}\right) + \frac{2a}{\alpha}\left(\frac{\beta - p}{\alpha}\right) + \frac{b}{\alpha}$$

$$= \frac{(\alpha\beta + 2a\beta + \alpha b)}{\alpha^2} - \frac{2(\alpha + a)}{\alpha^2}p$$

Thus setting $\frac{dP}{dp} = 0$ and solving for p, we obtain $p = \frac{(\alpha\beta + 2a\beta + \alpha b)}{2(\alpha + a)}$

Further, $$\frac{d^2P}{dp^2} = -\frac{2(\alpha + a)}{\alpha^2} < 0$$

Hence the maximum profit is obtained if the firm fixes the price at $p = \frac{(\alpha\beta + 2a\beta + \alpha b)}{2(\alpha + a)}$, and the corresponding output is:

$$x = \frac{1}{\alpha}\left[\beta - \frac{(\alpha\beta + 2a\beta + \alpha b)}{2(\alpha + a)}\right]$$

$$= \frac{1}{\alpha}\left[\frac{2\alpha\beta + 2a\beta - \alpha\beta - 2a\beta - \alpha b}{2(\alpha + a)}\right]$$

$$= \frac{1}{\alpha}\left[\frac{(\alpha\beta - \alpha b)}{2(\alpha + a)}\right]$$

$$= \frac{(\beta - b)}{2(\alpha + a)},$$

which is the same level of output as when the firm fixed the output.

Example 77. A radio manufacturer finds that he can sell x radios per week at ₹ p each, where $p = 2[100 - (x/4)]$. His cost of production of x radios per week is ₹ $\{120x + x^2/2\}$. Show that his profit is maximum when the production is 40 radios per week. Find also his maximum profit per week.

Solution. Here $p = 2\{100 - (x/4)\} = 200 - \frac{x}{2}$ and $C = 120x + \frac{x^2}{2}$

The revenue R is given by : $R = px = \left(200 - \frac{x}{2}\right)x = 200x - \frac{x^2}{2}$

The profit P is given by : $P = R - C = \left(200x - \frac{x^2}{2}\right) - \left(120x + \frac{x^2}{2}\right) = 80x - x^2$

First order condition for maximum profit gives $\frac{dP}{dx} = 0$ *i.e.,* $80 - 2x = 0$ or $x = 40$

Second order condition for maximum profit requires $\frac{d^2P}{dx^2} < 0$, *i.e.,* $-2 < 0$,

which is always satisfied. Hence maximum profit is obtained when the production is 40 radios per week and this profit is $80\ (40) - (40)^2 =$ ₹ 1600.

Example 78. There are 60 newly built apartments. At a rent of ₹ 4500 per month all will be occupied. However, one apartment will be vacant for each ₹ 150 increase in rent. An occupied apartment requires ₹ 600 per month for maintenance. Find the relationship between the profit and the number of unoccupied apartments. What is the number of vacant apartments for which profit is maximum? **[Delhi Univ. B. Com. (H) 2014]**

Solution. Let x be the number of vacant apartments, then the number of occupied apartments will be $(60 - x)$. Also, the rent for each occupied apartment will be $(4500 + 150x)$. Thus if R is the revenue received from the occupied apartments, then

$$R = (60 - x)(4500 + 150x)$$

However, each occupied apartment requires ₹ 600 per month for maintenance, thus the total maintenance cost for all occupied apartments will be

$$C = (60 - x)\ 600$$

Hence the profit function P is given by

$$\begin{aligned} P &= R - C \\ &= (60 - x)(4500 + 150x) - (60 - x)\ 600 \\ &= (60 - x)[4500 + 150x - 600] \\ &= (60 - x)(3900 + 150x) \\ &= 2{,}34{,}000 + 5100x - 150x^2 \end{aligned}$$

That is, $$P = 2{,}34{,}000 + 5100x - 150x^2$$

$$\therefore \quad \frac{dP}{dx} = 5100 - 300x$$

For maximum profit, $\frac{dP}{dx} = 0$, *i.e.,* $5100 - 300x = 0$. This gives $x = 17$. Also

$$\frac{d^2P}{dx^2} = -300 < 0$$

Hence maximum profit is obtained when 17 apartments remain vacant.

Example 79. For a monopolist's product, the demand function is $p = 50/\sqrt{x}$ and the average cost function is $AC = 0.50 + (1000/x)$. Find the profit maximizing price and output. At this level, show marginal revenue is equal to marginal cost.

[Delhi Univ. B. Com. (H) 2019 (Modified)]

Solution. The revenue function is :

$$R = px = (50/\sqrt{x})\,x = 50\sqrt{x}$$

If C denotes the cost function, then

$$AC = \frac{C}{x} = 0.50 + \frac{1000}{x}$$

$$\Rightarrow \qquad C = 0.50x + 1000$$

Hence the profit function is given by

$$P = R - C = 50\sqrt{x} - 0.50x - 1000$$

$$\Rightarrow \qquad \frac{dP}{dx} = \frac{25}{\sqrt{x}} - 0.50$$

For maximum profit, we have

$$\frac{dP}{dx} = 0 \;\; i.e., \;\; \frac{25}{\sqrt{x}} - 0.50 = 0 \;\Rightarrow\; \sqrt{x} = 50 \text{ or } x = 2500$$

Also $$\frac{d^2P}{dx^2} = -\frac{25}{2}x^{-3/2},$$

which is negative when $x = 2500$. Hence maximum profit is obtained when $x = 2500$ and the corresponding price is $p = 50/\sqrt{2500} = 50/50 = 1$. We'll now show that $MR = MC$ at $x =$ 2500. We have

$$MR = \frac{25}{\sqrt{x}} \quad \text{and} \quad MC = 0.50$$

Therefore, when $x = 2{,}500$, $MR = \dfrac{25}{\sqrt{2500}} = \dfrac{25}{50} = 0.50$, and $MC = 0.50$ and the result follows.

Example 80. A company makes a profit of ₹ 5 on each unit of its product it sells. If it spends an amount of ₹ A per week on advertising, then the number of items per week it sells is given by $x = 2000\,(1 - e^{-kA})$, where $k = 0.001$. Find the value of A that will maximize the net profit. ***[Delhi Univ. B. Com. (H) 1994, 1997, 2011]***

Solution. Let P denote the net profit. Then P is given by

$$P = 5x - A = 10{,}000\,(1 - e^{-kA}) - A$$

$$\Rightarrow \qquad \frac{dP}{dA} = 10{,}000\,ke^{-kA} - 1 = 10e^{-0.001A} - 1 \qquad (\because k = 0.001)$$

For maximum net profit, $$\frac{dP}{dA} = 0 \quad i.e., \quad 10e^{-0.001A} - 1 = 0 \;\Rightarrow\; e^{-0.001A} = 0.1$$

$\Rightarrow \quad -0.001A \log_{10} e = -1 \quad \Rightarrow \quad (0.001)(0.4343)\, A = 1 \Rightarrow \quad A =$ Rs. 2302.55 (app.)

Further $$\frac{d^2P}{dA^2} = -0.01e^{-0.001A},$$

which is always negative. Thus P is maximum when $A =$ ₹ 2302.55.

Example 81. A manufacturer plans to sell a new product at a price of ₹ 150 per unit. He estimates that if he spends x thousand rupees on development and y thousand rupees on promotion, approximately $\frac{320y}{y+2} + \frac{160x}{x+4}$ units of the product will be sold. The cost of production is ₹ 50 per unit. If the manufacturer has a total of ₹ 8000 to spend on development and promotion, how should this money be allocated to generate the largest possible profit ?

Solution. It is given that each unit costs the manufacturer ₹ 50 and can be sold at ₹ 150, therefore the profit per unit is ₹ 100. If he spends x thousand rupees on development and y thousand rupees on promotion, the number of units sold is given by

$$\frac{320y}{y+2} + \frac{160x}{x+4}$$

Hence the total profit is given by

$$100\left[\frac{320y}{y+2} + \frac{160x}{x+4}\right]$$

However, this profit is achieved by making an expenditure of $1000\,(x + y)$ on development and promotion. Thus net profit, P, is given by

$$P = 100\left[\frac{320y}{y+2} + \frac{160x}{x+4}\right] - 1000\,(x + y)$$

But $1000(x + y) = 8000$ (given), *i.e.*, $x + y = 8$. Using this equation, P can be expressed explicitly in terms of x :

$$P = \left[320.\frac{8-x}{10-x} + 160\frac{x}{x+4}\right] - 8000$$

$$\Rightarrow \quad \frac{dP}{dx} = 100\left[320\frac{(10-x)(-1)-(8-x)(-1)}{(10-x)^2} + 160\frac{(x+4).1 - x.1}{(x+4)^2}\right]$$

$$= 100\left[320\frac{-2}{(10-x)^2} + 160\frac{4}{(x+4)^2}\right]$$

For P to be maximum, $\frac{dP}{dx} = 0$. That is,

$$\Rightarrow \quad \frac{1}{(10-x)^2} = \frac{1}{(x+4)^2} \quad \Rightarrow \quad 10 - x = \pm (x+4), \quad i.e., \quad x = 3$$

Also $\frac{d^2P}{dx^2} = 100\left[320\,\frac{-4}{(10-x)^3} + 160\,\frac{-8}{(x+4)^2}\right] = -1{,}28{,}000\left[\frac{1}{(10-x)^3} + \frac{8}{(x+4)^2}\right]$,

which is negative when $x = 3$. Thus P is maximum when $x = 3$. Substituting this value of x in $x + y = 8$, we obtain $y = 5$. Hence the manufacturer should allocate ₹ 3,000 on development and ₹ 5,000 on promotion.

Example 82. The production function of a commodity is given by $Q = 40F + 3F^2 - \frac{1}{3}F^3$, where Q is the total output and F is the input.

(i) Find the number of units of input required to give maximum output.

(ii) Find the maximum value of marginal product.

(iii) Verify that when the average product is maximum, it is equal to marginal product.

Solution. The production function is given by : $Q = 40F + 3F^2 - \frac{1}{3}F^3$

(i)
$$\frac{dQ}{dF} = 40 + 6F - F^2$$

For Q to be maximum, $\frac{dQ}{dF} = 0$. This implies

$$40 + 6F - F^2 = 0 \quad \Rightarrow \quad F = 10 \qquad \textit{(on simplification)}$$

Also, $\frac{d^2Q}{dF^2} = 6 - 2F$, which is negative when $F = 10$. Thus Q is maximum when $F = 10$.

(ii)
$$\text{Marginal product} = MQ = \frac{dQ}{dF} = 40 + 6F - F^2$$

$$\Rightarrow \quad \frac{dMQ}{dQ} = 6 - 2F$$

Setting $\frac{dMQ}{dQ} = 0$ gives $F = 3$. Also $\frac{d^2MQ}{dQ^2} = -2 < 0$. Thus MQ is maximum when $F = 3$ and $\max(MQ) = (MQ)_{F=3} = 49$.

(iii) Average Product $= AQ = \dfrac{Q}{F} = 40 + 3F - \dfrac{1}{3}F^2$

$$\Rightarrow \qquad \frac{dAQ}{dF} = 3 - \frac{2}{3}F$$

Setting $\dfrac{dAQ}{dF} = 0$ gives $F = \dfrac{9}{2}$. Also $\dfrac{d^2AQ}{dF^2} = -\dfrac{2}{3} < 0$. Thus AQ is maximum when $F = \dfrac{9}{2}$. We now verify that $AQ = MQ$ at $F = \dfrac{9}{2}$. We have

$$(AQ)_{F=\frac{9}{2}} = 40 + 3\left(\frac{9}{2}\right) - \frac{1}{3}\left(\frac{9}{2}\right)^2 = \frac{187}{4} \quad \text{and} \quad (MQ)_{F=\frac{9}{2}} = 40 + 6\left(\frac{9}{2}\right) - \left(\frac{9}{2}\right)^2 = \frac{187}{4}$$

Thus $$AQ = MQ \quad \text{at} \quad F = \frac{9}{2}.$$

Optimal Trade-in Time. Car owners are often faced with the following problem : When is the best time for the car to be traded in, *i.e.,* replaced ?

There are two major considerations that form the framework of determining optimal replacement time. The first is the estimated cost of repairs needed to maintain the car; the second is the replacement cost of the car. We express these costs as functions of time 't'. Our problem is to determine t for which the average cost of repair and replacement is a minimum.

Example 83. The average cost per repair of a car is found to be ₹ 320 and the number of repairs is given by $\frac{1}{2}t^{3/2}$, where t is time in months since the purchase of the car. The replacement cost is estimated to be ₹ 80,000. Find the optimum replacement period.

[*Delhi Univ. B. Com. (H) 1990*]

Solution. Total number of repairs in t months $= \frac{1}{2}t^{3/2}$

$\therefore$ The average number of repairs per month $= \frac{1}{2}t^{1/2}$

The average cost per repair = ₹ 320

$\therefore$ The average repair cost per month $= 320\left(\frac{1}{2}t^{1/2}\right) = 160t^{1/2}$

The cost of replacement = ₹ 80,000

$\therefore$ The average replacement cost per month $= \dfrac{80{,}000}{t}$

Hence the average cost of repair and replacement per month is given by

$$C(t) = 160t^{1/2} + \frac{80{,}000}{t} \quad \Rightarrow \quad C'(t) = 80t^{-1/2} - \frac{80{,}000}{t^2}$$

Our problem is to determine t for which $C(t)$ is a minimum.

For $C(t)$ to be minimum, $C'(t) = 0$. That is,

$$80t^{-1/2} - \frac{80,000}{t^2} = 0 \quad \Rightarrow \quad 80t^{3/2} - 80,000 = 0, \quad i.e., \quad t^{3/2} = 1000 \quad \Rightarrow \quad t = 100$$

Also, $$C''(t) = -\frac{40}{t^{3/2}} + \frac{1,60,000}{t^3} \quad \Rightarrow \quad C''(100) = -0.04 + 0.16 = 0.12 > 0$$

Hence the optimum replacement period is $t = 100$ months.

Example 84. A piece of machinery costs ₹ 12,000. The total cost of operation from the time of purchase up to a time t is given by the function $20t^2 + 15t$. If the machine is sold as scrap after t years, the resale value is given by the function $6880 - 60t^2$. Find the optimum time for replacement of the machine.

[*Delhi Univ. B. Com. (H) 1993, 2000, 2012, 2019, 2022 (Modified)*]

Solution. The original cost of the machine $=$ ₹ 12,000

Its scrap value after t years $= 6880 - 60t^2$

$\therefore$ Total depreciation $=$ original cost – scrap value

$= 12,000 - (6880 - 60t^2) = 5120 + 60t^2$

and hence the average depreciation per year $= \dfrac{5120}{t} + 60t$

Further, the total cost of operation $= 20t^2 + 15t$

$\therefore$ the average operation cost per year $= 20t + 15$

Hence the average cost of depreciation and operation per year is given by

$$C(t) = \frac{5120}{t} + 60t + 20t + 15 = \frac{5120}{t} + 80t + 15$$

Our problem is to find t so that $C(t)$ is minimum. For $C(t)$ to be minimum, we set

$$C'(t) = 0, \quad i.e., \quad -\frac{5120}{t^2} + 80 = 0 \quad \Rightarrow \quad t^2 = \frac{5120}{80} = 64 \quad \Rightarrow \quad t = 8$$

Also $$C''(t) = \frac{10,240}{t^3} \quad \Rightarrow \quad C''(8) = 20 > 0$$

Thus $C(t)$ is minimum if $t = 8$ and hence the optimum time for replacement is 8 years.

Example 85. A major airline purchases a small plane at a cost of ₹ 5,20,00,000. The company estimates that average capital cost and average operating cost are a function of x, the number of hours of flight time. The salvage value of the plane (in rupees) is expressed by the function $S(x) = 4,20,00,000 - 10,000\,x$. Average operating cost per hour is given by the function: $OC(x) = 800 + 0.40x$. Determine how many hours a plane should be flown before replacement if the objective is to minimize the sum of average capital and average operating cost per hour?

[*Delhi Univ. B. Com. (H) 2009*]

Solution. Average capital cost, $C(x)$, per hour equals the purchase cost minus the salvage cost, all divided by the number of hours of flight time. That is,

$$C(x) = \frac{5,20,00,000 - (4,20,00,000 - 10,000\ x)}{x} = \frac{1,00,00,000}{x} + 10,000$$

The sum of average capital cost and average operating cost is:

$$f(x) = C\ (x) + OC\ (x) = \frac{1,00,00,000}{x} + 10,000 + 800 + 0.40x$$

$$= \frac{1,00,00,000}{x} + 10,800 + 0.40x \qquad ...(1)$$

Our problem is to find x so that $f(x)$ is minimum. For $f(x)$ to be minimum. we set $f'(x) = 0$. That is,

$$-\frac{1,00,00,000}{x^2} + 0.40 = 0 \Rightarrow x^2 = \frac{1,00,00,000}{0.40} = 2,50,00,000 \Rightarrow x = 5,000$$

Also, $f''(x) = \dfrac{2,00,00,000}{x^3} > 0$ when $x = 5000$

Thus $f(x)$ is minimum when $x = 5000$. In other words, the plane should be flown 5000 hours before replacement. The minimum cost per hour is obtained by substituing $x = 5000$ in (1) :

$$f(5000) = ₹\ 14,800 \qquad \textit{(on simplification)}$$

EXERCISE 6.4

1. The demand equation for a manufacturer's product is $p = (80 - x)/4$, where x is the number of units and p is price per unit. At what value of x will there be maximum revenue ? What is the maximum revenue ?
2. For a monopolist's product, the demand function is $x = 10,000e^{-0.02p}$. Find the value of p for which maximum revenue is obtained.
3. A firm's demand function is $x = ae^{-bp}$, where p is price per unit, x is quantity demanded and a, b are positive constants. Find marginal revenue as a function of x. Also, find the price and quantity where total revenue is maximum. ***[Delhi Univ. B.Com. (H) 2005]***
4. A company notices that higher sale of a particular item which it produces are achieved by lowering the price charged. As a result the total revenue from the sales at first rises as the number of units sold increases, reaches a maximum and then falls off. This pattern of the total revenue is described by the relation $R = 40,00,000 - (x - 2,000)^2$, where R is the total revenue and x is the number of units sold.

   (*i*) Find what number of units sold maximizes total revenue.

(*ii*) What is the amount of maximum revenue ?

(*iii*) What would be the total revenue if 2500 units are sold ?

5. A company charges ₹ 15,000 for a television set on orders of 60 or less sets. The charge is reduced on every set by ₹ 100 per set for each set ordered in excess of 60. Find the largest size order the company should allow so as to receive a maximum revenue.
6. A company charges ₹ 550 for a transister set on orders of 50 or less sets. The charge is reduced by ₹ 5 per set for each set ordered in excess of 50. Find the largest size order the company should allow so as to receive a maximum revenue.
7. The owner of a 40 unit motel can rent all the units nightly at ₹ 1000 per night. However, for each ₹ 100 per night increase in rent, 2 units will be vacant with no possibility of filling them. How many units should he rent per night and at what rate to maximize his daily income ?
8. A steel plant is capable of producing x tons per day of a low-grade steel and y tons per day of a high-grade steel, where $y = \dfrac{40 - 5x}{10 - x}$. If the fixed market price of low-grade steel is half of that of high-grade steel, show that about $5\frac{1}{2}$ tons of low-grade steel are produced per day for maximum revenue. **[*Delhi Univ. B. Com. (H) 1981*]**
9. A firm, with fixed plant, supplies of raw materials, etc., produces two goods X and Y in amounts related by the function $y = f(x)$. The market prices of the two goods are fixed at p_x and p_y. Show that the outputs for maximum total revenue are such that $-f'(x)$ equals the ratio of p_x to p_y.
10. The manufacturing cost of an item consists of ₹ 400 as overheads, material cost ₹ 3 per item and the labour cost $x^2/4$ for x - items produced. Find how many items be produced to have the average cost as minimum.
11. A manufacturer's total cost function is given by $C = \dfrac{x^2}{4} + 3x + 400$, where x is the number of units produced. At what level of output will average cost be a minimum ? What is this minimum?
12. Let the cost function of a firm be given by $C = 300x - 10x^2 + \frac{1}{3}x^3$, where C stands for cost and x for output. Calculate

    (*i*) the output at which marginal cost is minimum,

    (*ii*) the output at which average cost is minimum,

    (*iii*) the output at which average cost is equal to marginal cost.
13. The total cost of daily output of x tons of coal is ₹ $\left(\dfrac{1}{10}x^3 - 3x^2 + 50x\right)$. Find the output level at which the average cost is minimum. Verify that at this level, average cost = marginal cost.
14. If x be the number of workers employed, the average cost of production is given by

$$AC = \frac{3}{2(x-4)} + 24x,$$

show that $x = 4.25$ will make the expression minimum. In the interest of the management will you then advise to employ four or five workers ? Give reasons for your answer.

15. A firm produces x units of output per week at a total cost of ₹ $\left(\frac{1}{3}x^3 - x^2 + 5x + 3\right)$. Find the output levels at which the marginal cost and the average variable cost attain their respective minima.

16. The cost of fuel in running a vehicle is proportional to the square of the speed and is ₹ 45 per hour for a speed of 15 km per hour. Other expenses amount to ₹ 500 per hour. Find the most economical speed for the vehicle and also the minimum cost for a distance of 100 km.

[Delhi Univ. B. Com. (H) 2002]

17. The cost of fuel in running an engine is proportional to the square of the speed in kms per hour, and is ₹ 48 per hour when the speed is 16 kms. Other costs amount to ₹ 300 per hour. Find the most economical speed.

18. The cost of fuel for running a train is proportional to the cube of the speed in kms per hour, and is ₹ 64 per hour when the speed is 12 kms. Other costs amount to ₹ 2000 per hour. Find the most economical speed of the train in running a distance of 100 km.

19. Show using calculus that profit of a monopolist is maximum when :

(*i*) $MR = MC$; and

(*ii*) Slope of MR < Slope of MC. **[Delhi Univ. B. Com. (H) 2002, 2003, 2014]**

20. A firm has found from past experience that its profit in terms of number of units x produced is given by $P(x) = -\frac{x^3}{3} + 729x - 2500, \quad 0 \le x \le 35$. Compute

(*i*) the value of x that maximizes the profit,

(*ii*) the profit per unit of the product when this maximum level is achieved.

21. A radio manufacturer produces x sets per week at a total cost of ₹ $(x^2 + 78x + 2000)$. He is a monopolist and the demand function for his product is $x = (600 - p)/8$, when the price is ₹ p per set. Show that maximum net revenue (*i.e.*, profit) is obtained when 29 sets are produced per week. What is the monopoly price ?

22. A radio manufacturer produces x sets per week at a total cost of ₹ $\left[\frac{x^2}{25} + 3x + 100\right]$. He is a monopolist and the demand for his product is $x = 75 - 3p$, where p is the price in rupees per set. Show that the maximum net revenue is obtained when about 30 sets are produced per week. What is the monopoly price ?

23. A monopolist has a demand curve $x = 219 - 3p$ and average cost curve $AC = 8 + \left(\frac{x}{75}\right)$ where p is the price per kg and x is the number of kg of output. Determine the output at which the maximum profit can be earned. **[Delhi Univ. B. Com. (H) 2019]**

24. The total cost of a firm is $C(x) = 1000 + \frac{1}{2}(x/50)^2$, where C is the total cost, x is the output, and price under pure competition is fixed at ₹ 2 per unit. Find the number of units produced that will maximize profit. What is the amount of this maximum profit?

25. Find the profit maximizing output level, given $x = 240 - 10p$ and $AC = 10 + \frac{x}{25}$, where x represents the units of output, p represents price, and AC represents average cost.

26. If a monopolist has a total cost of $C = ax^2 + bx + c$ and if the demand law is $p = \beta - \alpha x^2$, show that the output for maximum net revenue is $x = \frac{\sqrt{a^2 + 3\alpha(\beta - b)} - a}{3\alpha}$.

27. The demand of a monopolist's market is $p = \beta - \alpha x$ and he produces an output of x units at a total cost of $C = ax^3 - bx^2 + cx + d$. Show that the output for maximum net revenue is the positive root of $3ax^2 - 2(b - \alpha)x - (\beta - c) = 0$.

28. The quantity sold (q) and the price (p) are related by $q = ae^{-bp}$ and the production cost is given by $C = l + mq$, where a, b, l and m are positive constants. Find the optimal price which maximizes the profit.

29. For XYZ manufacturing Co., total fixed costs are ₹ 1200, material and labour costs combined are ₹ 2 per unit, and the demand equation is $p = 100/\sqrt{x}$. What level of output will maximize profit? Show that this occurs when marginal revenue is equal to marginal cost. What is the price at profit maximization?

30. Each day a firm makes x tons of chemical A ($x \leq 4$) and y tons of chemical B with $y = (24 - 6x)/(5 - x)$. The profit on chemical A is ₹ 2000 per ton and on B it is ₹ 1000 per ton. How much of chemical A should be produced per day to maximize profit?

31. The production function for a commodity is given by $Q = aF + bF^2 + cF^3$, where Q is output and F is input. Verify that when average product is maximum, it is equal to marginal product. Are there any restrictions for values of a, b and c?

32. The relationship between sales, $TR(x)$, and advertising cost, ₹ x, is given by

$$TR(x) = \frac{32{,}000x}{500 + x}$$

It is known that gross profit is 25 per cent of the sales. Determine (*i*) the corresponding net profit as a function of x, $P(x)$, (*ii*) the value of x which maximizes $P(x)$, and (*iii*) maximized $P(x)$.

33. For a company, the gross profit per unit of its product sold is ₹ 8 and the number of units it can sell per month is dependent on the amount spent on the advertisement given by $x = 9000(1 - e^{-.002A})$, where x is the number of units sold per month when it spends ₹ A per month on advertisement. Find the value of A which maximizes the net profit.

[Delhi Univ. B.Com. (H) 2005]

34. The proportion of people responding to an advertisement of a new product in a certain marketing area is given by $(1 - e^{-0.3t})$, where t is the number of days the advertisement runs. The marketing area contains 4,00,000 potential customers and each response to the advertisement results in an average profit of ₹ 2. The profit is exclusive of advertisement cost. The fixed cost of producing the advertising is ₹ 30,000 and the variable cost is ₹ 4000 for each day the advertisement runs. How many days should the advertisement run to maximize total net profit?

[Delhi Univ. B. Com. (H) 1992]

Hint : The net profit is given by

$$P(t) = 8{,}00{,}000\,(1 - e^{-0.3t}) - (30{,}000 + 4000t)$$

$$\Rightarrow \quad P'(t) = 2{,}40{,}000e^{-0.3t} - 4000$$

Setting $P'(t) = 0$ gives $e^{0.3t} = 60 \Rightarrow t = 14$ (app.) *(on simplification)*

Further, $P''(t) = -72{,}000e^{-0.3t} < 0$.

35. A manufacturer of a product finds that for the first 500 units that are produced and sold, the profit is ₹ 50 per unit. The profit on each of the units beyond 500 is decreased by ₹ 0.10 times the number of additional units produced. What level of output will maximize profit ?

36. The average cost per repair of a television is found to be ₹ 300 and the number of repairs is given by $t^{4/3}$, where t is time in years since the purchase of the television. The replacement cost is estimated to be ₹ 4500. Find the optimal replacement period.

37. The average cost per repair of a scooter is found to be ₹ 50 and the number of repairs is given by $t^{3/2}$, where t is time in months since the purchase of the scooter. The replacement cost is estimated to be ₹ 25,000. Find the optimum replacement period. **[*Delhi Univ. B. Com. (H) 2014*]**

38. The average cost per repair of a car is found to be ₹ 400 and the numbner of repairs is given by $\frac{1}{2}t^{3/2}$, where t is time in months since the purchase of the car. The replacement cost is estimated to be ₹ 1,00,000. Find the optimum replacement period.

39. A piece of machinery costs ₹ 10,000. The total cost of operation from the time of purchase up to a time t is given by the function $10t^2 + 100t$. If the machine is sold as scrap after t years, the resale value is given by the function $7840 - 50t^2$. Find the optimum time for the replacement of the machine.

40. A piece of machinery costs ₹ 15,000. The total cost of operation from the time of purchase upto a time t is given by the function $30t^2 + 20t$. If the machine is sold as scrap after t years, the resale value is given by the function $8520 - 50t^2$. Find the optimum time for the replacement of the machine.

ANSWERS

1. 40, 400 **2.** 50 **3.** $MR = \frac{1}{b}\left[\log\left(\frac{a}{x}\right) - 1\right]$; $x = \frac{a}{e}$ and $p = \frac{1}{b}$

4. (*i*) 2000 (*ii*) 40,00,000 (*iii*) 37,50,000

5. 105 **6.** 80 **7.** 30 units per night, ₹ 1500

10. 40 **11.** $x = 40$, min $AC = 23$ **12.** (*i*) 10 (*ii*) 15 (*iii*) 15

13. 15 **14.** 5 workers **15.** 1 and 3/2

16. 50 kms/hr, ₹ 2000 **17.** 40 kms/hr **18.** 30 kms/hr

20. (*i*) 27 (*ii*) ₹ 393.40 **21.** ₹ 368

22. ₹ 15 **23.** $x = 93.75$ **24.** 5000 units ; ₹ 4000

25. 50 **28.** $p = \frac{1}{b} + m$ **29.** $x = 625,\ p = 4$

30. $5 - \sqrt{3}$ **31.** b positive, c negative

32. (*i*) $P(x) = \frac{8000x}{500 + x} - x$ (*ii*) ₹ 1500 (*iii*) ₹ 4500 **33.** ₹ 2484.92

34. 14 days **35.** $x = 750$ **36.** 17.38 years **37.** 100 months

38. 100 months **39.** 6 years **40.** 9 years

6.12 INVENTORY CONTROL

Another application of max-min theory can be observed by looking at the classical inventory model. We will use this model to determine the **economic order quantity**, *i.e.,* the order quantity such that the inventory costs are at a minimum. We abbreviate economic order quantity by *EOQ*.

Inventory costs are made up of three types of costs – purchasing cost, ordering cost and holding cost. These are described below:

Purchasing Cost. It is the cost incurred to purchase the item. It includes the price of the item *plus* any taxes and transportation costs. If the item is produced by the firm itself, then the full cost involved would be called the **production cost**.

Ordering Cost. It represents the cost which incurred each time an order is placed. It includes the labour to prepare the order, telephone calls, postage, travel, and any other direct cost. If the item is produced internally, ordering cost would include all set up costs.

Holding Cost. It is the cost of keeping an item in stock. It includes the physical storage, maintenance, the cost of capital, insurance, deterioration and depreciation.

A simple *EOQ* model involves the following basic assumptions:

(*a*) There is no *uncertainty* - demand is assumed to be constant and uniform.

(*b*) The lead time is zero - the lead time is the time between placing an order and receiving the goods.

(*c*) All costs are constant.

To develop a formula for *EOQ*, the purchasing cost can be omitted, since it is constant. The only costs to be considered are the ordering cost and the holding cost. Thus

Total inventory cost = ordering cost + holding cost

Looking first at ordering cost:

Annual ordering cost = number of orders × cost per order

If the total annual demand is D units per year and we are ordering Q units on each occasion, then number of orders per year = D/Q. It follows that if C_0 is the cost of one order, then

$$\text{Annual ordering cost} = \frac{D}{Q} C_0$$

Next consider annual holding cost. This would be

Annual holding cost = average inventory × holding cost per unit per year.

The inventory level starts at a peak of Q units and steadily declines to reach at 0, at which time a new order for Q units is placed. When the order is received, it brings up the level back and cycle is repeated. (see Fig. 6.16).

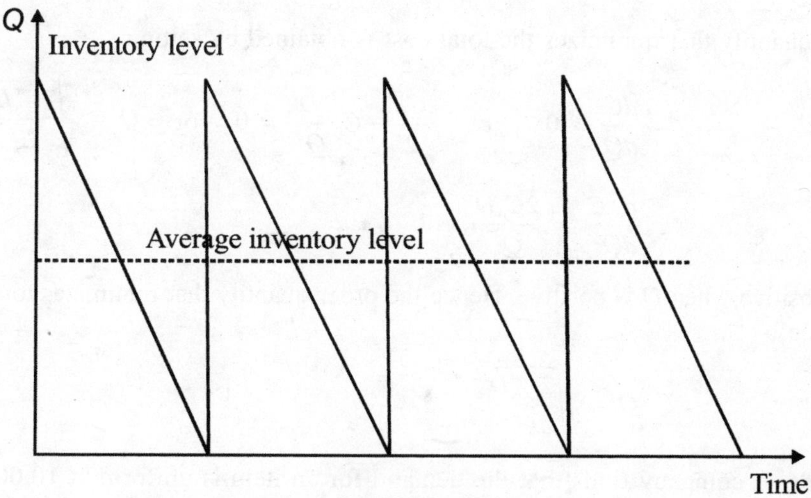

Fig. 6.16

It may be observed that the average inventory level is $Q/2$ units, since the inventory level varies uniformly between 0 and Q units. Hence letting C_h represent the holding cost per unit per year, we obtain

$$\text{Annual holding cost} = \frac{Q}{2} C_h$$

Thus the total inventory cost can be expressed as:

$$C = \frac{D}{Q} C_0 + \frac{Q}{2} C_h$$

We observe that as the size of the order increases, the total annual ordering cost decreases whereas the total annual holding cost increases. We are faced with a trade-off of costs such that total cost, C, will be at a minimum. These costs are shown in Figure 6.17.

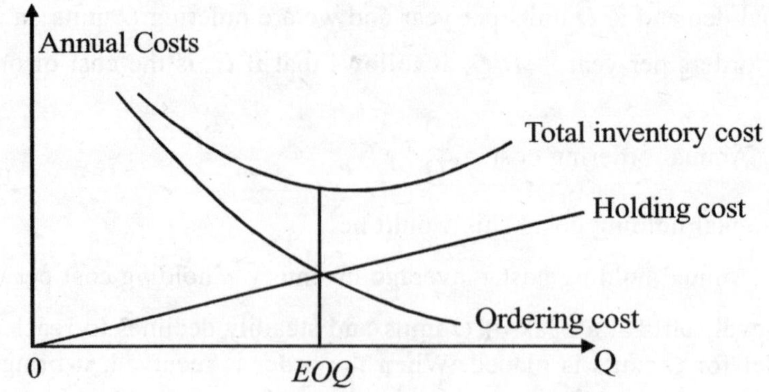

Fig. 6.17

The order quantity that minimizes the total cost is obtained by setting

$$\frac{dC}{dQ} = 0 \quad i.e., \quad \frac{C_h}{2} - C_0 \frac{D}{Q^2} = 0 \quad \text{or} \quad Q = \sqrt{\frac{2C_0 D}{C_h}}$$

Also $$\frac{d^2C}{dQ^2} = \frac{2C_0 D}{Q^3},$$

which is positive when Q is positive. Hence the order quantity that minimizes total inventory cost is given by

$$Q = \sqrt{\frac{2C_0 D}{C_h}}.$$

Example 86. A company finds that the demand for an item is uniform at 10,000 units per year. It costs ₹ 50 each time to place an order, regardless of the quantity order. If the holding cost is ₹ 1 per unit per year, find the economic order quantity.

Solution. Let the economic order quantity be Q, *i.e.*, company is ordering Q units at a time. Since the demand for the item is uniform at 10,000 units per year, therefore

$$\text{Number of orders per year} = \frac{10,000}{Q}$$

The cost of ordering remains fixed at ₹ 50 per order.

$$\therefore \quad \text{Annual ordering cost} = \frac{10,000}{Q} \times 50 = \frac{5,00,000}{Q}$$

The annual holding cost is:

$$\text{Annual holding cost} = \text{average inventory} \times \text{holding cost per unit per year} = \frac{Q}{2} \times 1 = \frac{Q}{2}$$

Hence the total inventory cost per year is:

$$C = \frac{5,00,000}{Q} + \frac{Q}{2}$$

$$\therefore \qquad \frac{dC}{dQ} = -\frac{5,00,000}{Q^2} + \frac{1}{2}$$

Setting $\frac{dC}{dQ} = 0$ gives $Q^2 = 10,00,000 \quad \Rightarrow \quad Q = 1000$

To determine if this value of Q minimizes C, we shall examine second-derivative test :

$$\frac{d^2C}{dQ^2} = \frac{10,00,000}{Q^3},$$

which is positive for $Q = 1000$. Therefore, the company must order 1000 units at a time to minimize its total inventory cost per year.

Example 87. A firm requires 10,000 units of a material per annum. The cost of purchasing is ₹ 1 per unit, the cost of replenishment of stock of material is ₹ 25 and the cost of storing material is 12.5 per cent per annum of the average rupee inventory. Find the optimum order size and the corresponding total cost using calculus.

[*Delhi Univ. B. Com. (H) 1998, 2006, 2014*]

Solution. Let the optimum order size be Q units. That is, Q represent the number of units ordered at a time. Since the demand is 10,000 units per year, therefore

$$\text{Number of orders per year} = \frac{10,000}{Q}$$

and hence

$$\text{Total annual ordering cost} = \frac{10,000}{Q} \times 25 = \frac{2,50,000}{Q}$$

Further

Total annual holding cost = cost of storing per unit × average number of units in stock

$$= (12.5\% \text{ of ₹ } 1) \times \frac{Q}{2} = \frac{125}{1000} \times \frac{Q}{2} = \frac{Q}{16}$$

and finally

Total annual purchasing cost = Total annual demand × cost of purchasing per unit

$$= 10,000 \times 1 = 10,000$$

$\therefore$ Total annual inventory cost is : $\quad C = \frac{2,50,000}{Q} + \frac{Q}{16} + 10,000 \qquad \text{... (1)}$

Differentiating C *w.r.t.* Q, we obtain

$$\frac{dC}{dQ} = -\frac{2{,}50{,}000}{Q^2} + \frac{1}{16}$$

Setting $\quad \dfrac{dC}{dQ} = 0$ gives $Q^2 = 40{,}00{,}000 \quad$ or $\quad Q = 2000$

Also, $\quad \dfrac{d^2C}{dQ^2} = \dfrac{5{,}00{,}000}{Q^3}$,

which is positive and hence C is minimum if the order size is 2000 units. The corresponding cost per annum is obtained by substituting $Q = 2000$ in Eq. (1):

$$C = \frac{2{,}50{,}000}{2000} + \frac{2000}{16} + 10{,}000 = ₹\,10{,}250$$

Example 88. A chemical company acquires its requirement in 20 orders of the size 250 drums each. Each drum costs ₹ 80. The replenishment cost is ₹ 40 irrespective of the order size, and the annual carrying cost is 2% of the average rupee inventory. Find how much can company save if it buys according to economic order quantity using calculus.

[*Delhi Univ. B. Com. (H) 1999, 2017 (modified)*]

Solution. Annual demand : $D = 20 \times 250 = 5000$ units

Let the economic order quantity be Q units. That is, Q represents the number of units ordered at a time.

Then

$$\text{Number of orders per year} = \frac{5000}{Q}$$

The replenishment cost is fixed at ₹ 40 per order irrespective of the order size.

$$\therefore \quad \text{Total annual ordering cost} = \frac{5000}{Q} \times 40 = \frac{2{,}00{,}000}{Q}$$

The annual holding cost is:

Annual holding cost = cost of storing one unit × average number of units in stock

$$= (2\% \text{ of } ₹80) \times \frac{Q}{2} = 0.80Q$$

Total annual purchasing cost = Total annual demand × cost of purchasing per unit

$$= 5000 \times 80 = 4{,}00{,}000$$

Hence the total annual inventory cost is:

$$C = \frac{2{,}00{,}000}{Q} + 0.80Q + 4{,}00{,}000$$

Differentiating C *w.r.t.* Q, we obtain $\dfrac{dC}{dQ} = -\dfrac{2,00,000}{Q^2} + 0.80$

Setting $\dfrac{dC}{dQ} = 0$ gives $Q^2 = 2,50,000$ or $Q = 500$

Also, $$\frac{d^2C}{dQ^2} = \frac{4,00,000}{Q^3} > 0 \text{ when } Q = 500. \text{ Hence } EOQ = 500 \text{ units.}$$

Further, $$C\,(500) = \frac{2,00,000}{500} + 0.80(500) + 4,00,000 = ₹\ 4,00,800$$

and $$C\,(250) = \frac{2,00,000}{250} + 0.80\,(250) + 4,00,000 = ₹\ 4,01,000$$

Hence annual saving = ₹ 4,01,000 – ₹ 4,00,800 = ₹ 200.

Example 89. A firm produces 36,000 items per year. It costs ₹ 250 to make the machine run regardless of the number (x) of items produced in a run. The cost of storage is 50 paise per item per year on average inventory $x/2$ in hand. The cost of material per item is ₹ 5. Find the economic lot size x. ***[Delhi Univ. B. Com. (H) 1987]***

Solution. Since the firm is producing 36,000 items per year at the rate of x items per run, therefore the number of production runs per year is $\dfrac{36,000}{x}$. Thus

$$\text{Total annual set up cost} = \frac{36,000}{x} \times 250 = \frac{90,00,000}{x}$$

Further,

$$\text{Total annual storage cost} = \text{cost of storing per item} \times \text{average number of items in stock}$$

$$= 0.50 \times \frac{x}{2} = 0.25x$$

and finally

$$\text{Total annual material cost} = \text{total annual production} \times \text{cost of material per item}$$

$$= 36,000 \times 5 = 1,80,000$$

Hence the total annual inventory cost is:

$$C = \frac{90,00,000}{x} + 0.25x + 1,80,000$$

Differentiating C w.r.t. x, we obtain

$$\frac{dC}{dx} = -\frac{90,00,000}{x^2} + 0.25$$

Setting $\frac{dC}{dx} = 0$ gives $x^2 = 3,60,00,000$ or $x = 6000$

Also, $\frac{d^2C}{dx^2} = \frac{1,80,00,000}{x^3} > 0$ when $x = 6000$. Thus C is minimum if the lot size is 6000 units.

Example 90. A factory follows an economic order quantity system for maintaining stocks of one of its component requirements. The annual demand is for 24,000 units, the cost of placing an order is ₹ 300, and the component cost is ₹ 60 per unit. The factory has imputed 24% as the inventory carrying rate. Determine the *EOQ* and the optimal interval for placing orders, assuming a year is equivalent to 360 days. **[*Delhi Univ. B. Com. (H) 2011*]**

Solution. Let economic order quantity (*EOQ*) be Q units. Then the total annual cost is:

$$C(Q) = \text{Total ordering cost} + \text{Total carrying cost} + \text{Total purchasing cost}$$

$$= \frac{24,000}{Q} \times 300 + \frac{24}{100} \times 60 \times \frac{Q}{2} + 24,000 \times 60$$

$$= \frac{72,00,000}{Q} + 7.2Q + 14,40,000$$

Differentiating C w.r.t. Q, we obtain

$$C'(Q) = -\frac{72,00,000}{Q^2} + 7.2$$

Setting $C'(Q) = 0$ gives $Q^2 = 10,00,000$ or $Q = 1000$

Also $C''(Q) = \frac{1,44,00,000}{Q^3} > 0$ when $Q = 1000$. Hence $EOQ = 1000$ units.

$$\text{Number of orders in a year} = \frac{24,000}{1000} = 24$$

$$\text{Optimal interval between placing orders} = \frac{360}{24} = 15 \text{ days}$$

Example 91. A publishing house purchases 2000 units of a particular item per year at a unit cost of ₹ 20. The ordering cost per order is ₹ 50 and the inventory cost is 25% of the value of the average inventory. Find the optimal order quantity and the minimum total cost.

If the supplier offers a discount of 3% on purchases in lots of 1000, should the offer be accepted ? **[*Delhi Univ. B. Com. (H) 2012, 2018*]**

Solution. Let the optimal order quantity be Q units. Then the total annual inventory cost is :

$$C(Q) = \text{Total ordering cost} + \text{Total holding cost} + \text{Total purchasing cost}$$

$$= \frac{2000}{Q} \times 50 + \frac{25}{100} \times 20 \times \frac{Q}{2} + 2000 \times 20$$

$$= \frac{1,00,000}{Q} + \frac{5}{2}Q + 40,000$$

$$\Rightarrow \quad C'(Q) = -\frac{1,00,000}{Q^2} + \frac{5}{2}$$

Setting $C'(Q) = 0$ gives $Q^2 = 40,000$ or $Q = 200$

Also $C''(Q) = \dfrac{2,00,000}{Q^3} > 0$ when $Q = 200$

Thus C is minimum when $Q = 200$. The corresponding annual inventory cost is:

$$C(200) = ₹\ 41,000$$

If the supplier offers a discount of 3% on purchases in lots of 1000, then the total annual inventory cost is :

$$C = \frac{2000}{1000} \times 50 + \frac{1}{4} \times \frac{1000}{2} \times 19.40 + 2000 \times 19.40$$

$$= 100 + 2425 + 38,800 = ₹\ 41,325$$

which is greater than ₹ 41,000 and hence the offer should not be accepted.

Example 92. A company uses components at the rate of 500 a month which are bought at the cost of ₹ 12 each from the supplier. The cost of placing an order is ₹ 200, irrespective of the quantity ordered. The total holding cost is made up of the capital cost of 10% per annum of the value of stock plus 30 paise per item per annum for insurance plus 60 paise per item per annum for storage plus 30 paise per item for deterioration. Determine the number of components the company should order, the number of orders per year and the total annual cost of the inventory.

Solution. Annual demand = 500 × 12 = 6000 units

Let Q represent the number of units ordered at a time. Then

$$\text{Number of orders per year} = \frac{6000}{Q}$$

The cost of placing an order remains fixed at ₹ 200 per order.

$$\therefore \quad \text{Total annual ordering cost} = \frac{6000}{Q} \times 200 = \frac{12,00,000}{Q}$$

Further,

Total annual holding cost $= \text{Cost of holding per unit} \times \text{average number of units in stock}$

$$= (0.10 \times 12 + 0.30 + 0.60 + 0.30) \times \frac{Q}{2}$$

$$= 2.40 \times \frac{Q}{2} = 1.20Q$$

and finally,

Total annual purchasing cost $= \text{Total annual demand} \times \text{cost of purchasing per item}$

$$= 6000 \times 12 = 72{,}000$$

$\therefore$ Total annual inventory cost is :

$$C = \frac{12{,}00{,}000}{Q} + 1.20Q + 72{,}000 \qquad \text{...(1)}$$

$$\Rightarrow \qquad \frac{dC}{dQ} = -\frac{12{,}00{,}000}{Q^2} + 1.2$$

Setting $\frac{dC}{dQ} = 0$ gives $Q^2 = 10{,}00{,}000$ or $Q = 1000$

Also, $\frac{d^2C}{dQ^2} = \frac{24{,}00{,}000}{Q^3} > 0$ when $Q = 1000$

Hence, the company must order 1000 units at a time to minimize its total inventory cost per year.

$$\text{The number of orders per year} = \frac{6000}{Q} = \frac{6000}{1000} = 6$$

The total annual inventory cost is obtained by substituting $Q = 1000$ in Equation (1) :

$$C = \frac{12{,}00{,}000}{1000} + 1.20 \times 1000 + 72{,}000 = ₹74{,}400.$$

EXERCISE 6.5

1. Suppose that the cost of holding one item in inventory is ₹ 0.50 and that the cost of placing an order is ₹ 100. Also assume that the annual demand for the item is 4900 units. Find the economic order quantity.
2. The demand for components of a particular type is 300 units per year. Each order, regardless of the size of order, incurs a cost of ₹ 40. The cost of holding a piece of component for a year is reckoned to be ₹ 0.60. Determine the economic order quantity.
3. A company annually uses 8000 units of an item which are bought at the cost of ₹ 40 each from the supplier. It costs ₹ 80 each time to place an order, regardless of the quantity ordered. The holding cost is estimated to be 20% of the value of average inventory. Determine the order quantity that minimizes total annual cost.

4. A company annually uses 10,000 units of an item which are bought at the cost of ₹ 5 each from the supplier. It costs ₹ 25 each time to place an order, regardless of the quantity ordered. The storage cost is 10% of the value of average inventory. Find the optimum order size.

[Delhi Univ. B. Com. (H) 2010, 2014]

5. A company annually uses 10,000 units of an item which are bought at the cost of ₹ 5 each from the supplier. It costs ₹ 25 each time to place an order, regardless of the quantity ordered. The holding cost is estimated to be 10% of the value of average inventory. Find the optimum order size and the corresponding total cost using calculus.

If the supplier offers a discount of 5% on purchases in lots of 2000, should the offer be accepted?

6. A bicycle manufacturer buys 6000 tyres in a year from a distributor. The ordering cost is ₹ 20 per order, the storage cost is 96 paisa per tyre per year. Each tyre costs ₹ 25. Suppose that the tyres are used at a constant rate throughout the year. Using calculus, find the economic order quantity (*EOQ*). If the tyre requirement increases fourfold and other conditions remain the same, what will be the new (*EOQ*) ? **[Delhi Univ. B. Com. (H) 2001]**

7. A firm requires 5000 units of a material per annum. The cost of purchasing is ₹ 30 per unit and the cost of placing on order is ₹ 20, irrespective of the quantity ordered. If the storage cost is 80 paisa per item per year on average inventory in hand, find the economic order quantity (*EOQ*). If the material requirement increases fourfold and other conditions remain the same, what will be the new (*EOQ*)?

8. A firm has to produce 1,44,000 units of an item per year. The cost of material is ₹ 5 per unit and cost of storage is 50 paise per item per year on average inventory $x/2$ in hand. If it costs ₹ 90 to make the factory ready for a production run, irrespective of the units x produced in a run, find the economic lot size x. **[Delhi Univ. B. Com. (H) 1987]**

9. A firm has to produce 1,44,000 units of an item during a year. The production is done in lots and it costs ₹ 60 to make the factory ready for a production run, irrespective of the units produced in a run. If the storage cost is 50 paise per unit per year, find the most economic lot size.

[Delhi Univ. B. Com. (H) 1989]

10. A firm's annual sales are s units of a product which the firm buys from a supplier. If the replenishment cost is ₹ r per order and the holding cost ₹ h per unit per year, find the economic order quantity using calculus. **[Delhi Univ. B. Com. (H) 1991]**

11. A manufacturing company has a contract to supply 4000 units of an item per year at a uniform rate. The storage cost per unit per year amounts to ₹ 50 and the set up cost per production run is ₹ 160. If the production run can be started instantaneously and shortages are not permitted, determine the number of units which should be produced per run to minimize the total cost. Also calculate the cost. **[Delhi Univ. B. Com. (H) 1995]**

12. A company annually produces and sells 6000 units of a product. Sales are uniformly distributed throughout the year. It costs ₹ 30 to make the machine run regardless of the number of units produced in a run. The production cost of each unit is ₹ 5 and carrying cost is estimated to be 20% of the value of the average inventory. Find the economic lot size.

13. A manufacturer has to produce annually 1000 units of a product that is sold at a uniform rate during the year. The production cost of each unit is ₹ 10 and carrying cost is estimated to be

12.8% of the value of average inventory. Set up cost per production run is ₹ 40. Find the economic lot size.

14. A departmental store sells 2500 top-of-the-line refrigerators per year, with sales occurring at a relatively constant rate. The annual holding cost per refrigerator is ₹ 10. To re-order there is a basic ₹ 20 service fee per order plus a ₹ 9 insurance charge for each refrigerator ordered. How many times per year and in what lot size should the store re-order to minimize the total inventory cost ?

Hint: Total annual inventory cost is :

$$C = \frac{2500}{Q} \times 20 + 10 \times \frac{Q}{2} + 2500 \times 9$$

15. A firm acquires its annual requirements in 5 orders of 1000 units each at present. If per order cost is ₹ 40, irrespective of the order size, holding cost is 10% of the average rupee inventory and the cost per unit is ₹ 100, find Economic Order Quantity (EOQ) using calculus. Also find annual saving by switching to EOQ system. ***[Delhi Univ. B. Com. (H) 2017]***

16. A firm has to produce 14,400 units of an item per year. The production cost of each unit is ₹ 5 and the carrying cost is estimated to be 10% of the value of the average inventory. It costs ₹ 25 to make the machine run regardless of the number of units produced in a run. Find the economic lot size. If currently, the firm produces in the lots of 1,000 units, find how much can the firm save if it produces according to the economic lot size.

ANSWERS

1. 1400 **2.** 200 **3.** 400 **4.** 1000

5. 1000; ₹ 50,500; yes **6.** 500; 1000 **7.** 500; 1000 **8.** $x = 7200$

9. 5879 **10.** $\sqrt{\frac{2rs}{h}}$ **11.** 160; ₹ 8,000 **12.** 600

13. 250 **14.** 25; 100 **15.** 200; ₹ 800 **16.** 1200; ₹ 10

CHAPTER

7

PARTIAL DIFFERENTIATION

OBJECTIVES

After studying the material in this chapter, you should be able to:

- *Understand the meaning of partial differentiation.*
- *Evaluate first and second-order partial derivatives of functions of more than o n e independent variable.*
- *Understand the concepts of homogeneous functions and Euler's theorem.*
- *Find total differentials of first and second order.*
- *Examine a function of two variables for local maxima and local minima.*

7.1 INTRODUCTION

Our treatment of differentiation so far has been limited to functions of one independent variable. In this chapter we extend the concept of differentiation of functions with more than one independent variable. These functions occur frequently in business and economic analysis as well as other areas of study. For example, a firm's output may depend on its taxes, labour costs, cost of warehouse, space and so on. Similarly, the price of a commodity may depend on many variables, such as the demand for that commodity, the cost of labour, competitor's price, maintenance and insurance. The point to bear in mind is that in most applications, a quantity depends on more than one quantity.

In this chapter we will usually restrict our study to the case of two independent variables. However, this can often be easily extended to the case where there are more than two independent variables.

7.2 FUNCTIONS OF TWO VARIABLES

If to each point (x, y) in a certain subset of the two-dimensional plane there corresponds onc and only value of z, then the correspondence is called a *function of two variables.* Just as in the one independent variable case, this correspondence is often represented by an equation. However, the equation is now of the form

$$z = f(x, y),$$

where z is the dependent variable, and x and y are the independent variables. The set of all allowable values for the independent variables constitutes the domain of the function. For example, the equation

$$z = f(x, y) = \frac{x^2 + y^2}{y - 2}$$

defines z as a function of x and y. Because the denominator is zero when $y = 2$, the domain of f is all ordered pairs (x, y) such that $y \neq 2$. Some function values are

$$f(1, 3) = \frac{(1)^2 + (3)^2}{3 - 2} = 10, \quad f(2, 0) = \frac{(2)^2 + (0)^2}{0 - 2} = -2$$

Turning to another example, let us define z by $z^2 = x^2 + y^2$, then for $x = 3$ and $y = 4$, we have $z^2 = 25$. Consequently, $z = \pm 5$. Thus with the ordered pair (3,4) we cannot associate exactly one value of z. Hence z is not a function of x and y.

The following example gives an applied situation that involves two independent variables.

Example 1. A company produces two types of products, A and B. Product A gives a profit of ₹ 50 for each unit produced and sold. Product B gives a profit of ₹ 40 for each unit produced and sold. State the profit function. Also find the profit when the number of units of A produced and sold is 5 and the number of units of B produced and sold is 10.

Solution. Let

x = the number of units of A produced and sold,

y = the number of units of B produced and sold.

Then $P = 50x + 40y$,

where P denotes profit. Here x and y are independent variables and P is the dependent variable. The profit function is of the form $P = f(x, y)$, where $f(x, y) = 50x + 40y$. When $x = 5$ and $y = 10$, then

$$P = f(5, 10) = 50\,(5) + 40\,(10) = ₹\ 650.$$

7.3 PARTIAL DERIVATIVES

In the earlier chapters we studied the derivative as the rate of change of functions of the type $y = f(x)$ – that is, functions of one variable only. This section is concerned with the generalization of some of the previous concepts to functions of two variables.

For example, consider a function of two independent variables written in the form $z = f(x, y)$. We might wish to study the rate at which z changes as x changes if y is held constant.

By holding y as a constant, $z = f(x, y)$ essentially becomes a function of x alone and we can calculate its derivative with respect x. This derivative is called the *partial derivative of* z *(or* f*) with respect to* x and is denoted by

$$\frac{\partial z}{\partial x}, \quad \frac{\partial}{\partial x}[f(x, y)], \; f_x(x, y) \quad \text{or} \quad z_x\,(x, y)$$

Similarly, the partial derivative of z (or f) with respect to y is denoted by

$$\frac{\partial z}{\partial y}, \quad \frac{\partial}{\partial y}[f(x, y)], \; f_y(x, y) \quad \text{or} \quad z_y\,(x\; y)$$

and gives the rate of change of z with respect to y, with x held constant.

Partial derivative of $z = f(x, y)$ with respect to x evaluated at the point (x_0, y_0) is denoted by

$$\left(\frac{\partial z}{\partial x}\right)_{(x_0,\, y_0)}, \; f_x\,(x_0, y_0) \quad \textit{or} \quad z_x\,(x_0, y_0)$$

Similarly, partial derivative of $z = f(x, y)$ with respect to y evaluated at (x_0, y_0) is denoted by

$$\left(\frac{\partial z}{\partial y}\right)_{(x_0,\, y_0)}, \; f_y\,(x_0, y_0) \quad \textit{or} \quad z_y\,(x_0, y_0)$$

Example 2. If $z = 2x^2y + 5xy - y^3$, find $\dfrac{\partial z}{\partial x}$ and $\dfrac{\partial z}{\partial y}$.

Solution. To find $\dfrac{\partial z}{\partial x}$, we treat y as a constant and differentiate z with respect to x. Thus

$$\frac{\partial z}{\partial x} = 4xy + 5y$$

To find $\dfrac{\partial z}{\partial y}$, we treat x as a constant and differentiate z with respect to y. Thus

$$\frac{\partial z}{\partial y} = 2x^2 + 5x - 3y^2.$$

Example 3. If $z = \log(x^2 + y^2)$, find $\dfrac{\partial z}{\partial x}$ and $\dfrac{\partial z}{\partial y}$.

Solution. Treating y as a constant and differentiating z with respect to x, we obtain

$$\frac{\partial z}{\partial x} = \frac{1}{x^2 + y^2}\,\frac{\partial}{\partial x}\left(x^2 + y^2\right) = \frac{2x}{x^2 + y^2}$$

Similarly, $$\frac{\partial z}{\partial y} = \frac{2y}{x^2 + y^2}.$$

Example 4. If $z = \log (y/x)$, find $\frac{\partial z}{\partial x}$ and $\frac{\partial z}{\partial y}$. Also verify that $x\frac{\partial z}{\partial x} + y\frac{\partial z}{\partial y} = 0$.

Solution. (*i*) Treating y as a constant, $\frac{\partial z}{\partial x} = \frac{1}{y/x}\left(-\frac{y}{x^2}\right) = -\frac{1}{x}$

Treating x as a constant, $\frac{\partial z}{\partial y} = \frac{1}{y/x}\left(\frac{1}{x}\right) = \frac{1}{y}$

$$\therefore \quad x\frac{\partial z}{\partial x} + y\frac{\partial z}{\partial y} = x\left(-\frac{1}{x}\right) + y\left(\frac{1}{y}\right) = -1 + 1 = 0.$$

Example 5. Find the first-order partial derivatives of z if

(*i*) $z = y^2 e^{5x^2 + y^3}$ (*ii*) $z = x^y y^x$

Solution. (*i*) $z = y^2 e^{5x^2 + y^3}$

$$\therefore \quad \frac{\partial z}{\partial x} = y^2 e^{5x^2 + y^3} \frac{\partial}{\partial x}(5x^2 + y^3) = y^2 e^{5x^2 + y^3}(10x) = 10xy^2 e^{5x^2 + y^3}$$

$$\frac{\partial z}{\partial y} = y^2 \frac{\partial}{\partial y}\left(e^{5x^2 + y^3}\right) + e^{5x^2 + y^3}\frac{\partial}{\partial y}(y^2)$$

$$= y^2 e^{5x^2 + y^3}(3y^2) + e^{5x^2 + y^3}(2y)$$

$$= 3y^4 e^{5x^2 + y^3} + 2y\, e^{5x^2 + y^3} = (3y^3 + 2)y\, e^{5x^2 + y^3}$$

(*ii*) $z = x^y y^x$

$$\Rightarrow \quad \frac{\partial z}{\partial x} = x^y \frac{\partial}{\partial x}(y^x) + y^x \frac{\partial}{\partial x}(x^y)$$

$$= x^y (y^x \log y) + y^x (yx^{y-1}) \qquad (y \text{ is treated as a constant})$$

$$= x^y y^x \log y + y^{x+1} x^{y-1}$$

$$= x^{y-1} y^x (x \log y + y)$$

By symmetry, $\frac{\partial z}{\partial y} = y^{x-1} x^y (y \log x + x)$

Note. The concept of partial derivatives can be extended in a natural way to functions of more than two variables. For example, if $z = f(x_1, x_2, ..., x_n)$ is a function of n variables, then to find $\dfrac{\partial z}{\partial x_1}$, treat $x_2, x_3, ..., x_n$ as constants and differentiate z with respect to x_1. To find $\dfrac{\partial z}{\partial x_2}$, treat $x_1, x_3, ..., x_n$ as constants and differentiate z with respect to x_2, and so on. It should be noted when we take a partial derivative with respect to a particular independent variable, all other independent variables are held constants. For example, if $z = 2x_1^2 x_2 + x_1 x_2^2 + x_3^2$ then by treating x_2 and x_3 as constants and differentiating z with respect to x_1, we obtain

$\dfrac{\partial z}{\partial x_1} = 4x_1x_2 + x_2^2$. Similarly, $\dfrac{\partial z}{\partial x_2} = 2x_1^2 + 2x_1x_2$ and $\dfrac{\partial z}{\partial x_3} = 2x_3$.

Example 6. If $u = \log(x^2 + y^2 + z^2)$, show that $x\dfrac{\partial u}{\partial x} + y\dfrac{\partial u}{\partial y} + z\dfrac{\partial u}{\partial z} = 2.$

Solution. We have

$$\frac{\partial u}{\partial x} = \frac{1}{x^2 + y^2 + z^2}\frac{\partial}{\partial x}(x^2 + y^2 + z^2) = \frac{1}{x^2 + y^2 + z^2}(2x) = \frac{2x}{x^2 + y^2 + z^2}$$

Similarly,

$$\frac{\partial u}{\partial y} = \frac{2y}{x^2 + y^2 + z^2} \quad \text{and} \quad \frac{\partial u}{\partial z} = \frac{2z}{x^2 + y^2 + z^2}$$

$$\therefore \quad x\frac{\partial u}{\partial x} + y\frac{\partial u}{\partial y} + z\frac{\partial u}{\partial z} = \frac{2x^2}{x^2 + y^2 + z^2} + \frac{2y^2}{x^2 + y^2 + z^2} + \frac{2z^2}{x^2 + y^2 + z^2}$$

$$= \frac{2x^2 + 2y^2 + 2z^2}{x^2 + y^2 + z^2} = 2.$$

EXERCISE 7.1

1. Find the first-order partial derivatives of the following functions :

(*i*) $2x^3y - xy^2 + y^4$ (*ii*) $\log(x^2 - y^2)$ (*iii*) $(5x + 4y)^3$ (*iv*) $e^{x^2 + xy + y^2}$

2. Find $\dfrac{\partial z}{\partial x}$ and $\dfrac{\partial z}{\partial y}$ for each of the following functions :

(*i*) $z = e^{xy} + \log(xy)$ (*ii*) $z = xe^{x^2 - y^2}$ (*iii*) $z = (x^2 + y^3)^4$ (*iv*) $z = \dfrac{x^2 + y^2}{xy}$

3. In the following problems, find f_x and f_y.

(i) $f(x, y) = 3x^2y^2 + x^5 + 3y^2$ (ii) $f(x, y) = \dfrac{x^2}{y^3} + \dfrac{y^2}{x^3}$

(iii) $f(x, y) = \sqrt{xy}$ (iv) $f(x, y) = 5x \log (x^2 + y)$

4. (i) If $f(x, y) = 2x^3 - 11x^2y + 3y^3$, prove that $x f_x(x, y) + y f_y(x, y) = 3 f(x, y)$.

(ii) If $z = 3x^4 - x^2y^2 + xy^3$, prove that $x\dfrac{\partial z}{\partial x} + y\dfrac{\partial z}{\partial y} = 4z$.

5. (i) If $z = f(y/x)$, show that $x\dfrac{\partial z}{\partial x} + y\dfrac{\partial z}{\partial y} = 0$.

(ii) If $z = x^2 f(y/x)$, show that $x\dfrac{\partial z}{\partial x} + y\dfrac{\partial z}{\partial y} = 2z$.

6. (i) If $z = \log(e^x + e^y)$, prove that $\dfrac{\partial z}{\partial x} + \dfrac{\partial z}{\partial y} = 1$.

(ii) If $z = \dfrac{e^{xy}}{e^x + e^y}$, prove that $\dfrac{\partial z}{\partial x} + \dfrac{\partial z}{\partial y} = (x + y - 1) z$.

7. In the following problems, find f_x, f_y and f_z.

(i) $f(x, y, z) = x^2y + yz^3 + xyz$ (ii) $f(x, y, z) = (xy + yz) / xz$.

8. (i) If $u = x^2y + y^2z + z^2x$, show that $\dfrac{\partial u}{\partial x} + \dfrac{\partial u}{\partial y} + \dfrac{\partial u}{\partial z} = (x + y + z)^2$.

(ii) If $u = x^2 (y - z) + y^2 (z - x) + z^2 (x - y)$, show that $\dfrac{\partial u}{\partial x} + \dfrac{\partial u}{\partial y} + \dfrac{\partial u}{\partial z} = 0$.

ANSWERS

1. (i) $\dfrac{\partial z}{\partial x} = 6x^2y - y^2$, $\dfrac{\partial z}{\partial y} = 2x^3 - 2xy + 4y^3$ (ii) $\dfrac{\partial z}{\partial x} = \dfrac{2x}{x^2 - y^2}$, $\dfrac{\partial z}{\partial y} = -\dfrac{2y}{x^2 - y^2}$

(iii) $\dfrac{\partial z}{\partial x} = 15 (5x + 4y)^2$, $\dfrac{\partial z}{\partial y} = 12 (5x + 4y)^2$

(iv) $\dfrac{\partial z}{\partial x} = e^{x^2 + xy + y^2} (2x + y)$, $\dfrac{\partial z}{\partial y} = e^{x^2 + xy + y^2} (x + 2y)$

2. (i) $\dfrac{\partial z}{\partial x} = ye^{xy} + \dfrac{1}{x}$, $\dfrac{\partial z}{\partial y} = xe^{xy} + \dfrac{1}{y}$ (ii) $\dfrac{\partial z}{\partial x} = (2x^2 + 1)e^{x^2 - y^2}$, $\dfrac{\partial z}{\partial y} = -2xye^{x^2 - y^2}$

(iii) $\dfrac{\partial z}{\partial x} = 8x (x^2 + y^3)^3$, $\dfrac{\partial z}{\partial y} = 12y^2 (x^2 + y^3)^3$ (iv) $\dfrac{\partial z}{\partial x} = \dfrac{1}{y} - \dfrac{y}{x^2}$, $\dfrac{\partial z}{\partial y} = -\dfrac{x}{y^2} + \dfrac{1}{x}$

3. (i) $f_x = 6xy^2 + 5x^4,\ f_y = 6x^2y + 6y$ (ii) $f_x = \dfrac{2x}{y^3} - \dfrac{3y^2}{x^4},\ f_y = -\dfrac{3x^2}{y^4} + \dfrac{2y}{x^3}$

(iii) $f_x = \dfrac{1}{2}\sqrt{\dfrac{y}{x}},\quad f_y = \dfrac{1}{2}\sqrt{\dfrac{x}{y}}$ (iv) $f_x = \dfrac{10x^2}{x^2+y} + 5\log(x^2+y),\ f_y = \dfrac{5x}{x^2+y}$

7. (i) $f_x = 2xy + yz,\ f_y = x^2 + z^3 + xz,\ f_z = 3yz^2 + xy$ (ii) $f_x = -\dfrac{y}{x^2},\ f_y = \dfrac{1}{z} + \dfrac{1}{x},\ f_z = -\dfrac{y}{z^2}$

7.4 SECOND ORDER PARTIAL DERIVATIVES

If $z = f(x, y)$, then not only is z a function of x and y, but also $\dfrac{\partial z}{\partial x}$ and $\dfrac{\partial z}{\partial y}$ are each functions of x and y. Thus we can form partial derivatives of $\dfrac{\partial z}{\partial x}$ and $\dfrac{\partial z}{\partial y}$ just as we formed those of z to obtain the following second-order partial derivatives of z :

(i) $\dfrac{\partial}{\partial x}\left(\dfrac{\partial z}{\partial x}\right)$, denoted by $\dfrac{\partial^2 z}{\partial x^2}$ or f_{xx} (ii) $\dfrac{\partial}{\partial y}\left(\dfrac{\partial z}{\partial x}\right)$, denoted by $\dfrac{\partial^2 z}{\partial y \partial x}$ or f_{xy}

(iii) $\dfrac{\partial}{\partial x}\left(\dfrac{\partial z}{\partial y}\right)$, denoted by $\dfrac{\partial^2 z}{\partial x \partial y}$ or f_{yx} (iv) $\dfrac{\partial}{\partial y}\left(\dfrac{\partial z}{\partial y}\right)$ denoted by $\dfrac{\partial^2 z}{\partial y^2}$ or f_{yy}

***Note* 1.** To find $\dfrac{\partial^2 z}{\partial x^2}$ $\left(\text{respectively } \dfrac{\partial^2 z}{\partial y^2}\right)$ we take two successive derivatives, each time treating y (respectively x) as a constant. Similarly, to find $\dfrac{\partial^2 z}{\partial y \partial x}$ $\left(\text{respectively } \dfrac{\partial^2 z}{\partial x \partial y}\right)$, we take two successive derivatives, whereas, in the first differentiation y (respectively x) is treated as a constant and in the second differentiation x (respectively y) is treated as a constant.

***Note* 2.** The derivatives $\dfrac{\partial^2 z}{\partial x\, \partial y}$ and $\dfrac{\partial^2 z}{\partial y\, \partial x}$ are called **mixed or cross partial derivatives.**

Usually, but not always $\dfrac{\partial^2 z}{\partial x\, \partial y} = \dfrac{\partial^2 z}{\partial y\, \partial x}$. That is, the order of partial differentiation is immaterial. We assume that this is the case for all functions that we consider.

The following examples illustrate this point.

Example 7. Find all the second-order partial derivatives of $z = 2x^2y^2 + 3xy^2 + y^3$.

Solution. We have $\dfrac{\partial z}{\partial x} = 4xy^2 + 3y^2$ and $\dfrac{\partial z}{\partial y} = 4x^2y + 6xy + 3y^2$

$$\therefore \quad \frac{\partial^2 z}{\partial x^2} = \frac{\partial}{\partial x}\left(\frac{\partial z}{\partial x}\right) = \frac{\partial}{\partial x}(4xy^2 + 3y^2) = 4y^2$$

$$\frac{\partial^2 z}{\partial y\, dx} = \frac{\partial}{\partial y}\left(\frac{\partial z}{\partial x}\right) = \frac{\partial}{\partial y}(4xy^2 + 3y^2) = 8xy + 6y$$

$$\frac{\partial^2 z}{\partial x\, \partial y} = \frac{\partial}{\partial x}\left(\frac{\partial z}{\partial y}\right) = \frac{\partial}{\partial x}(4x^2y + 6xy + 3y^2) = 8xy + 6y$$

$$\frac{\partial^2 z}{\partial y^2} = \frac{\partial}{\partial y}\left(\frac{\partial z}{\partial y}\right) = \frac{\partial}{\partial y}(4x^2y + 6xy + 3y^2) = 4x^2 + 6x + 6y$$

Notice that $\dfrac{\partial^2 z}{\partial y\, \partial x} = \dfrac{\partial^2 z}{\partial x\, dy} = 8xy + 6y.$

Example 8. Find the four second order partial derivatives of $z = \dfrac{x}{y} + \dfrac{y}{x}$.

Solution. $\dfrac{\partial z}{\partial x} = \dfrac{1}{y} - \dfrac{y}{x^2}$ and $\dfrac{\partial z}{\partial y} = -\dfrac{x}{y^2} + \dfrac{1}{x}$

$$\therefore \quad \frac{\partial^2 z}{\partial x^2} = \frac{\partial}{\partial x}\left(\frac{\partial z}{\partial x}\right) = \frac{2y}{x^3}$$

$$\frac{\partial^2 z}{\partial y\, \partial x} = \frac{\partial}{\partial y}\left(\frac{\partial z}{\partial x}\right) = -\frac{1}{y^2} - \frac{1}{x^2} = -\frac{(x^2 + y^2)}{x^2y^2}$$

$$\frac{\partial^2 z}{\partial x\, \partial y} = \frac{\partial}{\partial x}\left(\frac{\partial z}{\partial y}\right) = -\frac{1}{y^2} - \frac{1}{x^2} = -\frac{(x^2 + y^2)}{x^2y^2}$$

$$\frac{\partial^2 z}{\partial y^2} = \frac{\partial}{\partial y}\left(\frac{\partial z}{\partial y}\right) = \frac{2x}{y^3}.$$

Example 9. (*i*) If $z = \log(x^2 + y^2)$, show that $\dfrac{\partial^2 z}{\partial x^2} + \dfrac{\partial^2 z}{\partial y^2} = 0$.

(*ii*) If $z = f(x + ay) + g(x - ay)$, show that $\dfrac{\partial^2 z}{\partial y^2} = a^2 \dfrac{\partial^2 z}{\partial x^2}$.

Solution. (*i*) $\dfrac{\partial z}{\partial x} = \dfrac{1}{x^2+y^2}\dfrac{\partial}{\partial x}(x^2+y^2) = \dfrac{2x}{x^2+y^2}$

$$\frac{\partial z}{\partial y} = \frac{1}{x^2+y^2}\frac{\partial}{\partial y}(x^2+y^2) = \frac{2y}{x^2+y^2}$$

$$\frac{\partial^2 z}{\partial x^2} = \frac{\partial z}{\partial x}\left(\frac{\partial z}{\partial x}\right) = \frac{(x^2+y^2)\dfrac{\partial}{\partial x}(2x) - 2x\dfrac{\partial}{\partial x}(x^2+y^2)}{(x^2+y^2)^2}$$

$$= \frac{(x^2+y^2)\,2 - 2x\,(2x)}{(x^2+y^2)^2} = \frac{2y^2-2x^2}{(x^2+y^2)^2}$$

$$\frac{\partial^2 z}{\partial y^2} = \frac{\partial z}{\partial y}\left(\frac{\partial z}{\partial y}\right) = \frac{(x^2+y^2)\dfrac{\partial}{\partial y}(2y) - 2y\dfrac{\partial}{\partial y}(x^2+y^2)}{(x^2+y^2)^2}$$

$$= \frac{(x^2+y^2)\,2 - 2y\,(2y)}{(x^2+y^2)^2} = \frac{2x^2-2y^2}{(x^2+y^2)^2}$$

$$\therefore \quad \frac{\partial^2 z}{\partial x^2} + \frac{\partial^2 z}{\partial y^2} = \frac{2y^2-2x^2}{(x^2+y^2)} + \frac{2x^2-2y^2}{(x^2+y^2)^2} = \frac{2y^2-2x^2+2x^2-2y^2}{(x^2+y^2)^2} = 0.$$

(*ii*) Given $\quad z = f(x+ay) + g(x-ay)$

$$\therefore \quad \frac{\partial z}{\partial x} = f'(x+ay)\,.\,1 + g'(x-ay)\,.\,1 = f'(x+ay) + g'(x-ay),$$

where $f'(x+ay)$ denotes the derivative of $f(x+ay)$ with respect to $(x+ay)$.

Similarly, $\quad \dfrac{\partial z}{\partial y} = f'(x+ay).\,a + g'(x-ay)\,.\,(-a) = af'(x+ay) - ag'(x-ay)$

Now $\quad \dfrac{\partial^2 z}{\partial x^2} = \dfrac{\partial}{\partial x}\left(\dfrac{\partial z}{\partial x}\right) = f''(x+ay) + g''(x-ay)$

and $\quad \dfrac{\partial^2 z}{\partial y^2} = \dfrac{\partial}{\partial y}\left(\dfrac{\partial z}{\partial y}\right) = a^2 f''(x+ay) + a^2 g''(x-ay)$

$$= a^2\,[f''(x+ay) + g''(x-ay)] = a^2\,\frac{\partial^2 z}{\partial x^2}$$

That is, $\quad \dfrac{\partial^2 z}{\partial y^2} = a^2\,\dfrac{\partial^2 z}{\partial x^2}.$

7.5 HOMOGENEOUS FUNCTIONS AND EULER'S THEOREM

Many of the functions that are useful in economic analysis share the property of being homogeneous.

Definition. A function $z = f(x, y)$ is said to be **homogeneous of degree** n (n being a constant) if, for any real number λ,

$$f(\lambda x, \lambda y) = \lambda^n f(x, y)$$

That is, if both x and y are multiplied by the same real number, then the resulting function value is a power of the number times the function value $f(x, y)$

For example, if $\quad f(x, y) = 2x^2y + xy^2 - y^3$, then

$$\begin{aligned} f(\lambda x, \lambda y) &= 2(\lambda x)^2(\lambda y) + (\lambda x)(\lambda y)^2 - (\lambda y)^3 \\ &= 2\lambda^3 x^2 y + \lambda^3 x y^2 - \lambda^3 y^3 \\ &= \lambda^3 (2x^2 y + x y^2 - y^3) \\ &= \lambda^3 f(x, y) \end{aligned}$$

Thus f is homogeneous of degree 3.

Notice that the function $f(x, y)$ is a polynomial in x and y such that the degree of each term is 3, which is the degree of homogeneity of the function. In general, we have the following remark for such functions.

***Remark* 1.** A polynomial function in two variables x and y is homogeneous if the degree of each term in the polynomial is same. In that case, the degree of homogeneity is the common degree of each term. In particular, a polynomial function in two variables containing only one term is always homogeneous. For example, the function $z = f(x,y) = x^2 y^3$ is homogeneous of degree 5.

We next give an example of a homogeneous function which is not a polynomial.

Example 10. If $z = f(x, y) = \dfrac{x^3 + y^3}{x - y}$, show that z is homogeneous of degree 2.

Solution. We have $\quad f(\lambda x, \lambda y) = \dfrac{(\lambda x)^3 + (\lambda y)^3}{\lambda x - \lambda y} = \dfrac{\lambda^3 x^3 + \lambda^3 y^3}{\lambda x - \lambda y}$

$$= \frac{\lambda^3 \left(x^3 + y^3\right)}{\lambda (x - y)} = \lambda^2 \left(\frac{x^3 + y^3}{x - y}\right) = \lambda^2 f(x, y)$$

Hence $z = f(x, y)$ is homogeneous of degree 2.

***Remark* 2.** A rational function (*i.e.* quotient of two polynomials) in two variables is homogeneous if both numerator and denominator are separately homogeneous. In that case, degree of homogeneity of the given function is degree of numerator minus degree of denominator. To quote some example, each of the following functions is homogeneous:

(i) $z = \dfrac{ax^2 + 2hxy + by^2}{cx + dy}$: *homogeneous of degree* 1

(ii) $z = \dfrac{x + y}{x - y}$: *homogeneous of degree* 0

(iii) $z = \dfrac{x^2 + y^2}{x^3 - y^3}$: *homogeneous of degree* – 1

(iv) $z = \dfrac{x^2 + y^2}{xy}$: *homogeneous of degree* 0

Example 11. Show that each of the following functions is homogeneous of degree 1.

(i) $f(x, y) = \left[ax^{-\beta} + by^{-\beta}\right]^{-1/\beta}$ (ii) $f(x, y) = \sqrt{ax^2 + 2h\,xy + by^2}$

Solution. (i)
$$f(\lambda x, \lambda y) = \left[a(\lambda x)^{-\beta} + b(\lambda y)^{-\beta}\right]^{-1/\beta}$$
$$= \left[a\lambda^{-\beta}x^{-\beta} + b\lambda^{-\beta}y^{-\beta}\right]^{-1/\beta}$$
$$= \left[\lambda^{-\beta}\left(ax^{-\beta} + by^{-\beta}\right)\right]^{-1/\beta}$$
$$= \left(\lambda^{-\beta}\right)^{-1/\beta}\left[ax^{-\beta} + by^{-\beta}\right]^{-1/\beta}$$
$$= \lambda\left(ax^{-\beta} + by^{-\beta}\right)^{-1/\beta} = \lambda f(x, y)$$

Hence $f(x, y)$ is homogeneous of degree 1.

(ii)
$$f(\lambda x, \lambda y) = \sqrt{a(\lambda x)^2 + 2h(\lambda x)(\lambda y) + b(\lambda y)^2}$$
$$= \sqrt{a\lambda^2x^2 + 2h\lambda^2xy + b\lambda^2y^2}$$
$$= \sqrt{\lambda^2\left(ax^2 + 2h\,xy + by^2\right)}$$
$$= \lambda\sqrt{ax^2 + 2hxy + by^2} = \lambda f(x, y)$$

Hence $f(x, y)$ is homogeneous of degree 1.

***Note* 1.** Homogeneous functions of degree one are also called **linear homogeneous functions.**

***Note* 2.** The definition of homogeneous functions can be extended without difficulty to the case of functions of more than 2 variables. The function $z = f(x_1, x_2, \ldots.., x_m)$ is homogeneous of degree n if

$$f(\lambda x_1, \lambda x_2, \ldots.., \lambda x_m) = \lambda^n f(x_1, x_2, \ldots.., x_m)$$

for any real number λ.

***Note* 3.** If $z = f(x, y)$ is a homogeneous function of degree n, then it can be expressed in the form

$$z = x^n \varphi(y/x).$$

For, since $f(\lambda x, \lambda y) = \lambda^n f(x, y)$ for any λ, we have

$$f\left(1, \frac{y}{x}\right) = \frac{1}{x^n} f(x, y) \text{ for } \lambda = \frac{1}{x} \quad \Rightarrow \quad z = f(x, y) = x^n f\left(1, \frac{y}{x}\right) = x^n \varphi(y/x).$$

Euler's Theorem. If $z = f(x, y)$ is a homogeneous function of degree n, then

$$x\frac{\partial z}{\partial x} + y\frac{\partial z}{\partial y} = nz$$

Proof. Since z is a homogeneous function of degree n, it can be expressed as

$$z = x^n \varphi(y/x)$$

Differentiating z with respect to x treating y as a constant, we obtain

$$\begin{aligned}\frac{\partial z}{\partial x} &= x^n \varphi'(y/x)(-y/x^2) + nx^{n-1}\varphi(y/x) \\ &= -x^{n-2}y\varphi'(y/x) + nx^{n-1}\varphi(y/x)\end{aligned}$$

Differentiating z with respect to y treating x as a constant, we obtain

$$\frac{\partial z}{\partial y} = x^n \varphi'(y/x)(1/x) = x^{n-1}\varphi'(y/x)$$

$$\begin{aligned}\therefore \quad x\frac{\partial z}{\partial x} + y\frac{\partial z}{\partial y} &= x[-x^{n-2}y\varphi'(y/x) + n\,x^{n-1}\varphi(y/x)] + y[x^{n-1}\varphi'(y/x)] \\ &= -x^{n-1}y\varphi'(y/x) + nx^n\varphi(y/x) + x^{n-1}y\varphi'(y/x) \\ &= nx^n\varphi(y/x) = nz\end{aligned}$$

Hence proved.

Corollary. If $z = f(x, y)$ is a homogeneous function of degree n, then

$$x^2\frac{\partial^2 z}{\partial x^2} + 2xy\frac{\partial^2 z}{\partial x \partial y} + y^2\frac{\partial^2 z}{\partial y^2} = n(n-1)z$$

Proof. Since z is a homogeneous function of degree n, we have

$$x\frac{\partial z}{\partial x} + y\frac{\partial z}{\partial y} = nz \quad ...(1)$$

Differentiating equation (1) with respect to x treating y as a constant, we obtain

$$x\frac{\partial^2 z}{\partial x^2} + \frac{\partial z}{\partial x} + y\frac{\partial^2 z}{\partial x \partial y} = n\frac{\partial z}{\partial x}$$

$$\Rightarrow \quad x\frac{\partial^2 z}{\partial x^2} + y\frac{\partial^2 z}{\partial x \partial y} = (n-1)\frac{\partial z}{\partial x}$$

Multiplying both sides of this equation by x gives

$$x^2\frac{\partial^2 z}{\partial x^2} + xy\frac{\partial^2 z}{\partial x \partial y} = (n-1)\,x\frac{\partial z}{\partial x} \quad ...(2)$$

Now, differentiating equation (1) with respect to y treating x as a constant, we obtain

$$x\frac{\partial^2 z}{\partial x \partial y} + y\frac{\partial^2 z}{\partial y^2} + \frac{\partial z}{\partial y} = n\frac{\partial z}{\partial y} \qquad \left(\textit{assuming}\ \frac{\partial^2 z}{\partial x \partial y} = \frac{\partial^2 z}{\partial y \partial x}\right)$$

$$\Rightarrow \quad x\frac{\partial^2 z}{\partial x \partial y} + y\frac{\partial^2 z}{\partial y^2} = (n-1)\frac{\partial z}{\partial y}$$

Multiplying both sides of this equation by y gives

$$xy\frac{\partial^2 z}{\partial x \partial y} + y^2\frac{\partial^2 z}{\partial y^2} = (n-1)\,y\frac{\partial z}{\partial y} \quad ...(3)$$

Adding equations (2) and (3), we get

$$x^2\frac{\partial^2 z}{\partial x^2} + 2xy\frac{\partial^2 z}{\partial x \partial y} + y^2\frac{\partial^2 z}{\partial y^2} = (n-1)\left[x\frac{\partial z}{\partial x} + y\frac{\partial z}{\partial y}\right] = (n-1)\,nz = n\,(n-1)\,z$$

i.e.,

$$x^2\frac{\partial^2 z}{\partial x^2} + 2xy\frac{\partial^2 z}{\partial x \partial y} + y^2\frac{\partial^2 z}{\partial y^2} = n(n-1)z.$$

***Remark* 3.** If $z = f(x_1, x_2, ..., x_m)$ is a homogeneous function of degree n, then $x_1\frac{\partial z}{\partial x_1} + x_2\frac{\partial z}{\partial x_2} + ... + x_m\frac{\partial z}{\partial x_m} = nz$. This is Euler's theorem for a homogeneous function of m independent variables.

Example 12. Verify Euler's theorem for

(*i*) $z = \dfrac{x^2+y^2}{xy}$ (*ii*) $z = ax^2 + 2hxy + by^2$ (*iii*) $z = x^2 e^{y/x}$

Solution. (*i*) Given $z = \dfrac{x^2+y^2}{xy} = \dfrac{x}{y} + \dfrac{y}{x}$

Obviously, z is homogeneous of degree 0, therefore to verify Euler's theorem, we must verify that

$$x\frac{\partial z}{\partial x} + y\frac{\partial z}{\partial y} = 0.z = 0$$

Now $\dfrac{\partial z}{\partial x} = \dfrac{1}{y} - \dfrac{y}{x^2}$ and $\dfrac{\partial z}{\partial y} = -\dfrac{x}{y^2} + \dfrac{1}{x}$

$$\therefore \quad x\frac{\partial z}{\partial x} + y\frac{\partial z}{\partial y} = x\left[\frac{1}{y} - \frac{y}{x^2}\right] + y\left[-\frac{x}{y^2} + \frac{1}{x}\right] = \frac{x}{y} - \frac{y}{x} - \frac{x}{y} + \frac{y}{x} = 0.$$

(*ii*) We have $z = ax^2 + 2hxy + by^2$, a homogeneous function of degree 2. We, therefore, must verify that

$$x\frac{\partial z}{\partial x} + y\frac{\partial z}{\partial y} = 2z$$

Now $\dfrac{\partial z}{\partial x} = 2ax + 2hy$ and $\dfrac{\partial z}{\partial y} = 2hx + 2by$

$$\therefore \quad x\frac{\partial z}{\partial x} + y\frac{\partial z}{\partial y} = x(2ax + 2hy) + y(2hx + 2by)$$

$$= 2ax^2 + 2hxy + 2hxy + 2by^2$$

$$= 2(ax^2 + 2hxy + by^2) = 2z.$$

(*iii*) We have $z = x^2 e^{y/x} = f(x, y)$ (*say*)

Hence $f(\lambda x, \lambda y) = (\lambda x)^2 e^{\lambda y/\lambda x} = \lambda^2 x^2 e^{y/x} = \lambda^2 f(x, y)$

Thus $z = f(x, y)$ is homogeneous of degree 2. Thus to verify Euler's theorem, we must verify that

$$x\frac{\partial z}{\partial x} + y\frac{\partial z}{\partial y} = 2z$$

But $\dfrac{\partial z}{\partial x} = x^2 e^{y/x}(-y/x^2) + 2x\, e^{y/x} = -y e^{y/x} + 2x\, e^{y/x}$

and $$\frac{\partial z}{\partial y} = x^2 e^{y/x}(1/x) = xe^{y/x}$$

$$\therefore \quad x\frac{\partial z}{\partial x} + y\frac{\partial z}{\partial y} = x(-y\, e^{y/x} + 2xe^{y/x}) + y\,(x\, e^{y/x})$$

$$= -xy\, e^{y/x} + 2x^2\, e^{y/x} + xye^{y/x} = 2x^2\, e^{y/x} = 2z.$$

Example 13. If $z = ax^{\alpha}\, y^{\beta}$, verify that

(*i*) $x\dfrac{\partial z}{\partial x} + y\dfrac{\partial z}{\partial y} = (\alpha + \beta)\, z$ (*ii*) $x^2\dfrac{\partial^2 z}{\partial x^2} + 2xy\dfrac{\partial^2 z}{\partial x\,\partial y} + y^2\dfrac{\partial^2 z}{\partial y^2} = (\alpha + \beta)(\alpha + \beta - 1)\, z$

Solution. (*i*) We have $z = ax^{\alpha}\, y^{\beta}$

$$\therefore \quad \frac{\partial z}{\partial x} = \alpha\, a\, x^{\alpha - 1}\, y^{\beta} \quad \text{and} \quad \frac{\partial z}{\partial y} = \beta\, a\, x^{\alpha}\, y^{\beta - 1}$$

Hence $$x\frac{\partial z}{\partial y} + y\frac{\partial z}{\partial y} = x(\alpha\, a\, x^{\alpha-1}\, y^{\beta}) + y\,(\beta\, a\, x^{\alpha} y^{\beta - 1})$$

$$= \alpha\, a\, x^{\alpha} y^{\beta} + \beta\, a\, x^{\alpha}\, y^{\beta}$$

$$= (\alpha + \beta)\, ax^{\alpha}\, y^{\beta} = (\alpha + \beta)\, z.$$

(*ii*) Now $$\frac{\partial^2 z}{\partial x^2} = \frac{\partial}{\partial x}\left(\frac{\partial z}{\partial x}\right) = \frac{\partial}{\partial x}(\alpha\, a\, x^{\alpha - 1} y^{\beta}) = \alpha\,(\alpha - 1)\, a\, x^{\alpha - 2} y^{\beta}$$

$$\frac{\partial^2 z}{\partial x\,\partial y} = \frac{\partial}{\partial x}\left(\frac{\partial z}{\partial y}\right) = \frac{\partial}{\partial x}(\beta\, ax^{\alpha}\, y^{\beta - 1}) = \alpha\, \beta\, ax^{\alpha - 1}\, y^{\beta - 1}$$

$$\frac{\partial^2 z}{\partial y^2} = \frac{\partial}{\partial y}\left(\frac{\partial z}{\partial y}\right) = \frac{\partial}{\partial y}(\beta a x^{\alpha}\, y^{\beta - 1}) = \beta(\beta - 1)\, a x^{\alpha}\, y^{\beta - 2}$$

$$\therefore \quad x^2\frac{\partial^2 z}{\partial x^2} + 2xy\frac{\partial^2 z}{\partial x\,\partial y} + y^2\frac{\partial^2 z}{\partial y^2}$$

$$= x^2\,[\alpha\,(\alpha - 1)\, a\, x^{\alpha - 2} y^{\beta}] + 2xy\,[\alpha\,\beta a\, x^{\alpha - 1} y^{\beta - 1}] + y^2\,[\beta\,(\beta - 1)\, ax^{\alpha} y^{\beta - 2}]$$

$$= \alpha\,(\alpha - 1)\, a\, x^{\alpha} y^{\beta} + 2\,\alpha\,\beta\, a\, x^{\alpha} y^{\beta} + \beta\,(\beta - 1) a\, x^{\alpha} y^{\beta}$$

$$= [\alpha\,(\alpha - 1) + 2\,\alpha\beta + \beta\,(\beta - 1)]\, ax^{\alpha}\, y^{\beta}$$

$$= [(\alpha^2 + 2\,\alpha\beta + \beta^2) - (\alpha + \beta)]z = [(\alpha + \beta)^2 - (\alpha + \beta)]z$$

$$= (\alpha + \beta)(\alpha + \beta - 1)\, z\,.$$

EXERCISE 7.2

1. Find the four second-order partial derivatives of the following functions :

(*i*) $f(x, y) = 2x^2y^3$ (*ii*) $f(x, y) = e^{2x-2y}$ (*iii*) $f(x, y) = 2x^3 - 5xy + y^3$

(*iv*) $f(x, y) = xe^{xy} - y^2$ (*v*) $f(x, y) = \log(x^2y^4)$ (*vi*) $f(x, y) = y^2e^x + \log(xy)$

2. Verify that $\dfrac{\partial^2 z}{\partial x \partial y} = \dfrac{\partial^2 z}{\partial y \partial x}$ for

(*i*) $z = 3x^3y + 4x^2y^2 - y^3$ (*ii*) $z = \dfrac{xy}{\sqrt{1 + x^2 + y^2}}$ (*iii*) $z = \log(2x^2 + 3y^4)$

3. (*i*) If $z = x^3 + y^3 - 3x^2y$, verify that $x^2\dfrac{\partial^2 z}{\partial x^2} + 2xy\dfrac{\partial^2 z}{\partial x \partial y} + y^2\dfrac{\partial^2 z}{\partial y^2} = 6z.$

(*ii*) If $z = \sqrt{ax^2 + 2hxy + by^2}$, verify that $x^2\dfrac{\partial^2 z}{\partial x^2} + 2xy\dfrac{\partial^2 z}{\partial x \partial y} + y^2\dfrac{\partial^2 z}{\partial y^2} = 0..$

4. (*i*) If $u = \dfrac{1}{\sqrt{x^2 + y^2 + z^2}}$, prove that $\dfrac{\partial^2 u}{\partial x^2} + \dfrac{\partial^2 u}{\partial y^2} + \dfrac{\partial^2 u}{\partial z^2} = 0.$

(*ii*) If $u = (ax + by)^{-1}$, prove that $\dfrac{\partial^2 u}{\partial x^2} + \dfrac{\partial^2 u}{\partial y^2} = \dfrac{2(a^2 + b^2)}{(ax + by)^3}.$

5. Verify Euler's theorem for each of the following functions :

(*i*) $z = x^2 + y^2$ (*ii*) $z = \dfrac{\sqrt{x} + \sqrt{y}}{\sqrt{x} - \sqrt{y}}$ (*iii*) $z = \dfrac{xy}{x + y}$

(*iv*) $z = x^n \log(y/x)$ (*v*) $z = x^3 + y^3 + x^2y$ (*vi*) $z = ax^3 + bx^2y + cxy^2 + dy^3$

ANSWERS

1. (*i*) $f_{xx} = 4y^3$, $f_{xy} = 12xy^2$, $f_{yx} = 12xy^2$, $f_{yy} = 12x^2y$

(*ii*) $f_{xx} = 4e^{2x-2y}$, $f_{xy} = -4e^{2x-2y}$, $f_{yx} = -4e^{2x-2y}$, $f_{yy} = 4e^{2x-2y}$

(*iii*) $f_{xx} = 12x$, $f_{xy} = -5$, $f_{yx} = -5$, $f_{yy} = 6y$

(*iv*) $f_{xx} = (xy + 2)ye^{xy}$, $f_{xy} = (xy + 2)xe^{xy}$, $f_{yx} = (xy + 2)xe^{xy}$, $f_{yy} = x^3e^{xy} - 2$

(*v*) $f_{xx} = -\dfrac{2}{x^2}$, $f_{xy} = 0$, $f_{yx} = 0$, $f_{yy} = -\dfrac{4}{y^2}$

(*vi*) $f_{xx} = y^2e^x - \dfrac{1}{x^2}$, $f_{xy} = 2ye^x$, $f_{yx} = 2ye^x$, $f_{yy} = 2e^x - \dfrac{1}{y^2}$

7.6 TOTAL DIFFERENTIAL

It may be recalled that if $y = f(x)$ is a differentiable function of x, then the derivative, $\frac{dy}{dx}$, of y with respect to x is given by

$$\frac{dy}{dx} = \lim_{\Delta x \to 0} \frac{\Delta y}{\Delta x},$$

where Δy denotes the increment in y due to an arbitrary small increment Δx in the variable x. Hence

$$\frac{\Delta y}{\Delta x} = \frac{dy}{dx} \quad \text{approximately when } \Delta x \text{ is small}$$

i.e.,

$$\Delta y = \frac{dy}{dx}\Delta x \text{ approximately when } \Delta x \text{ is small}$$

In other words, if x changes by Δx, then the change in $y, \Delta y$, is approximately $\frac{dy}{dx}$ times the change in x. Thus the expression

$$\Delta y = \frac{dy}{dx}\Delta x$$

gives the approximate increment in y for an arbitrary small increment Δx in x.

We now generalize this notion to the case of a function $z = f(x, y)$ of two variables. The partial derivative, $\frac{\partial z}{\partial x}$, gives the rate at which z changes when x changes (y remaining constant). Hence, if $\Delta_x z$ denotes the increment in z due to an arbitrary small increment Δx in the variable x from a point (x, y), then

$$\Delta_x z = \frac{\partial z}{\partial x}\Delta x \text{ approximately}$$

In the same way, if $\Delta_y z$ denotes the increment in z due to an arbitrary small increment Δy in the variable y from the point (x, y), then

$$\Delta_y z = \frac{\partial z}{\partial y}\Delta y \text{ approximately}$$

Thus the increment in z when x varies alone is represented approximately by $\frac{\partial z}{\partial x} \Delta x$ and the increment when y varies alone is represented approximately by $\frac{\partial z}{\partial y} \Delta y$. There remains the important problem of expressing the increment in z when x and y vary together. It has been established (under some continuity conditions) that the increment in the function $z = f(x, y)$ corresponding to arbitrary small increments Δx and Δy in x and y is approximately

$$\Delta z = \frac{\partial z}{\partial x} \Delta x + \frac{\partial z}{\partial y} \Delta y$$

This is called the **differential** of z and is denoted by dz. For convenience, we denote the arbitrary increment Δx by dx and call it the **differential** of the independent variable x. Similarly, Δy is written dy and called the **differential** of y. The differential of the dependent variable z is then defined in terms of the independent differentials by the formula

$$dz = \frac{\partial z}{\partial x} dx + \frac{\partial z}{\partial y} dy$$

It must always be remembered, in using this formula, that dx and dy are no more than arbitrary increments in the independent variables. An alternative notation for the differential of z is :

$$df = f_x \, dx + f_y \, dy$$

The expression $dz = \frac{\partial z}{\partial x} dx + \frac{\partial z}{\partial y} dy$ is often called the **total** or **complete** differential of the function $z = f(x, y)$ of first order.

Example 14. Find the total differential of the function $z = f(x, y) = 2x^2 y + y^3$.

Solution. We have

$$\frac{\partial z}{\partial x} = 4xy \quad and \quad \frac{\partial z}{\partial y} = 2x^2 + 3y^2$$

Hence the total differential of z is

$$dz = \frac{\partial z}{\partial x} dx + \frac{\partial z}{\partial y} dy = (4xy)\,dx + (2x^2 + 3y^2)\,dy$$

Remark. The total differential $dz = \frac{\partial z}{\partial x} dx + \frac{\partial z}{\partial y} dy$ is defined for a function of two **independent** variables x and y. We state, without proof, that exactly the same formula holds for the differential of a function in which the variables x and y are not independent. Hence, the differential of the function $z = f(x, y)$ is given by $dz = \frac{\partial z}{\partial x} dx + \frac{\partial z}{\partial y} dy$, whether the variables x and y are independent or not.

7.7 SECOND-ORDER TOTAL DIFFERENTIALS

The second-order total differential of a function of two variables is defined and obtained from the first order differential. If $z = f(x, y)$, the first-order total differential of z is

$$dz = f_x\, dx + f_y\, dy$$

The second-order total differential of z, denoted by $d^2 z$,is given by $d^2 z = d(dz)$. Thus

$$d^2 z = d(f_x\, dx + f_y\, dy)$$

$$= \frac{\partial}{\partial x}(f_x\, dx + f_y\, dy)\, dx + \frac{\partial}{\partial y}(f_x\, dx + f_y\, dy)\, dy$$

$$= [f_{xx}\, dx + f_x \frac{\partial}{\partial x}(dx) + f_{yx}\, dy + f_y \frac{\partial}{\partial x}(dy)]dx + [f_{xy}\, dx + f_x \frac{\partial}{\partial y}(dx) + f_{yy}\, dy + f_y \frac{\partial}{\partial y}(dy)]dy$$

However, dx and dy are considered as constants, so

$$\frac{\partial}{\partial x}(dx) = 0, \quad \frac{\partial}{\partial x}(dy) = 0, \quad \frac{\partial}{\partial y}(dx) = 0, \quad \frac{\partial}{\partial y}(dx) = 0$$

$$\therefore \quad d^2 z = (f_{xx}\, dx + f_{xy}\, dy)\, dx + (f_{xy}\, dx + f_{yy}\, dy)\, dy \quad (\text{assuming } f_{xy} = f_{yx})$$

$$= f_{xx}\,(dx)^2 + f_{xy}\, dx\, dy + f_{xy}\, dx\, dy + f_{yy}\,(dy)^2$$

$$= f_{xx}\,(dx)^2 + 2 f_{xy}\, dx\, dy + f_{yy}\,(dy)^2$$

Hence the second-order total differential of z is :

$$d^2 z = f_{xx}\,(dx)^2 + 2 f_{xy}\, dx\, dy + f_{yy}\,(dy)^2.$$

Example 15. Find the first-and second-order total differentials of the function :

$$z = f(x, y) = 7y \log(1 + x)$$

[*Delhi Univ. B. Com. (H) 1992*]

Solution. We have

$$f_x = \frac{\partial z}{\partial x} = \frac{7y}{1+x} \quad \text{and} \quad f_y = \frac{\partial z}{\partial y} = 7 \log(1 + x)$$

Thus the first-order total differential of z is :

$$dz = f_x\,dx + f_y\,dy$$
$$= \frac{7y}{1+x}\,dx + 7\log(1+x)\,dy$$

Also $f_{xx} = \frac{\partial^2 z}{\partial x^2} = -\frac{7y}{(1+x)^2}$, $f_{xy} = \frac{\partial^2 z}{\partial y\,\partial x} = \frac{7}{1+x}$ *and* $f_{yy} = \frac{\partial^2 z}{\partial y^2} = 0$

Hence the second-order total differential of z is :

$$d^2 z = f_{xx}\,(dx)^2 + 2 f_{xy}\,dx\,dy + f_{yy}\,(dy)^2$$
$$= -\frac{7y}{(1+x)^2}(dx)^2 + 2\left(\frac{7}{1+x}\right)dx\,dy + (0)\,(dy)^2$$
$$= -\frac{7y}{(1+x)^2}(dx)^2 + \left(\frac{14}{1+x}\right)dx\,dy.$$

Example 16. Calculate the first and second order total differentials of the function :

$$z = 3x^2 + xy - 2y^2$$ **[*Delhi Univ. B. Com. (H) 1995*]**

Solution. We have $\frac{\partial z}{\partial x} = 6x + y$ and $\frac{\partial z}{\partial y} = x - 4y$

The first-order total differential of z is :

$$dz = \frac{\partial z}{\partial x}\,dx + \frac{\partial z}{\partial y}\,dy = (6x + y)\,dx + (x - 4y)\,dy$$

Further, $\frac{\partial^2 z}{\partial x^2} = 6$, $\frac{\partial^2 z}{\partial x\,\partial y} = 1$ and $\frac{\partial^2 z}{\partial y^2} = -4$

The second-order total differential of z is :

$$d^2 z = \frac{\partial^2 z}{\partial x^2}(dx)^2 + 2\frac{\partial^2 z}{\partial x\,\partial y}\,dx\,dy + \frac{\partial^2 z}{\partial y^2}(dy)^2$$
$$= 6(dx)^2 + 2\,dx\,dy - 4(dy)^2.$$

7.8 MAXIMA AND MINIMA OF FUNCTIONS OF TWO VARIABLES

In an earlier chapter we considered the problem of finding local extrema (local maxima, local minima) of a function of one independent variable. We now study the analogous problem of local extrema for functions of two variables.

Definition. A function $z = f(x, y)$ is said to have a **local maximum** at the point (a, b) if

$$f(a, b) \geq f(x, y)$$

for all (x, y) that are sufficiently close to (a, b). A **local minimum** is defined similarly.

A local maximum or a local minimum is referred to as a **local extremum.**

It may be recalled that we used derivatives to find local maxima and local minima of functions involving a single independent variable. For example, it was proved that if $y = f(x)$ has a local maximum or a local minimum at $x = a$, then $f'(a) = 0$. Similarly, if $z = f(x, y)$ has a local maximum or local minimum at (a, b), then the two first-ordered partial derivatives of z are zero at (a, b). That is,

$$f_x(a, b) = 0 \quad \text{and} \quad f_y(a, b) = 0$$

Definition. A point (a, b) for which $f_x(a, b) = f_y(a, b) = 0$ is called a **critical point** of $z = f(x, y)$.

Just as in the case of function of one variable, a critical point can give rise to a local maximum, a local minimum or nothing.

Definition. A critical point that is neither a local maximum nor a local minimum is called a **saddle point.**

We now state a test (omitting the proof), similar to the second-derivative test previously studied, that gives conditions which determine whether a critical point yields a local maximum, a local minimum or a saddle point.

Second Derivative Test for Local Extrema

Given $\quad z = f(x, y)$

Step 1. *Find $f_x(x, y)$ and $f_y(x, y)$.*

Step 2. *Find the critical point (s) by solving $f_x(x, y) = 0$ and $f_y(x, y) = 0$. Let (a, b) be a critical point.*

Step 3. *Find $f_{xx}(x, y)$, $f_{xy}(x, y)$ and $f_{yy}(x, y)$.*

Step 4. *Define $A = f_{xx}(a, b)$, $B = f_{xy}(a, b)$ and $C = f_{yy}(a, b)$. Consider $AC - B^2$. Then*

(i) If $AC - B^2 > 0$ and $A < 0$, f has a local maximum at (a, b),

(ii) If $AC - B^2 > 0$ and $A > 0$, f has a local minimum at (a, b),

(iii) If $AC - B^2 < 0$, f has a saddle point at (a, b),

(iv) If $AC - B^2 = 0$, no conclusion can be drawn and further analysis is required.

Note. It may be noticed that when $AC - B^2 > 0$, then neither A nor C can be zero. Further, both A and C will be positive or negative together.

The following examples illustrate the procedure :

Example 17. Examine the following functions for local maxima and minima:

(*i*) $z = f(x, y) = 6xy - x^3 - y^3$ (*ii*) $z = f(x, y) = 10x + 20y - 2xy - 3x^2 - 2y^2$

Solution. (*i*) Given $z = f(x, y) = 6xy - x^3 - y^3$

$\therefore \quad f_x = 6y - 3x^2$ and $f_y = 6x - 3y^2$

For critical points, we have

$$f_x = 0, \quad i.e, \quad 6y - 3x^2 = 0$$

and

$$f_y = 0, \quad i.e, \quad 6x - 3y^2 = 0$$

Solving these for x and y, we find that critical points are (0, 0) and (2, 2).

Now $\quad f_{xx} = -6x, \quad f_{xy} = 6, \quad f_{yy} = -6y$

At (0, 0), $\quad A = f_{xx}(0, 0) = 0, \quad B = f_{xy}(0, 0) = 6, \quad C = f_{yy}(0, 0) = 0$

Thus $AC - B^2 = -36 < 0$. Hence (0, 0) is a saddle point.

At (2, 2), $A = f_{xx}(2, 2) = -12, \quad B = f_{xy}(2, 2) = 6, \quad C = f_{yy}(2, 2) = -12$

Thus $AC - B^2 = 144 - 36 = 108 > 0$. Since at (2, 2), $A = -12 < 0$ and $AC - B^2 > 0$, it follows that there is a local maximum at (2, 2). At this point the value of the function is :

$$f(2, 2) = 6(2)(2) - (2)^3 - (2)^3 = 8.$$

(*ii*) Given $\quad z = f(x, y) = 10x + 20y - 2xy - 3x^2 - 2y^2$

$$\therefore \quad f_x = \frac{\partial z}{\partial x} = 10 - 2y - 6x \quad \text{and} \quad f_y = \frac{\partial z}{\partial y} = 20 - 2x - 4y$$

For critical points, we solve the equations $f_x = 0, f_y = 0$, which are

$$10 - 2y - 6x = 0 \quad i.e. \quad 6x + 2y = 10$$

$$20 - 2x - 4y = 0 \quad i.e. \quad 2x + 4y = 20$$

On solving, we get (0, 5) as the only critical point. Now

$$f_{xx} = \frac{\partial^2 z}{\partial x^2} = -6, \quad f_{xy} = \frac{\partial^2 z}{\partial y \partial x} = -2, \quad f_{yy} = \frac{\partial^2 z}{\partial y^2} = -4$$

At (0, 5), $\quad A = -6, \quad B = -2, \quad C = -4$

Thus $AC - B^2 = 24 - 4 = 20 > 0$. Since $A = -6 < 0$, there is a local maximum at (0, 5). The value of the function at this point is :

$$f(0, 5) = 10(0) + 20(5) - 2(0)(5) - 3(0)^2 - 2(5)^2 = 50.$$

Example 18. Examine the function $z = f(x, y) = -x^3 + 27x - 4y^2$ for local extrema.

Solution. We have $\quad \dfrac{\partial z}{\partial x} = -3x^2 + 27$ and $\dfrac{\partial z}{\partial y} = -8y$

For critical points, we solve the following system for x and y:

$$-3x^2 + 27 = 0$$
$$-8y = 0$$

This gives $x = \pm 3$ and $y = 0$. Thus (3, 0) and (– 3, 0) are the critical points. Now

$$\frac{\partial^2 z}{\partial x^2} = -6x, \quad \frac{\partial^2 z}{\partial x \partial y} = 0, \quad \frac{\partial^2 z}{\partial y^2} = -8$$

At (3, 0), $A = -18$, $B = 0$, $C = -8$, so that $AC - B^2 = 144 > 0$.

Since $A = -18 < 0$, it follows that (3, 0) is a local maximum point.

At (– 3, 0) $A = 18$, $B = 0$, $C = -8$, so that $AC - B^2 = -144 < 0$.

Since $AC - B^2 < 0$, it follows that (– 3, 0) is a saddle point.

EXERCISE 7.3

1. Find the first - order total differentials of the following functions :

(*i*) $z = x^3 + y^3 + 3xy$ (*ii*) $z = e^{x^2 + y^2}$ (*iii*) $z = e^{y/x}$

(*iv*) $z = \log\left(\dfrac{x+y}{x-y}\right)$ (*v*) $z = \log(x^2 + y^2)$ (*vi*) $z = \dfrac{x^2}{x - y + 1}$

2. Find the first-and second-order total differentials of the following functions:

(*i*) $z = x^2 y^2$ (*ii*) $z = x^3 + y^3$ (*iii*) $z = x + y + \log(xy)$

3. Examine the following functions for local maxima and minima:

(*i*) $f(x, y) = 2xy - x^3 - y^3$ (*ii*) $f(x, y) = xy^2 - 6x^2 - 3y^2$

(*iii*) $f(x, y) = x^2 + y^2 - 3x$ (*iv*) $f(x, y) = xy(1 - x - y)$

(*v*) $f(x, y) = 3x^2 - y^2 + x^3$ (*vi*) $f(x, y) = y^2 - x^2$

(*vii*) $f(x, y) = x^2 + 3y^2 + 4x - 9y + 3$ (*viii*) $f(x, y) = x^2 + y^2 + xy - 9x + 1$

4. Show that the function $f(x, y) = 3x^2 + y^2 - 3xy$ has a local minimum at (0, 0).

ANSWERS

1. (*i*) $dz = (3x^2 + 3y)\, dx + (3y^2 + 3x)\, dy$ (*ii*) $dz = 2e^{x^2 + y^2}(x\, dx + y\, dy)$

(*iii*) $dz = \dfrac{e^{y/x}}{x^2}[-y\, dx + x\, dy]$ (*iv*) $dz = \dfrac{2}{x^2 - y^2}[-y\, dx + x\, dy]$

(*v*) $dz = \dfrac{2}{x^2 + y^2}[x\, dx + y\, dy]$ (*vi*) $dz = \dfrac{x}{(x - y + 1)^2}[(x - 2y + 2)\, dx + x\, dy]$

2. (*i*) $dz = 2xy(y\,dx + x\,dy)$, $d^2z = 2y^2(dx)^2 + 8xy\,dx\,dy + 2x^2(dy)^2$

(*ii*) $dz = 3x^2\,dx + 3y^2\,dy$, $d^2z = 6x(dx)^2 + 6y(dy)^2$

(*iii*) $dz = \left(1 + \frac{1}{x}\right)dx + \left(1 + \frac{1}{y}\right)dy$, $d^2z = -\frac{1}{x^2}(dx)^2 - \frac{1}{y^2}(dy)^2$

3. (*i*) Local maximum at (2/3, 2/3)

(*ii*) Local maximum at (0, 0), saddle points : (3, 6) and (3, – 6)

(*iii*) Local minimum at (3/2, 0)

(*iv*) Local maximum at (1/3, 1/3), saddle points at (0, 0), (1, 0) and (0, 1)

(*v*) Local maximum at (– 2, 0) and saddle point at (0, 0)

(*vi*) Saddle point at (0, 0)

(*vii*) Local minimum at (– 2, 3/2)

(*viii*) Local minimum at (6, – 3).

CHAPTER

8

APPLICATIONS OF PARTIAL DIFFERENTIATION

OBJECTIVES

After studying the material in this chapter, you should be able to:

- *Discuss the concept of partial marginal demand functions.*
- *Find the nature of commodities.*
- *Find partial elasticities for a demand function of two variables.*
- *Determine the average and marginal productivities of factors of production.*
- *Determine the nature of returns to scale for a given production function.*
- *Understand the law of diminishing returns to factors.*
- *Evaluate the marginal rate of technical substitution and the elasticity of substitution for a given production function.*
- *Solve applied optimization problems.*

8.1 INTRODUCTION

In this chapter several applications of partial differentiation to business and economics are discussed. For example, if L units of labour and K units of capital are used to produce P units of a product, then the function $P = f(L, K)$ is called a *production function* and its partial derivatives are called *marginal productivity functions.* The concept of partial derivatives is also used to find the nature of commodities. Another use of partial derivatives is to solve applied extrema problems. For example, a monopolist producing two products with known market demands might be interested to find how many items of each be produced to have the joint profit maximum.

8.2 DEMAND ANALYSIS

Sometimes two commodities may be related so that the changes in the price of one of them can affect the demand for the other. For example, cameras and film are related commodities:

an increase in the price of film will make picture-taking more expensive and hence the demand for cameras will decrease.

Let X_1 and X_2 be two related commodities, then the demand for each product is dependent on the prices of both. Suppose x_1 and x_2 are the quantities demanded for X_1 and X_2 respectively, and p_1 and p_2 are their respective prices. Then both x_1 and x_2 are functions of p_1 and p_2 :

$$x_1 = f(p_1, p_2), \text{ demand function for } X_1$$

$$x_2 = g(p_1, p_2), \text{ demand function for } X_2$$

We now introduce the notion of partial marginal demand functions.

(*i*) **The partial marginal demand for X_1 with respect to p_1** is defined as $\dfrac{\partial x_1}{\partial p_1}$. This represents the rate of change of x_1 with respect to p_1 when p_2 is held fixed.

(*ii*) **The partial marginal demand for X_1 with respect to p_2** is defined as $\dfrac{\partial x_1}{\partial p_2}$. This is the rate of change of x_1 with respect to p_2 when p_1 is held fixed.

(*iii*) **The partial marginal demand for X_2 with respect to p_1** is defined as $\dfrac{\partial x_2}{\partial p_1}$. This gives the rate of change of x_2 with respect to p_1 when p_2 is held fixed.

(*iv*) **The partial marginal demand for X_2 with respect to p_2** is defined as $\dfrac{\partial x_2}{\partial p_2}$. This gives the rate of change of x_2 with respect to p_2 when p_1 is held fixed.

Remark. Under normal conditions, if the price of X_2 is fixed and the price of X_1 increases then the quantity of X_1 demanded will decrease. Hence $\dfrac{\partial x_1}{\partial p_1} < 0$. Similarly, $\dfrac{\partial x_2}{\partial p_2} < 0$.

However, $\dfrac{\partial x_1}{\partial p_2}$ and $\dfrac{\partial x_2}{\partial p_1}$ may be either positive or negative.

Example 1. The demand functions for two commodities X and Y are given by

$$x = 5q^2 - 3pq \quad \text{and} \quad y = 8p^2 - pq$$

respectively, where x and y are the quantities demanded for X and Y respectively, and p and q are their respective prices. Find the four partial marginal demands when $p = 50$ and $q = 70$. Interpret the results.

Solution. Partial marginal demand for X with respect to $p = \dfrac{\partial x}{\partial p} = -3q$

Partial marginal demand for X with respect to $q = \dfrac{\partial x}{\partial q} = 10q - 3p$

Evaluating when $p = 50$ and $q = 70$, we obtain

$$\frac{\partial x}{\partial p} = -210 \quad \text{and} \quad \frac{\partial x}{\partial q} = 550$$

Thus, if $p = 50$ and $q = 70$, increasing p to 51 and holding q at 70 will decrease demand for X by 210 units. But, if q is increased to 71 while p is held at 50, the demand for X increases by 550 units. Further,

Partial marginal demand for Y with respect to $p = \dfrac{\partial y}{\partial p} = 16p - q$

Partial marginal demand for Y with respect to $q = \dfrac{\partial y}{\partial q} = -p$

Evaluating when $p = 50$ and $q = 70$, we obtain

$$\frac{\partial y}{\partial p} = 730 \quad \text{and} \quad \frac{\partial y}{\partial q} = -50$$

Thus, if $p = 50$ and $q = 70$, increasing p to 51 and holding q at 70 will increase demand for Y by 730 units. But if q is increased to 71 while p is held at 50, the demand for Y decreases by 50 units.

Nature of Commodities. Two commodities are said to be **complementary** if an increase in the price of one of them causes a decrease in the demand for the other, provided the price of the other commodity is held fixed. For example, car and petrol are complementary commodities.

Clearly, two commodities X_1 and X_2 are complementary if and only if $\dfrac{\partial x_1}{\partial p_2}$ and $\dfrac{\partial x_2}{\partial p_1}$ are both negative.

Proceeding to a different situation, two commodities are said to be **competitive** or **substitutes** if an increase in the price of one of them causes an increase in the demand for the other, provided the price of the other commodity is held fixed. For example, Tea and Coffee are substitutes.

Clearly, two commodities X_1 and X_2 are competitive if and only if $\dfrac{\partial x_1}{\partial p_2}$ and $\dfrac{\partial x_2}{\partial p_1}$ are both positive.

These concepts are best illustrated with the help of following examples.

Example 2. The demand functions for two commodities X_1 and X_2 are given by

$$x_1 = \frac{100}{p_1^2 p_2} \quad \text{and} \quad x_2 = \frac{50}{p_1 p_2^2}$$

respectively. Find the four partial marginal demand functions. Also determine whether X_1 and X_2 are complementary or competitive.

Solution. Partial marginal demand for X_1 with respect to $p_1 = \dfrac{\partial x_1}{\partial p_1} = -\dfrac{200}{p_1^3 p_2}$

Partial marginal demand for X_1 with respect to $p_2 = \dfrac{\partial x_1}{\partial p_2} = -\dfrac{100}{p_1^2 p_2^2}$

Partial marginal demand for X_2 with respect to $p_1 = \dfrac{\partial x_2}{\partial p_1} = -\dfrac{50}{p_1^2 p_2^2}$

Partial marginal demand for X_2 with respect to $p_2 = \dfrac{\partial x_2}{\partial p_2} = -\dfrac{100}{p_1 p_2^3}$

Since p_1 and p_2 represent prices, they are both positive. Hence $\dfrac{\partial x_1}{\partial p_2}$ and $\dfrac{\partial x_2}{\partial p_1}$ are both negative. We thus conclude that X_1 and X_2 are complementary commodities.

Example 3. The demand functions for two commodities X_1 and X_2 are each a function of the prices of X_1 and X_2 and are given by

$$x_1 = p_1^{-.2} e^{.4p_2} \quad \text{and} \quad x_2 = p_2^{-.5} e^{.6p_1}$$

respectively. Find the four partial marginal demand functions. Also discuss whether X_1 and X_2 are complementary or substitutes.

Solution. We have

$$\frac{\partial x_1}{\partial p_1} = -.2p_1^{-1.2} e^{.4p_2} < 0, \qquad \frac{\partial x_1}{\partial p_2} = .4p_1^{-.2} e^{.4p_2} > 0,$$

$$\frac{\partial x_2}{\partial p_1} = .6p_2^{-.5} e^{.6p_1} > 0, \qquad \frac{\partial x_2}{\partial p_2} = -.5p_2^{-1.5} e^{.6p_1} < 0$$

Since $\dfrac{\partial x_1}{\partial p_2}$ and $\dfrac{\partial x_2}{\partial p_1}$ are both positive, X_1 and X_2 are substitutes.

Partial Elasticities of Demand. We now define partial elasticities of demand on the basis of formula derived for price elasticity of demand:

(*i*) **The (direct) partial elasticity of demand for X_1 with respect to p_1**, denoted by η_{11}, is defined as

$$\eta_{11} = \frac{p_1}{x_1}\frac{\partial x_1}{\partial p_1}$$

(*ii*) **The (cross) partial elasticity of demand for X_1 with respect to p_2**, denoted by η_{12}, is defined as

$$\eta_{12} = \frac{p_2}{x_1}\frac{\partial x_1}{\partial p_2}$$

(*iii*) **The (cross) partial elasticity of demand for X_2 with respect to p_1**, denoted by η_{21}, is defined as

$$\eta_{21} = \frac{p_1}{x_2}\frac{\partial x_2}{\partial p_1}$$

(*iv*) **The (direct) partial elasticity of demand for X_2 with respect to p_2**, denoted by η_{22}, is defined as

$$\eta_{22} = \frac{p_2}{x_2}\frac{\partial x_2}{\partial p_2}$$

Loosely speaking, η_{11} (respectively η_{22}) can be interpreted as the ratio of a percentage change in the quantity of X_1 (respectively X_2) demanded to a percentage change in the price of X_1 (respectively X_2) when the price of X_2 (respectively X_1) is fixed. Similarly, η_{12} (respectively η_{21}) can be loosely interpreted as the ratio of a percentage change in the quantity of X_1 (respectively X_2) demanded to a percentage of change in the price of X_2 (respectively X_1) when the price of X_1 (respectively X_2) is fixed.

It should be noted that if η_{12} and η_{21} are both positive, commodities are competitive, and if η_{12} and η_{21} both are negative, commodities are complementary.

Example 4. The demand functions of two commodities, X_1 and X_2, are $x_1 = p_1^{-1.4}p_2^{0.6}$ and $x_2 = p_1^{0.5}p_2^{-1.2}$ respectively, where x_1 and x_2 are the quantities demanded of X_1 and X_2 respectively, and p_1 and p_2 are their respective prices. Find the four partial elasticities of demand and determine whether the commodities are competitive or complementary.

[*Delhi Univ. B.Com. (H) 2013*]

Solution. We have

$$\frac{\partial x_1}{\partial p_1} = -1.4p_1^{-2.4}p_2^{0.6}, \qquad \frac{\partial x_1}{\partial p_2} = 0.6p_1^{-1.4}p_2^{-0.4}$$

$$\frac{\partial x_2}{\partial p_1} = 0.5 p_1^{-0.5} p_2^{-1.2}, \qquad \frac{\partial x_2}{\partial p_2} = -1.2 p_1^{0.5} p_2^{-2.2}$$

The (direct) partial elasticity of demand of X_1 with respect to p_1 is

$$\eta_{11} = \frac{p_1}{x_1}\frac{\partial x_1}{\partial p_1} = \frac{p_1}{p_1^{-1.4} p_2^{0.6}}(-1.4 p_1^{-2.4} p_2^{0.6}) = -1.4$$

The (cross) partial elasticity of demand of X_1 with respect to p_2 is

$$\eta_{12} = \frac{p_2}{x_1}\frac{\partial x_1}{\partial p_2} = \frac{p_2}{p_1^{-1.4} p_2^{0.6}}(0.6 p_1^{-1.4} p_2^{-0.4}) = 0.6$$

The (cross) partial elasticity of demand of X_2 with respect to p_1 is

$$\eta_{21} = \frac{p_1}{x_2}\frac{\partial x_2}{\partial p_1} = \frac{p_1}{p_1^{0.5} p_2^{-1.2}}(0.5 p_1^{-0.5} p_2^{-1.2}) = 0.5$$

The (direct) partial elasticity of demand of X_2 with respect to p_2 is

$$\eta_{22} = \frac{p_2}{x_2}\frac{\partial x_2}{\partial p_2} = \frac{p_2}{p_1^{0.5} p_2^{-1.2}}(-1.2 p_1^{0.5} p_2^{-2.2}) = -1.2$$

Since η_{12} and η_{21} are both positive, we conclude that X_1 and X_2 are competitive.

Example 5. The joint demand functions of two products are $x_1 = 2p_1^{-0.6} p_2^{0.8}$ and $x_2 = 3p_1^{0.7} p_2^{-0.5}$, where x_1 and and x_2 are the units demanded of the two products when their prices are ₹ p_1 and ₹ p_2 per unit respectively. Find the four price elasticities at the particular prices p_1 and p_2. Also find the percentage change in the quantity demanded when

(*i*) p_1 increases by 2% from the existing level.

(*ii*) p_2 decreases by 2% from the existing level. **[*Delhi Univ. B.Com. (H) 2005*]**

Solution. The joint demand functions of two products are :

$$x_1 = 2p_1^{-0.6} p_2^{0.8} \quad \text{and} \quad x_2 = 3p_1^{0.7} p_2^{-0.5}$$

$$\Rightarrow \quad \frac{\partial x_1}{\partial p_1} = -1.2\, p_1^{-1.6} p_2^{0.8}, \qquad \frac{\partial x_1}{\partial p_2} = 1.6 p_1^{-0.6} p_2^{-0.2}$$

$$\frac{\partial x_2}{\partial p_1} = 2.1\, p_1^{-0.3} p_2^{-0.5}, \qquad \frac{\partial x_2}{\partial p_2} = -1.5 p_1^{0.7} p_2^{-1.5}$$

The four partial price elasticities of demand are calculated as follows :

$$\eta_{11} = \frac{p_1}{x_1}\frac{\partial x_1}{\partial p_1} = \frac{p_1}{2p_1^{-0.6} p_2^{0.8}} \times -1.2 p_1^{-1.6} p_2^{0.8} = -0.6$$

$$\eta_{12} = \frac{p_2}{x_1}\frac{\partial x_1}{\partial p_2} = \frac{p_2}{2p_1^{-0.6} p_2^{0.8}} \times 1.6 p_1^{-0.6} p_2^{-0.2} = 0.8$$

$$\eta_{21} = \frac{p_1}{x_2}\frac{\partial x_2}{\partial p_1} = \frac{p_1}{3p_1^{0.7}\ p_2^{-0.5}} \times 2.1p_1^{-0.3}\ p_2^{-0.5} = 0.7$$

$$\eta_{22} = \frac{p_2}{x_2}\frac{\partial x_2}{\partial p_2} = \frac{p_2}{3p_1^{0.7}\ p_2^{-0.5}} \times -1.5p_1^{0.7}\ p_2^{-1.5} = -0.5$$

(*i*) When p_1 increases by 2%, x_1 decreases by 1.2% and x_2 increases by 1.4%.

(*ii*) When p_2 decreases by 2%, x_1 decreases by 1.6% and x_2 increases by 1%.

Example 6. Given the demand functions of two commodities as

$$Q_1 = 2000 + \frac{400}{p_1 + 3} - 50\,p_2 \quad \text{and} \quad Q_2 = 2000 - 100p_2 + \frac{500}{p_1 + 4}$$

(*i*) Find the nature of commodities.

(*ii*) Calculate the four partial elasticities of demand at $p_1 = 5$ and $p_2 = 1$.

[*Delhi Univ. B.Com. (H) 2001*]

Solution. Given: $Q_1 = 2000 + \dfrac{400}{p_1 + 3} - 50\,p_2$ and $Q_2 = 2000 - 100p_2 + \dfrac{500}{p_1 + 4}$

$$\therefore \quad \frac{\partial Q_1}{\partial p_1} = -\frac{400}{(p_1 + 3)^2}, \quad \frac{\partial Q_1}{\partial p_2} = -50, \quad \frac{\partial Q_2}{\partial p_1} = -\frac{400}{(p_1 + 4)^2}, \quad \frac{\partial Q_2}{\partial p_2} = -100$$

(*i*) Since $\dfrac{\partial Q_1}{\partial p_2}$ and $\dfrac{\partial Q_2}{\partial p_1}$ are both negative, the commodities are *complementary.*

(*ii*) The (direct) partial elasticity of Q_1 with respect to p_1 is:

$$\eta_{11} = \frac{p_1}{Q_1}\frac{\partial Q_1}{\partial p_1} = \frac{p_1}{2000 + \dfrac{400}{p_1 + 3} - 50p_2} \times \frac{-400}{(p_1 + 3)^2}$$

At $p_1 = 5, p_2 = 1, \eta_{11} = \dfrac{5}{2000 + \dfrac{400}{8} - 50} \times \dfrac{-400}{64} = -\dfrac{1}{64}$

The (cross) partial elasticity of Q_1 with respect to p_2 is:

$$\eta_{12} = \frac{p_2}{Q_1}\frac{\partial Q_1}{\partial p_2} = \frac{p_2}{2000 + \dfrac{400}{p_1 + 3} - 50p_2} \times -50$$

At $p_1 = 5, p_2 = 1,$ $$\eta_{12} = \frac{1}{2000 + \frac{400}{8} - 50} \times -50 = -\frac{1}{40}$$

The (cross) partial elasticity of Q_2 with respect to p_1 is:

$$\eta_{21} = \frac{p_1}{Q_2}\frac{\partial Q_2}{\partial p_1} = \frac{p_1}{2000 - 100p_2 + \frac{500}{p_1 + 4}} \times \frac{-500}{(p_1 + 4)^2}$$

At $p_1 = 5, p_2 = 1,$ $$\eta_{21} = \frac{5}{2000 - 100 + \frac{500}{9}} \times \frac{-500}{81} = -\frac{25}{1584}$$

The (direct) partial elasticity of Q_2 with respect to p_2 is:

$$\eta_{22} = \frac{p_2}{Q_2}\frac{\partial Q_2}{\partial p_2} = \frac{p_2}{2000 - 100p_2 + \frac{500}{p_1 + 4}} \times -100$$

At $p_1 = 5, p_2 = 1,$ $$\eta_{22} = \frac{1}{2000 - 100 + \frac{500}{9}} \times -100 = -\frac{9}{176}.$$

Example 7. The demand functions for two commodities X_1 and X_2 in terms of their respective prices p_1 and p_2 are given by

$$x_1 = p_1^{-a_1} e^{b_1 p_2 + c_1} \quad \text{and} \quad x_2 = p_2^{-a_2} e^{b_2 p_1 + c_2}$$

where a_1, a_2, b_1, b_2 and c_1, c_2 are constants.

(*i*) Find the four partial marginal demand functions.

(*ii*) Determine the conditions so that the commodities are (*a*) competitive (*b*) complementary.

(*iii*) Show that the 'direct' partial elasticities of demand are independent of the prices, while the 'cross' partial elasticities of demand are determined in signs by the constants b_1 and b_2 respectively. **[*Delhi Univ. B. Com. (H) 2017, 2019*]**

Solution. (*i*) The partial marginal demand of X_1 with respect to p_1 is :

$$\frac{\partial x_1}{\partial p_1} = -a_1 p_1^{-a_1 - 1} e^{b_1 p_2 + c_1} = -\frac{a_1 x_1}{p_1}$$

The partial marginal demand of X_1 with respect to p_2 is :

$$\frac{\partial x_1}{\partial p_2} = b_1 p_1^{-a_1} e^{b_1 p_2 + c_1} = b_1 x_1$$

The partial marginal demand of X_2 with respect to p_1 is :

$$\frac{\partial x_2}{\partial p_1} = b_2\, p_2^{-a_2} e^{b_2 p_1 + c_2} = b_2\, x_2$$

The partial marginal demand of X_2 with respect to p_2 is :

$$\frac{\partial x_2}{\partial p_2} = -a_2\, p_2^{-a_2-1} e^{b_2 p_1 + c_2} = -\frac{a_2\, x_2}{p_2}$$

(*ii*) The commodities X_1 and X_2 will be competitive or complementary depending upon whether $\frac{\partial x_1}{\partial p_2}$ and $\frac{\partial x_2}{\partial p_1}$ are both positive or negative. But $\frac{\partial x_1}{\partial p_2} = b_1\, x_1$ and $\frac{\partial x_2}{\partial p_1} = b_2\, x_2$.

Since x_1 and x_2 are both positive, therefore the signs of $\frac{\partial x_1}{\partial p_2}$ and $\frac{\partial x_2}{\partial p_1}$ depend on the signs of b_1 and b_2 respectively. We thus conclude that X_1 and X_2 will be competitive if both b_1 and b_2 are positive, and complementary if both b_1 and b_2 are negative.

(*iii*) The two direct partial elasticities of demand are:

$$\eta_{11} = \frac{p_1}{x_1}\frac{\partial x_1}{\partial p_1} = \frac{p_1}{x_1}\left(-\frac{a_1\, x_1}{p_1}\right) = -a_1 \quad \text{and} \quad \eta_{22} = \frac{p_2}{x_2}\frac{\partial x_2}{\partial p_2} = \frac{p_2}{x_2}\left(-\frac{a_2\, x_2}{p_2}\right) = -a_2$$

Since a_1 and a_2 are constants, it follows that η_{11} and η_{22} are independent of prices.

The two cross partial elasticities of demand are:

$$\eta_{12} = \frac{p_2}{x_1}\frac{\partial x_1}{\partial p_2} = \frac{p_2}{x_1}(b_1\, x_1) = b_1\, p_2 \quad \text{and} \quad \eta_{21} = \frac{p_1}{x_2}\frac{\partial x_2}{\partial p_1} = \frac{p_1}{x_2}(b_2\, x_2) = b_2\, p_1$$

Since p_1 and p_2 are both positive, we conclude that η_{12} and η_{21} are determined in sign by the constants b_1 and b_2 respectively.

***Remark* 1.** We have defined partial elasticities for a demand function of two variables. Actually the concept of partial elasticities can be extended to demand functions of more than two variables. For example, if the demand for any commodity X_r is a function of all market prices:

$$x_r = f_r\,(p_1, p_2, \ldots, p_n) \qquad (r = 1, 2, \ldots n)$$

then the partial elasticity of X_r with respect to p_s, denoted by η_{rs}, is defined by

$$\eta_{rs} = \frac{p_s}{x_r}\frac{\partial x_r}{\partial p_s}$$

***Remark* 2.** The concept of partial elasticities of demand can be generalized in a natural way to arbitrary functions of n variables. For example, with $z = f\,(x_1, x_2, \ldots, x_n)$, the partial elasticity of z with respect to x_r is defined to be $\frac{x_r}{z}\frac{\partial z}{\partial x_r}$. In particular, if the supply function for a commodity X_r is given by $x_r = f_r(p_1, p_2, \ldots, p_n)$, where x_r is the quantity of X_r supplied when the price per unit of X_1 is p_1, the price per unit of X_2 is p_2, and so on, then the partial elasticity of supply of X_r with respect to p_s is given by $\frac{p_s}{x_r}\frac{\partial x_r}{\partial p_s}$.

Example 8. If the supply function $x = f(p_1, p_2, \ldots, p_m)$ is homogeneous of degree n, show that sum of the partial price elasticities of supply equals n. (x denotes the quantity supplied of a particular commodity and $p_1, p_2, \ldots, p_m$ are the prices of the different commodities).

[*Delhi Univ. B.Com. (H) 1991, 1997*]

Solution. Let η_r denote the partial price elasticity of supply for x with respect to p_r ($r = 1, 2, \ldots, m$). Then

$$\eta_r = \frac{p_r}{x}\frac{\partial x}{\partial p_r}$$

Since $x = f(p_1, p_2, \ldots, p_m)$ is homogeneous of degree n, therefore by Euler's theorem, we have

$$p_1\frac{\partial x}{\partial p_1} + p_2\frac{\partial x}{\partial p_2} + \ldots + p_m\frac{\partial x}{\partial p_m} = nx$$

Dividing both sides of this equation by x, we obtain

$$\frac{p_1}{x}\frac{\partial x}{\partial p_1} + \frac{p_2}{x}\frac{\partial x}{\partial p_2} + \ldots + \frac{p_m}{x}\frac{\partial x}{\partial p_m} = n$$

i.e.,
$$\eta_1 + \eta_2 + \ldots + \eta_m = n$$

In other words, the sum of the partial price elasticities of supply equals n.

Example 9. Demand function for a commodity X is given by

$$D(x) = 300 - \frac{p_x^2}{2} + \frac{p_y}{50} + \frac{Y}{20},$$

where p_x is the price of the commodity, p_y is the price of a related commodity and Y the income of the consumer. Find the cross elasticity and income elasticity of demand for X when $p_x = 10$, $p_y = 15$ and $Y = 300$. **[*Delhi Univ. B.Com. (H) 1988*]**

Solution. The cross partial elasticity of demand for X with respect to p_y is given by

$$\eta_{x,p_y} = \frac{p_y}{D(x)}\frac{\partial D(x)}{\partial p_y} = \frac{p_y}{300 - \frac{p_x^2}{2} + \frac{p_y}{50} + \frac{Y}{20}} \times \frac{1}{50}$$

Thus evaluating when $p_x = 10$, $p_y = 15$ and $Y = 300$, we obtain $\eta_{x,p_y} = \frac{3}{2653}$.

Income elasticity of demand for X is given by

$$\eta_{x,Y} = \frac{Y}{D(x)}\frac{\partial D(x)}{\partial Y} = \frac{Y}{300 - \frac{p_x^2}{2} + \frac{p_y}{50} + \frac{Y}{20}} \times \frac{1}{20}$$

Thus evaluating when $p_x = 10$, $p_y = 15$ and $Y = 300$, we obtain $\eta_{x,Y} = \frac{150}{2653}$.

Example 10. Find the partial elasticities of $z = x^2 e^y$.

Solution. Partial elasticity of z *w.r.t.* x is : $\eta_{z,x} = \frac{x}{z}\frac{\partial z}{\partial x} = \frac{x}{x^2 e^y} 2x e^y = 2$

Partial elasticity of z *w. r. t.* y is : $\eta_{z,y} = \frac{y}{z}\frac{\partial z}{\partial y} = \frac{y}{x^2 e^y} x^2 e^y = y.$

EXERCISE 8.1

1. The demand functions for two commodities X and Y are $x = \frac{10}{p_1^2 p_2}$ and $y = \frac{5}{p_1 p_2^2}$ respectively, where p_1 and p_2 are the prices of the commodities. Find $\frac{\partial x}{\partial p_1}, \frac{\partial x}{\partial p_2}, \frac{\partial y}{\partial p_1}$ and $\frac{\partial y}{\partial p_2}$ and interpret them in economic sense. Examine whether the two commodities are complementary or competitive.
2. Discuss whether the two goods are substitutes or complementary on the basis of demand functions as given below:

$$x_1 = p_1^{-.4} e^{.2p_2}, \quad x_2 = p_2^{-.6} e^{.5p_1},$$

where p_1 and p_2 denote the prices and x_1, x_2 are the quantities of the two goods respectively. **[*Delhi Univ. B.Com. (H) 1985*]**

3. For the demand functions of two commodities given below, find the four partial elasticities of demand with respect to price and indicate whether the commodities are competitive or complementary:

$$x_1 = 2p_1^{-0.6} p_2^{0.8}, \qquad x_2 = 3p_1^{0.7} p_2^{-0.5}$$

(where x_1 and x_2 are the quantities demanded of the two commodities at prices p_1 and p_2 respectively.) **[*Delhi Univ. B.Com. (H) 1996*]**

4. If x_1 and p_1 are the demand and price of tea, and x_2 and p_2 are demand and price of coffee and the demand functions are given by

$$x_1 = p_1^{-1.7} p_2^{0.6} \quad \text{and} \quad x_2 = p_1^{0.4} p_2^{-0.8},$$

calculate the two cross elasticities of demand and point out whether the commodities are competitive or complementary. **[*Delhi Univ. B.Com. (H) 2014*]**

5. The demand functions for commodities X_1 and X_2 are each a function of the prices of X_1 and X_2 and are given by $x_1 = \dfrac{4}{p_1^2 p_2}$ and $x_2 = \dfrac{16}{p_1 p_2^2}$, respectively. Find the four partial marginal demand functions and determine whether X_1 and X_2 are competitive, complementary or neither. Also determine the four partial elasticities of demand.

6. The demand functions of two commodities X_1 and X_2 are given as

$$x_1 = 1000 + \frac{200}{p_1 + 2} - 20p_2 \quad \text{and} \quad x_2 = 1000 + \frac{500}{p_1 + 4} - 50p_2$$

(*i*) Find the four partial marginal demand functions and determine whether the commodities are competitive or complementary.

(*ii*) Calculate the four partial elasticities of demand at $p_1 = 3$ and $p_2 = 1$.

7. The demand functions for two related goods X_r and X_s are given by :

$$x_r = 4\, p_r^{-2} p_s^2 \qquad \text{and} \qquad x_s = 4\, p_s^{-3} p_r^2,$$

where p_r, p_s are the prices of X_r and X_s respectively. Find the four partial elasticities of demand.

8. When the price of an umbrella is p rupees, x umbrellas are demanded, and when the price of a rain coat is q rupees, y rain coats are demanded. The respective demand functions are :

$$x = 4\, e^{-p/100q} \qquad \text{and} \qquad y = 8\, e^{-q/200p}$$

(*a*) Show that the two commodities are substitutes.

(*b*) Find the four partial elasticities of demand.

9. The demand functions for two commodities X_1 and X_2 in terms of their respective prices p_1 and p_2 are given by:

$$x_1 = p_1^{-0.7}\, e^{0.5p_2 + 3} \qquad \text{and} \qquad x_2 = p_2^{-0.6}\, e^{0.4p_1 + 7}$$

Find the four partial elasticities of demand and determine whether the commodities are competitive or complementary. Also, show that the direct price elasticities of demand are independent of prices whereas cross-price elasticities of demand depend on the price of the other commodity.

[Delhi Univ. B.Com. (H) 2022]

10. Two commodities' joint demand functions are:

$$x_1 = p_1^{-a_{11}}\, e^{a_{12}p_2 + a_1} \qquad \text{and} \qquad x_2 = p_2^{-a_{22}}\, e^{a_{21}p_1 + a_2}$$

where x_1 and x_2 are the units demanded when prices are p_1 and p_2 per unit respectively. Show that direct price elasticities of demand are independent of prices whereas cross price elasticities of demand depend on the price of other commodity. Also find the condition when both the commodities are : (*i*) Competitive (*ii*) Complementary. **[Delhi Univ. B.Com. (H) 1999]**

11. The demand D of passenger automobiles is given by $D = 0.90\, I^{1.1}\, p^{-0.7}$, where I is the income and p is the price per car. Find (*i*) the price elasticity of demand and (*ii*) the income elasticity of demand. **[Delhi Univ. B.Com. (H) 1981]**

12. A demand function is given by $q_r = 5y^2 + \dfrac{2}{p_r} - 4p_s$, where q_r is the quantity demanded, p_r the price of the commodity, p_s the price of another related commodity and y the income. Find the partial direct and cross elasticities of demand and also the income elasticity of demand.

13. The demand function for mutton is : $Q_M = 4850 - 5P_M + 1.5P_C + 0.1Y$. Find (*i*) the income elasticity of demand, and (*ii*) the cross price elasticity of demand for mutton at Y (income) = ₹ 1000, P_M (price of mutton) = ₹ 200, P_C (price of chicken) = ₹ 100.

[*Delhi Univ. B.A. Eco. (Hons.) 1985*]

14. Find the partial elasticities of the function $z = (x+y)\,e^{x+y}$.

ANSWERS

1. $\dfrac{\partial x}{\partial p_1} = -\dfrac{20}{p_1^3 p_2}$, $\dfrac{\partial x}{\partial p_2} = -\dfrac{10}{p_1^2 p_2^2}$, $\dfrac{\partial y}{\partial p_1} = -\dfrac{5}{p_1^2 p_2^2}$, $\dfrac{\partial y}{\partial p_2} = -\dfrac{10}{p_1 p_2^3}$; complementary

2. substitutes

3. $\eta_{11} = -0.6$, $\eta_{12} = 0.8$, $\eta_{21} = 0.7$, $\eta_{22} = -0.5$; competitive

4. $\eta_{12} = 0.6$, $\eta_{21} = 0.4$; competitive

5. $\dfrac{\partial x_1}{\partial p_1} = -\dfrac{8}{p_1^3 p_2}$, $\dfrac{\partial x_1}{\partial p_2} = -\dfrac{4}{p_1^2 p_2^2}$, $\dfrac{\partial x_2}{\partial p_1} = -\dfrac{16}{p_1^2 p_2^2}$, $\dfrac{\partial x_2}{\partial p_2} = -\dfrac{32}{p_1 p_2^3}$; complementary

$\eta_{11} = -2$, $\eta_{12} = -1$, $\eta_{21} = -1$, $\eta_{22} = -2$

6. (*i*) $\dfrac{\partial x_1}{\partial p_1} = -\dfrac{200}{(p_1+2)^2}$, $\dfrac{\partial x_1}{\partial p_2} = -20$, $\dfrac{\partial x_2}{\partial p_1} = -\dfrac{500}{(p_1+4)^2}$, $\dfrac{\partial x_2}{\partial p_2} = -50$; complementary

(*ii*) At $p_1 = 3$, $p_2 = 1$, $\eta_{11} = -\dfrac{2}{85}$, $\eta_{12} = -\dfrac{1}{51}$, $\eta_{21} = -\dfrac{30}{1001}$, $\eta_{22} = -\dfrac{7}{143}$

7. $\eta_{rr} = -2$, $\eta_{rs} = 2$, $\eta_{sr} = 2$, $\eta_{ss} = -3$

8. (*b*) $\eta_{xp} = -\dfrac{p}{100q}$, $\eta_{xq} = \dfrac{p}{100q}$, $\eta_{yp} = \dfrac{q}{200p}$, $\eta_{yq} = -\dfrac{q}{200p}$

9. $\eta_{11} = -0.7$, $\eta_{12} = 0.5p_2$, $\eta_{21} = 0.4p_1$, $\eta_{22} = -0.6$; Competitive

10. (*i*) Competitive if a_{12} and a_{21} are positive (*ii*) Complementary if a_{12} and a_{21} are negative

11. (*i*) -0.7 (*ii*) 1.1

12. $\eta_{q_r\,p_r} = -\dfrac{2}{p_r\,q_r}$, $\eta_{q_r\,p_s} = -\dfrac{4p_s}{q_r}$, $\eta_{q_r\,y} = \dfrac{10y^2}{q_r}$

13. (*i*) $\dfrac{1}{41}$ (*ii*) $\dfrac{3}{82}$ **14.** $\dfrac{x(x+y+1)}{x+y}$, $\dfrac{y(x+y+1)}{x+y}$

8.3 THE PRODUCTION FUNCTION

The production of a commodity depends on various factors of production such as labour, capital, machinery, land, and so on. For simplicity, we assume that production depends only on labour and capital. Suppose the producer uses L units of labour and K units of capital, then the function $P = f(L, K)$ gives the amount of output P and is called a **production function.** An important production function in economics is the **Cobb-Douglas production function** defined by

$$P = f(L, K) = a\, L^{\alpha} K^{1-\alpha},$$

where a and α are constants.

If $P = f(L, K)$ is a production function, the **marginal productivity of labour,** denoted by MP_L, is defined by $\frac{\partial P}{\partial L}$. It represents the rate of change of P with respect to L when K is held fixed. Likewise, the **marginal productivity of capital,** denoted by MP_K, is defined by $\frac{\partial P}{\partial K}$. It represents the rate of change of P with respect to K when L is held fixed. The **average product of labour,** denoted by AP_L, is defined by $\frac{P}{L}$ (*i.e.,* the amount of product per unit of the input L). Similarly, the **average product of capital,** denoted by AP_K, is defined as $\frac{P}{K}$ (*i.e.*, the amount of product per unit of the input K). Both marginal and average products are functions of L and K.

Example 11. The production function of a firm is given by $P = 20\, L^{1/3} K^{2/3}$. Find the average and marginal products and evaluate them when $L = 8$ and $K = 27$. Interpret the results.

Solution. Average product of labour is : $AP_L = \dfrac{P}{L} = \dfrac{20L^{1/3}K^{2/3}}{L} = 20L^{-2/3}K^{2/3}$

Average product of capital is : $AP_K = \dfrac{P}{K} = \dfrac{20L^{1/3}K^{2/3}}{K} = 20L^{1/3}K^{-1/3}$

Evaluating when $L = 8$ and $K = 27$, we obtain $AP_L = 45$ and $AP_K = \dfrac{40}{3}$

Thus, if $L = 8$ and $K = 27$, then by holding K at 27, the output per unit of labour is 45 units. But, if L is held at 8, the output per unit of capital is 40/3 units. Now

marginal product of labour is : $MP_L = \dfrac{\partial P}{\partial L} = \dfrac{20}{3} L^{-2/3}\ K^{2/3}$, and

marginal product of capital is : $MP_K = \dfrac{\partial P}{\partial K} = \dfrac{40}{3} L^{1/3}\ K^{-1/3}$

Evaluating when $L = 8$ and $K = 27$, we obtain $MP_L = 15$ and $MP_K = \frac{80}{9}$

Thus, if $L = 8$ and $K = 27$, increasing L to 9 and holding K at 27 will increase output by 15 units. But if K is increased to 28 while L is held at 8, the output increases by $\frac{80}{9}$ units.

Example 12. The production function of a firm is given by $Q = 8LK - L^2 - K^2$, $L > 0, K > 0$. Find the marginal productivities of labour and capital. Also show that

$$L\frac{\partial Q}{\partial L} + K\frac{\partial Q}{\partial K} = 2Q$$

[*Delhi Univ. B.Com. (H) 1982*]

Solution. Given $Q = 8LK - L^2 - K^2$, we have

$$\text{marginal productivity of labour} = \frac{\partial Q}{\partial L} = 8K - 2L$$

and

$$\text{marginal productivity of capital} = \frac{\partial Q}{\partial K} = 8L - 2K$$

$$\therefore \quad L\frac{\partial Q}{\partial L} + K\frac{\partial Q}{\partial K} = L(8K - 2L) + K(8L - 2K)$$

$$= 8LK - 2L^2 + 8LK - 2K^2$$

$$= 2(8LK - L^2 - K^2) = 2Q.$$

Example 13. The production function for a commodity is :

$$Q = 10L - 0.1L^2 + 15K - 0.2K^2 + 2KL,$$

where L is labour, K is capital and Q is production .

(*i*) Calculate the marginal products of the two inputs when 10 units each of labour and capital are used.

(*ii*) If 10 units of capital are used, what is the upper limit for use of labour which a rational producer will never exceed? **[*Delhi Univ. B.Com. (H) 2008*]**

Solution. Given : $Q = 10L - 0.1L^2 + 15K - 0.2K^2 + 2KL$

(*i*) Marginal product of labour is : $MP_L = \frac{\partial Q}{\partial L} = 10 - 0.2L + 2K$

$\therefore$ When $L = 10$, $K = 10$, $MP_L = 10 - 2 + 20 = 28$

Marginal product of capital is : $MP_K = \frac{\partial Q}{\partial K} = 15 - 0.4K + 2L$

$\therefore$ When $L = 10,\ K = 10,$ $\qquad MP_K = 15 - 4 + 20 = 31$

(*ii*) If 10 units of capital are used, then

$$\begin{aligned} Q &= 10L - 0.1L^2 + 15 \times 10 - 0.2(10)^2 + 2 \times 10 \times L \\ &= 10L - 0.1L^2 + 150 - 20 + 20L \\ &= 30L - 0.1L^2 + 130 \end{aligned}$$

$$\Rightarrow \qquad \frac{dQ}{dL} = 30 - 0.2L$$

It is required to find L so that $\frac{dQ}{dL} \geq 0$. That is,

$$30 - 0.2L \geq 0 \quad \Rightarrow \quad 0.2L \leq 30 \quad \text{or} \quad L \leq \frac{30}{0.2} = 150$$

Thus the upper limit for use of labour which a rational producer will never exceed is 150.

Example 14. The output at a certain plant is $Q = f(x, y) = 0.08x^2 + 0.12xy + 0.03y^2$ units per day, where x is the number of hours of skilled labour and y is the number of hours of unskilled labour used. Currently 80 hours of skilled labour and 200 hours of unskilled labour are used each day. Use total differential of Q to estimate the change in the output that will result if an additional hour of skilled labour is used along with an additional 2 hours of unskilled labour.

Solution. The total differential of Q is given by

$$dQ = \frac{\partial Q}{\partial x} dx + \frac{\partial Q}{\partial y} dy$$

But $\qquad \frac{\partial Q}{\partial x} = 0.16x + 0.12y \quad$ and $\quad \frac{\partial Q}{\partial y} = 0.12x + 0.06y$

$$\therefore \qquad dQ = (0.16x + 0.12y)\,dx + (0.12x + 0.06y)\,dy \qquad \text{...(1)}$$

We have to evaluate dQ when $x = 80,\quad y = 200,\quad dx = 1$ and $dy = 2$.

Substituting these values in (1), the total change in Q is given by

$$\begin{aligned} dQ &= (0.16 \times 80 + 0.12 \times 200)(1) + (0.12 \times 80 + 0.06 \times 200)(2) \\ &= (12.8 + 24) + (9.6 + 12)(2) = 36.8 + 21.6 \times 2 = 36.8 + 43.2 = 80. \end{aligned}$$

The Linear Homogeneous Production Function. Many of the production functions that are useful in economic analysis share the property of being homogeneous. Of particular interest are the production functions which are homogeneous of degree 1 (also called linearly homogeneous). The Cobb-Douglas production referred to earlier is linearly homogeneous in the form in which it is usually written. Homogeneous production functions of degree 1 have many interesting properties.

1. *Linearly homogeneous production functions exhibit constant returns to scale, i.e.,* a proportional increase in all inputs leads to the same proportional increase in output.

For, if $P = f(L, K)$ is such a function, then from the definition,

$$f(\lambda L, \lambda K) = \lambda f(L, K) \qquad (\lambda \text{ is a constant})$$

Thus, if both inputs are doubled, then

$$f(2L, 2K) = 2f(L, K),$$

and output is doubled. Similarly, if both inputs are tripled, output is tripled, etc.

2. *The average and marginal products depend on the ratio of factors used in the production for linear homogeneous production functions.*

For, if $P = f(L, K)$ is a linear homogeneous production function, then

$$f(\lambda L, \lambda K) = \lambda f(L, K),$$

where λ is any constant. Setting $\lambda = \frac{1}{L}$, we obtain

$$f\left(1, \frac{K}{L}\right) = \frac{1}{L} f(L, K)$$

$$\Rightarrow \qquad P = f(L, K) = L f\left(1, \frac{K}{L}\right) = L\,\varphi\left(\frac{K}{L}\right) \qquad \text{...(1)}$$

$$\Rightarrow \qquad AP_L = \frac{P}{L} = \varphi\left(\frac{K}{L}\right), \text{ a function of } \frac{K}{L} \text{ only.}$$

Similarly, by setting $\lambda = \frac{1}{K}$, it can be seen that the average product of capital, AP_K, depends on the ratio of factors used.

To show that the marginal product of labour depends on the ratio, we differentiate equation (1) partially with respect to L:

$$\frac{\partial P}{\partial L} = L\,\varphi'\left(\frac{K}{L}\right)\left(-\frac{K}{L^2}\right) + \varphi\left(\frac{K}{L}\right) = -\left(\frac{K}{L}\right)\varphi'\left(\frac{K}{L}\right) + \varphi\left(\frac{K}{L}\right),$$

which clearly depends on the ratio $\left(\frac{K}{L}\right)$. Similarly, it can be verified that the marginal product of capital depends on the ratio of factors used.

3. *For a linear homogeneous production function, the total product is equal to labour times the marginal productivity of labour plus capital times the marginal productivity of capital, i.e., total product will get exhausted if each factor is paid a price according to its marginal product. This is a well-known property in theory of production under the name of "Adding-up Theorem".*

This property follows from Euler's theorem: $L\dfrac{\partial P}{\partial L} + K\dfrac{\partial P}{\partial K} = P$, and the fact that $\dfrac{\partial P}{\partial L}$ is the marginal productivity of labour and $\dfrac{\partial P}{\partial K}$ is the marginal productivity of capital.

The above properties are best illustrated by the following examples.

Example 15. For the following production function, show that the marginal products of the factors depend only on the ratio of factors

$$x = f(l, k) = \frac{2hlk - \alpha l^2 - \beta k^2}{cl + dk}$$

Also, determine the degree of homogeneity and verify Euler's theorem for the function.

[*Delhi Univ. B.Com. (H) 1990, 1993, 2000*]

Solution. Marginal product of labour is:

$$MP_l = \frac{\partial x}{\partial l} = \frac{(cl + dk)(2hk - 2\alpha l) - (2hlk - \alpha l^2 - \beta k^2)(c)}{(cl + dk)^2}$$

Dividing numerator and denominator by k^2, we obtain

$$MP_l = \frac{2\left(c\dfrac{l}{k} + d\right)\left[h - \alpha\left(\dfrac{l}{k}\right)\right] - \left[2h\left(\dfrac{l}{k}\right) - \alpha\left(\dfrac{l}{k}\right)^2 - \beta\right](c)}{\left[c\left(\dfrac{l}{k}\right) + d\right]^2}$$

which clearly shows that MP_l depends only on the ratio $\dfrac{l}{k}$ of factors.

Marginal product of capital is:

$$MP_k = \frac{\partial x}{\partial k} = \frac{(cl + dk)(2hl - 2\beta k) - (2hlk - \alpha l^2 - \beta k^2)(d)}{(cl + dk)^2}$$

Dividing numerator and denominator by k^2, we obtain

$$MP_k = \frac{2\left(c\dfrac{l}{k} + d\right)\left[h\left(\dfrac{l}{k}\right) - \beta\right] - \left[2h\left(\dfrac{l}{k}\right) - \alpha\left(\dfrac{l}{k}\right)^2 - \beta\right](d)}{\left[c\left(\dfrac{l}{k}\right) + d\right]^2}$$

which shows that MP_k also depends on the ratio $\dfrac{l}{k}$ of factors.

To determine the degree of homogeneity, we consider

$$f(\lambda l, \lambda k) = \frac{2h(\lambda l)(\lambda k) - \alpha(\lambda l)^2 - \beta(\lambda k)^2}{c(\lambda l) + d(\lambda k)} = \frac{\lambda^2[2hlk - \alpha l^2 - \beta k^2]}{\lambda(cl + dk)} = \lambda f(l, k).$$

This shows that $x = f(l, k)$ is homogeneous of degree 1.

Finally, to verify Euler's theorem, we'll show that $l\frac{\partial x}{\partial l} + k\frac{\partial x}{\partial k} = x$.

Consider

$$l\frac{\partial x}{\partial l} + k\frac{\partial x}{\partial k} = l\left\{\frac{(cl + dk)(2hk - 2\alpha l) - (2hlk - \alpha l^2 - \beta k^2)(c)}{(cl + dk)^2}\right\} +$$

$$k\left\{\frac{(cl + dk)(2hl - 2\beta k) - (2hlk - \alpha l^2 - \beta k^2)(d)}{(cl + dk)^2}\right\}$$

$$= \frac{(cl + dk)\left[2hlk - 2\alpha l^2 + 2hlk - 2\beta k^2\right] - \left[2hlk - \alpha l^2 - \beta k^2\right](cl + dk)}{(cl + dk)^2}$$

$$= \frac{(cl + dk)\left[2hlk - \alpha l^2 - \beta k^2\right]}{(cl + dk)^2} = \frac{2hlk - \alpha l^2 - \beta k^2}{cl + dk} = x.$$

Hence Euler's theorem is verified.

Example 16. Suppose a firm has a production function $x = [aL^4 + bK^4]^{½}$ where $0<a<1$, $0<b<1$. L and K are labour and capital respectively. Find the marginal product of labour and marginal product of capital and verify Euler's theorem. **[*Delhi Univ. B.Com. (H) 2017*]**

Solution. Given $x = f(L, K) = (aL^4 + bK^4)^{½}$

$$\text{Marginal product of labour} = \frac{\partial x}{\partial L} = \frac{1}{2}(aL^4 + bK^4)^{-½}\, 4aL^3 = \frac{2aL^3}{\sqrt{aL^4 + bK^4}}$$

$$\text{Marginal product of capital} = \frac{\partial x}{\partial K} = \frac{1}{2}(aL^4 + bK^4)^{-½}\, 4bK^3 = \frac{2bK^3}{\sqrt{aL^4 + bK^4}}$$

Verification of Euler's Theorem: To verify Euler's Theorem, let us first determine the degree of homogeneity. For this, consider

$$f(\lambda L, \lambda K) = \left[a(\lambda L)^4 + b(\lambda K)^4\right]^{½} = \left[\lambda^4(aL^4 + bK^4)\right]^{½}$$

$$= \lambda^2\left(aL^4 + bK^4\right)^{½} = \lambda^2 f(L, K)$$

This shows that $x = f(L, K)$ is homogeneous of degree 2 and hence we must verify that

$$L\frac{\partial x}{\partial L} + K\frac{\partial x}{\partial K} = 2x$$

Consider

$$LHS = L\frac{\partial x}{\partial L} + K\frac{\partial x}{\partial K} = L\left(\frac{2aL^3}{\sqrt{aL^4 + bK^4}}\right) + K\left(\frac{2bK^3}{\sqrt{aL^4 + bK^4}}\right)$$

$$= \frac{2a\,L^4}{\sqrt{aL^4 + bK^4}} + \frac{2b\,K^4}{\sqrt{aL^4 + bK^4}}$$

$$= \frac{2(aL^4 + bK^4)}{\sqrt{aL^4 + bK^4}} = 2\sqrt{aL^4 + bK^4} = 2x = RHS$$

Hence Euler's Theorem is verified.

Example 17. For the production function: $Q = f(L, K) = AK^{\alpha} L^{\beta} e^{a(K/L)}$ where A, α, β and a are positive constants, find the marginal products of labour and capital. Also verify Euler's theorem. **[*Delhi Univ. B.Com. (H) 1999*]**

Solution. Marginal product of labour is:

$$MP_L = \frac{\partial Q}{\partial L} = \beta\, A\, K^{\alpha} L^{\beta - 1}\, e^{a(K/L)} + A\, K^{\alpha} L^{\beta} e^{a(K/L)}\left(-\frac{aK}{L^2}\right)$$

$$= A\, K^{\alpha}\, L^{\beta - 1} e^{a(K/L)}\left(\beta - \frac{aK}{L}\right)$$

Marginal product of capital is:

$$MP_K = \frac{\partial Q}{\partial K} = \alpha\, A\, K^{\alpha - 1} L^{\beta}\, e^{a(K/L)} + A\, K^{\alpha} L^{\beta} e^{a(K/L)}\left(\frac{a}{L}\right)$$

$$= A\, K^{\alpha - 1}\, L^{\beta}\, e^{a(K/L)}\left(\alpha + \frac{aK}{L}\right)$$

Verification of Euler's theorem. To verify Euler's theorem, let us first determine the degree of homogeneity. For this, consider

$$f(\lambda L, \lambda K) = A\,(\lambda K)^{\alpha}\,(\lambda L)^{\beta}\, e^{a\,(\lambda K/\lambda L)} = \lambda^{\alpha + \beta} A\, K^{\alpha} L^{\beta}\, e^{a\,(K/L)} = \lambda^{\alpha + \beta} f(L, K)$$

This shows that $Q = f(L, K)$ is homogeneous of degree $\alpha + \beta$ and hence we must verify that

$$L\frac{\partial Q}{\partial L} + K\frac{\partial Q}{\partial K} = (\alpha + \beta)\, Q$$

Consider

$$L\frac{\partial Q}{\partial L} + K\frac{\partial Q}{\partial K} = L\left[AK^{\alpha} L^{\beta - 1} e^{a(K/L)}\left(\beta - \frac{aK}{L}\right)\right] + K\left[AK^{\alpha - 1} L^{\beta} e^{a(K/L)}\left(\alpha + \frac{aK}{L}\right)\right]$$

$$= AK^{\alpha} L^{\beta} e^{a(K/L)}\left[\beta - \frac{aK}{L} + \alpha + \frac{aK}{L}\right] = (\alpha + \beta)\, Q\,.$$

Hence Euler's theorem is verified.

Example 18. **For the linear homogeneous production function $z = \sqrt{x_1 x_2}$, where z is output and x_1, x_2 are inputs, show that the average and marginal products of inputs depend only on the ratio of the factors and verify Euler's theorem for this production function.**

[*Delhi Univ. B.Com. (H) 2011*]

Solution. Average product of input x_1 is: $AP_{x_1} = \dfrac{z}{x_1} = \dfrac{\sqrt{x_1 x_2}}{x_1} = \sqrt{\dfrac{x_2}{x_1}}$

Average product of input x_2 is: $AP_{x_2} = \dfrac{z}{x_2} = \dfrac{\sqrt{x_1 x_2}}{x_2} = \sqrt{\dfrac{x_1}{x_2}}$.

The expressions for AP_{x_1} and AP_{x_2} clearly indicate that they depend only on the ratio of factors.

Marginal product of input x_1 is: $MP_{x_1} = \dfrac{\partial z}{\partial x_1} = \dfrac{1}{2}\sqrt{\dfrac{x_2}{x_1}}$

Marginal product of input x_2 is: $MP_{x_2} = \dfrac{\partial z}{\partial x_2} = \dfrac{1}{2}\sqrt{\dfrac{x_1}{x_2}}$

The expressions for marginal products clearly indicate that they depend only on the ratio of factors. Now

$$x_1 \frac{\partial z}{\partial x_1} + x_2 \frac{\partial z}{\partial x_2} = x_1\left(\frac{1}{2}\sqrt{\frac{x_2}{x_1}}\right) + x_2\left(\frac{1}{2}\sqrt{\frac{x_1}{x_2}}\right) = \frac{1}{2}\sqrt{x_1 x_2} + \frac{1}{2}\sqrt{x_1 x_2} = \sqrt{x_1 x_2} = z$$

Thus $\quad x_1 \dfrac{\partial z}{\partial x_1} + x_2 \dfrac{\partial z}{\partial x_2} = z.$

Hence Euler's theorem is verified.

Example 19. **The following is a linear homogeneous production function, where X, L, K represent output, labour, and capital, respectively:**

$$X = \sqrt{aL^2 + 2hLK + bK^2}$$

Show that the sum of L times the marginal product of labour and K times the marginal product of capital equals total product. **[*Delhi Univ. B.Com. (H) 2012*]**

Solution. Marginal product of labour is:

$$MP_L = \frac{\partial X}{\partial L} = \frac{1}{2}(aL^2 + 2hLK + bK^2)^{-1/2}(2aL + 2hK)$$

$$= \frac{(aL + hK)}{\sqrt{aL^2 + 2hLK + bK^2}}$$

Marginal product of capital is:

$$MP_K = \frac{\partial X}{\partial K} = \frac{1}{2}(aL^2 + 2hLK + bK^2)^{-1/2}(2hL + 2bK)$$

$$= \frac{(hL + bK)}{\sqrt{aL^2 + 2hLK + bK^2}}$$

Now $$LMP_L + KMP_K = L\frac{aL + hK}{\sqrt{aL^2 + 2hLK + bK^2}} + K\frac{hL + bK}{\sqrt{aL^2 + 2hLK + bK^2}}$$

$$= \frac{aL^2 + hLK + hLK + bK^2}{\sqrt{aL^2 + 2hLK + bK^2}}$$

$$= \frac{aL^2 + 2hLK + bK^2}{\sqrt{aL^2 + 2hLK + bK^2}} = X.$$

That is, L times the marginal product of labour plus K times the marginal product of capital equals total product.

Diminishing Returns to Factors. As we have mentioned already, production of any commodity requires many different types of inputs. These inputs are called the factors of production. The production function provides only a technical relationship between quantities of inputs and quantities of output and does not take into account the prices of either the factors or the products.

Economic theory looks to two kinds of input-output relations in production functions. One is the relation where the input of only one factor is varied while the other factors of production are fixed. In the other relation, all of the factors are variable.

When a firm increases output by holding all inputs constants but one, the gains in output are subject to the *law of diminishing returns to inputs,* also known as the *law of variable proportions,* which states that *when total output or production of a commodity is increased by adding units of a variable input while the quantities of other inputs are held fixed, the increases in total production become, after some point, smaller and smaller.* Observe that total production increases but at a diminishing rate. Thus the law of diminishing returns states that as additional units of a variable input are added to a given quantity of fixed inputs, the marginal product of the variable input will eventually decline (or diminish).

Returns to Scale. When a firm increases output by using more of all inputs, the input-output relation is one of a *scale. Returns to scale* can be increasing, constant and decreasing. As discussed before, if the increase in output is proportional to the increase in the quantities of the inputs, returns to scale are said to be *constant.* If instead, the increase in output is more than proportional, returns to scale are *increasing.* And if the increase in output is less than proportional, returns to scale are *decreasing.*

Remark. Returns to scale can easily be determined for homogeneous production functions. For example, if $P = f(L, K)$ is a homogeneous production of degree n, then

$$f(\lambda L, \lambda K) = \lambda^n f(L, K),$$

where λ is a constant. From this equation, it follows that there are increasing, decreasing or constant returns to scale according as the degree of homogeneity, n, is greater than, less than or equal to one.

Example 20. Show that the production function

$$x = f(l, k) = 2\sqrt{lk}$$

(where x, l and k are the units of output, labour and capital respectively) gives constant returns to scale and diminishing returns to inputs. **[*Delhi Univ. B.Com. (H) 1992*]**

Solution. Replacing l by $\lambda\, l$ and k by $\lambda\, k$ in the given production function, we obtain

$$f(\lambda l, \lambda k) = 2\sqrt{(\lambda l)(\lambda k)} = 2\sqrt{\lambda^2\, lk} = \lambda 2\sqrt{lk} = \lambda f(l, k)$$

i.e. $$f(\lambda l, \lambda k) = \lambda f(l, k)$$

This shows that the given production function is homogeneous of degree 1 and hence there are constant returns to scale.

To see that the production function gives diminishing returns to inputs, we'll show that marginal products of inputs are decreasing functions. Now

$$MP_l = \text{marginal product of input } l = \frac{\partial x}{\partial l} = \sqrt{\frac{k}{l}},$$

and $$MP_k = \text{marginal product of input } k = \frac{\partial x}{\partial k} = \sqrt{\frac{l}{k}}$$

Differentiating MP_l partially with respect to l gives

$$\frac{\partial}{\partial l}(MP_l) = -\frac{1}{2}\frac{\sqrt{k}}{l^{3/2}} < 0$$

Thus, if l is increased while k is held fixed, marginal product of labour will diminish. Similarly, differentiating MP_k partially with respect to k gives

$$\frac{\partial}{\partial k}(MP_k) = -\frac{1}{2}\frac{\sqrt{l}}{k^{3/2}} < 0$$

Thus, if k is increased while l is held fixed, marginal product of capital will diminish.

Example 21. Suppose a firm has production function $Q = AK^{\alpha}L^{1-\alpha}$ where $A > 0$ and $0 < \alpha < 1$. Show that production function has diminishing returns to inputs.

[*Delhi Univ. B.Com. (H) 2010, 2018*]

Solution. The production function is given by $Q = AK^{\alpha} L^{1-\alpha}$, $A > 0$ and $0 < \alpha < 1$

$$\Rightarrow \quad MP_L = \frac{\partial Q}{\partial L} = A(1-\alpha) K^{\alpha} L^{-\alpha} \quad \text{and} \quad MP_K = \frac{\partial Q}{\partial K} = A\alpha K^{\alpha-1} L^{1-\alpha}$$

$$\Rightarrow \quad \frac{\partial MP_L}{\partial L} = -A\alpha(1-\alpha)K^{\alpha} L^{-\alpha-1}$$

and
$$\frac{\partial MP_K}{\partial K} = A\alpha(\alpha-1)K^{\alpha-2}L^{1-\alpha} = -A\alpha(1-\alpha)K^{\alpha-2}L^{1-\alpha}$$

Since $A > 0$ and $0 < \alpha < 1$, $\dfrac{\partial MP_L}{\partial L}$ and $\dfrac{\partial MP_K}{\partial K}$ are both negative and hence production function shows diminishing returns to inputs.

Example 22. A production function is given by $Q = f(L, K) = 27L^{2/3} K^{1/3}$, where L is labour and K is capital.

(*i*) Find the behaviour of the marginal product of each factor.

(*ii*) What is the nature of returns to scale ?

(*iii*) What is the total reward of labour and capital if each factor is paid a price equal to its marginal product ? **[*Delhi Univ. B.Com. (H) 2009*]**

Solution. (*i*) Differentiating the production function Q partially with respect to L gives

$$MP_L = \frac{\partial Q}{\partial L} = 18L^{-1/3} K^{1/3}$$

Differentiating MP_L partially with respect to L, we obtain

$$\frac{\partial}{\partial L}(MP_L) = -6L^{-4/3} K^{1/3} < 0,$$

which shows that the marginal product of labour decreases as L increases while K is held fixed. Similarly,

$$MP_K = \frac{\partial Q}{\partial K} = 9L^{2/3} K^{-2/3}$$

and
$$\frac{\partial}{\partial K}(MP_K) = -6L^{2/3} K^{-5/3} < 0,$$

which shows that the marginal product of capital decreases as K increases while L is held fixed.

(*ii*) We have
$$f(\lambda L, \lambda K) = 27(\lambda L)^{2/3}(\lambda K)^{1/3} = 27\lambda^{2/3} L^{2/3} \lambda^{1/3} K^{1/3}$$
$$= \lambda(27 L^{2/3} K^{1/3}) = \lambda f(L, K)$$

Hence $Q = f(L, K)$ is homogeneous of degree 1 and the production function exhibits constant returns to scale.

(*iii*) If each factor is paid a price equal to its marginal product, then we have

$$L\frac{\partial Q}{\partial L} + K\frac{\partial Q}{\partial K} = L\,(18L^{-1/3}\,K^{1/3}) + K\,(9L^{2/3}\,K^{-2/3})$$

$$= 18L^{2/3}\,K^{1/3} + 9L^{2/3}\,K^{1/3}$$

$$= 27L^{2/3}\,K^{1/3} = Q,$$

which shows that the total product is exhausted.

Example 23. The production function of a firm is given by $x = f(l, k) = Al^{\alpha}\,k^{\beta}$, where $\alpha + \beta < 1$, and l and k denote inputs of labour and capital respectively. Show that there are decreasing returns to scale and deduce that the total product is greater than l times the marginal product of labour plus k times the marginal product of capital.

Solution. Replacing l by λl and k by λk in the production function, we obtain

$$f(\lambda l, \lambda k) = A(\lambda l)^{\alpha}\,(\lambda k)^{\beta} = \lambda^{\alpha+\beta}\,Al^{\alpha}\,k^{\beta} = \lambda^{\alpha+\beta}\,f(l, k)$$

i.e., $$f(\lambda l, \lambda k) = \lambda^{\alpha+\beta}\,f(l, k)$$

That means $x = f(l, k)$ is homogeneous of degree $\alpha + \beta$. However, $\alpha + \beta < 1$, therefore there are decreasing returns to scale. Applying Euler's theorem, we get

$$l\frac{\partial x}{\partial l} + k\frac{\partial x}{\partial k} = (\alpha+\beta)\,x < x \qquad (\because \alpha + \beta < 1)$$

Since $\frac{\partial x}{\partial l}$ is the marginal product of labour and $\frac{\partial x}{\partial k}$ is the marginal product of capital, we conclude that the total product is greater than l times the marginal product of labour plus k times the marginal product of capital.

Example 24. At a certain factory, daily output is $Q = 60\,L^{1/3}\,K^{1/2}$ units, where L indicates the size of labour force and K denotes capital investment. Use concept of total differential to estimate the percentage by which daily output change if capital investment and labour force are both increased by 3%. What is nature of return to scale in this case?

[*Delhi Univ. B.Com. (H) 2010*]

Solution. The daily output is given by : $Q = 60\,L^{1/3}\,K^{1/2}$

$$\Rightarrow \quad \frac{\partial Q}{\partial L} = 60 \times \frac{1}{3}L^{-2/3}\,K^{1/2} = 20L^{-2/3}\,K^{1/2} \quad \text{and} \quad \frac{\partial Q}{\partial K} = 60 \times \frac{1}{2}\,L^{1/3}\,K^{-1/2} = 30L^{1/3}\,K^{-1/2}$$

The total differential of Q is given by

$$dQ = \frac{\partial Q}{\partial L}dL + \frac{\partial Q}{\partial K}dK = \left(20L^{-2/3}K^{1/2}\right)dL + \left(30L^{1/3}K^{-1/2}\right)dK \quad \ldots\ldots(1)$$

The problem is to find dQ when $dL = 0.03L$ and $dK = 0.03K$. Substituting these values of dL and dK in (1), we obtain

$$dQ = (20L^{-2/3}K^{1/2})(0.03L) + (30L^{1/3}K^{-1/2})(0.03K)$$

$$= 0.6L^{1/3}K^{1/2} + 0.9L^{1/3}K^{1/2} = 1.5L^{1/3}K^{1/2} = (0.025)60L^{1/3}K^{1/2} = 0.025Q$$

This shows that output is increased by 2.5%. That is, when labour force and capital investment are both increased by 3%, output increases by 2.5%. Hence production shows decreasing returns to scale.

EXERCISE 8.2

1. If a manufacturer's production function is defined by $P = 20L^{0.7}K^{0.3}$, determine the marginal productivities of labour and capital and evaluate them when $L = 1$ and $K = 1$. Interpret the results.
2. For the production function given by $P = 20LK - 2L^2 - 4K^2 + 800$, find the marginal productivities of labour and capital respectively.
3. For the production function $Q = f(L, K)$ given implicitly, in the form $f(L, K, Q) = 0$, by the equation

$$6Q^3 - Q^2 - 6L - 24K + L^2 + 4K^2 + 50 = 0,$$

find the marginal productivities of labour (L) and capital (K).

4. The production function of a firm is given by $Q = 4L^{3/4}K^{1/4}$, $L > 0$, $K > 0$. Find the marginal productivities of Labour (L) and Capital (K). Also show that $L\frac{\partial Q}{\partial L} + K\frac{\partial Q}{\partial K} = Q$.

[Delhi Univ. B.Com. (H) 1982]

5. Given the production function $x = \dfrac{2HKL - AL^2 - BK^2}{CL + DK}$, where H, A, B, C, D are positive constants, x denotes output and L, K denote labour and capital inputs respectively.

   (*i*) Find the average and marginal products of factors of production.

   (*ii*) Verify Euler's theorem for the production function. **[Delhi Univ. B. Com. (H) 1989]**

6. Define a homogeneous function. Find out the degree of homogeneity of the following production function :

$$Q^{-\beta} = aK^{-\beta} + bL^{-\beta}$$

where Q represents output, L and K denote factors of production labour and capital respectively.

[Delhi Univ. B. Com. (H) 2007]

7. State Euler's theorem. Compute the degree of homogeneity and verify Euler's theorem for the production function $x = f(l, k) = (al^4 + bk^4)^{1/2}$. **[Delhi Univ. B. Com. (H) 1996]**

8. The production function of a firm is given by :

$$Q = f(L, K) = A[\alpha K^{-\beta} + (1-\alpha)L^{-\beta}]^{-1/\beta}, \quad L > 0, K > 0$$

Find the marginal productivities of Labour (L) and Capital (K). Also, determine the degree of homogeneity and verify Euler's theorem for the function.

9. For the Cobb-Douglas production function $Q = f(L, K) = AL^{\alpha} K^{\beta}$, show that there are increasing, decreasing, or constant returns to scale according as $(\alpha + \beta) > 1, < 1$, or $= 1$.

[Delhi Univ. B. Com. (H) 1989]

10. A firm's production function is : $Q = f(L, K) = AL^{\alpha} K^{\beta}$. Under what conditions will it give increasing returns to scale and diminishing returns to inputs, if possible.

[Delhi Univ. B. Com. (H) 1998, 2004]

11. Define the degree of homogeneity of a function and state Euler's theorem. Show that the production function $q = f(l, k) = \sqrt{lk}$, (where q is total output obtained by using l and k units of labour and capital respectively) gives constant returns to scale and diminishing returns to inputs.

[Delhi Univ. B. Com. (H) 2002]

12. Suppose a firm has a production function $Q = AK^{\alpha} L^{1-\alpha}$ where $A > 0$ and $0 < \alpha < 1$. Show that production function has diminishing returns to inputs and constant returns to scale.

[Delhi Univ. B. Com. (H) 2018]

13. For the Cobb-Douglas production function : $Q = AL^{\alpha} K^{\beta}$, where A, α and β are positive constants and $\alpha + \beta = 1$, show that

(*i*) $\dfrac{\partial Q}{\partial L} = \dfrac{\alpha Q}{L}$ and $\dfrac{\partial Q}{\partial K} = \dfrac{\beta Q}{K}$ (*ii*) $L\dfrac{\partial Q}{\partial L} + K\dfrac{\partial Q}{\partial K} = Q$

(*iii*) There are diminishing returns to inputs.

14. A production function is given by $Q = AL^{3/8} K^{3/8}$, where L is labour and K is capital.

(*i*) What is the nature of returns to scale ?

(*ii*) What is the total reward of labour and capital if each factor is paid a price equal to its marginal product ? Show that total product is not exhausted.

[Delhi Univ. B.Com. (H) 2014]

15. A production function is given by $Q = AL^{1/3} K^{1/3}$, where L is labour and K is capital.

(*i*) Find the behaviour of marginal product of each factor.

(*ii*) What is the nature of returns to scale ?

(*iii*) Show that the total product is not exhausted if each factor is paid a price equal to its marginal product. **[Delhi Univ. B.Com. (H) 2012]**

16. At a certain factory, daily output is $Q = f(L, K) = 40L^{3/4} K^{1/2}$, where L indicates the size of labour force and K denotes the capital investment. Use calculus to estimate the percentage by which the daily output will increase when capital investment and labour force are increased by 4%. What is the nature of returns to scale in this case?

17. For the Cobb-Douglas production function $Q = AL^{\alpha} K^{1-\alpha}$, show that the average and marginal products of inputs depend only on the ratio of the factors used.

18. For the production function $Q = \dfrac{LK}{L+K}$, show that the average and marginal products of inputs depend only on the ratio of the factors used.

19. Given a linear homogeneous production function $Z = AL^{\alpha} K^{\beta} P^{\gamma}$, where L, K, P stand for factor quantities and A is a constant, show that

(*i*) the sum of marginal products of factors each multiplied by its respective quantity equals the total output,

(*ii*) the sum of production elasticities with respect to the factors is unity, and

(*iii*) In (*i*) and (*ii*) above, consider how these results change if the given production function is not linear homogeneous but homogeneous of degree n.

[Delhi Univ. B. Com. (H) 1983]

ANSWERS

1. $(MP_L)_{L=1, K=1} = 14$ which means if L is increased to 2 while K is held at 1, the output increases by 14 units.
$(MP_K)_{L=1, K=1} = 6$ which means if K is increased to 2 while L is held at 1, the output increases by 6 units.

2. $MP_L = 20K - 4L, \quad MP_K = 20L - 8K$

3. $MP_L = \dfrac{3-L}{9Q^2 - Q}, \quad MP_K = \dfrac{12-4K}{9Q^2 - Q}$ **4.** $MP_L = 3L^{-1/4} K^{1/4}, \quad MP_K = L^{3/4} K^{-3/4}$

6. 1 **10.** $0 < \alpha, \beta < 1$ and $\alpha + \beta > 1$

14. (*i*) Decreasing returns to scale (*ii*) $\dfrac{3}{4}Q$

15. (*i*) Marginal product of each factor is a decreasing function (*ii*) Decreasing returns to scale.

16. The daily output is increased by 5%; increasing returns to scale

19. (*iii*) nZ; n

8.4 CONSTANT PRODUCT CURVES OR ISOQUANTS

If there are two inputs L and K, production function $P = f(L, K)$ can be represented in a diagram. Put the quantity of labour on the horizontal axis and the quantity of capital on the vertical axis, as in Figure 8.1. Each point in this diagram indicates a particular combination of labour and capital that might be used in production. Any one combination of labour and capital will produce a particular level of output.

When all the different points that lead to the production of the same level of output are connected, they form a curve known as **a constant product curve** or **an isoquant.** The word *isoquant* simply means equal quantities. Since each point in the labour capital diagram represents some level of output, there is an isoquant going through every point in the diagram. Some isoquants are drawn in Figure 8.1 Isoquants toward the northeast represent higher levels of output. The isoquant labelled 300, for example, shows all the different combination of labour and capital that will produce exactly 300 units of output.

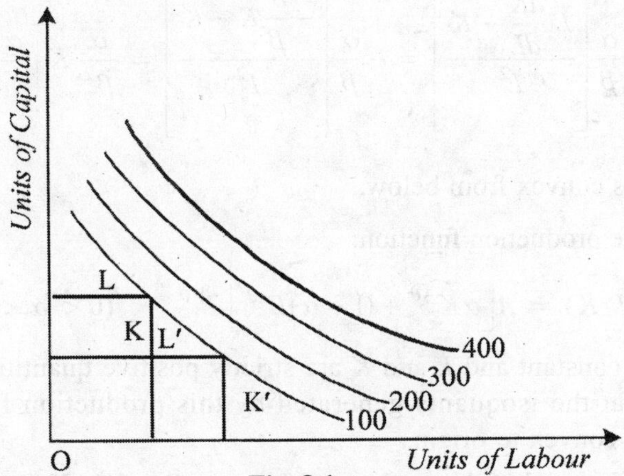

Fig. 8.1

The constant product curves have the following properties:

1. *The constant product curves are downward sloping.*
2. *The constant product curves are convex to the origin.*
3. *The constant product curves never intersect each other.*
4. *Higher isoquants represent higher levels of output.*

Example 25. Show that the isoquants generated by the Cobb-Douglas production function $X = f(L, K) = AL^{\alpha} K^{\beta}$ are negatively sloped and convex from below where A, α and β are positive constants. **[Delhi Univ. B.Com. (H) 2004, 2019]**

Solution. The total differential of the production function $X = f(L, K) = AL^{\alpha} K^{\beta}$ is :

$$dX = \frac{\partial X}{\partial L} dL + \frac{\partial X}{\partial K} dK$$

But $dX = 0$ for movements along an isoquant.

$$\therefore \quad \frac{\partial X}{\partial L} dL + \frac{\partial X}{\partial K} dK = 0 \quad \Rightarrow \quad \frac{dK}{dL} = -\frac{\partial X / \partial L}{\partial X / \partial K}$$

$$\text{But} \quad \frac{\partial X}{\partial L} = \alpha A L^{\alpha - 1} K^{\beta} \quad \text{and} \quad \frac{\partial X}{\partial K} = \beta A L^{\alpha} K^{\beta - 1}$$

$$\therefore \quad \frac{dK}{dL} = -\frac{\alpha A L^{\alpha - 1} K^{\beta}}{\beta A L^{\alpha} K^{\beta - 1}} = -\left(\frac{\alpha}{\beta}\right)\left(\frac{K}{L}\right) < 0$$

Thus each isoquant is downward sloping. Further,

$$\frac{d^2 K}{dL^2} = -\frac{\alpha}{\beta}\left[\frac{L\frac{dK}{dL} - K}{L^2}\right] = -\frac{\alpha}{\beta}\left[\frac{-\frac{\alpha}{\beta}K - K}{L^2}\right] = \frac{\alpha}{\beta^2} K\left[\frac{\alpha + \beta}{L^2}\right] > 0$$

Hence each isoquant is convex from below.

Example 26. Given the production function:

$$Q = f(L, K) = A\left[\alpha K^{-\theta} + (1 - \alpha)L^{-\theta}\right]^{-1/\theta} \qquad (0 < \alpha < 1, \quad -1 < \theta \neq 0)$$

where A is a positive constant and L and K are strictly positive quantities of two factors of production. Show that the isoquants generated by this production function are always negatively sloped and convex to origin.

Solution. The total differential of the production function $Q = f(L, K)$ is:

$$dQ = \frac{\partial Q}{\partial L} dL + \frac{\partial Q}{\partial K} dK$$

However, $dQ = 0$ for movements along an isoquant.

$$\therefore \quad 0 = \frac{\partial Q}{\partial L} dL + \frac{\partial Q}{\partial K} dK \quad \Rightarrow \quad \frac{dK}{dL} = -\frac{\partial Q / \partial L}{\partial Q / \partial K}$$

But $$\frac{\partial Q}{\partial L} = A\left(-\frac{1}{\theta}\right)\left[\alpha K^{-\theta} + (1-\alpha)L^{-\theta}\right]^{-\frac{1}{\theta}-1}(1-\alpha)(-\theta)L^{-\theta-1}$$

$$= A(1-\alpha)L^{-\theta-1}\left[\alpha K^{-\theta} + (1-\alpha)L^{-\theta}\right]^{-\frac{1}{\theta}-1}$$

Similarly, $$\frac{\partial Q}{\partial K} = A\alpha K^{-\theta-1}\left[\alpha K^{-\theta} + (1-\alpha)L^{-\theta}\right]^{-\frac{1}{\theta}-1}$$

$$\therefore \quad \frac{dK}{dL} = -\frac{\partial Q/\partial L}{\partial Q/\partial K} = -\frac{A(1-\alpha)L^{-\theta-1}\left[\alpha K^{-\theta} + (1-\alpha)L^{-\theta}\right]^{-\frac{1}{\theta}-1}}{A\alpha K^{-\theta-1}\left[\alpha K^{-\theta} + (1-\alpha)L^{-\theta}\right]^{-\frac{1}{\theta}-1}}$$

$$= -\left(\frac{1-\alpha}{\alpha}\right)\left(\frac{K}{L}\right)^{\theta+1}, \qquad \text{....(1)}$$

which is negative under given conditions. Thus each isoquant is downward sloping. Further,

$$\frac{d^2K}{dL^2} = -\left(\frac{1-\alpha}{\alpha}\right)(\theta+1)\left(\frac{K}{L}\right)^{\theta}\left[\frac{L\frac{dK}{dL} - K}{L^2}\right]$$

$$= -\left(\frac{1-\alpha}{\alpha}\right)(\theta+1)\left(\frac{K}{L}\right)^{\theta}\frac{1}{L^2}\left[-\left(\frac{1-\alpha}{\alpha}\right)\left(\frac{K}{L}\right)^{\theta+1}L - K\right] \qquad \textit{(using (1))}$$

$$= \left(\frac{1-\alpha}{\alpha}\right)(\theta+1)\left(\frac{K}{L}\right)^{\theta}\frac{K}{L^2}\left[\left(\frac{1-\alpha}{\alpha}\right)\left(\frac{K}{L}\right)^{\theta} + 1\right]$$

which is positive under given conditions. Hence each isoquant is convex to origin.

Marginal Rate of Technical Substitution. *The marginal rate of technical substitution of L for K, denoted by MRTS, is defined as the number of units of K that a producer is willing to sacrifice for an additional unit of L, so as to maintain the same level of output* (*i.e.*, remain on the same isoquant). The marginal rate of technical substitution between L and K is numerically equal to the negative of the slope of an isoquant at a point. If, for a given level to output P_0, the isoquant is given by the equation

$$P_0 = f(L, K),$$

the slope of the tangent at any point on the isoquant is negative. Minus one times the slope of the isoquant is the marginal rate of technical substitution (*MRTS*). Thus

$$MRTS = -\frac{dK}{dL}$$

The total differential of the production function $P = f(L, K)$ is

$$dP = f_L\, dL + f_K\, dK.$$

Here, f_L and f_K, the partial derivatives of P with respect to L and K, are the marginal productivities of labour and capital. But $dP = 0$ for movements along an isoquant. Therefore

$$0 = f_L\, dL + f_K\, dK$$

$$\Rightarrow \qquad f_L\, d_L = -f_K\, dK \quad i.e. \quad -\frac{dK}{dL} = \frac{f_L}{f_K} \quad \text{or} \quad MRTS = \frac{f_L}{f_K}$$

Thus the marginal rate of technical substitution between labour and capital is equal to the ratio of their marginal productivities.

Example 27. For the Cobb-Douglas production function given by $P = AL^{\alpha} K^{1-\alpha}$, find the marginal rate of technical substitution between the factors.

Solution. The marginal product of labour is:

$$MP_L = \frac{\partial P}{\partial L} = \alpha\, AL^{\alpha-1} K^{1-\alpha}$$

The marginal product of capital is:

$$MP_K = \frac{\partial P}{\partial K} = (1-\alpha)\, AL^{\alpha} K^{-\alpha}$$

The marginal rate of technical substitution (*MRTS*) is the ratio of the marginal product of labour to the marginal product of capital. Thus

$$MRTS = \frac{MP_L}{MP_K} = \frac{\alpha\, AL^{\alpha-1} K^{1-\alpha}}{(1-\alpha)\, AL^{\alpha} K^{-\alpha}} = \frac{\alpha}{(1-\alpha)}\left(\frac{K}{L}\right).$$

Example 28. Production function for a commodity is: $x = 2HLK - AL^2 - BK^2$, find out at any point on constant product curve, the marginal rate of substitution of L for K.

[Delhi Univ. B. Com. (H) 1983]

Solution. The marginal rate of substitution of L for K at any point on the constant product curve is numerically equal to the negative of the slope of the tangent at the point. That is,

$$MRTS = -\frac{dK}{dL} = \frac{\partial x / \partial L}{\partial x / \partial K}$$

Now $$\frac{\partial x}{\partial L} = 2HK - 2AL \quad \text{and} \quad \frac{\partial x}{\partial K} = 2HL - 2BK$$

$$\therefore \qquad MRTS = \frac{2HK - 2AL}{2HL - 2BK} = \frac{HK - AL}{HL - BK}.$$

Example 29. For the production function given by

$$x = \sqrt{2Hab - Aa^2 - Bb^2},$$

show that the marginal rate of technical substitution between the factors depends only on the ratio $b : a$ of the factors used.

Solution. Marginal product of a is:

$$MP_a = \frac{\partial x}{\partial a} = \frac{1}{2\sqrt{2Hab - Aa^2 - Bb^2}} . (2Hb - 2Aa) = \frac{(Hb - Aa)}{x}$$

Marginal product of b is :

$$MP_b = \frac{\partial x}{\partial b} = \frac{1}{2\sqrt{2Hab - Aa^2 - Bb^2}} . (2Ha - 2Bb) = \frac{(Ha - Bb)}{x}$$

$$\therefore \qquad MRTS = \frac{MP_a}{MP_b} = \frac{(Hb - Aa)/x}{(Ha - Bb)/x} = \frac{Hb - Aa}{Ha - Bb} = \frac{H(b/a) - A}{H - B(b/a)},$$

which show that $MRTS$ depends only on (b/a).

8.5 ELASTICITY OF SUBSTITUTION

Under normal situation, the constant product curves are convex to the origin. Thus the value of $MRTS$ must increase as L decreases (and K increases) along the constant product curve. The interesting point now is to determine how fast $MRTS$ increases, *i.e.*, to measure the "elasticity" of substitution.

Definition. The **elasticity of substitution** between L and K, denoted by σ, is defined as the ratio of the proportionate change in (K/L) to the proportionate change in the marginal rate of technical substitution ($MRTS$). That is,

$$\sigma = \frac{d(K/L)/K/L}{d(MRTS)/MRTS}$$

Denoting $MRTS$ by r, we have

$$\sigma = \frac{\frac{L}{K} d\left(\frac{K}{L}\right)}{\frac{1}{r} dr} \qquad \text{...(1)}$$

But $$d\left(\frac{K}{L}\right) = \frac{LdK - KdL}{L^2} \qquad \text{...(2)}$$

and $$dr = \frac{\partial r}{\partial L} dL + \frac{\partial r}{\partial K} dK \qquad \text{...(3)}$$

Now $$\frac{dK}{dL} = -\frac{f_L}{f_K} = -r \quad \left(\text{Since } r = \frac{f_L}{f_K}\right) \quad \Rightarrow \quad dK = -rdL$$

Substituting this value of dK in (2) and (3), we obtain

$$d\left(\frac{K}{L}\right) = -\frac{Lr + K}{L^2} dL \quad \text{and} \quad dr = -\left(r\frac{\partial r}{\partial K} - \frac{\partial r}{\partial L}\right) dL$$

Substituting these values in (1), we obtain

$$\sigma = \frac{r(Lr + K)}{LK\left(r\dfrac{\partial r}{\partial K} - \dfrac{\partial r}{\partial L}\right)} \qquad ...(4)$$

This gives the value of σ in terms of r and its partial derivatives. The value of σ can also be expressed in terms of the production function and its partial derivatives. We have

$$\frac{\partial r}{\partial L} = \frac{\partial}{\partial L}\left(\frac{f_L}{f_K}\right) = \frac{f_K f_{LL} - f_L f_{LK}}{f_K^2} \quad \textit{and} \quad \frac{\partial r}{\partial K} = \frac{\partial}{\partial K}\left(\frac{f_L}{f_K}\right) = \frac{f_K f_{LK} - f_L f_{KK}}{f_K^2}$$

Substituting these values in (4), we obtain

$$\sigma = -\frac{f_L f_K (L f_L + K f_K)}{LK(f_{LL} f_K^2 - 2 f_{LK} f_L f_K + f_{KK} f_L^2)} \qquad \textit{(on simplification)}$$

$$= \frac{f_L f_K (L f_L + K f_K)}{LKT}, \qquad ...(5)$$

where $T = -(f_{LL} f_K^2 - 2 f_{LK} f_L f_K + f_{KK} f_L^2)$.

***Remark* 1.** When the production function $P = f(L, K)$ is linear and homogeneous, σ becomes of simpler form:

$$\sigma = \frac{f_L f_K}{P f_{LK}} = \frac{\dfrac{\partial P}{\partial L}\dfrac{\partial P}{\partial K}}{P\dfrac{\partial^2 P}{\partial L\,\partial K}}$$

To get this formula, we have by Euler's Theorem

$$L f_L + K f_K = P = f(L, K)$$

Differentiating this equation partially with respect to L and K, we obtain

$$L f_{LL} + f_L + K f_{LK} = f_L \quad \Rightarrow \quad f_{LL} = -\frac{K}{L} f_{LK}$$

and

$$L f_{LK} + K f_{KK} + f_K = f_K \quad \Rightarrow \quad f_{KK} = -\frac{L}{K} f_{LK}$$

$\therefore$

$$T = -(f_{LL} f_K^2 - 2 f_{LK} f_L f_K + f_{KK} f_L^2)$$

$$= \frac{f_{LK}(L f_L + K f_K)^2}{LK} \qquad \textit{(on simplification)}$$

Hence

$$\sigma = \frac{f_L f_K}{f_{LK}(L f_L + K f_K)} = \frac{f_L f_K}{P f_{LK}}.$$

***Remark* 2.** We referred to the production function to introduce (in a concrete manner) the concepts of marginal rate of technical substitution and the elasticity of substitution, but the concepts are applicable to any other function of two variables.

Example 30. For the Cobb-Douglas production function $Q = AK^{\alpha} L^{1-\alpha}$, find the elasticity of substitution by using the formula $\sigma = \dfrac{\dfrac{\partial Q}{\partial L}\dfrac{\partial Q}{\partial K}}{Q\dfrac{\partial^2 Q}{\partial L\,\partial K}}$. **[*Delhi Univ. B. Com. (H) 1984*]**

Solution. We have $\dfrac{\partial Q}{\partial L} = (1-\alpha)\,AK^{\alpha} L^{-\alpha}$, $\quad \dfrac{\partial Q}{\partial K} = \alpha\, AK^{\alpha-1} L^{1-\alpha}$

and
$$\frac{\partial^2 Q}{\partial L\,\partial K} = \alpha(1-\alpha)AK^{\alpha-1}L^{-\alpha}$$

$$\therefore \qquad \sigma = \frac{\dfrac{\partial Q}{\partial L}\dfrac{\partial Q}{\partial K}}{Q\dfrac{\partial^2 Q}{\partial L\,\partial K}} = \frac{(1-\alpha)\,AK^{\alpha}\,L^{-\alpha}\;\alpha\,AK^{\alpha-1}\,L^{1-\alpha}}{AK^{\alpha}\,L^{1-\alpha}\;\alpha\,(1-\alpha)\,AK^{\alpha-1}\,L^{-\alpha}} = 1.$$

Example 31. Find the marginal rate of technical substitution and the elasticity of substitution for the following production function:

$$x = f(l, k) = \left[\alpha k^{-\theta} + (1-\alpha)\, l^{-\theta}\right]^{-\frac{1}{\theta}}$$

(where x is the total output obtained by using l and k units of labour and capital respectively)

[*Delhi Univ. B. Com. (H) 2003, 2006, 2011, 2018*]

Solution. We have $x = f(l, k) = [\alpha\, k^{-\theta} + (1-\alpha)\, l^{-\theta}]^{-1/\theta}$

The marginal product of labour is :

$$MP_l = \frac{\partial x}{\partial l} = -\frac{1}{\theta}\left[\alpha\, k^{-\theta} + (1-\alpha)\, l^{-\theta}\right]^{-\frac{1}{\theta}-1}(1-\alpha)(-\theta)\, l^{-\theta-1}$$

$$= (1-\alpha)\, l^{-\theta-1}\left[\alpha\, k^{-\theta} + (1-\alpha)\, l^{-\theta}\right]^{\frac{-1-\theta}{\theta}}$$

The marginal product of capital is :

$$MP_k = \frac{\partial x}{\partial k} = -\frac{1}{\theta}\left[\alpha\, k^{-\theta} + (1-\alpha)\, l^{-\theta}\right]^{-\frac{1}{\theta}-1}\alpha(-\theta)\, k^{-\theta-1}$$

$$= \alpha\, k^{-\theta-1}\left[\alpha\, k^{-\theta} + (1-\alpha)\, l^{-\theta}\right]^{\frac{-1-\theta}{\theta}}$$

$$\Rightarrow \qquad r = MRTS = \frac{MP_l}{MP_k} = \left(\frac{1-\alpha}{\alpha}\right)\left(\frac{k}{l}\right)^{\theta+1}$$

To find the elasticity of substitution, we make use of the following formula :

$$\sigma = \frac{r\,d\,(k/l)}{k/l\;dr}$$

Now $$dr = (\theta+1)\left(\frac{1-\alpha}{\alpha}\right)\left(\frac{k}{l}\right)^{\theta} d\,(k/l)$$

$$\therefore \qquad \sigma = \frac{\left(\frac{1-\alpha}{\alpha}\right)\left(\frac{k}{l}\right)^{\theta+1} d\,(k/l)}{\frac{k}{l}(\theta+1)\left(\frac{1-\alpha}{\alpha}\right)\left(\frac{k}{\alpha}\right)^{\theta} d\,(k/l)} = \frac{1}{\theta+1}.$$

Example 32. Show that the following production function has the constant elasticity of substitution:

$$x = f\,(l, k) = ak^{\theta}\,(1-a)\,l^{-1/\theta},$$

where x is total output obtained by using l and k units of labour and capital respectively and a and θ are constants. **[Delhi Univ. B. Com. (H) 2002]**

Solution. The elasticity of substitution is given by :

$$\sigma = \frac{f_l / f_k}{k/l}\cdot\frac{d\,(k/l)}{d(f_l/f_k)}$$

Now, $$f_l = -\frac{1}{\theta}ak^{\theta}\,(1-a)\,l^{-\frac{1}{\theta}-1} \quad \text{and} \quad f_k = \theta\,ak^{\theta-1}(1-a)\,l^{-\frac{1}{\theta}}$$

$$\therefore \qquad \frac{f_l}{f_k} = -\frac{1}{\theta^2}\cdot\frac{k}{l} \quad \Rightarrow \quad d\left(\frac{f_l}{f_k}\right) = -\frac{1}{\theta^2}\,d\left(\frac{k}{l}\right)$$

Hence $$\sigma = \frac{-\frac{1}{\theta^2}\left(\frac{k}{l}\right)}{\frac{k}{l}}\cdot\frac{d\left(\frac{k}{l}\right)}{-\frac{1}{\theta^2}\,d\left(\frac{k}{l}\right)} = 1.$$

Example 33. If r denotes the marginal rate of technical substitution between the factors for the production function $x = f\,(a, b)$, show that the elasticity of substitution (σ) between the factors is given by

$$\sigma = \frac{r}{ab} \cdot \frac{ar+b}{r\dfrac{\partial r}{\partial b} - \dfrac{\partial r}{\partial a}}$$

Hence or otherwise, find the value of σ for the production function $x = 20 - a^{-1} - b^{-1}$.

Solution. By definition,

$$\sigma = \frac{\dfrac{d(b/a)}{b/a}}{dr/r} = \frac{r}{b/a} \cdot \frac{d(b/a)}{dr}, \qquad \text{...(1)}$$

where $\quad r = MRTS = -\dfrac{db}{da}.$

Now $\quad d(b/a) = \dfrac{a\,db - b\,da}{a^2} = \dfrac{-ra\,da - b\,da}{a^2} = -\left(\dfrac{ra+b}{a^2}\right)da \qquad \text{...(2)}$

and $\quad dr = \dfrac{\partial r}{\partial a}da + \dfrac{\partial r}{\partial b}db = \dfrac{\partial r}{\partial a}da - r\dfrac{\partial r}{\partial a}da = -\left(r\dfrac{\partial r}{\partial b} - \dfrac{\partial r}{\partial a}\right)da \qquad \text{...(3)}$

Substituting these values in (1), we obtain

$$\sigma = \frac{r}{ab} \cdot \frac{ar+b}{r\dfrac{\partial r}{\partial b} - \dfrac{\partial r}{\partial a}} \qquad \text{...(4)}$$

To derive the value of σ for the function $x = 20 - a^{-1} - b^{-1}$, we have

$$\frac{\partial x}{\partial a} = a^{-2}, \qquad \frac{\partial x}{\partial b} = b^{-2},$$

and hence $\quad r = \dfrac{\partial x/\partial a}{\partial x/\partial b} = \dfrac{a^{-2}}{b^{-2}} = \left(\dfrac{b}{a}\right)^2$

$$\therefore \quad \frac{\partial r}{\partial a} = 2(b/a)(-b/a^2) = -2(b^2/a^3)$$

and $\quad \dfrac{\partial r}{\partial b} = 2(b/a)(1/a) = 2(b/a^2)$

Substituting the values of r, $\dfrac{\partial r}{\partial a}$ and $\dfrac{\partial r}{\partial b}$ in (4), we obtain

$$\sigma = \frac{(b/a)^2}{ab} \cdot \frac{a(b/a)^2 + b}{(b/a)^2\, 2(b/a^2) + 2(b^2/a^3)} = \frac{1}{2} \qquad \textit{(on simplification)}$$

EXERCISE 8.3

1. For the production function given by $x = f(a,b) = 2\,Hab - Aa^2 - Bb^2$. show that the marginal rate of technical substitution between the factors depends only on the ratio b/a of the factors used.

2. If $x = f(a, b)$ is a linear homogeneous production function, show that the marginal rate of substitution between the factors depends only on the ratio b/a of the factors used. Verify the same for the production function given by

$$x = f(a, b) = \frac{2\,Hab - Aa^2 - Bb^2}{Ca + Db}$$

3. For the production function given by $x = \sqrt{2\,Hlk - Al^2 - Bk^2}$, show that the elasticity of substitution (σ) is : $\sigma = \dfrac{Hx^2}{(H^2 - AB)lk} - 1.$

4. Show that the production function $Q^{-\beta} = aK^{-\beta} + bL^{-\beta}$ is homogeneous of degree one, where Q represents output; L and K denote factors of production, Labour and Capital respectively. Also obtain elasticity of substitution (σ) between factors defined by $\sigma = f_L f_K / Q f_{LK}$, where

$$f_L = \frac{\partial Q}{\partial L}, \quad f_K = \frac{\partial Q}{\partial K}, \quad f_{LK} = \frac{\partial^2 Q}{\partial L\, \partial K}$$

[*Delhi Univ. B. Com. (H) 1985, 2022*]

5. Find the marginal rate of technical substitution and the elasticity of substitution for the production function : $Q = f(L, K) = \dfrac{LK}{L + K}$ **[*Delhi Univ. B. Com. (H) 1998*]**

6. Find the elasticity of substitution for the production function:

$$Q = 75(0.3K^{-0.4} + 0.7L^{-0.4})^{-2.5}$$ **[*Delhi Univ. B. A. (Econ. Hons.) 1986*]**

7. Find the elasticity of substitution for the production function:

$$Q = \left(\frac{0.2}{K} + \frac{0.8}{L}\right)^{-1}$$ **[*Delhi Univ. B.Com. (H) 2019*]**

8. Show that the production function given below has constant elasticity of substitution :

$$x = f(l, k) = \left[\alpha k^{-\theta} + (1 - \alpha) l^{-\theta}\right]^{-1/\theta},$$

where x is the total output obtained by using l and k units of labour and capital respectively and α and θ are constants. **[*Delhi Univ. B.Com. (H) 2001*]**

9. Find the elasticity of substitution for the production function:

$$x = f(a, b) = 20 - a^{-2} - b^{-2}$$

10. If the production function is : $X = 2ALK - BL^2 - CK^2$, find $\dfrac{dK}{dL}$ and $\dfrac{d^2K}{dL^2}$. Under what condition the isoquants are downward sloping and convex to the origin ?

[Delhi Univ. B. Com. (H) 2008]

11. Show that the isoquants generated by the production function $Q = f(L, K) = \dfrac{LK}{L+K}$ are always negatively sloped and convex from below.

ANSWERS

4. $\sigma = \dfrac{1}{\beta + 1}$ **5.** $MRTS = (K/L)^2$ and $\sigma = 1/2$ **6.** $\sigma = 5/7$

7. $\sigma = 1/2$ **8.** $\sigma = \dfrac{1}{\theta + 1}$ **9.** $\sigma = 1/3$

10. $\dfrac{dK}{dL} = -\dfrac{AK - BL}{AL - CK}$, $\dfrac{d^2K}{dL^2} = \dfrac{(AL - CK)\left(B - A\dfrac{dK}{dL}\right) - (BL - AK)\left(A - C\dfrac{dK}{dL}\right)}{(AL - CK)^2}$

8.6 APPLIED OPTIMIZATION PROBLEMS

This section is concerned with the applications of maxima and minima of functions of two variables to applied business problems. For example, a firm producing two types of products might be interested to find the number of units produced of each in order to minimize the total joint-cost. Similarly, a monopolist practicing price discrimination by selling the same product in two different markets at different prices might be interested in finding the prices to the charged in each market so as to maximize his total profit.

Example 34. (*Minimizing Joint-Cost*) The joint cost function of a firm producing two products is given by

$$C = C(x, y) = 6x^2 - 9x - 3xy - 7y + 5y^2 + 20,$$

where x and y denote their units. Find the values of x and y that minimize $C(x, y)$. Also find the minimum cost. **[Delhi Univ. B.Com. (H) 2014]**

Solution. To minimize C with respect to x and y, we set its partial derivatives equal to 0. Thus

$$\frac{\partial C}{\partial x} = 12x - 9 - 3y = 0 \qquad ...(1)$$

$$\frac{\partial C}{\partial y} = -3x - 7 + 10y = 0 \qquad ...(2)$$

Solving (1) and (2) for x and y yield $x = 1$ and $y = 1$.

Also $$\frac{\partial^2 C}{\partial x^2} = 12, \quad \frac{\partial^2 C}{\partial x\, \partial y} = -3, \quad \frac{\partial^2 C}{\partial y^2} = 10$$

$\therefore$ At (1, 1) $$A = \frac{\partial^2 C}{\partial x^2} = 12, \quad B = \frac{\partial^2 C}{\partial x \partial y} = -3, \quad C = \frac{\partial^2 C}{\partial y^2} = 10$$

$$\Rightarrow \qquad AC - B^2 = (12)(10) - (-3)^2 = 111 > 0$$

Since $A = 12 > 0$, we conclude that C is minimum when $x = 1$ and $y = 1$. And the minimum cost is :

$$C(1, 1) = 6(1)^2 - 9(1) - 3(1)(1) - 7(1) + 5(1)^2 + 20 = 12.$$

Example 35. The demand function and the total cost function of a monopolist are as follows:

$$p = 100 - 3q + 4A^{1/2} \quad \text{and} \quad C = 4q^2 + 10q + A,$$

where A is the level of advertisement expenditure. Final the values of A, p and q to maximize profit. **[*Delhi Univ. B.Com. (H) 2005, 2008, 2018*]**

Solution. The demand function and the total cost functions of a monopolist are :

$$p = 100 - 3q + 4A^{1/2} \quad \text{and} \quad C = 4q^2 + 10q + A$$

The revenue function R is :

$$R = pq = 100q - 3q^2 + 4qA^{1/2}$$

and the profit function P is :

$$P = R - C = 100\,q - 3q^2 + 4q\,A^{1/2} - 4q^2 - 10q - A$$
$$= 90q - 7q^2 + 4qA^{1/2} - A$$

$$\Rightarrow \qquad \frac{\partial P}{\partial q} = 90 - 14q + 4\sqrt{A} \quad \text{and} \quad \frac{\partial P}{\partial A} = 2qA^{-1/2} - 1$$

To maximize P, we set its partial derivatives equal to 0.

Thus $$\frac{\partial P}{\partial q} = 0 \quad \Rightarrow \quad 90 - 14q + 4\sqrt{A} = 0 \qquad ...(1)$$

and $$\frac{\partial P}{\partial A} = 0 \quad \Rightarrow \quad 2qA^{-1/2} - 1 = 0 \qquad ...(2)$$

Solving Equations (1) and (2) for q and A, we get $q = 15$ and $A = 900$. Substituting the values of q and A in the demand function, we get $p = 175$.

Also, $$\frac{\partial^2 P}{\partial q^2} = -14, \quad \frac{\partial^2 P}{\partial q \partial A} = \frac{2}{\sqrt{A}}, \quad \text{and} \quad \frac{\partial^2 P}{\partial A^2} = -qA^{-3/2}$$

$$\therefore \text{ At}(q = 15,\ A = 900), \quad A = -14, \quad B = \frac{1}{15}, \quad C = -\frac{1}{1800}$$

$$\Rightarrow \qquad AC - B^2 = \frac{14}{1800} - \frac{1}{225} = \frac{14-8}{1800} = \frac{6}{1800} = \frac{1}{300} > 0$$

and since $A = -14$ is negative, we conclude that P is maximum when $q = 15$, $p = 175$ and $A = 900$.

Two Product Firm. Sometimes a monopolist or a firm produces two products with known market demands. The firm, in that case, is interested to find how many items of each be produced to have the joint profit maximum. We shall illustrate this with the help of following examples.

Example 36. A manufacturer is planning to sell a new product at a price of ₹ 350 per unit and estimates that if x thousand rupees is spent on development and y thousand rupees is spent on advertisement, consumer will buy approximately $\left[\frac{250y}{y+2} + \frac{100x}{x+5}\right]$ units of the product. If manufacturing costs for this product are ₹ 150 per unit, how much should the manufacturer spend on development and how much on advertisement to generate the largest profit.

Solution. It is given that each unit costs the manufacturer ₹ 150 and can be sold for ₹ 350, therefore the profit per unit is ₹ 200. Hence the total profit is given by

$$200\left(\frac{250y}{y+2} + \frac{100x}{x+5}\right)$$

However, this profit is achieved by making an expenditure of 1000 $(x + y)$ on development and advertisement. Thus the net profit, P, is given by

$$P = 200\left(\frac{250y}{y+2} + \frac{100x}{x+5}\right) - 1000\,(x + y)$$

$$\Rightarrow \qquad \frac{\partial P}{\partial x} = 200\left[\frac{(x+5)(100) - 100x}{(x+5)^2}\right] - 1000 = \frac{1,00,000}{(x+5)^2} - 1000$$

and $$\frac{\partial P}{\partial y} = 200\left[\frac{(y+2)(250) - 250y}{(y+2)^2}\right] - 1000 = \frac{1,00,000}{(y+2)^2} - 1000$$

To maximize P, we set its partial derivatives equal to zero. Thus

$$\frac{1,00,000}{(x+5)^2} - 1000 = 0 \quad \text{and} \quad \frac{1,00,000}{(y+2)^2} - 1000 = 0$$

$\Rightarrow \quad x = 5 \quad$ and $\quad y = 8$ *(on simplification)*

Also,
$$\frac{\partial^2 P}{\partial x^2} = -\frac{2,00,000}{(x+5)^3}, \quad \frac{\partial^2 P}{\partial x \partial y} = 0, \quad \frac{\partial^2 P}{\partial y^2} = -\frac{2,00,000}{(y+2)^3}$$

$\therefore \quad$ At $x = 5,\ y = 8,\quad A = -200,\quad B = 0,\quad C = -200$

$\Rightarrow \quad AC - B^2 = (-200)(-200) = 40,000 > 0$ and since $A < 0$, we conclude that profit is maximum when $x = 5$ and $y = 8$.

Hence the manufacturer should allocate ₹ 5000 on development and ₹ 8000 on advertisement to get maximum net profit.

Example 37. A firm produces two items X_1 and X_2. The market prices are given by $p_1 = 100 - 2x_1$ and $p_2 = 125 - 3x_2$. The cost of production is $12x_1 + 11x_2 + 4x_1x_2$ for producing X_1 and X_2 items. Find how many items of each should be produced to have the joint profit maximum ? ***[Delhi Univ. B.Com. (H) 2008, 2012, 2014]***

Solution. Let R_1 and R_2 denote the revenues obtained from X_1 and X_2 respectively. Then

$$R_1 = x_1 p_1 = x_1(100 - 2x_1) = 100x_1 - 2x_1^2$$

and
$$R_2 = x_2 p_2 = x_2(125 - 3x_2) = 125x_2 - 3x_2^2$$

Thus the joint revenue, R, is given by

$$R = R_1 + R_2 = 100x_1 - 2x_1^2 + 125x_2 - 3x_2^2$$

Let C denote the total cost function, then

$$C = 12x_1 + 11x_2 + 4x_1x_2$$

Therefore, the profit function of the firm is :

$$P = R - C$$
$$= (100x_1 - 2x_1^2 + 125x_2 - 3x_2^2) - (12x_1 + 11x_2 + 4x_1x_2)$$
$$= 88x_1 + 114x_2 - 2x_1^2 - 4x_1x_2 - 3x_2^2$$

$$\therefore \quad \frac{\partial P}{\partial x_1} = 88 - 4x_1 - 4x_2 \quad \text{and} \quad \frac{\partial P}{\partial x_2} = 114 - 4x_1 - 6x_2$$

Setting these equal to zero, we obtain

$$88 - 4x_1 - 4x_2 = 0 \qquad ...(1)$$

$$114 - 4x_1 - 6x_2 = 0 \qquad ...(2)$$

Solving (1) and (2) for x_1 and x_2, we obtain $x_1 = 9$ and $x_2 = 13$.

Also $$\frac{\partial^2 P}{\partial x_1^2} = -4, \quad \frac{\partial^2 P}{\partial x_1 \partial x_2} = -4, \quad \frac{\partial^2 P}{\partial x_2^2} = -6$$

Therefore, at (9, 13), we have $A = -4$, $B = -4$, $C = -6$. Thus $AC - B^2 = 8 > 0$, and since $A = -4 < 0$, we conclude that P is maximum when $x_1 = 9$ and $x_2 = 13$.

Example 38. A multiproduct firm produces two commodities X_1 and X_2 whose prices per unit are 12 and 18 respectively. Assuming the firm's cost function as :

$$C = 2x_1^2 + x_1x_2 + 2x_2^2$$

Obtain the optimum levels of its products which maximize profit.

[*Delhi Univ. B.Com. (H) 1985*]

Solution. Let x_1 and x_2 denote the number of units produced and sold of each commodity. Then the revenue function of the firm is :

$$R = 12x_1 + 18x_2$$

Also, the total cost function is:

$$C = 2x_1^2 + x_1x_2 + 2x_2^2$$

$\therefore$ The profit function of the firm is given by

$$P = R - C = (12x_1 + 18x_2) - (2x_1^2 + x_1x_2 + 2x_2^2)$$

$$= 12x_1 + 18x_2 - 2x_1^2 - x_1x_2 - 2x_2^2$$

Differentiating P partially with respect to x_1 and x_2, we obtain

$$\frac{\partial P}{\partial x_1} = 12 - 4x_1 - x_2 \quad \text{and} \quad \frac{\partial P}{\partial x_2} = 18 - x_1 - 4x_2$$

To maximize P, we set its partial derivatives equal to zero. Thus

$$12 - 4x_1 - x_2 = 0 \qquad ...(1)$$

$$18 - x_1 - 4x_2 = 0 \qquad ...(2)$$

Solving (1) and (2) for x_1 and x_2, we obtain $x_1 = 2$ and $x_2 = 4$,

Also, $\dfrac{\partial^2 P}{\partial x_1^2} = -4, \quad \dfrac{\partial^2 P}{\partial x_1 \partial x_2} = -1, \quad \dfrac{\partial^2 P}{\partial x_2^2} = -4$

$\therefore$ At (2,4) $A = -4, \quad B = -1, \quad C = -4$

Hence $AC - B^2 = 15 > 0$, and since $A = -4 < 0$, we thus conclude that profit is maximum when $x_1 = 2$ and $x_2 = 4$.

Example 39. A monopolist firm produces commodities of two types X_1 and X_2 at constant average cost ₹ 2.50 and ₹ 3.00 per item respectively. If p_1 and p_2 are the prices charged and the market demands are $x_1 = 5(p_2 - p_1)$ and $x_2 = 32 + 5p_1 - 10p_2$, find the prices of two commodities for maximum joint monopoly profit. **[Delhi Univ. B.Com. (H) 1984, 2019]**

Solution. Let R denote the total revenue function, then

$$R = p_1x_1 + p_2x_2 = p_1(5p_2 - 5p_1) + p_2(32 + 5p_1 - 10p_2)$$

$$= 10p_1p_2 - 5p_1^2 - 10p_2^2 + 32p_2$$

Let C_1 denote the cost of producing x_1 units of X_1 and let C_2 denote the cost of producing x_2 units of x_2, then we have

$$\frac{C_1}{x_1} = 2.50 \Rightarrow C_1 = 2.5x_1 \quad \text{and} \quad \frac{C_2}{x_2} = 3.00 \Rightarrow C_2 = 3x_2$$

Thus the joint cost function of the firm is :

$$C = C_1 + C_2 = 2.5x_1 + 3x_2 = 2.5(5p_2 - 5p_1) + 3(32 + 5p_1 - 10p_2)$$
$$= 2.5p_1 - 17.5p_2 + 96$$

Hence the joint profit is given by

$$P = R - C = (10p_1p_2 - 5p_1^2 - 10p_2^2 + 32p_2) - (2.5p_1 - 17.5p_2 + 96)$$

$$= 10p_1p_2 - 5p_1^2 - 10p_2^2 - 2.5p_1 + 49.5p_2 - 96$$

Setting partial derivatives of P equal to 0, we obtain

$$\frac{\partial P}{\partial p_1} = 10p_2 - 10p_1 - 2.5 = 0 \quad \text{...(1)}$$

$$\frac{\partial P}{\partial p_2} = 10p_1 - 20p_2 + 49.5 = 0 \quad \text{...(2)}$$

Solving (1) and (2) for p_1 and p_2 yield $p_1 = 4.45$ and $p_2 = 4.70$.

Now $\dfrac{\partial^2 P}{\partial p_1^2} = -10, \quad \dfrac{\partial^2 P}{\partial p_1 \partial p_2} = 10, \quad \dfrac{\partial^2 P}{\partial p_2^2} = -20$

Thus at (4.45, 4.70), $A = -10$, $B = 10$, $C = -20$. Hence $AC - B^2 = 100 > 0$ and since $A = -10 < 0$, we conclude that P is maximum when $p_1 = 4.45$ and $p_2 = 4.70$.

Discriminating Monopoly. When a firm charges a different price to different consumers for the same product, the firm is practicing *price discrimination.* Suppose a monopolist is practicing price discrimination by selling the same product in two separate markets at different prices. Let x_1 be the number of units sold in market *I*, where the demand function is $p_1 = f(x_1)$, and let x_2 be the number of units sold in market *II*, where the demand function is $p_2 = g(x_2)$. Then the revenue functions for the two markets are :

$$R_1 = x_1 p_1 = x_1 f(x_1) \quad \text{and} \quad R_2 = x_2 p_2 = x_2 g(x_2)$$

Assume that all units are produced at one plant and let the cost function for producing x (= $x_1 + x_2$) units be $C = C(x)$.

The monopolist's profit is : $P = R_1 + R_2 - C$,

which can be maximized by applying the usual technique of maxima and minima of functions of two variables.

Let us consider some examples.

Example 40. Suppose a monopolist is practicing price discrimination in the sale of a product by charging different prices in two separate markets.In market *I* the demand function is $p_1 = 100 - x_1$, and in *II* it is $p_2 = 84 - x_2$, where x_1 and x_2 are the quantities sold per week in *I* and *II*, and p_1 and p_2 are the respective prices per unit. If the monopolist's cost function is $C = 600 + 4(x_1 + x_2)$, how much should be sold in each market to maximize profit ? What selling prices give this maximum profit ? Find the maximum profit.

Solution. Let R_1 and R_2 denote the revenue functions in the two markets. Then

$$R_1 = p_1 x_1 = (100 - x_1)\,x_1 = 100x_1 - x_1^2$$

and
$$R_2 = p_2 x_2 = (84 - x_2)\,x_2 = 84x_2 - x_2^2$$

The cost function is given to be : $C = 600 + 4(x_1 + x_2)$

Therefore, the monopolist's profit function is :

$$\begin{aligned} P &= R_1 + R_2 - C \\ &= (100x_1 - x_1^2) + (84x_2 - x_2^2) - (600 + 4x_1 + 4x_2) \\ &= 96x_1 - x_1^2 + 80x_2 - x_2^2 - 600 \qquad ...(1) \end{aligned}$$

To maximize P, we set its partial derivatives equal to zero. That is,

$$\frac{\partial P}{\partial x_1} = 96 - 2x_1 = 0 \quad \Rightarrow \quad x_1 = 48$$

and
$$\frac{\partial P}{\partial x_2} = 80 - 2x_2 = 0 \quad \Rightarrow \quad x_2 = 40$$

Also,
$$\frac{\partial^2 P}{\partial x_1^2} = -2, \quad \frac{\partial^2 P}{\partial x_1\,\partial x_2} = 0, \quad \frac{\partial^2 P}{\partial x_2^2} = -2$$

At (48, 40), we have $A = -2$, $B = 0$, $C = -2$. Thus, $AC - B^2 = 4 > 0$, and since $A < 0$, we conclude that the monopolist's profit is maximum when quantities sold per week in markets *I* and *II* are 48 units and 40 units respectively. Substituting $x_1 = 48$ and $x_2 = 40$ in the demand functions, we obtain the corresponding selling prices:

$$p_1 = 100 - 48 = 52 \quad \text{and} \quad p_2 = 84 - 40 = 44$$

Substituting $x_1 = 48$ and $x_2 = 40$ into Equation (1), we obtain the maximum profit:

$$P = 96\,(48) - (48)^2 + 80\,(40) - (40)^2 - 600 = 3304.$$

Example 41. A discriminating monopolist, serving two isolated markets with demand elasticities η_1 and η_2, sells quantities x_1 and x_2 of his product, at prices p_1 and p_2 respectively. Show that for maximum profit,

$$p_1 / p_2 = \left(1 - \frac{1}{\eta_2}\right) \Big/ \left(1 - \frac{1}{\eta_1}\right)$$

Hence deduce that the price will be higher in the market where the demand is less elastic.

Solution. Let R_1 and R_2 denote the revenue functions in the two markets, then

$$R_1 = p_1 x_1 \quad \text{and} \quad R_2 = p_2 x_2$$

Keep in mind that R_1 is a function of x_1 and R_2 is a function of x_2. Let $C = C(x)$ denote the cost function for producing $x\ (= x_1 + x_2)$ units. Then the monopolist's profit P is

$$P = R_1 + R_2 - C$$

Differentiating P partially with respect to x_1, we obtain

$$\frac{\partial P}{\partial x_1} = \frac{dR_1}{dx_1} + 0 - \frac{\partial C}{\partial x_1} = \frac{dR_1}{dx_1} - \frac{dC}{dx}\frac{\partial x}{\partial x_1} \qquad \textit{(Chain Rule)}$$

$$= \frac{dR_1}{dx_1} - \frac{dC}{dx} \qquad \textit{(since } x = x_1 + x_2\textit{)}$$

$$= MR_1 - MC$$

Similarly, differentiating P partially with respect to x_2 gives

$$\frac{\partial P}{\partial x_2} = MR_2 - MC$$

To maximize P, we set $\frac{\partial P}{\partial x_1}$ and $\frac{\partial P}{\partial x_2}$ equal to zero. Thus, we have

$$\frac{\partial P}{\partial x_1} = MR_1 - MC = 0 \quad \Rightarrow \quad MR_1 = MC \qquad \text{...(1)}$$

and

$$\frac{\partial P}{\partial x_2} = MR_2 - MC = 0 \quad \Rightarrow \quad MR_2 = MC \qquad \text{...(2)}$$

From Equations (1) and (2), we get $MR_1 = MC = MR_2$. Hence, to maximize profit it is necessary to charge prices so that the marginal revenues in both markets will be the same.

But $$MR_1 = p_1\left(1-\frac{1}{\eta_1}\right) \quad \text{and} \quad MR_2 = p_2\left(1-\frac{1}{\eta_2}\right)$$

$\therefore$ For maximum profit, it is necessary that

$$p_1\left(1-\frac{1}{\eta_1}\right) = p_2\left(1-\frac{1}{\eta_2}\right) \quad \Rightarrow \quad \frac{p_1}{p_2} = \frac{\left(1-\frac{1}{\eta_2}\right)}{\left(1-\frac{1}{\eta_1}\right)}$$

Now, assume that the demand is less elastic in market *I*, *i.e.*, $\eta_1 < \eta_2$.

$$\Rightarrow \quad \frac{1}{\eta_1} > \frac{1}{\eta_2} \quad i.e. \quad -\frac{1}{\eta_1} < -\frac{1}{\eta_2} \quad \Rightarrow \quad \left(1-\frac{1}{\eta_1}\right) < \left(1-\frac{1}{\eta_2}\right)$$

Thus $$\frac{\left(1-\frac{1}{\eta_2}\right)}{\left(1-\frac{1}{\eta_1}\right)} > 1 \quad \Rightarrow \quad \frac{p_1}{p_2} > 1 \quad \text{or} \quad p_1 > p_2 .$$

Hence the price will be higher in the market where the demand is less elastic.

Example 42. A monopolist charges different prices in the two markets where his demand functions are $x_1 = 21 - 0.1p_1$ and $x_2 = 50 - 0.4p_2$, p_1 and p_2, being prices and x_1 and x_2 quantities demanded. His total cost function is $TC = 10x + 2000$, where x is total output. Find the prices that the monopolist should charge to maximize his profit. Also, verify that higher price will be charged in the market having the lower price elasticity of demand.

[*Delhi Univ. B.Com. (H) 1990, 2017 (modified)*]

Solution. Let R_1 and R_2 denote the revenue functions in the two markets, then

$$R_1 = p_1x_1 = p_1(21 - 0.1p_1) = 21p_1 - 0.1p_1^2$$

and $$R_2 = p_2x_2 = p_2(50 - 0.4p_2) = 50p_2 - 0.4p_2^2$$

The total cost function is:

$$\begin{aligned} TC &= 10x + 2000 = 10(x_1 + x_2) + 2000 \\ &= 10[(21 - 0.1p_1) + (50 - 0.4p_2)] + 2000 \\ &= 2710 - p_1 - 4p_2 \end{aligned}$$

∴ The monopolist's profit function is:

$$P = R_1 + R_2 - TC$$

$$= (21\,p_1 - 0.1\,p_1^2) + (50\,p_2 - 0.4\,p_2^2) - (2710 - p_1 - 4\,p_2)$$

$$= 22p_1 + 54p_2 - 0.1p_1^2 - 0.4p_2^2 - 2710$$

$$\therefore \qquad \frac{\partial P}{\partial p_1} = 22 - 0.2p_1 \quad \text{and} \quad \frac{\partial P}{\partial p_2} = 54 - 0.8p_2$$

For maximum profit, it is necessary that $\dfrac{\partial P}{\partial p_1} = 0$ and $\dfrac{\partial P}{\partial p_2} = 0$. That is,

$$22 - 0.2p_1 = 0 \quad \Rightarrow \quad p_1 = 110 \text{ and } 54 - 0.8p_2 = 0 \quad \Rightarrow \quad p_2 = 67.5$$

Also, $\dfrac{\partial^2 P}{\partial p_1^2} = -0.2, \quad \dfrac{\partial^2 P}{\partial p_1\,\partial p_2} = 0, \quad \dfrac{\partial^2 P}{\partial p_2^2} = -0.8$

∴ At (110, 67.5), $A = -0.2 \quad B = 0, \quad C = -0.8$. Thus $AC - B^2 = 0.16 > 0$, and since $A < 0$, we conclude that profit is maximum when the price charged in the market *I* is 110 and that in the market *II* is 67.5.

We'll now show that the price elasticity of demand in the market *I* is less than the price elasticity of demand in the market *II*. If η_1 and η_2 are the price elasticities of demand in markets *I* and *II* respectively, then

$$\eta_1 = -\frac{p_1}{x_1}\frac{d\,x_1}{d\,p_1} = -\frac{110}{[21 - 0.1\,(110)]}(-0.1) = \frac{11}{10} \qquad ...(1)$$

and
$$\eta_2 = -\frac{p_2}{x_2}\frac{d\,x_2}{d\,p_2} = -\frac{67.5}{[50 - 0.4\,(67.5)]}(-0.4) = \frac{27}{23} \qquad ...(2)$$

From Equations (1) and (2), we see that $\eta_1 < \eta_2$.

Example 43. A monopolist discriminates prices between two markets and the average revenue functions for the two markets are :

$$AR_1 = 53 - 4Q_1 \quad \text{and} \quad AR_2 = 29 - 3Q_2$$

and his total cost function is $C = 20 + 5Q$. Find the profit maximizing outputs and prices in the two markets. What are these values when there is no price discrimination?

[*Delhi Univ. B.Com. (H) 2010*]

Solution. Let R_1 and R_2 denote the revenue functions in the two markets, then

$$R_1 = 53Q_1 - 4Q_1^2 \quad \text{and} \quad R_2 = 29Q_2 - 3Q_2^2$$

$\therefore$ $$R = R_1 + R_2 = 53Q_1 + 29Q_2 - 4Q_1^2 - 3Q_2^2$$

Also, $$C = 20 + 5Q = 20 + 5Q_1 + 5Q_2$$

$\Rightarrow$ $$P = R - C = 53Q_1 + 29Q_2 - 4Q_1^2 - 3Q_2{}^2 - 20 - 5Q_1 - 5Q_2$$

$$= 48Q_1 + 24Q_2 - 4Q_1^2 - 3Q_2^2 - 20$$

$\therefore$ $$\frac{\partial P}{\partial Q_1} = 48 - 8Q_1 \quad \text{and} \quad \frac{\partial P}{\partial Q_2} = 24 - 6Q_2$$

To maximize P, we set these partial derivatives equal to zero. Thus, we have

$$\frac{\partial P}{\partial Q_1} = 48 - 8Q_1 = 0 \quad \Rightarrow \quad Q_1 = 6$$

and $$\frac{\partial P}{\partial Q_2} = 24 - 6Q_2 = 0 \quad \Rightarrow \quad Q_2 = 4$$

Also, $$\frac{\partial^2 P}{\partial Q_1^2} = -8, \qquad \frac{\partial^2 P}{\partial Q_1 \partial Q_2} = 0, \qquad \frac{\partial^2 P}{\partial Q_2^2} = -6$$

$\therefore$ At (6, 4), $A = -8, \quad B = 0, \quad C = -6$

Thus $AC - B^2 = 48 > 0$ and since $A = -8 < 0$, we conclude that P is maximum when $Q_1 = 6$ and $Q_2 = 4$. The corresponding prices are then $p_1 = 29$ and $p_2 = 17$. When there is no discrimination, $p_1 = p_2 = p$, (say)

$\therefore$ $$Q_1 = \frac{53}{4} - \frac{p}{4} \quad \text{and} \quad Q_2 = \frac{29}{3} - \frac{p}{3}$$

$\Rightarrow$ $$Q = Q_1 + Q_2 = \frac{53}{4} - \frac{p}{4} + \frac{29}{3} - \frac{p}{3} = \frac{275}{12} - \frac{7p}{12}$$

$\Rightarrow$ $$P = R - C = pQ - C = \frac{275p}{12} - \frac{7p^2}{12} - 20 - \frac{1375}{12} + \frac{35p}{12}$$

$\Rightarrow$ $$\frac{dP}{dp} = \frac{275}{12} - \frac{14p}{12} + \frac{35}{12}$$

For P to be maximum, $\dfrac{dP}{dp} = 0$. This gives

$$\frac{275}{12} - \frac{14p}{12} + \frac{35}{12} = 0 \quad \Rightarrow \quad p = \frac{310}{14} = 22.14$$

Also, $\dfrac{d^2P}{dp^2} = -\dfrac{14}{12} < 0$. Thus profit P is maximum when $p = 22.14$ and

$Q_1 = \dfrac{53}{4} - \dfrac{155}{28} = 7.714$ and $Q_2 = \dfrac{29}{3} - \dfrac{155}{21} = 2.285.$

EXERCISE 8.4

1. The production of a certain commodity requires x machine hours and y man hours, and the cost of production per unit is given by

$$C = 2x^3 - 6xy + y^2 + 500$$

Determine x and y in order to minimize C.

2. Given the following least squares function :

$$S = f(a, b) = [50 - (5a + b)]^2 + [30 - (10a + b)]^2 + [20 - (15a + b)]^2$$

You are required to determine the values of a and b which will minimize the sum of squares, S. Also calculate the minimum value of S. ***[Delhi Univ. B. Com. (H) 1994]***

3. A company manufactures two types of typewriters—electrical (E) and manual (M). The revenue function of the company, in units of thousands, is :

$$R = 8E + 5M + 2EM - E^2 - 2M^2 + 20,$$

where E represents the units of electrical typewriter in thousands and M represents manual typewriter in thousands. Determine the quantity of electrical and manual typewriters which lead to maximum revenue. Also calculate the maximum revenue. ***[Delhi Univ. B. Com. (H) 1995]***

4. A firm sells two products A and B. Their joint demand functions are

$$x_1 = 175 - 4p_1 - p_2 \quad \text{and} \quad x_2 = 90 - 2p_1 - 3p_2$$

where x_1 and x_2 are the units demanded of two products when their market prices are ₹ p_1 and ₹ p_2 per unit respectively. Determine the prices which should be charged to maximize total revenue of the two products. Also, find the maximized revenue.

[Delhi Univ. B.Com. (H) 2005]

5. The demand functions of two commodities X_1 and X_2 are $p_1 = 16 - x_1^2$ and $p_2 = 8 - x_2$ and the joint cost function is $C = 10 + 4x_1 + 2x_2$. Determine the quantities that maximize the profit of the monopolist and also find the maximum profit.

6. A monopolist sells two products X and Y for which the demand is $x = 50 - 0.5p_x$ and $y = 76 - p_y$. The combined cost function is $C = 3x^2 + 2xy + 2y^2 + 55$. Find

(*i*) the profit maximizing level of output and price for each product, and

(*ii*) the maximum profit. ***[Delhi Univ. B. Com. (H) 1993]***

7. A monopolist produces two commodities X_1 and X_2 which have the following demand functions:

$$p_1 = 100 - 2x_1 \quad \text{and} \quad p_2 = 80 - x_2,$$

where p_1 and p_2 are their respective prices and x_1 and x_2 are their respective quantities. If the total cost function of the monopolist is $TC = 20\ (x_1 + x_2)$, find the prices and quantities that would maximize his profit. **[Delhi Univ. B. Com. (H) 1984]**

8. A firm sells two products. The annual total revenue R behaves as a function of the number of units sold. Specifically,

$$R = 400x - 4x^2 + 1960y - 8y^2$$

where x and y equal, respectively, the number of units sold of each product. The cost of producing two products is :

$$C = 100 + 2x^2 + 4y^2 + 2xy$$

(*i*) Determine the number of units which should be produced and sold in order to maximize annual profit.

(*ii*) What is total revenue equal to at the output level determined in (*i*) ?

(*iii*) What is the maximum profit ? **[Delhi Univ. B. Com. (H) 2000]**

9. The demand functions of two competitive products are: $x_1 = 11 - 2p_1 - 2p_2$ and $x_2 = 16 - 2p_1 - 3p_2$, and the average costs of production of the products are constants, 3 and 1 respectively. Determine prices and quantities that maximize the profit.

10. A monopolist sells two competitive products, X_1 and X_2, for which the demand functions are: $x_1 = 1 - 2p_1 + 4p_2$ and $x_2 = 11 + 2p_1 - 6p_2$. If the constant average cost of producing a unit of X_1 is 4 and for X_2 it is 1, how many units of X_1 and X_2 should be sold to maximize the monopolist's profit?

11. Two commodities X and Y are produced by a firm at constant average cost of ₹ 2 and ₹ 4 per item respectively. The market demands are : $x = 50\ (p_y - p_x)$ and $y = 320 + 50p_x - 100p_y$, where p_x and p_y are the prices charged. If the firm is a monopolist, find the prices of the two commodities for the maximum joint profit. **[Delhi Univ. B. Com. (H) 1987]**

12. Assume that a monopolist sells his product in two separate markets and the demand functions are : $p_1 = 12 - x_1, p_2 = 20 - 3x_2$. The total cost function for his product is given by: $C = 3 + 2x$, where $x = x_1 + x_2$. Determine the prices that the monopolist will charge and the quantities he will sell in each of the markets to maximize his profit.

13. A firm producing two commodities X_1 and X_2 has the total cost function

$$C = x_1^2 + 2x_1x_2 + 3x_2^2,$$

where x_1, x_2 are the quantities of the two commodities and 8 and 10 are the prices per unit of these commodities. Determine the optimum levels of products which maximize profit.

14. A discriminating monopolist can separate his consumers into two distinct markets with the following demand functions :

Market I: $Q_1 = 16 - 0.2P_1$ *Market II*: $Q_2 = 180 - 2P_2$

Assume that the monopolist's total cost function takes the form $TC - 20Q - 20 = 0$, where Q is the total output. Obtain the total profit function and determine the prices he would charge in the two markets to maximize the profit. What is the total profit ? Do you agree that the price charged in the market with a higher elasticity of demand would be higher ? Show by calculation.

[*Delhi Univ. B. Com. (H) 1994, 2017*]

15. A discriminating monopolist, serving two sub-markets, faces the demand functions $x_1 = 16 - 0.2p_1$ and $x_2 = 9 - 0.05p_2$. If the monopolist's total cost function is given by $C = 20 + 20(x_1 + x_2)$, find the prices that he should charge in the two markets to maximize his profit.

16. A monopolist charges different prices in the two markets where his demand functions are $p_1 = 180 - 15x_1$ and $p_2 = 380 - 20x_2$. His total cost function is $C = 150 + 120x$, where x is the total output. Find the prices that the monopolist should charge to maximize his profit. Also, verify that higher price will be charged in the market having the lower price elasticity of demand.

[*Delhi Univ. B.Com. (H) 2022*]

ANSWERS

1. $x = 3, \quad y = 9$

2. $a = -3, \quad b = \dfrac{190}{3}$ and min $S = \dfrac{50}{3}$

3. $E = 10.5, \quad M = 6.5$ and max $R = 78.25$

4. $p_1 =$ ₹ 20, $p_2 =$ ₹ 5 and $R_{max} =$ ₹1925

5. $x_1 = 2, \quad x_2 = 3$ and maximum profit = 15

6. (*i*) $x = 8, \quad y = 10, \quad p_x = 84, \quad p_y = 66$ (*ii*) maximum profit = 725

7. $x_1 = 20, \quad x_2 = 30, \quad p_1 = 60, \quad p_2 = 50$

8. (*i*) $x = 20, \quad y = 80$ (*ii*) 1,12,000 (*iii*) 82,300

9. $p_1 = 1.75, \quad p_2 = 3, x_1 = 1.5, \quad x_2 = 3.5$

10. $x_1 = 4/3, \quad x_2 = 3$

11. $p_x = 4.2, \quad p_y = 5.2$

12. $x_1 = 5, \quad x_2 = 3, p_1 = 7, \quad p_2 = 11$

13. $x_1 = 3.5, \quad x_2 = 0.5$

14. $P_1 = 50, \quad P_2 = 55$ and max profit = 6138, No

15. $p_1 = 50, \quad p_2 = 100$

16. $p_1 = 150, \quad p_2 = 250$

CHAPTER

9

INTEGRATION

OBJECTIVES

After studying the material in this chapter, you should be able to:

- *Understand the meaning of the term indefinite integral.*
- *Distinguish between differentiation and integration.*
- *Know various techniques of integration.*
- *Evaluate definite integrals.*
- *Discuss the important properties of the definite integral.*
- *Find areas under curves and between curves.*

9.1 INTRODUCTION

In this chapter the process of taking derivatives is reversed. Essentially, we are starting with the derivative and asking what function has this derivative. This reverse process is called *integration.*

$$f \xrightarrow{\text{differentiation}} f' \text{ (derivative)}$$

$$\text{(integral) } f \xleftarrow{\text{integration}} f'.$$

9.2 THE INDEFINITE INTEGRAL

Given a function f, if F is a function such that

$$F'(x) = f(x)$$

then F is called an **antiderivative** of f. Thus an antiderivative of f is simply a function whose derivative is f.

For example, since $\frac{d}{dx}(x^3) = 3x^2$, x^3 is an antiderivative of $3x^2$. However, it is not the only antiderivative of $3x^2$. Since

$$\frac{d}{dx}(x^3+2) = 3x^2 \quad \text{and} \quad \frac{d}{dx}(x^3-7) = 3x^2,$$

both $x^3 + 2$ and $x^3 - 7$ are also antiderivatives of $3x^2$. It can be shown that any antiderivative of $3x^2$ must have the form $x^3 + c$. In general, if $F(x)$ is an antiderivative of $f(x)$, then for any constant c. we have

$$\frac{d}{dx}[F(x)+C] = \frac{d}{dx}[F(x)] + 0 = f(x)$$

Therefore, $F(x) + c$ is also an antiderivative of $f(x)$.

This motivates the following definition.

Definition. If the function $F(x)$ is an antiderivative of $f(x)$, then the **indefinite integral** or simply **integral** of $f(x)$ is defined as $F(x) + c$, where c is an arbitrary constant. It is denoted by $\int f(x)\,dx$.

The symbol $\int$ is called the **integral sign,** the function $f(x)$ is called the **integrand,** and c is called the **constant of integration.** The process of finding the integral of a function is called **integration.**

Example 1. (*i*) $\int 3\,dx = 3x + c$, since $\frac{d}{dx}(3x) = 3$.

(*ii*) $\int x^2\,dx = \frac{x^3}{3} + c$, since $\frac{d}{dx}\left(\frac{x^3}{3}\right) = x^2$.

Basic Integration Formulas. A very brief list of some basic integration formulas is given below. These formulas can easily be verified by using the differentiation formulas.

1. $\int k\,dx = kx + c$, k a constant

2. $\int x^n\,dx = \frac{x^{n+1}}{n+1} + c$, $n \neq -1$

3. $\int \frac{1}{x}\,dx = \log x + c$

4. $\int e^x\,dx = e^x + c$

5. $\int a^x\,dx = \frac{a^x}{\log a} + c$

6. $\int k\,f(x)\,dx = k\int f(x)\,dx$, k a constant

7. $\int [f(x) \pm g(x)]\,dx = \int f(x)\,dx \pm \int g(x)\,dx$

Remark. Formula 7 can be extended to any number of sums or differences.

The use of above formulas is best illustrated with the help of following examples.

Example 2. Find the following indefinite integrals:

(*i*) $\int 1\,dx$ (*ii*) $\int x^3\,dx$ (*iii*) $\int \frac{1}{\sqrt{x}}\,dx$

(*iv*) $\int 2^x\,dx$ (*v*) $\int 5x\,dx$ (*vi*) $\int \frac{1}{3x^2}\,dx$

Solution. (*i*) $\int 1\,dx = 1x + c = x + c$

Note. Usually, we write $\int 1\,dx$ as $\int dx$.

(*ii*) $$\int x^3\,dx = \frac{x^{3+1}}{3+1} + c = \frac{x^4}{4} + c$$

(*iii*) $$\int \frac{1}{\sqrt{x}}\,dx = \int x^{-1/2}\,dx = \frac{x^{-1/2+1}}{-\frac{1}{2}+1} + c = 2\sqrt{x} + c$$

(*iv*) $$\int 2^x\,dx = \frac{2^x}{\log 2} + c$$

(*v*) $$\int 5x\,dx = 5\int x\,dx = 5.\frac{x^2}{2} = \frac{5}{2}x^2 + c$$

(*vi*) $$\int \frac{1}{3x^2}\,dx = \frac{1}{3}\int \frac{1}{x^2}\,dx = \frac{1}{3}\int x^{-2}\,dx = \frac{1}{3}\cdot\frac{x^{-2+1}}{-2+1} = -\frac{1}{3x} + c$$

Sometimes, in order to apply the basic integration formula, it is necessary to simplify an integrand, as is shown in the next example.

Example 3. Find the following integrals:

(*i*) $\int (e^{a \log x} + e^{x \log a})\,dx$ **[*Delhi Univ. B.Com.(H) 1989*]**

(*ii*) $\int \sqrt{x}\,(x+3)\,dx$ (*iii*) $\int (x^2+5)(x-3)\,dx$ (*iv*) $\int \frac{(x^3+2x^2+3x)}{\sqrt{x}}\,dx$

Solution.

(*i*) $\int (e^{a \log x} + e^{x \log a})\, dx = \int \left(e^{\log x^a} + e^{\log a^x}\right) dx$

$= \int (x^a + a^x)\, dx$ (since $e^{\log t} = t$)

$= \dfrac{x^{a+1}}{a+1} + \dfrac{a^x}{\log a} + c$

(*ii*) $\int \sqrt{x}\,(x+3)\, dx = \int (x^{3/2} + 3x^{1/2})\, dx$

$= \dfrac{x^{5/2}}{\frac{5}{2}} + 3 \cdot \dfrac{x^{3/2}}{\frac{3}{2}} + c = \dfrac{2}{5}x^{5/2} + 2x^{3/2} + c$

(*iii*) $\int (x^2+5)(x-3)\, dx = \int (x^3 - 3x^2 + 5x - 15)\, dx$

$= \dfrac{x^4}{4} - 3 \cdot \dfrac{x^3}{3} + 5 \cdot \dfrac{x^2}{2} - 15x + c = \dfrac{x^4}{4} - x^3 + \dfrac{5}{2}x^2 - 15x + c$

(*iv*) $\int \dfrac{x^3 + 2x^2 + 3x}{\sqrt{x}}\, dx = \int (x^{5/2} + 2x^{3/2} + 3x^{1/2})\, dx$

$= \dfrac{x^{7/2}}{\frac{7}{2}} + 2 \cdot \dfrac{x^{5/2}}{\frac{5}{2}} + 3 \cdot \dfrac{x^{3/2}}{\frac{3}{2}} + c = \dfrac{2}{7}x^{7/2} + \dfrac{4}{5}x^{5/2} + 2x^{3/2} + c$

Although there are many antiderivatives of a function, we may want to find a *particular* one that satisfies a certain condition. The following example will illustrate.

Example 4. In each of the following problems, find y subject to the given conditions :

(*i*) $\dfrac{dy}{dx} = 3x^2 + 2x$; given that $y = 4$ when $x = 1$.

(*ii*) $\dfrac{dy}{dx} = e^x + 1$; given that $y = 3$ when $x = 0$.

Solution. (*i*) Since $\dfrac{dy}{dx} = 3x^2 + 2x$, y is an antiderivative of $(3x^2 + 2x)$. Thus

$$y = \int (3x^2 + 2x)\, dx = x^3 + x^2 + c$$

Thus any function of the form

$$y = x^3 + x^2 + c \qquad ...(1)$$

satisfies $\frac{dy}{dx} = 3x^2 + 2x$. To determine the particular one for which $y = 4$ when $x = 1$, we substitute $y = 4$ and $x = 1$ into Equation (1). Thus

$$4 = (1)^3 + (1)^2 + c = 2 + c \quad \Rightarrow \quad c = 2$$

$$\therefore \quad y = x^3 + x^2 + 2$$

(*ii*) Since $\frac{dy}{dx} = e^x + 1$, y is an antiderivative of $e^x + 1$. Thus

$$y = \int (e^x + 1)\, dx \quad \Rightarrow \quad y = e^x + x + c \qquad ...(2)$$

Since $y = 3$ when $x = 0$, from Equation (2), we obtain

$$3 = e^0 + 0 + c = 1 + c \quad \Rightarrow \quad c = 2$$

$$\therefore \quad y = e^x + x + 2.$$

EXERCISE 9.1

1. Find the following indefinite integrals:

(*i*) $\int 3\, dx$ (*ii*) $\int \log 2\, dx$ (*iii*) $\int x^7\, dx$

(*iv*) $\int 5x^{-7}\, dx$ (*v*) $\int \frac{1}{2x^{3/2}}\, dx$ (*vi*) $\int 5^x\, dx$

(*vii*) $\int dt$ (*viii*) $\int (5 + e)\, dx$ (*ix*) $\int \left(\frac{e^x}{3} + 2x\right) dx$

(*x*) $\int \left(\frac{x^3}{3} - \frac{3}{x^3}\right) dx$ (*xi*) $\int 0\, dx$ (*xii*) $\int (7 - 3t - 2t^2)\, dt$

2. Find the following integrals:

(*i*) $\int \left(6 - \frac{5}{4}z^2 + 2e^z\right) dz$ (*ii*) $\int x^4(x^3 + 3x^2 - 7)\, dx$ (*iii*) $\int \frac{(x^3 + 2x + 5)}{\sqrt{x}}\, dx$

(*iv*) $\int \frac{(x+1)^2}{\sqrt{x}}\, dx$ (*v*) $\int \left(5x + \frac{4}{x} + 3e^x\right) dx$ (*vi*) $\int x(x+2)(x-3)\, dx$

(*vii*) $\int \sqrt{x}\left(2\sqrt{x} + \frac{1}{\sqrt{x}}\right) dx$ (*viii*) $\int \frac{x^4 + 2x^2 + 1}{x^2} dx$ (*ix*) $\int \left(\sqrt{x} + \frac{1}{\sqrt{x}}\right)^3 dx$

(*x*) $\int \frac{x^3 - 2}{x^2} dx$ (*xi*) $\int (e^5 + e^x)\, dx$ (*xii*) $\int \frac{x^2 (4 - x)^2}{\sqrt{x}} dx$

3. In each of the following problems, find y subject to the given conditions.

(*i*) $\frac{dy}{dx} = \frac{1}{2\sqrt{x}}$; given that $y = 1$ when $x = 1$.

(*ii*) $\frac{dy}{dx} = 6x^2 + 2x + 3$; given that $y = 4$ when $x = 0$.

ANSWERS

1. (*i*) $3x + c$ (*ii*) $(\log 2)\, x + c$ (*iii*) $\frac{x^8}{8} + c$

(*iv*) $-\frac{5}{6}x^{-6} + c$ (*v*) $-x^{-1/2} + c$ (*vi*) $\frac{5^x}{\log 5} + c$

(*vii*) $t + c$ (*viii*) $5x + ex + c$ (*ix*) $\frac{e^x}{3} + x^2 + c$

(*x*) $\frac{x^4}{12} + \frac{3}{2x^2} + c$ (*xi*) c (*xii*) $7t - \frac{3t^2}{2} - \frac{2t^3}{3} + c$

2. (*i*) $6z - \frac{5z^3}{12} + 2e^z + c$ (*ii*) $\frac{x^8}{8} + \frac{3x^7}{7} - \frac{7x^5}{5} + c$

(*iii*) $\frac{2}{7}x^{7/2} + \frac{4}{3}x^{3/2} + 10x^{1/2} + c$ (*iv*) $\frac{2}{5}x^{5/2} + 2x^{1/2} + \frac{4}{3}x^{3/2} + c$

(*v*) $\frac{5}{2}x^2 + 4 \log x + 3e^x + c$ (*vi*) $\frac{x^4}{4} - \frac{x^3}{3} - 3x^2 + c$

(*vii*) $x^2 + x + c$ (*viii*) $\frac{x^3}{3} + 2x - \frac{1}{x} + c$

(*ix*) $\frac{2}{5}x^{5/2} - 2x^{-1/2} + 2x^{3/2} + 6x^{1/2} + c$ (*x*) $\frac{x^2}{2} + 2x^{-1} + c$

(*xi*) $e^5 x + e^x + c$ (*xii*) $\frac{32}{5}x^{5/2} + \frac{2}{9}x^{9/2} - \frac{16}{7}x^{7/2} + c$

3. (*i*) $y = \sqrt{x}$ (*ii*) $y = 2x^3 + x^2 + 3x + 4$.

9.3 TECHNIQUES OF INTEGRATION

There exist many functions whose integral cannot be found directly by applying the basic integration formulas. As such some techniques must be developed for integrating more complicated functions. Of many techniques, we shall discuss only the following:

(*a*) *Integration by substitution.*

(*b*) *Integration by parts.*

(*c*) *Integration by partial fractions.*

(a) Integration by Substitution. Many a times an integrand is reduced to a standard form by making a suitable substitution. This is illustrated in the following examples.

Example 5. Evaluate $\int \sqrt{x^2+1}\ 2x\,dx$

Solution . Let $u = x^2 + 1$ $\quad\therefore\quad$ $du = 2x\,dx$

Upon substitution, we obtain

$$\int \sqrt{x^2+1}\ 2x\,dx = \int \sqrt{u}\,du = \int u^{1/2}du = \frac{2}{3}u^{3/2} + c$$

Replacing u by $x^2 + 1$, we get

$$\int \sqrt{x^2+1}\ 2x\,dx = \frac{2}{3}(x^2+1)^{3/2} + c\,.$$

Example 6. Find the following integrals:

(*i*) $\int (2x+3)^2\,dx$ (*ii*) $\int e^{3x+4}\,dx$ (*iii*) $\int \frac{1}{5x-1}dx$

Solution . (*i*) Given $\int (2x+3)^2\,dx$

Let $u = 2x + 3$. Then $du = 2\,dx$.

$$\therefore\quad \int (2x+3)^2\,dx = \int \frac{(2x+3)^2}{2}.2\,dx = \frac{1}{2}\int u^2\,du = \frac{1}{2}.\frac{u^3}{3} + c = \frac{1}{6}(2x+3)^3 + c$$

(*ii*) Given $\int e^{3x+4}\,dx$

Let $u = 3x + 4$. Then $du = 3\,dx$.

$$\therefore\quad \int e^{3x+4}\,dx = \int \frac{e^{3x+4}}{3}\cdot 3\,dx = \frac{1}{3}\int e^u\,du = \frac{1}{3}e^u + c = \frac{e^{3x+4}}{3} + c\,.$$

(*iii*) Given $\int \frac{1}{5x-1}dx$

Let $u = 5x - 1$. Then $du = 5\,dx$.

$$\therefore \quad \int \frac{1}{5x-1}\,dx = \frac{1}{5}\int \frac{1}{5x-1}\cdot 5\,dx = \frac{1}{5}\int \frac{1}{u}\,du = \frac{1}{5}\log u + c = \frac{1}{5}\log(5x-1) + c.$$

An important result:

$$\text{If} \int f(x)\,dx = F(x) + c, \text{ then } \int f(ax+b)\,dx = \frac{F(ax+b)}{a} + c$$

Using this result, we immediately obtain the following formulas:

1. $\int (ax+b)^n\,dx = \frac{(ax+b)^{n+1}}{a(n+1)} + c, n \neq -1$
2. $\int \frac{1}{(ax+b)}\,dx = \frac{\log(ax+b)}{a} + c$
3. $\int e^{ax+b}\,dx = \frac{e^{ax+b}}{a} + c$
4. $\int a^{lx+m}\,dx = \frac{a^{lx+m}}{l(\log a)} + c$

Example 7. Evaluate the following integrals:

(*i*) $\int (5x-7)^3\,dx$ (*ii*) $\int \frac{1}{1-x}\,dx$

(*iii*) $\int e^{5x}\,dx$ (*iv*) $\int 2^{3-x}\,dx$

Solution . (*i*) $\int (5x-7)^3\,dx = \frac{(5x-7)^{3+1}}{3+1}\cdot\frac{1}{5} = \frac{1}{20}(5x-7)^4 + c$

(*ii*) $\int \frac{1}{1-x}\,dx = \frac{\log(1-x)}{-1} = -\log(1-x) + c$

(*iii*) $\int e^{5x}\,dx = \frac{e^{5x}}{5} + c$

(*iv*) $\int 2^{3-x}\,dx = \frac{2^{3-x}}{\log 2}\cdot\frac{1}{(-1)} = -\frac{2^{3-x}}{\log 2} + c$

Example 8. Evaluate the following integrals:

(*i*) $\int x^3 e^{x^4}\,dx$ (*ii*) $\int \frac{(2+\log x)^3}{x}\,dx$

(*iii*) $\int \frac{2x+1}{x^2+x}\,dx$ (*iv*) $\int \frac{x^2}{\sqrt[3]{2x^3+9}}\,dx$

Solution **.** (*i*) To find $\int x^3 e^{x^4}\, dx$, let $u = x^4$ so that $du = 4x^3\, dx$.

$$\therefore \quad \int x^3 e^{x^4}\, dx = \frac{1}{4}\int e^{x^4} . 4x^3\, dx$$

$$= \frac{1}{4}\int e^u\, du = \frac{1}{4}e^u + c = \frac{1}{4}e^{x^4} + c$$

(*ii*) To evaluate $\int \frac{(2+\log x)^3}{x}\, dx$, let $u = 2 + \log x$, then $du = \frac{1}{x}dx$

$$\therefore \quad \int \frac{(2+\log x)^3}{x}\, dx = \int u^3\, du = \frac{u^4}{4} + c = \frac{(2+\log x)^4}{4} + c$$

(*iii*) To evaluate $\int \frac{2x+1}{x^2+x}\, dx$, let $u = x^2 + x$, then $du = (2x+1)\, dx$.

$$\therefore \quad \int \frac{2x+1}{x^2+x}\, dx = \int \frac{1}{u}du = \log u + c = \log(x^2+x) + c$$

(*iv*) To find $\int \frac{x^2}{\sqrt[3]{2x^3+9}}\, dx$, let $u = 2x^3 + 9$, then $du = 6x^2\, dx$.

$$\therefore \quad \int \frac{x^2}{\sqrt[3]{2x^3+9}}\, dx = \frac{1}{6}\int \frac{1}{\sqrt[3]{u}}\, du = \frac{1}{6}\int u^{-1/3}\, du = \frac{1}{6}\cdot\frac{u^{2/3}}{\frac{2}{3}} + c = \frac{1}{4}(2x^3+9)^{2/3} + c.$$

Two Important Formulas :

1. $$\int f(x)^n f'(x)\, dx = \frac{f(x)^{n+1}}{n+1} + c, \; n \neq -1$$

2. $$\int \frac{f'(x)}{f(x)}\, dx = \log f(x) + c$$

Example 9. Find the following:

(*i*) $\int \frac{(\log x)^5}{x} dx$ (*ii*) $\int \frac{e^x + e^{-x}}{e^x - e^{-x}} dx$ (*iii*) $\int \frac{x}{2x^2+1} dx$

Solution **.** (*i*) Using Formula (1) with $f(x) = \log x$, we get

$$\int (\log x)^5 \frac{1}{x}\, dx = \frac{(\log x)^6}{6} + c.$$

(*ii*) Observe that $\frac{e^x + e^{-x}}{e^x - e^{-x}}$ is of the form $\frac{f'(x)}{f(x)}$, where $f(x) = e^x - e^{-x}$

Thus, applying Formula (2), we obtain

$$\int \frac{e^x + e^{-x}}{e^x - e^{-x}} dx = \log(e^x - e^{-x}) + c.$$

(*iii*) $\int \frac{x}{2x^2 + 1} dx = \frac{1}{4}\int \frac{4x}{2x^2 + 1} dx = \frac{1}{4}\log(2x^2 + 1) + c.$

Example 10. Solve: $\frac{dy}{dx} = x\sqrt{x^2 + 1}$; given that $y = 6$ at $x = 0$.

[*Delhi Univ. B. Com.(H) 1988*]

***Solution*.** Since $\frac{dy}{dx} = x\sqrt{x^2 + 1}$

$$\Rightarrow \qquad y = \int x\sqrt{x^2 + 1}\; dx = \frac{1}{2}\int (x^2 + 1)^{1/2} \cdot 2x\, dx = \frac{1}{2}\frac{(x^2 + 1)^{3/2}}{3/2} + c$$

Substituting $y = 6$ and $x = 0$ in the above equation, we get

$$6 = \frac{1}{3} + c \qquad \text{or} \qquad c = \frac{17}{3}$$

Thus
$$y = \frac{(x^2 + 1)^{3/2}}{3} + \frac{17}{3}.$$

(*b*) *Integration by Parts.* This technique is used to integrate the product of two functions. It can be stated as follows :

Integral of the product of two functions is equal to the first function times the integral of the second minus the integral of the product of the derivative of first and the integral of second. That is,

$$\int uv\,dx = u\int v\,dx - \int\left[\frac{du}{dx}.\int v\,dx\right]dx$$

Remark. While using the integration by parts formula, sometimes the "best choice" for u and v may not be obvious. In some cases one choice may be as good as another; in other cases only one choice may be suitable. Insight into making a good choice will come only with practice.

Example 11. Evaluate the following integrals: (*i*) $\int xe^x dx$ (*ii*) $\int x \log x\, dx$

***Solution*.** (*i*) Integrating by parts taking $u = x$, we get

$$\int xe^x\,dx = x\int e^x dx - \int\left[\frac{dx}{dx}\cdot\int e^x\,dx\right]dx$$

$$= xe^x - \int e^x\,dx = xe^x - e^x = e^x\,(x-1)+c\,.$$

(ii) Integrating by parts taking $u = \log x$, we get

$$\int x\log x\,dx = \log x\int x\,dx - \int\left[\frac{d}{dx}(\log x).\int x\,dx\right]dx$$

$$= \log x.\frac{x^2}{2} - \int\frac{1}{x}.\frac{x^2}{2}\,dx$$

$$= \frac{x^2\log x}{2} - \frac{1}{2}\int x\,dx$$

$$= \frac{x^2\log x}{2} - \frac{x^2}{4} + c$$

Example 12. Determine the following:

(*i*) $\int \log x\,dx$ (*ii*) $\int (\log x)^2\,dx$ (*iii*) $\int e^x\,(1+x)\log(xe^x)\,dx$

***Solution*.** (*i*) $\int \log x\,dx = \int \log x\,.1\,dx$ *(Note this step)*

$$= \log x\int 1\,dx - \int\left[\frac{d}{dx}(\log x)\cdot\int 1\,dx\right]dx$$

$$= \log x.x - \int\frac{1}{x}.\,x\,dx$$

$$= x\log x - \int 1.\,dx$$

$$= x\log x - x = x\,(\log x - 1) + c\,.$$

(*ii*) $\int (\log x)^2 dx = \int (\log x)^2\,.1\,dx$

$$= (\log x)^2\int 1\,dx - \int\left[\frac{d}{dx}(\log x)^2\,.\int 1\,dx\right]dx$$

$$= (\log x)^2\,.\,x - \int\frac{2\log x}{x}\,x\,dx$$

$$= x\,(\log x)^2 - 2\int \log x\,dx$$

$$= x(\log x)^2 - 2[x(\log x - 1)] + c \qquad \textit{(see part (i))}$$

$$= x(\log x)^2 - 2x \log x + 2x + c$$

(iii) Let $xe^x = u \quad \Rightarrow \quad e^x(1+x)\,dx = du$

$$\therefore \quad \int e^x (1+x) \log(xe^x)\,dx = \int \log u\,du$$

$$= u(\log u - 1) + c \qquad \textit{[see part (i)]}$$

$$= xe^x[\log(xe^x) - 1] + c.$$

Example 13. Determine the following integrals:

(*i*) $\int x^2 \log x\,dx$ (*ii*) $\int x\sqrt{x+1}\,dx$ (*iii*) $\int x^2 e^{5x}\,dx$

Solution. (*i*) $\int x^2 \log x\,dx = \log x \int x^2 dx - \int \left[\frac{d}{dx}(\log x) \cdot \int x^2\,dx\right] dx$

$$= \log x \cdot \frac{x^3}{3} - \int \frac{1}{x} \cdot \frac{x^3}{3}\,dx$$

$$= \frac{x^3 \log x}{3} - \frac{1}{3}\int x^2\,dx$$

$$= \frac{x^3 \log x}{3} - \frac{x^3}{9} + c$$

(*ii*) $\int x\sqrt{x+1}\,dx = x\int \sqrt{x+1}\,dx - \int \left[\frac{dx}{dx}\int \sqrt{x+1}\,dx\right] dx$

$$= x\,\frac{(x+1)^{3/2}}{\frac{3}{2}} - \int \frac{(x+1)^{3/2}}{\frac{3}{2}}\,dx$$

$$= \frac{2x(x+1)^{3/2}}{3} - \frac{2}{3}\int (x+1)^{3/2}\,dx$$

$$= \frac{2x(x+1)^{3/2}}{3} - \frac{2}{3}\,\frac{(x+1)^{5/2}}{\frac{5}{2}} + c$$

$$= \frac{2x(x+1)^{3/2}}{3} - \frac{4}{15}(x+1)^{5/2} + c$$

$$= \frac{2}{15}(3x-2)(x+1)^{3/2} + c.$$

(iii) $\displaystyle\int x^2\, x^{5x}\, dx = x^2\,\frac{e^{5x}}{5} - \int 2x \cdot \frac{e^{5x}}{5}\, dx$

$$= \frac{x^2\, e^{5x}}{5} - \frac{2}{5}\left[\frac{x\, e^{5x}}{5} - \int 1 \cdot \frac{e^{5x}}{5}\, dx\right]$$

$$= \frac{x^2\, e^{5x}}{5} - \frac{2}{25}\, x\, e^{5x} + \frac{2}{25}\int e^{5x}\, dx$$

$$= \frac{x^2\, e^{5x}}{5} - \frac{2}{25}\, x\, e^{5x} + \frac{2}{25}\,\frac{e^{5x}}{5} + c = \frac{(25x^2 - 10x + 2)\, e^{5x}}{125} + c.$$

(c) *Integration by Partial Fractions.* **This technique is used to integrate functions of the type $\dfrac{N(x)}{D(x)}$, where both $N(x)$ and $D(x)$ are polynomials. Such functions are called rational functions.** We may assume, without any loss of generality, that the numerator $N(x)$ and denominator $D(x)$ have no common polynomial factor and that the degree of $N(x)$ is less than the degree of $D(x)$. For, if the degree of numerator were greater than or equal to the degree of denominator, we could use long division to divide $N(x)$ by $D(x)$ so that

$$\frac{N(x)}{D(x)} = Q(x) + \frac{R(x)}{D(x)},$$

where $Q(x)$ is the quotient (a polynomial : easily integrable) and $R(x)$ is the remainder whose degree is less than the degree of $D(x)$.

The method of expressing $\dfrac{N(x)}{D(x)}$ into partial fractions depends on factors of the denominator. It is a fact that every polynomial $D(x)$ can be expressed as a product of linear and irreducible quadratic (one that can not be expressed as a product of two linear factors with real coefficients) factors with real coefficients. As such we have the following four cases to consider.

Case 1. All factors in the denominator are linear and non-repeated

To each linear non-repeated factor $ax + b$, there corresponds a fraction of the type

$$\frac{A}{ax + b},$$

where A is a constant.

The following examples will illustrate.

Example 14. Find $\int \frac{x}{(x+1)(x-2)} dx$.

Solution . Since the denominator consists only of linear factors, each factor occurring exactly once, we can write

$$\frac{x}{(x+1)(x-2)} = \frac{A}{x+1} + \frac{B}{x-2} \qquad \text{...(1)}$$

$$= \frac{A(x-2) + B(x+1)}{(x+1)(x-2)}$$

$$\Rightarrow \qquad x = A(x-2) + B(x+1) \qquad \text{...(2)}$$

By successively setting x in Equation (2) equal to any two different numbers, we can obtain a system of equations which can be solved for A and B. In particular, the work can be simplified by letting x be the roots of the denominator. Thus, for example, if $x = 2$, we have

$$2 = A(0) + B(3) \quad \Rightarrow \quad B = \frac{2}{3}$$

If $x = -1$, we have

$$-1 = A(-3) + B(0) \quad \Rightarrow \quad A = \frac{1}{3}$$

Thus Equation (1) becomes

$$\frac{x}{(x+1)(x-2)} = \frac{1/3}{x+1} + \frac{2/3}{x-2}$$

Hence

$$\int \frac{x}{(x+1)(x-2)} dx = \int \left[\frac{1/3}{x+1} + \frac{2/3}{x-2}\right] dx$$

$$= \frac{1}{3}\log(x+1) + \frac{2}{3}\log(x-2) + c$$

$$= \frac{1}{3}\left[\log(x+1) + \log(x-2)^2\right] + c$$

$$= \frac{1}{3}\log\left[(x+1)(x-2)^2\right] + c.$$

Note. There is an alternative method of finding the constants A and B. It involves comparing the coefficients of like powers of x on both sides. For example, comparing the coefficients of x and the constant terms on both sides of Equation (2), we obtain

$$1 = A + B \qquad \text{and} \qquad 0 = -2A + B$$

Solving gives $A = \frac{1}{3}$ and $B = \frac{2}{3}$.

Example 15. Find $\int \frac{1}{x - x^3}\, dx.$ **[Delhi Univ. B. Com. (H) 1982]**

Solution . Observe that the denominator of the integrand consists of linear, non-repeated factors:

$$x - x^3 = x\,(1 - x)\,(1 + x)$$

Therefore, we can write

$$\frac{1}{x - x^3} = \frac{A}{x} + \frac{B}{1-x} + \frac{C}{1+x}$$

$$\Rightarrow \qquad 1 = A\,(1 - x^2) + B\,(x)\,(1 + x) + C\,(x)\,(1 - x)$$

If $x = 0$, then $A = 1$. If $x = 1$, then $1 = 2B \Rightarrow B = \frac{1}{2}$ If $x = -1$, then $1 = -2C \Rightarrow C = -\frac{1}{2}$. Thus

$$\frac{1}{x - x^3} = \frac{1}{x} + \frac{1}{2} \cdot \frac{1}{1-x} - \frac{1}{2} \cdot \frac{1}{1+x}$$

Hence

$$\int \frac{1}{x - x^3}\, dx = \int \frac{1}{x}\, dx + \frac{1}{2} \int \frac{1}{1-x}\, dx - \frac{1}{2} \int \frac{1}{1+x}\, dx$$

$$= \log x + \frac{1}{2} \frac{\log(1-x)}{-1} - \frac{1}{2} \log(1+x)$$

$$= \log x - \frac{1}{2}\left[\log(1-x) + \log(1+x)\right]$$

$$= \log x - \frac{1}{2} \log\left(1 - x^2\right) = \log\left[\frac{x}{\sqrt{1 - x^2}}\right] + c.$$

Two Standard Formulas :

1. $\int \frac{1}{x^2 - a^2}\, dx = \frac{1}{2a} \log\left[\frac{x-a}{x+a}\right] + c$

2. $\int \frac{1}{a^2 - x^2}\, dx = \frac{1}{2a} \log\left[\frac{a+x}{a-x}\right] + c$

Example 16. Evaluate the following integrals :

(*i*) $\int \frac{x^2 - 1}{x^4 + 1}\, dx$ (*ii*) $\int \frac{3x^3 - 3x + 4}{x^2 - 1}\, dx$

***Solution*. (*i*)** $\displaystyle\int \frac{x^2-1}{x^4+1}\,dx = \int \frac{\left(1-\frac{1}{x^2}\right)}{x^2+\frac{1}{x^2}}\,dx = \int \frac{\left(1-\frac{1}{x^2}\right)}{\left(x+\frac{1}{x}\right)^2-2}\,dx$

Put $\quad x+\frac{1}{x} = t \quad \Rightarrow \quad \left(1-\frac{1}{x^2}\right)dx = dt$

Thus $\quad \displaystyle\int \frac{x^2-1}{x^4+1}\,dx = \int \frac{1}{t^2-\sqrt{2}^2}\,dt = \frac{1}{2\sqrt{2}}\log\frac{t-\sqrt{2}}{t+\sqrt{2}}$

$$= \frac{1}{2\sqrt{2}}\log\frac{x+\frac{1}{x}-\sqrt{2}}{x+\frac{1}{x}+\sqrt{2}} = \frac{1}{2\sqrt{2}}\log\left[\frac{x^2-\sqrt{2}\,x+1}{x^2+\sqrt{2}\,x+1}\right]+c.$$

(*ii*) To find $\displaystyle\int \frac{3x^3-3x+4}{x^2-1}\,dx$, observe that the degree of the numerator is greater than the degree of denominator. Therefore, we first divide numerator by denominator. Using long division, we obtain

$$\frac{3x^3-3x+4}{x^2-1} = 3x+\frac{4}{x^2-1}.$$

$$\therefore \quad \int \frac{3x^3-3x+4}{x^2-1}\,dx = \int 3x\,dx + 4\int \frac{1}{x^2-1}\,dx$$

$$= \frac{3}{2}x^2+4\cdot\frac{1}{2}\log\left[\frac{x-1}{x+1}\right]+c$$

$$= \frac{3}{2}x^2+2\log\left[\frac{x-1}{x+1}\right]+c.$$

Case 2. All factors in the denominator are linear but some of which are repeated.

To each linear factor $(ax+b)^r$ repeating r times, there will correspond the sum of r partial fractions as follows:

$$\frac{A_1}{ax+b}+\frac{A_2}{(ax+b)^2}+.....+\frac{A_r}{(ax+b)^r},$$

where $A_1, A_2, ..., A_r$ are constants.

Example 17. Evaluate $\int \frac{2x+1}{(x+2)(x-1)^2}\,dx$.

Solution. In the denominator of the given integrand, the factor $x + 2$ occurs exactly once and the factor $x - 1$ occurs twice. Thus we can write

$$\frac{2x+1}{(x+2)(x-1)^2} = \frac{A}{x+2} + \frac{B}{x-1} + \frac{C}{(x-1)^2}$$

$$\Rightarrow \quad 2x + 1 = A(x-1)^2 + B(x+2)(x-1) + C(x+2)$$

To find the constants A, B and C, let us choose $x = -2$, $x = 1$ and for convenience $x = 0$.

If $x = -2$, then $-3 = 9A \quad \Rightarrow \quad A = -\frac{1}{3}$

If $x = 1$, then $3 = 3C \quad \Rightarrow \quad C = 1$

If $x = 0$, then $1 = A - 2B + 2C$. Substituting the values of A and C and solving for B, we get $B = \frac{1}{3}$.

Hence $$\frac{2x+1}{(x+2)(x-1)^2} = -\frac{1}{3}\cdot\frac{1}{x+2} + \frac{1}{3}\cdot\frac{1}{x-1} + \frac{1}{(x-1)^2}$$

$$\therefore \quad \int \frac{2x+1}{(x+2)(x-1)^2}\,dx = -\frac{1}{3}\log(x+2) + \frac{1}{3}\log(x-1) - \frac{1}{(x-1)}$$

$$= \frac{1}{3}\log\left(\frac{x-1}{x+2}\right) - \frac{1}{x-1} + c$$

Example 18. Find $\int \frac{1}{(e^x-1)^2}\,dx$. **[*Delhi Univ. B. Com. (H) 1988*]**

***Solution*.** Let $e^x = t$. Then $e^x\,dx = dt$ or $dx = \frac{dt}{e^x} = \frac{dt}{t}$

$$\therefore \quad \int \frac{1}{(e^x-1)^2}\,dx = \int \frac{1}{t(t-1)^2}\,dt$$

Let $$\frac{1}{t(t-1)^2} = \frac{A}{t} + \frac{B}{t-1} + \frac{C}{(t-1)^2} \quad \Rightarrow \quad 1 = A(t-1)^2 + Bt(t-1) + Ct$$

Let us choose $t = 0$, $t = 1$ and $t = -1$.

If $t = 0$, then $A = 1$. If $t = 1$, then $C = 1$ If $t = -1$, then $1 = 4A + 2B - C$. Substituting the values of A and C and solving for B, we get $B = -1$.

Thus $$\frac{1}{t(t-1)^2} = \frac{1}{t} - \frac{1}{t-1} + \frac{1}{(t-1)^2}$$

$\Rightarrow$ $$\int \frac{1}{t(t-1)^2}\,dt = \log t - \log(t-1) - \frac{1}{t-1} + c = \log\frac{t}{t-1} - \frac{1}{t-1} + c$$

Thus $$\int \frac{1}{(e^x-1)^2}\,dx = \log\left(\frac{e^x}{e^x-1}\right) - \frac{1}{e^x-1} + c.$$

Case 3. The factors in the denominator are linear and irreducible quadratic but quadratic factors are non-repeated.

To each irreducible quadratic factor $ax^2 + bx + c$ that occurs exactly once in the denominator, there will correspond a partial fraction of the type

$$\frac{Ax + B}{ax^2 + bx + c}$$

Example 19. Find $\int \frac{x^3 - 5x}{(x^2-9)(x^2+1)}\,dx.$ **[*Delhi Univ. B. Com. (H) 1987*]**

***Solution*.** Let $$\frac{x^3 - 5x}{(x^2-9)(x^2+1)} = \frac{A}{x-3} + \frac{B}{x+3} + \frac{Cx + D}{x^2+1}$$

$\Rightarrow$ $$x^3 - 5x = A(x+3)(x^2+1) + B(x-3)(x^2+1) + (Cx+D)(x^2-9)$$

If $x = 3$, then $12 = 60A \Rightarrow A = \frac{1}{5}$. If $x = -3$, then $-12 = -60B \Rightarrow B = \frac{1}{5}$

To find the constants C and D, let us choose $x = 0$ and $x = 1$.

If $x = 0$, then $0 = 3A - 3B - 9D = -9D$ or $D = 0$.

If $x = 1$, then $-4 = 8A - 4B + (C + D)(-8)$. Substituting the values of A, B and D in the last equation and solving for C, we get $C = \frac{3}{5}$.

Thus $$\frac{x^3 - 5x}{(x^2-9)(x^2+1)} = \frac{1}{5}\frac{1}{x-3} + \frac{1}{5}\frac{1}{x+3} + \frac{3}{5}\frac{x}{x^2+1}$$

$\therefore$ $$\int \frac{x^3 - 5x}{(x^2-9)(x^2+1)}\,dx = \frac{1}{5}\int \frac{1}{x-3}\,dx + \frac{1}{5}\int \frac{1}{x+3}\,dx + \frac{3}{10}\int \frac{2x}{x^2+1}\,dx$$

$$= \frac{1}{5}\log(x-3) + \frac{1}{5}\log(x+3) + \frac{3}{10}\log(x^2+1)$$

$$= \frac{1}{5}\log(x^2-9)+\frac{3}{10}\log(x^2+1)+c.$$

Case 4. The factors in the denominator are linear and irreducible quadratic and some of the quadratic factors are repeated.

To each repeated quadratic factor $(ax^2 + bx + c)^r$ repeating r times, there will correspond a sum of r partial fractions of the type

$$\frac{A_1+B_1x}{ax^2+bx+c}+\frac{A_2+B_2x}{(ax^2+bx+c)^2}+\ldots\ldots+\frac{A_r+B_rx}{(ax^2+bx+c)^r}$$

Example 20. Evaluate $\int \frac{2x^4+9x^2+8}{x(x^2+2)^2}\,dx$.

Solution . The denominator of the given integral contains a linear factor which occurs exactly once and an irreducible quadratic factor which occurs twice. Thus we can write

$$\frac{2x^4+9x^2+8}{x(x^2+2)^2}=\frac{A}{x}+\frac{Bx+C}{x^2+2}+\frac{Dx+E}{(x^2+2)^2}$$

$$\Rightarrow \quad 2x^4+9x^2+8 = A(x^2+2)^2+(Bx+C)(x)(x^2+2)+(Dx+E)x$$

$$= (A+B)x^4+Cx^3+(4A+2B+D)x^2+(2C+E)x+4A$$

Equating like powers of x, we obtain

$$2 = A+B,\quad 0 = C,\quad 9 = 4A+2B+D,\quad 0 = 2C+E,\quad 8 = 4A$$

Solving gives $A = 2$, $B = 0$, $C = 0$, $D = 1$ and $E = 0$

$$\therefore \quad \frac{2x^4+9x^2+8}{x(x^2+2)^2}=\frac{2}{x}+\frac{x}{(x^2+2)^2}$$

$$\Rightarrow \quad \int\frac{2x^4+9x^2+8}{x(x^2+2)^2}\,dx = 2\int\frac{1}{x}\,dx+\int\frac{x}{(x^2+2)^2}\,dx$$

$$= 2\log x+\frac{1}{2}\int(x^2+2)^{-2}\,2x\,dx$$

$$= 2\log x+\frac{1}{2}\,\frac{(x^2+2)^{-2+1}}{(-2+1)}+c$$

$$= 2\log x-\frac{1}{2}\cdot\frac{1}{(x^2+2)}+c.$$

EXERCISE 9.2

1. Find the following integrals:

(i) $\int (x+5)^7\, dx$ (ii) $\int 2e^{3x+4}\, dx$ (iii) $\int \frac{1}{\sqrt{x-2}}\, dx$

2. Find the following integrals:

(i) $\int \frac{2x}{\sqrt{x^2+9}}\, dx$ (ii) $\int \frac{x}{\sqrt{a^2-x^2}}\, dx$ (iii) $\int x\sqrt{2x+3}\, dx$ (iv) $\int \frac{e^{\sqrt{x}}}{\sqrt{x}}\, dx$

3. Find the following integrals:

(i) $\int \frac{x^{n-1}}{\sqrt{a+bx^n}}\, dx$ (ii) $\int \frac{x^{e-1}+e^{x-1}}{x^e+e^x}\, dx$ (iii) $\int \frac{\sqrt{5+\log x}}{x}\, dx$

4. Find the following integrals:

(i) $\int xe^{-x}\, dx$ (ii) $\int xe^{3x}\, dx$ (ii) $\int (\log x)^2 dx$

5. Find the following integrals:

(i) $\int x^3 \log x\, dx.$ (ii) $\int \log(4x)\, dx$

6. Find the following integrals:

(i) $\int \frac{x}{(x-1)(2x+1)}\, dx$ (ii) $\int \frac{3x}{(x-1)(x-2)}\, dx$

7. Find the following integrals:

(i) $\int \frac{2x^2-5x-2}{(x-2)^2(x-1)}\, dx$ (ii) $\int \frac{3x^2+5x}{(x-1)(x+2)^2}\, dx$

8. Find the following integrals:

(i) $\int \frac{2x}{(x^2-1)(x^2+3)}\, dx$ (ii) $\int \frac{x^2+8}{x^3+4x}\, dx$

ANSWERS

1. (i) $\frac{(x+5)^8}{8}+c$ (ii) $\frac{2e^{3x+4}}{3}+c$ (iii) $2\sqrt{x-2}+c$

2. (i) $2\sqrt{x^2+9}+c$ (ii) $-\sqrt{a^2-x^2}+c$

(*iii*) $\frac{1}{2}\left[\frac{(2x+3)^{5/2}}{5}-(2x+3)^{3/2}\right]+c$ (*iv*) $2e^{\sqrt{x}}+c$

3. (*i*) $\frac{2\sqrt{a+bx^n}}{bn}+c$ (*ii*) $\frac{\log(x^e+e^x)}{e}+c$ (*iii*) $\frac{2(5+\log x)^{3/2}}{3}+c$

4. (*i*) $-xe^{-x}-e^{-x}+c$ (*ii*) $\frac{xe^{3x}}{3}-\frac{e^{3x}}{9}+c$ (*iii*) $x(\log x)^2-2x\log x+2x+c$

5. (*i*) $\frac{x^4\log x}{4}-\frac{x^4}{16}+c$ (*ii*) $x\log 4x-x+c$

6. (*i*) $\frac{1}{3}\log(x-1)+\frac{1}{6}\log(2x+1)+c$ (*ii*) $3\log\left[\frac{(x-2)^2}{x-1}\right]+c$

7. (*i*) $7\log(x-2)+\frac{4}{x-2}-5\log(x-1)+c$ (*ii*) $\frac{8}{9}\log(x-1)+\frac{19}{9}\log(x+2)+\frac{2}{3(x+2)}+c$

8. (*i*) $\frac{1}{4}\log\left[\frac{x^2-1}{x^2+3}\right]+c$ (*ii*) $2\log x-\frac{1}{2}\log(x^2+4)+c$

9.4 THE DEFINITE INTEGRAL

Let $F(x)$ be any antiderivative of $f(x)$, then for any two values of the independent variable x, say a and b, the difference $F(b)-F(a)$ is called the **definite integral** of $f(x)$ from a to b and is denoted by $\int_a^b f(x)\,dx$. Thus

$$\int_a^b f(x)\,dx = F(b)-F(a) \qquad \text{...(1)}$$

where $F(x)$ is any antiderivative of $f(x)$. The numbers a and b appearing with the integral sign $\int$ in Equation (1) are called the **limits of integration;** a is the lower limit and b is the upper limit. Usually, $F(b)-F(a)$ is abbreviated by writing $F(x)\Big|_a^b$.

The reason for using the term "definite integral" follows from the fact that the value of the definite integral is independent of the particular choice of the antiderivative of $f(x)$. For, if $F(x)+c$ is any other antiderivative of $f(x)$, then

$$\int_a^b f(x)\,dx = F(x)+c\Big|_a^b = [F(b)+c]-[F(a)+c] = F(b)-F(a),$$

which is same as before.

Thus to find $\int_a^b f(x)\, dx$, it suffices to find an antiderivative of $f(x)$, say $F(x)$ and subtract the value of $F(x)$ at the lower limit a from its value at the upper limit b.

Example 21. Find the following definite integrals:

(*i*) $\int_1^2 x\, dx$ (*ii*) $\int_0^1 \frac{1}{3x+2}\, dx$ (*iii*) $\int_0^1 e^{2t}\, dt$

Solution . (*i*) $\int_1^2 x\, dx = \frac{x^2}{2}\Big|_1^2 = \frac{(2)^2}{2} - \frac{(1)^2}{2} = \frac{3}{2}.$

(*ii*) $\int_0^1 \frac{1}{3x+2}\, dx = \frac{\log(3x+2)}{3}\Big|_0^1 = \frac{\log[3(1)+2]}{3} - \frac{\log[3(0)+2]}{3}$

$$= \frac{1}{3}[\log 5 - \log 2] = \frac{1}{3}\log\left(\frac{5}{2}\right).$$

(*iii*) $\int_0^1 e^{2t}\, dt = \frac{e^{2t}}{2}\Big|_0^1 = \frac{e^2}{2} - \frac{e^0}{2} = \frac{1}{2}(e^2 - 1).$

Properties of the Definite Integral

1. $\int_a^b f(x)\, dx = \int_a^b f(t)\, dt$

That is, the variable of integration is a "dummy variable" in the sense that any other variable produces the same result.

For Example, $\int_1^2 x\, dx = \int_1^2 t\, dt = \frac{3}{2}$

2. $\int_a^b f(x)\, dx = -\int_b^a f(x)\, dx$

That is, interchanging the limits of integration changes the integral's sign.

For Example, $\int_2^0 (5-x)\, dx = -\int_0^2 (5-x)\, dx$

3. $$\int_a^b f(x)\,dx = \int_a^c f(x)\,dx + \int_c^b f(x)\,dx, \quad a < c < b$$

That is, the interval over which a definite integral is to be evaluated may be subdivided.

4. *Geometrically,* $\int_a^b f(x)\,dx$ *can be interpreted as the area of the region bounded by the curve* $y = f(x)$*, the x-axis and the lines* $x = a$ *and* $x = b$*.* (Figure 9.1)

Similarly, $\int_c^d g(y)\,dy$ *can be interpreted as the area of the region bounded by the curve* $x = g(y)$*, the y-axis and the lines* $y = c$ *and* $y = d$ (Figure 9.2).

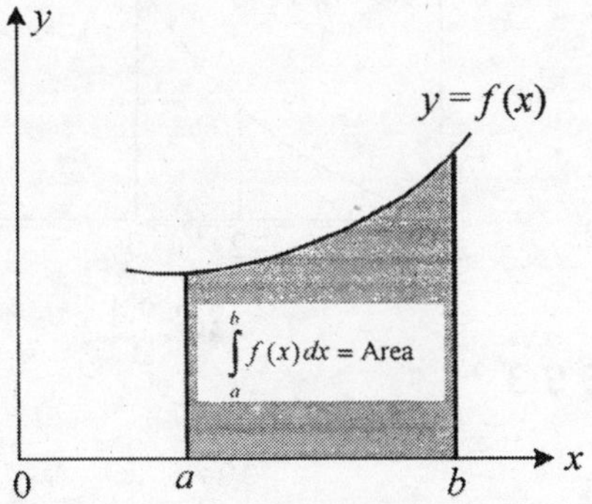

Fig. 9.1

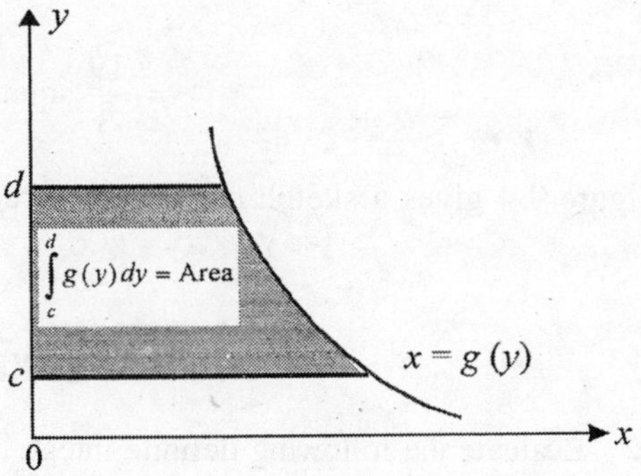

Fig. 9.2

Example 22. Use definite integral to find the area of the region bounded by the curve $y = |x|$, the *x*-axis and two lines $x = -2$ and $x = 1$.

Solution . A sketch of the region is given in Figure 9.3.

$$\text{Required Area} = \int_{-2}^{1} |x|\,dx$$

$$= \int_{-2}^{0} |x|\,dx + \int_{0}^{1} |x|\,dx$$

$$= \int_{-2}^{0} (-x)\,dx + \int_{0}^{1} x\,dx$$

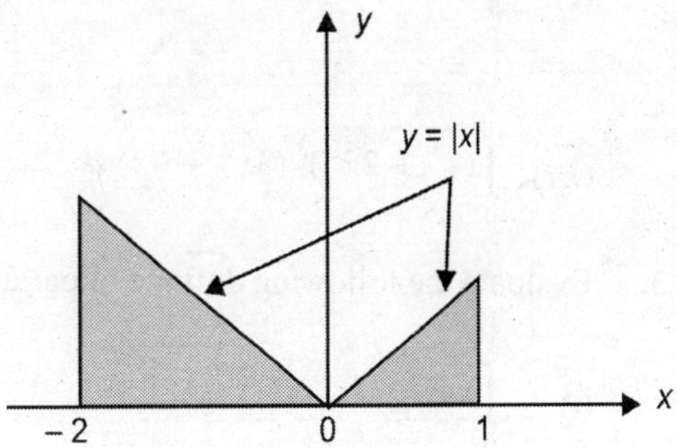

Fig. 9.3

$$= \left(-\frac{x^2}{2}\right)\Bigg|_{-2}^{0} + \left(\frac{x^2}{2}\right)\Bigg|_{0}^{1}$$

$$= \left[\left(-\frac{0^2}{2}\right)-\left(-\frac{(-2)^2}{2}\right)\right]+\left[\frac{(1)^2}{2}-\frac{(0)^2}{2}\right] = 2+\frac{1}{2}=\frac{5}{2} \text{ square units.}$$

Example 23. Find the area of the region bounded by the curve $y = x^2$, the x-axis and the lines $x = 2$ and $x = 3$.

Solution. By the property of the definite integral as an area, the required area bounded by the curve $y = x^2$, the x-axis and the lines $x = 2$ and $x = 3$ is given by

$$\text{Area} = \int_2^3 x^2\, dx = \left[\frac{x^3}{3}\right]_2^3 = \frac{3^3}{3}-\frac{2^3}{3}$$

$$= \frac{19}{3} \text{ square units.}$$

Figure 9.4 gives a sketch of the given region.

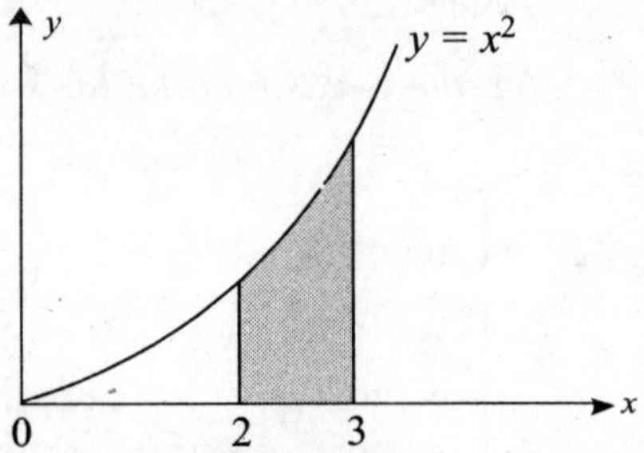

Fig. 9.4

EXERCISE 9.3

1. Evaluate the following definite integrals :

(*i*) $\int_0^4 5\, dx$ (*ii*) $\int_3^5 dx$ (*iii*) $\int_{-1}^1 (7x+2)\, dx$

2. Evaluate the following definite integrals :

(*i*) $\int_0^6 \sqrt{2x+4}\, dx$ (*ii*) $\int_{-2}^{-1} (3x^2 - x - 1)\, dx$

(*iii*) $\int_0^1 (x^3 + 2x^2)^3 (3x^2 + 4x)\, dx$ (*iv*) $\int_1^e \log x\, dx$

3. Evaluate the following definite integrals :

(*i*) $\int_0^1 xe^{-x^2}\, dx$ (*ii*) $\int_0^1 \frac{2-2x}{x^2+7x+12}\, dx$

4. Find the area of the region bounded by the curve $y = x^2$, the x-axis and the lines $x = 1$ and $x = 2$.

5. Find the area of the region bounded by the curve $y = e^x$, the x-axis and the lines $x = 0$ and $x = 3$.

ANSWERS

1. (*i*) 20 (*ii*) 2 (*iii*) 4 (*iv*) $\frac{98}{3}$ (*v*) 1 (*vi*) $\frac{5}{3}$

2. (*i*) $\frac{56}{3}$ (*ii*) $\frac{15}{2}$ (*iii*) $\frac{3}{4}$ (*iv*) 1

3. (*i*) $\frac{1}{2}\left(1 - e^{-1}\right)$ (*ii*) $8 \log\left(\frac{4}{3}\right) - 10 \log\left(\frac{5}{4}\right)$

4. $\frac{19}{3}$ square units **5.** $(e^3 - 1)$ square units

CHAPTER

10

APPLICATIONS OF INTEGRATION

OBJECTIVES

After studying the material in this chapter, you should be able to:

- *Find the cost function when marginal cost function is given.*
- *Find the revenue function and the demand function from a given marginal revenue function.*
- *Find the maximum profit if marginal revenue and marginal cost functions are given.*
- *Find the maximum profit over time if marginal revenue and marginal cost are given as functions of time.*
- *Derive the demand function when the price elasticity of demand is given.*
- *Evaluate consumers' surplus and producers' surplus.*

10.1 INTRODUCTION

Integration has a wide range of applications to a very large number of disciplines such as business and economics (marginal analysis, maximizing profit, consumer's surplus), biology (bacterial growth), geometry (finding the area under a curve), and to probability. In this chapter we shall discuss some of these applications.

10.2 APPLICATION TO MARGINAL ANALYSIS

In this section we investigate some of the applications of integration to marginal analysis. The first such application is that of finding the cost function when marginal cost function is given.

10.2.1 *To find the cost function and the average cost function from a given marginal cost function*

Since the marginal cost function MC is the derivative of the cost function C, therefore by definition C is an antiderivative of MC. That is,

$$C = \int MC \, dx + k,$$

where k is the constant of integration. However, if we are also given the fixed costs (*i.e.*, costs involved when $x = 0$) or the cost of production of a specific number of units of the commodity, we can determine the value of the constant of integration and thus find the particular cost function C. Once C is obtained, the average cost function can be obtained by using the equation

$$AC = \frac{C}{x}$$

Example 1. The marginal cost function of a firm is given by $MC = 3000\, e^{0.3x} + 50$, where x is quantity produced. If the fixed cost is ₹ 80,000, find the total cost of the firm.

[*Delhi Univ. B.Com. (H) 1990*]

***Solution*.** It is given that $\quad MC = 3000\, e^{0.3x} + 50$

Integrating with respect to x, we obtain

$$C = \int (3000 e^{0.3x} + 50)\, dx$$

$$= 3000 \frac{e^{0.3x}}{0.3} + 50x + k \qquad (k \text{ is a constant})$$

$$= 10{,}000\, e^{0.3x} + 50x + k$$

It is also given that the fixed cost is 80,000, *i.e.*, $C = 80{,}000$ when $x = 0$. Thus

$$80{,}000 = 10{,}000\, e^{(0.3)(0)} + 50\,(0) + k \quad \Rightarrow \quad k = 70{,}000$$

Therefore the total cost function of the firm is given by

$$C = 10{,}000\, e^{0.3x} + 50x + 70{,}000.$$

Example 2. The marginal cost function of manufacturing x pair of shoes is $6 + 10x - 6x^2$. The total cost of producing a pair of shoes is ₹ 1200. Find the total and average cost functions.

Solution. We are given : $\quad MC = 6 + 10x - 6x^2$

i.e., $$\frac{dC}{dx} = 6 + 10x - 6x^2$$

Integrating both sides with respect to x, we have

$$C = \int (6 + 10x - 6x^2)\, dx = 6x + 5x^2 - 2x^3 + k,$$

where k is the constant of integration. Since the cost of producing a pair of shoes is ₹ 1200, *i.e.*, $C = 1200$ when $x = 1$, therefore

$$1200 = 6(1) + 5(1)^2 - 2(1)^3 + k \quad \Rightarrow \quad k = 1191$$

Hence the total cost function is given by

$$C = 6x + 5x^2 - 2x^3 + 1191$$

The average cost function is given by

$$AC = \frac{C}{x} = 6 + 5x - 2x^2 + \frac{1191}{x}.$$

Example 3. Assume that the marginal cost in lakhs of rupees is given by

$$MC = 4 + 5x^2 + \frac{3}{2}e^{-x},$$

where x is the quantity produced. Find the total cost of production when $x = 2$, if fixed cost is ₹ 6 lakhs. **[*Delhi Univ. B.Com. (H) 1992*]**

Solution. The marginal cost is given to be

$$MC = 4 + 5x^2 + \frac{3}{2}e^{-x}$$

Integrating, we obtain

$$C = \int\left(4 + 5x^2 + \frac{3}{2}e^{-x}\right)dx = 4x + \frac{5}{3}x^3 - \frac{3}{2}e^{-x} + k,$$

where k is the constant of integration. Since $C = 6$, when $x = 0$, therefore

$$6 = -\frac{3}{2} + k \quad or \quad k = \frac{15}{2}$$

Hence $$C = 4x + \frac{5}{3}x^3 - \frac{3}{2}e^{-x} + \frac{15}{2},$$

which gives the cost of producing x units. In particular, when $x = 2$, we have

$$C = 8 + \frac{40}{3} - \frac{3}{2}e^{-2} + \frac{15}{2} = \frac{173}{6} - \frac{3}{2}e^{-2}.$$

Example 4. The marginal cost function of a firm is $MC = (\log x)^2$. Find the total cost of 100 units if the cost of producing one unit is ₹ 22. **[*Delhi Univ. B.Com. (H) 1999*]**

Solution. The marginal cost function is : $MC = (\log x)^2$

Integrating, we obtain $$C = \int(\log x)^2\, dx$$

$$= (\log x)^2\, x - \int\frac{2\log x}{x}\cdot x\, dx \qquad \textit{(integrating by parts)}$$

$$= x(\log x)^2 - 2\int \log x \, dx$$

$$= x(\log x)^2 - 2\left[\log x . x - \int \frac{1}{x} \cdot x \, dx\right] \quad \textit{(integrating by parts)}$$

$$= x(\log x)^2 - 2x \log x + 2x + k,$$

where k is the constant of integration. However, $C = 22$ when $x = 1$.

$$\therefore \quad 22 = 1(\log 1)^2 - 2(1)\log 1 + 2(1) + k \quad \Rightarrow \quad k = 20 \quad \text{(since } \log 1 = 0\text{)}$$

Hence the total cost function is given by: $C = x(\log x)^2 - 2x \log x + 2x + 20$.

In particular, the cost of producing 100 units is given by

$$C(x = 100) = 100(\log 100)^2 - 200 \log 100 + 200 + 20 = ₹1419.7252$$

Example 5. If the marginal cost function is given by $MC = 3(3x+4)^{-1/2}$ and the fixed cost is 2, find the average cost for 7 units of output.

Solution. The marginal cost function is : $MC = 3(3x+4)^{-1/2}$

Integrating, we obtain $C = \int 3(3x+4)^{-1/2} dx = 2(3x+4)^{1/2} + k$,

where k is the constant of integration. However, $C = 2$ when $x = 0$,

$$\therefore \quad 2 = 2[3(0)+4]^{1/2} + k \quad \Rightarrow \quad k = -2$$

Hence $\quad C = 2(3x+4)^{1/2} - 2$

The average cost function is given by

$$AC = \frac{C}{x} = \frac{2(3x+4)^{1/2}}{x} - \frac{2}{x}$$

In particular, when $x = 7$, $AC = \dfrac{2(25)^{1/2}}{7} - \dfrac{2}{7} = \dfrac{8}{7}$.

Example 6. If C is the total cost of an output x, it is known that the marginal cost MC always equals average cost AC. Show that C is a constant multiple of x.

Solution . We have $MC = AC$

i.e., $$\frac{dC}{dx} = \frac{C}{x} \quad \textit{or} \quad \frac{dC}{C} = \frac{dx}{x}$$

Integrating both sides with respect to x gives

$$\log C = \log x + \log k,$$

where k is an arbitrary constant. Thus

$$\log C = \log(kx) \quad \Rightarrow \quad C = kx,$$

which shows that C is a constant multiple of x.

Example 7. The marginal cost is given by $MC = 150 + e^{.5x}$, where x is the number of units of a product in hundreds. If x increases from 200 to 400, find the total increase in cost.

Solution. The marginal cost is :

$$MC = 150 + e^{.5x}, \quad \text{i.e.,} \quad \frac{dC}{dx} = 150 + e^{.5x}$$

We have to find total increase in cost when x increases from 200 to 400. That is, we have to find $C(4) - C(2)$. But

$$C(4) - C(2) = \int_2^4 \frac{dC}{dx}\,dx = \int_2^4 (150 + e^{.5x})\,dx$$

$$= (150x + 2e^{.5x})\Big|_2^4$$

$$= (600 + 2e^2) - (300 + 2e)$$

$$= 300 + 2e^2 - 2e.$$

Example 8. The cost of a truck engine overhaul is ₹ A, and the operating costs are at the rate of ₹ $a + bx$ per km, where the engine has run x kms since the last overhaul. If the engine is overhauled at M kms, find the total cost function C.

[*Delhi Univ. B.Com. (H) 2000*]

Solution. The total cost function, C, is given by

$$C = \text{Total engine overhauled cost} + \text{Total operating cost}$$

$$= A + \int_0^M (a + bx)\,dx$$

$$= A + \left| ax + \frac{bx^2}{2} \right|_0^M$$

$$= A + aM + \frac{bM^2}{2}.$$

EXERCISE 10.1

1. The marginal cost of a product is given by $MC = 0.003x^2 - 0.6x + 30$. Find the total cost function, given that the fixed cost is ₹ 42.
2. The marginal cost of production is found to be $MC = 1000 - 20x + x^2$, where x is the number of units produced. The fixed cost of production is ₹ 9000. Find the cost function.

[*Delhi Univ. B.Com. (H) 1982*]

3. If the marginal cost function is $MC = 4 - 2x + x^2$, find the total and the average cost functions, given that the fixed cost is ₹100. **[Delhi Univ. B.Com. (H) 1983]**

4. The marginal cost of production is found to be $MC = 2000 - 40x + 3x^2$, where x is the number of units produced. The fixed cost of production is ₹ 18,000. Find the total cost function and the average cost function. **[Delhi Univ. B.Com. (H) 1984]**

5. The marginal cost of a product is given by $MC = 10 - 0.01x + 0.0009x^2$. Find the total cost function and the average cost function if the cost of producing 10 units is given by ₹ 105.

6. The additional cost (in lakhs of rupees) of producing a motor car is given by $6 + 4x^2 + 1.5e^{-x}$, where x is the quantity produced. Determine the total cost of producing 5 motor cars if fixed cost is ₹ 7 lakhs. It is given that $e^{-5} = 0.006$. **[Delhi Univ. B.Com. (H) 2005(SOL)]**

7. The marginal cost function is given by $MC = 2e^{0.001x}$. Find the total cost function if fixed cost is ₹ 2000.

8. The marginal cost of a product is given by $MC = 2 + 3e^x$ where x is the number of units produced. Find the total cost of production when $x = 4$, if the fixed cost is ₹ 200.

9. The marginal cost function of a firm is $MC = (\log x)^2$. Find the total cost function if the cost of producing one unit is ₹ 20.

10. The marginal cost function of a firm is $MC = 3 + 2 \log x$. Find the total cost function when the cost of producing 1 unit is ₹ 21.

11. If the marginal cost function is given by $MC = 2\,(2x + 9)^{-1/2}$ and the fixed cost is ₹ 4, find the average cost for 8 units of output.

12. If the marginal cost function at x units of output is $\dfrac{a}{\sqrt{ax + b}}$ and if the cost of zero output is zero, find the total cost as a function of x. **[Delhi Univ. B.Com. (H) 1984]**

13. The marginal cost function of a product is given by $MC = \dfrac{x}{\sqrt{x^2 + 400}}$. Find the total cost function and the average cost function if the fixed cost is ₹ 900.

14. A factory manufacturing heavy machinery determined its marginal cost by the function $MC = x\sqrt{x + 1}$ and the cost is ₹ 7800 when 3 machines are produced. Find the cost function.

15. If the marginal cost of a product is equal to half its average cost, show that fixed cost is zero. If the cost of producing 9 units of the product is ₹ 60, find the cost function.

16. A manufacturer's marginal cost function is $\dfrac{dC}{dx} = 0.2x + 3$. If C is in rupees, determine the cost involved to increase the production from 60 to 70 units.

17. A manufacturer's marginal cost function is $\dfrac{dC}{dx} = \dfrac{500}{\sqrt{2x + 25}}$. If C is in rupees, determine the cost involved to increase the production from 100 to 300 units.

18. A manufacturer's marginal cost function is : $MC = 0.003x^2 - 0.6x + 40$, where x is the number of units of a product. If x increases from100 to 200 units, find the total increase in cost. **[Delhi Univ. B.Com. (H) 2007]**

19. The marginal cost function of a product is $MC = 3000e^{0.3x} + 100$, where x is the quantity produced. Determine the corresponding total and average cost functions when the fixed cost is ₹ 60,000. Also, find the total cost of first 10 units and the additional cost when output is increased from 10 to 20 units using definite integral.

ANSWERS

1. $C = 0.001x^3 - 0.3x^2 + 30x + 42$ **2.** $C = 1000x - 10x^2 + \frac{x^3}{3} + 9000$

3. $C = 4x - x^2 + \frac{x^3}{3} + 100, \quad AC = 4 - x + \frac{x^2}{3} + \frac{100}{x}$

4. $C = 2000x - 20x^2 + x^3 + 18{,}000, \quad AC = 2000 - 20x + x^2 + \frac{18{,}000}{x}$

5. $C = 10x - 0.005x^2 + 0.0003x^3 + 5.2, \quad AC = 10 - 0.005x + 0.0003x^2 + \frac{5.2}{x}$

6. 205.16 lakhs of rupees **7.** $C = 2000e^{0.001x}$ **8.** $205 + 3e^4$

9. $C = x(\log x)^2 - 2x\log x + 2x + 18$ **10.** $C = x + 2x\log x + 20$ **11.** 1

12. $C = 2\sqrt{ax+b} - 2\sqrt{b}$ **13.** $C = \sqrt{x^2+400} + 880, \quad AC = \frac{\sqrt{x^2+400}}{x} + \frac{880}{x}$

14. $C = \frac{2}{15}(x+1)^{3/2}(3x-2) + \frac{1{,}16{,}888}{15}$ **15.** $C = 20\sqrt{x}$ **16.** ₹ 160

17. ₹ 5000 **18.** ₹ 2000

19. $C = 10{,}000e^{0.3x} + 100x + 50{,}000, \quad AC = 10{,}000\frac{e^{0.3x}}{x} + 100 + \frac{50{,}000}{x}$

$C(10) = ₹\ 2{,}51{,}860, \quad C(20) - C(10) = ₹\ 38{,}34{,}440$

10.2.2 *To find the total revenue function and the demand function from a given marginal revenue function*

The marginal revenue MR is the derivative of the revenue function R. Therefore, to find R, we integrate the marginal revenue function. Thus

$$R = \int MR\,dx + k,$$

where k is the constant of integration to be determined by using the fact that $R = 0$, when $x = 0$.

Once R is obtained, the corresponding demand function can be obtained by using the equation $px = R$, or $p = R/x$.

Example 9. Given $MR = 2 - 6x$, where MR denotes marginal revenue, find the demand law.

Solution. Since MR is the derivative of the revenue function R, we have

$$R = \int (2 - 6x)\,dx = 2x - 3x^2 + k,$$

where k is the constant of integration. However, $R = 0$, when $x = 0$.

$$\therefore \qquad 0 = 0 + k \quad \Rightarrow \quad k = 0$$

Thus the revenue function is given by : $R = 2x - 3x^2$.

But $R = px$, where p is the price per unit. Thus

$$px = 2x - 3x^2 \quad \Rightarrow \quad p = \frac{2x - 3x^2}{x} = 2 - 3x,$$

which is the required demand function.

Example 10. Marginal revenue function of a firm is $\dfrac{ab}{(x-b)^2} - c$. Prove that the demand law is

$p = \dfrac{a}{b - x} - c$. **[*Delhi Univ. B.Com. (H) 1986, 2006 (SOL)*]**

Solution. If R denotes the revenue function, then we are given that

$$\frac{dR}{dx} = \frac{ab}{(x-b)^2} - c$$

Integrating both sides with respect to x gives

$$R = \int \left[\frac{ab}{(x-b)^2} - c\right] dx = -\frac{ab}{x-b} - cx + k,$$

where k is an arbitrary constant. However, $R = 0$, when $x = 0$.

$$\therefore \qquad 0 = -\frac{ab}{-b} - 0 + k \quad \Rightarrow \quad k = -a$$

Hence the revenue function is

$$R = -\frac{ab}{x-b} - cx - a$$

But $R = px$, where p is the price per unit. Thus

$$px = -\frac{ab}{x-b} - cx - a$$

$$\Rightarrow \qquad p = -\frac{ab}{x(x-b)} - c - \frac{a}{x}$$

$$= \frac{-ab - ax + ab}{x(x-b)} - c$$

$$= \frac{a}{b-x} - c,$$

which is the required demand law.

Example 11. The marginal revenue (in thousand of rupees) function for a particular commodity is $4 + e^{-0.03x}$, where x denotes the number of units sold. Determine the total revenue from the sale of 100 units. It is given that $e^{-3} = 0.05$ (app.)

[*Delhi Univ. B.Com. (H) 2019*]

Solution. If R denotes the revenue function, then we are given that

$$MR = 4 + e^{-0.03x}$$

$$\Rightarrow \quad R = \int (4 + e^{-0.03x})\, dx = 4x - \frac{e^{-0.03x}}{.03} + k,$$

where k is the constant of integration. However, $R = 0$ when $x = 0$.

$$\therefore \quad 0 = 0 - \frac{1}{.03} + k \quad \Rightarrow \quad k = 33.33$$

Hence the revenue from the sale of x units is given by

$$R(x) = 4x - \frac{1}{0.03} e^{-0.03x} + 33.33$$

In particular, the revenue from the sale of 100 units is:

$$R(100) = 400 - \frac{e^{-3}}{0.03} + 33.33$$

$$= 400 - \frac{0.05}{0.03} + 33.33$$

$= 431.67$ thousands of rupees = ₹ 4,31,670

ALITER:

Total revenue from the sale of 100 units $= \int_0^{100} MR\, dx = \int_0^{100} \left(4 + e^{-0.03x}\right) dx$

$$= \left[4x - \frac{e^{-0.03x}}{0.03}\right]_0^{100}$$

$$= \left[400 - \frac{e^{-3}}{0.03}\right] - \left[-\frac{1}{0.03}\right]$$

$$= 400 - 1.66 + 33.33$$

$$= 431.67 \text{ thousands of rupees.}$$

Example 12. Given the marginal revenue function $MR = \dfrac{4}{(2x+3)^2} - 1$, show that the average revenue function is $p = \dfrac{4}{6x+9} - 1$. **[*Delhi Univ. B.Com. (H) 2018*]**

Solution. Given : $\quad MR = \dfrac{4}{(2x+3)^2} - 1$

$$\therefore \quad R = \int \left[\frac{4}{(2x+3)^2} - 1 \right] dx = = 4 \cdot \frac{(2x+3)^{-1}}{(-1)2} - x + k,$$

where k is an arbitrary constant. Thus

$$\therefore \quad R = -\frac{2}{2x+3} - x + k$$

However, $R = 0$ when $x = 0$.

$$\therefore \quad 0 = -\frac{2}{2(0)+3} - 0 + k \quad \Rightarrow \quad k = \frac{2}{3}$$

Hence the revenue function is given by

$$R = -\frac{2}{2x+3} - x + \frac{2}{3}$$

The average revenue function is given by

$$p = AR = \frac{R}{x} = -\frac{2}{x(2x+3)} - 1 + \frac{2}{3x} = \frac{-6+4x+6}{3x(2x+3)} - 1 = \frac{4}{6x+9} - 1.$$

Thus $p = \dfrac{4}{6x+9} - 1$ is the desired average revenue function.

Example 13. A firm's marginal revenue function is $MR = 20e^{-x/10}\left(1 - \dfrac{x}{10}\right)$. Find the corresponding demand function. **[*Delhi Univ. B.Com. (H) 1998*]**

Solution. We have $\quad MR = 20e^{-x/10}\left(1 - \dfrac{x}{10}\right)$

Integrating both sides with respect to x, we obtain

$$R = \int 20e^{-x/10}\left(1-\frac{x}{10}\right)dx$$

$$= 20\int e^{-x/10}\,dx - 2\int xe^{-x/10}\,dx$$

$$= 20\int e^{-x/10}\,dx - 2\left[-10x\,e^{-x/10}\,dx - \int -10e^{-x/10}\,dx\right]$$

$$= 20\int e^{-x/10}\,dx + 20x\,e^{-x/10} - 20\int e^{-x/10}\,dx$$

$$= 20xe^{-x/10} + k,$$

where k is an arbitrary constant. However, $R = 0$, when $x = 0$.

$\therefore \quad 0 = 0 + k \quad \Rightarrow \quad k = 0$

Thus the revenue function is given by

$$R = 20xe^{-x/10}$$

But $R = px$, where p is the price per unit. Hence

$$px = 20xe^{-x/10} \quad \Rightarrow \quad p = 20e^{-x/10},$$

which is the required demand function.

Example 14. A manufacturer's marginal revenue function is $MR = 175 - 2x - 0.3x^2$. Find the increase in the manufacturer's total revenue if production is increased from 10 to 20 units.

Solution. The marginal revenue is :

$$MR = 175 - 2x - 0.3x^2, \quad \textit{i.e.,} \quad \frac{dR}{dx} = 175 - 2x - 0.3x^2$$

We have to find total increase in revenue when x increases from 10 to 20 units. That is, we have to find $R(20) - R(10)$. But

$$R(20) - R(10) = \int_{10}^{20} \frac{dR}{dx}\,dx = \int_{10}^{20} \left(175 - 2x - 0.3x^2\right) dx$$

$$= \left[175x - x^2 - 0.1x^3\right]_{10}^{20}$$

$$= \left[175(20) - (20)^2 - 0.1(20)^3\right] - \left[175(10) - (10)^2 - 0.1(10)^3\right]$$

$$= (3500 - 400 - 800) - (1750 - 100 - 100) = 750.$$

EXERCISE 10.2

1. If the marginal revenue function for a commodity is $MR = 9 - 4x^2$, find the demand function. **[Delhi Univ. B.Com. (H) 1983]**

2. If the marginal revenue function for output x is given by $MR = \frac{6}{(x+2)^2} + 5$, find the total revenue function and the demand function. **[Delhi Univ. B.Com. (H) 2006]**

3. If the marginal revenue function is $\frac{ab}{(x+b)^2} - c$, find the total revenue function and show that $p = \frac{a}{x+b} - c$ is the demand function (where p is price, x is quantity demanded and a, b and c are constants.) **[Delhi Univ. B.Com. (H) 1989]**

4. A firm has the marginal revenue function given by $MR = \frac{a}{(x+b)^2} - c$, where x is the output and a, b and c are constants. Show that the demand function is given by $x = \frac{a}{b(p+c)} - b$.

5. A firm has a marginal revenue function given by $MR = \frac{a}{x+b} - c$, where x is the output and a, b and c are constants. Show that the demand function is given by $p = \frac{a}{x}\log\left(\frac{x+b}{b}\right) - c$.

6. If the marginal revenue function of a firm is given by $MR = 25e^{-x/400}\left(1 - \frac{x}{400}\right)$, show that the corresponding demand function is $p = 25e^{-x/400}$, where p is price and x is quantity.

7. A firm's marginal revenue function is $MR = 10e^{-x/400}\left(1 - \frac{x}{400}\right)$, show that the corresponding demand function is $x = 400 \log (10/p)$, where p is price and x is quantity.

8. A manufacturer's marginal revenue function is $MR = 275 - x - 0.3x^2$. Find the increase in the manufacturer's total revenue if production is increased from 10 to 20 units.

9. A manufacturer's marginal revenue function is $MR = 200 - 3x^2$. Find the increase in the manufacturer's total revenue if production is increased from 5 to 10 units.

10. The sole producer of a product has determined that the marginal revenue function is $MR = 100 - 3x^2$. Determine the point elasticity of demand for the product when $x = 5$.

ANSWERS

1. $p = 9 - \frac{4}{3}x^2$ **2.** $R = \frac{3x}{x+2} + 5x, \ p = \frac{3}{x+2} + 5$

3. $R = \frac{ax}{x+b} - cx$ **8.** 1900 **9.** 125 **10.** $\frac{3}{2}$

10.2.3 *To find the maximum profit if marginal revenue and marginal cost functions are given*

If P denotes the profit function, then

$$\frac{dP}{dx} = \frac{d}{dx}(R - C) = \frac{dR}{dx} - \frac{dC}{dx} = MR - MC$$

Integrating both sides with respect to x gives

$$P = \int \frac{dP}{dx} = \int (MR - MC)\, dx + k,$$

where k is the constant of integration. However, if we are given an additional information, such as fixed cost or loss at zero level of output, we can determine the constant k. Once P is known, it can be maximized by using the concept of maxima and minima.

Remark. There is an alternative method for finding the maximum profit that avoids using the concept of maxima and minima. It uses the fact that maximum profit is obtained when "$MR = MC$". The maximum profit is then given by

$$P_{\max} = \int_0^{x_0} (MR - MC)\, dx, \qquad \text{...(1)}$$

where x_0 is obtained by solving $MR = MC$. Actually equation (1) gives the additional profit only. So loss at zero level of output must be subtracted from this for obtaining the actual profit.

The following examples will illustrate.

Example 15. The marginal cost $C'(x)$ and the marginal revenue $R'(x)$ are given by

$$C'(x) = 20 + \frac{x}{20} \quad \text{and} \quad R'(x) = 30$$

The fixed cost is ₹ 200. Determine the maximum profit and the number of items produced for this profit. **[*Delhi Univ. B.Com. (H) 1987*]**

Solution. Let P denote the profit function, then

$$\frac{dP}{dx} = \frac{dR}{dx} - \frac{dC}{dx} = 30 - 20 - \frac{x}{20} = 10 - \frac{x}{20}$$

Integrating both sides with respect to x, we get

$$P = \int\left(10 - \frac{x}{20}\right)dx = 10x - \frac{x^2}{40} + k,$$

where k is the constant of integration. However, we are given that fixed cost is 200. That is,

$$P = -200 \text{ when } x = 0 \quad \therefore \quad -200 = k$$

Thus $\quad P = 10x - \dfrac{x^2}{40} - 200$, which is the profit function

For P to be maximum, it is necessary that $\dfrac{dP}{dx} = 0$, *i.e.*, $10 - \dfrac{x}{20} = 0 \Rightarrow x = 200$

Also $\dfrac{d^2P}{dx^2} = -\dfrac{1}{20} < 0$. Thus the output $x = 200$ indeed gives a maximum profit. The maximum profit is given by

$$P = 10(200) - \frac{(200)^2}{40} - 200 = 800.$$

Hence maximum profit is ₹ 800 which is obtained when 200 items are produced.

Example 16. The XYZ Co. has approximated the marginal revenue function for one of its product by $MR = 20x - 2x^2$ and marginal cost $MC = 81 - 16x + x^2$. Determine the profit maximizing output and the total profit at the optimum level, assuming fixed cost as zero.

[*Delhi Univ. B.Com. (H) 2006, 2010, 2019*]

Solution. We are given : $\quad MR = 20x - 2x^2 \quad$ and $\quad MC = 81 - 16x + x^2$

For maximum profit, $\quad MR = MC \quad$ and $\quad \dfrac{dMR}{dx} < \dfrac{dMC}{dx}$

Now $\quad MR = MC \Rightarrow 20x - 2x^2 = 81 - 16x + x^2$

$\Rightarrow \quad 3x^2 - 36x + 81 = 0 \quad \Rightarrow \quad x = 3$ or $x = 9$

Also, $\quad \dfrac{dMR}{dx} = 20 - 4x \quad$ and $\quad \dfrac{dMC}{dx} = -16 + 2x$

At $x = 3$, $\quad \dfrac{dMR}{dx} = 8 \quad$ and $\quad \dfrac{dMC}{dx} = -10$

$\therefore \quad \dfrac{dMR}{dx}$ is not less than $\dfrac{dMC}{dx}$ at $x = 3$

Thus the value $x = 3$ is rejected. However,

at $x = 9$, $\frac{dMR}{dx} = -16$ and $\frac{dMC}{dx} = 2$

$$\therefore \quad \frac{dMR}{dx} < \frac{dMC}{dx} \quad \text{at} \quad x = 9$$

$\Rightarrow$ Profit is maximum when $x = 9$ and the maximum profit is :

$$P_{max} = \int_0^9 (MR - MC)\,dx = \int_0^9 (20x - 2x^2 - 81 + 16x - x^2)\,dx$$

$$= \int_0^9 (-3x^2 + 36x - 81)\,dx = \left| -x^3 + 18x^2 - 81x \right|_0^9 = 0.$$

Example 17. The marginal cost of producing x units of a commodity in a day is given as $MC = 16x - 1591$. The selling price is fixed at ₹ 9 per unit and the fixed cost is ₹ 1800 per day. Determine (*i*) cost function (*ii*) revenue function, (*iii*) profit function, and (*iv*) maximum profit that can be obtained in one day. ***[Delhi Univ. B.Com. (H) 1989]***

Solution. (*i*) The marginal cost function is : $MC = 16x - 1591$

Integrating, we obtain $\quad C = \int (16x - 1591)\,dx = 8x^2 - 1591x + k$,

where k is an arbitrary constant. However, since the fixed cost is ₹ 1800, *i.e.*, $C = 1800$ when $x = 0$, we have

$$1800 = 8(0)^2 - 1591(0) + k \quad \Rightarrow \quad k = 1800$$

Therefore the cost function is given by

$$C = 8x^2 - 1591x + 1800$$

(*ii*) Since the selling price is fixed at ₹ 9 per unit, the revenue function is given by

$$R = 9x$$

(*iii*) The profit function is given by

$$P = R - C = 9x - (8x^2 - 1591x + 1800) = -8x^2 + 1600x - 1800$$

(*iv*) For maximum profit, it is necessary that

$$\frac{dP}{dx} = 0 \quad i.e., \quad -16x + 1600 = 0 \quad \Rightarrow \quad x = 100$$

Also $$\frac{d^2P}{dx^2} = -16 < 0$$

This shows that profit is maximum when $x = 100$. The corresponding maximum profit is :

$$P = (-8)(100)^2 + 1600(100) - 1800 = ₹\ 78{,}200.$$

Example 18. A company suffers a loss of ₹ 121.50 if one of its products does not sell at all. Marginal revenue and Marginal cost functions for the product are given by

$$MR = 30 - 6x \quad \text{and} \quad MC = -24 + 3x$$

Determine the total profit function, the break-even points and the profit maximizing level of output.

Solution. If P denotes the profit function, then

$$\frac{dP}{dx} = MR - MC = (30 - 6x) - (-24 + 3x) = 54 - 9x$$

Integrating both sides with respect to x, we have

$$P = \int (54 - 9x)\, dx = 54x - \frac{9}{2}x^2 + k,$$

where k is an arbitrary constant. Since the company suffers a loss of ₹ 121.50 if the product does not sell at all, *i.e.*, $P = -121.50$ when $x = 0$.

$$\therefore \quad -121.50 = 54(0) - \frac{9}{2}(0)^2 + k \quad \Rightarrow \quad k = -121.50$$

Thus the profit function is given by

$$P = 54x - \frac{9}{2}x^2 - 121.50$$

For break-even points, $P = 0$. That is,

$$54x - \frac{9}{2}x^2 - 121.50 = 0 \quad \Rightarrow \quad x^2 - 12x + 27 = 0 \quad \Rightarrow \quad x = 3 \text{ and } x = 9$$

Thus the break-even points are $x = 3$ and $x = 9$.

For P to be maximum, it is necessary that

$$\frac{dP}{dx} = 0 \quad i.e., \quad 54 - 9x = 0 \quad \Rightarrow \quad x = 6$$

Also,
$$\frac{d^2P}{dx^2} = -9 < 0$$

This shows that profit is maximum when $x = 6$.

10.2.4 *To find the maximum profit over time if marginal revenue and marginal cost are given as functions of time*

In many business operations such as oil drilling, mining and other depletion operations, the initial revenue rate is generally higher than the revenue rate after a period of time has passed. That is, revenue rate, as a function of time, is a decreasing function. However, the cost rate of such operations generally increases with time because of inflation and other reasons. That is, cost rate, as a function of time, is an increasing function. In such operations, the management is faced with the problem of determining the time t^* that maximizes the profit function $P(t)$.

To solve such a problem, we denote the revenue and the cost function by $R(t)$ and $C(t)$, respectively, where t denotes time. This representation of revenue and cost differs from the usual economic definitions of price per unit times number of units, and cost per unit times number of units. The derivatives $R'(t)$ and $C'(t)$, taken with respect to time, represent revenue and cost as time rates. We, further, assume that the revenue rate is greater than the cost rate at the beginning of the business operation under consideration. Also, as time goes on, we assume the cost rate increases to the revenue rates, and thereafter exceeds it. The optimum time at which the business operation should continue is that point of time where the rates are equal. That is, the optimum time t^* is given by $R'(t) = C'(t)$.

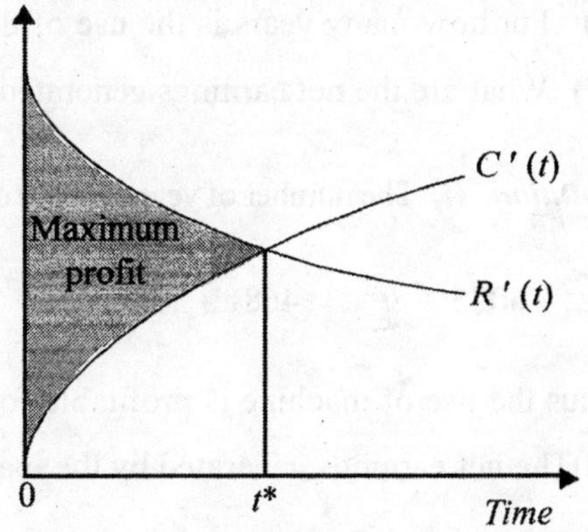

Fig. 10.1

The maximum profit is obtained when $t = t^*$ and the maximum profit is :

$$P(t^*) = \int_0^{t^*} [R'(t) - C'(t)]\, dt$$

Geometrically, the maximum profit $P(t^*)$ is the area enclosed by the graphs of $R'(t)$ and $C'(t)$ from $t = 0$ to $t = t^*$. See Figure 10.1.

Example 19. The revenue and the cost rate of G-B Oil Company are, respectively,

$$R'(t) = 12 - 3t^{1/3} \quad \text{and} \quad C'(t) = 4 + t^{1/3}$$

where t is measured in years and R and C are measured in millions of rupees. Determine how long the operation should continue and the profit that can be generated during this period.

[*Delhi Univ. B.Com. (H) 2022*]

Solution. The optimum time t^* at which the business operation should continue is given by

$$R'(t) = C'(t)$$

i.e., $12 - 3t^{1/3} = 4 + t^{1/3} \Rightarrow 8 = 4t^{1/3}$ or $t^{1/3} = 2 \Rightarrow t^* = 8$ years

The profit $P(t^*)$ that can be generated during this period is given by

$$P(t^*) = \int_0^8 [R'(t) - C'(t)]\, dt$$

$$= \int_0^8 \left[(12 - 3t^{1/3}) - (4 + t^{1/3})\right] dt = \int_0^8 (8 - 4t^{1/3})\, dt$$

$$= \left|8t - 3t^{4/3}\right|_0^8 = 16 \text{ million rupees.}$$

Example 20. Suppose that when it is t years old, an industrial machine generates revenue at the rate of $R'(t) = 6025 - 8t^2$ rupees per year and results in costs that accumulate at the rate of $C'(t) = 4681 + 13t^2$ rupees per year.

(*i*) For how many years is the use of the machine profitable ?

(*ii*) What are the net earnings generated by the machine during its period of profitability ?

[Delhi Univ. B.Com. (H) 2001, 2018]

Solution. (*i*) The number of years the use of machine will generate profit is given by $R'(t) = C'(t)$.

i.e., $$6025 - 8t^2 = 4681 + 13t^2 \Rightarrow 1344 - 21t^2 = 0 \Rightarrow t^2 = \frac{1344}{21} = 64 \Rightarrow t = 8$$

Thus the use of machine is profitable for 8 years.

(*ii*)The net earnings generated by the machine during its period of profitability is :

$$\int_0^8 [R'(t) - C'(t)]\,dt = \int_0^8 (1344 - 21t^2)\,dt = \left|1344t - 7t^3\right|_0^8 = ₹7168.$$

Example 21. Suppose that x years from now, one investment plan will be generating profit at the rate of $P_1(x) = 50 + x^2$ rupees per year, while a second plan will be generating profit at the rate of $P_2(x) = 200 + 5x$ rupees per year. For how many years will the second plan be more profitable one? Compute the net excess profit if invested in the second plan for the period of time in part (*i*) **[Delhi Univ. B.Com. (H) 2010]**

Solution. Second plan will be more profitable until $P_1(x) = P_2(x)$. That is,

$$50 + x^2 = 200 + 5x \Rightarrow x^2 - 5x - 150 = 0 \Rightarrow x = 15$$

The net excess profit that can be generated during this period is given by

$$\int_0^{15} [P_2(x) - P_1(x)]\,dx = \int_0^{15} \left[(200 + 5x) - (50 + x^2)\right]dx$$

$$= \int_0^{15} (150 + 5x - x^2)\,dx = \left|150x + \frac{5}{2}x^2 - \frac{x^3}{3}\right|_0^{15} = ₹1687.50.$$

EXERCISE 10.3

1. At the production level x, the marginal cost of production of a firm is ₹ $(4 + 0.11x)$ and the marginal revenue is ₹ 15. The overhead costs are ₹ 100. Find the maximum profit the firm can earn and the level of production for this profit. **[Delhi Univ. B.Com. (H) 1988]**

2. The marginal cost of production is found to be $MC = 1000 - 20x + x^2$, where x is the number of units produced. The fixed cost of production is ₹ 9000. Find the cost function. If the manufacturer fixes the price per unit at ₹ 3400, find the profit function and the sales volume that yields maximum profit.
3. The marginal cost (MC) and marginal revenue (MR) of a firm are given as $MC = 4 + .08x$ and $MR = 12$, where x denotes the level of output. Compute the maximum profit of the firm given that the initial fixed cost is ₹ 100. **[*Delhi Univ. B.Com. (H) 1985*]**
4. The marginal cost and marginal revenue of a firm are given as

$$MC = 40 + \frac{x}{8} \quad \text{and} \quad MR = 50 + \frac{x}{10},$$

where x denotes the level of output. The overhead costs are ₹ 500. Find the maximum profit the firm can earn and the level of production for this profit.

5. The marginal cost and marginal revenue of a firm are given as $MC = 3x^2 - 118x + 1315$ and $MR = 1000 - 4x$. Determine the maximized profit using definite integration, assuming fixed cost as ₹ 595.
6. The marginal cost of a product is given by $MC = 3x + 4$, where x is the number of units produced. The fixed cost of production is ₹ 10. Find the cost function. If the selling price is fixed at ₹ 40 per unit, find the revenue function and the maximum profit.
7. The marginal revenue and the marginal cost for an output x of a commodity are given as $MR = 5 - 4x + 3x^2$ and $MC = 3 + 2x$. If the fixed cost is zero, find the profit function and the profit when output is $x = 4$.
8. The marginal cost and marginal revenue of a firm are given as $MC = 14 + x$ and $MR = 36 - 24x + 3x^2$. Determine the maximized profit using definite integration, assuming fixed cost as zero.
9. A company suffers a loss of ₹ 1000 if its product does not sell at all. Marginal revenue and Marginal cost functions for the product are given by

$$MR = 50 - 4x \quad \text{and} \quad MC = -10 + x$$

Determine the total profit function, break-even points and the profit maximizing level of output. **[*Delhi Univ. B.Com. (H) 1989*]**

10. A company suffers a loss of ₹1000 if its product does not sell at all. The marginal revenue and marginal cost of the firm are given as $MR = 40 - 2x$ and $MC = x^2 - 62x + 40$, respectively. Find the profit function and the sales volume that yields maximum profit. Also, obtain the profit at this sales volume.
11. The marginal cost and marginal revenue of a firm are given as $MC = 15 - 2x - x^2$ and $MR = 25 - 5x - 2x^2$, where x denotes the level of output. Compute the maximum profit of the firm given that the initial fixed cost is zero.
12. A manufacturer estimates marginal revenue to be $10q^{-1/2}$ rupees per unit when the level of production is q units. The marginal cost has been found to be $0.4q$ rupees per unit. Suppose the manufacturer's profit is ₹ 520 when the level of production is 16 units. What is manufacturer's profit when level of production is 25 units ?

13. The revenue and the cost rate of Gold Star mining operations are, respectively,

$$R'(t) = 19 - t^{1/2} \text{ and } C'(t) = 3 + 3t^{1/2}$$

where t is measured in years and R and C are measured in millions of rupees. Determine how long the operation should continue and what the total profit will be at the end of the operation?

14. The revenue and the cost rate of G-B Oil Company are, respectively,

$$R'(t) = 9 - t^{1/3} \quad \text{and} \quad C'(t) = 1 + 3t^{1/3}$$

where t is measured in years and R and C are measured in millions of rupees. Determine how long the operation should continue and the profit that can be generated during this period.

15. Suppose that when an industrial machine is t years old, it generates revenue at the rate of $R'(t) = 50{,}000 - 200t^2$ rupees per year and results in costs that accumulate at the rate of $C'(t) = 20{,}000 + 100t^2$ rupees per year.

(*i*) For how many years is the use of machine profitable?

(*ii*) What are the net earnings generated by the machine during its period of profitability?

ANSWERS

1. Maximum profit = ₹ 450 and $x = 100$

2. $C = 1000x - 10x^2 + \frac{x^3}{3} + 9000$, Profit function : $P = 2400x + 10x^2 - \frac{x^3}{3} - 9000$; $x = 60$

3. Maximum profit = ₹ 300 and $x = 100$ **4.** $x = 400$, max P = ₹ 1500

5. $x = 35$, max. profit = ₹ 15,330

6. Revenue function : $R = 40x$, max. profit = ₹ 206

7. Profit function : $P = 2x - 3x^2 + x^3$; 24 **8.** 10.5

9. Profit function : $P = 60x - \frac{5}{2}x^2 - 1000$, No break-even point and $x = 12$

10. $P = -\frac{x^3}{3} + 30x^2 - 1000$; $x = 60$; max P = ₹35,000 **11.** $\frac{34}{3}$

12. ₹ 466.20 **13.** 16 years, 85.33 millions of rupees

14. 8 years, 16 millions of rupees **15.** 10 years, ₹ 2,00,000

10.2.5 *To find the demand function when the price elasticity of demand is given*

If x is the number of units demanded at price p per unit of a certain product, then we know that the price elasticity of demand is given by

$$\eta_d = -\frac{p}{x}\frac{dx}{dp}$$

The following examples illustrate how demand function could be obtained, using techniques of integration, if elasticity of demand is known.

Example 22. Derive the demand function which has the unit price elasticity of demand throughout. **[*Delhi Univ. B.Com. (H) 1983,1991*]**

Solution. It is given that

$$\eta_d = 1 \quad i.e., \quad -\frac{p}{x}\frac{dx}{dp} = 1 \quad \Rightarrow \quad \frac{dx}{x} + \frac{dp}{p} = 0$$

Integrating, we have

$$\log x + \log p = \log k,$$

where k is an arbitrary constant. Thus

$$\log (xp) = \log k$$

$$\Rightarrow \quad xp = k,$$

which is the required demand function.

Example 23. Obtain the demand function for a commodity whose elasticity of demand is given by $e_{xp} = a - bp$, where a, b are constants and p denotes the price per unit of the commodity. **[*Delhi Univ. B.Com. (H) 1985*]**

Solution. The elasticity of demand is given by

$$e_{xp} = a - bp$$

$$i.e., \quad -\frac{p}{x}\frac{dx}{dp} = a - bp$$

$$\Rightarrow \quad \frac{dx}{x} + \left(\frac{a}{p} - b\right)dp = 0$$

Integrating, we have

$$\log x + a \log p - bp = c,$$

where c is an arbitrary constant. Thus

$$\log x + \log p^a = bp + c,$$

$$\Rightarrow \quad \log (xp^a) = bp + c$$

$$\Rightarrow \quad xp^a = e^{bp + c}$$

$$\text{i.e.,} \quad x = p^{-a}e^{bp + c},$$

which is the required demand function.

Example 24. The price elasticity of demand for a commodity is p/x^3. Find the demand function if the demand is 3 when the price is 2.

Solution. We are given : $\eta_d = -\frac{p}{x}\frac{dx}{dp} = \frac{p}{x^3} \quad \Rightarrow \quad x^2\,dx + dp = 0.$

Integrating, we get $\quad \frac{x^3}{3} + p = k,$

where k is an arbitrary constant. However, it is given that $x = 3$ when $p = 2$.

Thus, $\quad \frac{(3)^3}{3} + 2 = k \quad \Rightarrow \quad k = 11$

$\therefore \quad \frac{x^3}{3} + p = 11 \quad \Rightarrow \quad p = 11 - \frac{x^3}{3}$, which is the required demand function.

Example 25. The price elasticity of demand for a commodity is $\eta_d = \frac{3p}{(p-1)(p+2)}$. Find the corresponding demand function if quantity demanded is 8 units when the price is ₹ 2.

[*Delhi Univ. B.Com. (H) 1997, 2011*]

Solution. Given : $\quad \eta_d = \frac{3p}{(p-1)(p+2)} \quad$ *i.e.,* $\quad -\frac{p}{x}\frac{dx}{dp} = \frac{3p}{(p-1)(p+2)}$

$$\Rightarrow \quad -\frac{dx}{x} = \frac{3\,dp}{(p-1)(p+2)}$$

$$\Rightarrow \quad \frac{dx}{x} + \frac{3\,dp}{(p-1)(p+2)} = 0$$

$$\Rightarrow \quad \frac{dx}{x} + \left(\frac{1}{p-1} - \frac{1}{p+2}\right)dp = 0 \quad \textit{(decomposing into partial fractions)}$$

Integrating, we get

$$\log x + \log(p-1) - \log(p+2) = \log k,$$

where k is an arbitary constant. Thus

$$\log\left[\frac{x(p-1)}{p+2}\right] = \log k \quad \Rightarrow \quad \frac{x(p-1)}{p+2} = k$$

However, it is given that $x = 8$ when $p = 2$.

$$\therefore \quad \frac{8(2-1)}{2+2} = k \quad \Rightarrow \quad k = 2$$

Thus the required demand function is given by $\frac{x(p-1)}{p+2} = 2$ or $x = \frac{2(p+2)}{p-1}$.

Example 26. The elasticity of cost is given by $\dfrac{5x}{2\,(5x+9)}$. Find the total cost function given that fixed cost is ₹ 18. **[*Delhi Univ. B.Com. (H) 2014, 2019*]**

Solution. The elasticity of cost is given by

$$\eta_C = \frac{5x}{2(5x+9)} \quad \textit{i.e.,} \quad \frac{x}{C}\frac{dC}{dx} = \frac{5x}{2(5x+9)}$$

$$\Rightarrow \qquad \frac{dC}{C} = \frac{5}{2}\cdot\frac{dx}{5x+9}$$

Integrating, we obtain

$$\log C = \frac{1}{2}\log(5x+9) + \log k,$$

where k is an arbitrary constant. Thus

$$\log C = \log\sqrt{5x+9} + \log k$$

$$\Rightarrow \qquad \log C = \log k\sqrt{5x+9} \quad \Rightarrow \quad C = k\sqrt{5x+9}$$

However, it is given that fixed cost is 18. That is, $C = 18$ when $x = 0$.

$$\therefore \qquad 18 = k\sqrt{0+9} \quad \Rightarrow \quad k = 6$$

Hence $\quad C = 6\sqrt{5x+9}$, which is the required cost function.

Example 27. The elasticity of a function $y = f(x)$ is $\dfrac{3x}{(x-1)(x+2)}$. Determine the function if $y = 6$ when $x = 4$.

Solution. By definition, the elasticity of a function $y = f(x)$ is $\dfrac{x}{y}\dfrac{dy}{dx}$. Thus we are given

$$\frac{x}{y}\frac{dy}{dx} = \frac{3x}{(x-1)(x+2)}$$

$$\Rightarrow \qquad \frac{dy}{y} - \frac{3dx}{(x-1)(x+2)} = 0$$

$$\textit{i.e.,} \qquad \frac{dy}{y} + \left(\frac{1}{x+2} - \frac{1}{x-1}\right)dx = 0 \qquad \textit{(by partial fraction)}$$

Integrating, we have

$$\log y + \log(x+2) - \log(x-1) = \log k,$$

where k is an arbitrary constant. Using properties of logarithm, we obtain

$$\log\left[\frac{y(x+2)}{x-1}\right] = \log k \quad \Rightarrow \quad \frac{y(x+2)}{x-1} = k$$

But we are given that $y = 6$ when $x = 4$.

$$\therefore \quad \frac{6(4+2)}{4-1} = k \quad \Rightarrow \quad k = 12$$

This gives $y = \dfrac{12(x-1)}{x+2}$, which is the required function.

10.2.6 ***To find consumption function if the marginal propensity to consume is given***

Recall that the marginal propensity to consume is defined as the rate of change of consumption with respect to income. Thus if $C = f(I)$ is the consumption function, where I is the total national income and C is the total national consumption, then the marginal propensity to consume (MPC) is given by

$$MPC = \frac{dC}{dI}$$

The following example illustrates how to find the consumption function if the marginal propensity to consume is given.

Example 28. The marginal propensity to consume is given by $\dfrac{dC}{dI} = \dfrac{1}{\sqrt{I}}$, where C is the consumption and I is the income. Find the consumption function if $C = 8$ when $I = 9$.

Solution. We have $\quad \dfrac{dC}{dI} = \dfrac{1}{\sqrt{I}}$

$$\Rightarrow \quad C = \int I^{-1/2}\, dI = 2\sqrt{I} + k,$$

where k is the constant of integration.

$$\because \quad C = 8 \text{ when } I = 9 \quad \therefore \quad 8 = 2\sqrt{9} + k \quad \Rightarrow \quad k = 2$$

Hence the consumption function is $C = 2\sqrt{I} + 2$.

Example 29. The marginal propensity to save is given by $\dfrac{dS}{dI} = \dfrac{1}{4} + \dfrac{1}{6\sqrt{I}}$, where S is savings and I is income. Find the total savings S if $S = 12$ when $I = 25$.

[Delhi Univ. B.Com. (H) 2007 (SOL)]

Solution. We are given : $\quad \dfrac{dS}{dI} = \dfrac{1}{4} + \dfrac{1}{6\sqrt{I}},$

$$\Rightarrow \qquad S = \int \left(\frac{1}{4} + \frac{1}{6\sqrt{I}} \right) dI = \frac{1}{4} I + \frac{1}{3} \sqrt{I} + k,$$

where k is the constant of integration.

$\because \quad S = 12$ when $I = 25 \quad \therefore \quad 12 = \frac{25}{4} + \frac{5}{3} + k \quad \Rightarrow \quad k = \frac{49}{12}$

Hence the total savings function is : $S = \frac{1}{4} I + \frac{1}{3} \sqrt{I} + \frac{49}{12}$.

EXERCISE 10.4

1. Obtain the demand function for a commodity for which price elasticity of demand is 'k' throughout. **[*Delhi Univ. B.Com. (H) 1994, 2014 (SOL*]**

2. Deduce a demand relation which shows a constant price elasticity of demand equal to 1/2. **[*Delhi Univ. B.Com. (H) 2007 (SOL)*]**

3. The price elasticity of demand for a commodity is p/x. Find the demand function if the demand is 3 when the price is 1.

4. Obtain the demand function for a commodity whose price elasticity of demand is given by $\frac{1}{kx}$, where k is constant.

5. The elasticity of demand of a commodity with respect to price is calculated to be $\frac{5p}{(p+3)(p-2)}$, where p is price. Find the demand function, if it is known that the quantity demanded is 5 units at $p = 3$. **[*Delhi Univ. B.Com. (H) 1996, 2022*]**

6. The elasticity of demand of a commodity with respect to price is calculated to be $\frac{p}{(p-1)(p-2)}$, where p is price. Find the demand function if it is known that the quantity demanded is 4 units at $p = 3$.

7. The elasticity of cost is given by $\eta_C = \frac{3x}{2(3x+4)}$. Find the total cost function given that the fixed cost is ₹ 18. **[*Delhi Univ. B.Com. (H) 2019 (Modified)*]**

8. If y is the number of incomes of ₹ x each, it is known that y decreases as x increases according to the law $\frac{dy}{dx} = -\frac{my}{x}$, where m is a given constant. Show that the dependence of y on x is $y = cx^{-m}$, where c is an arbitrary constant.

9. The marginal propensity to consume is given by $\frac{dC}{dI} = 0.5$, where C is the consumption and I is income. Find the consumption function if $C = 100$ when $I = 2$.

10. The marginal propensity to consume is given by $\frac{dC}{dI} = \frac{3}{4} - \frac{1}{2\sqrt{3I}}$, where consumption C is function of national income I. Determine the consumption function if it is known that consumption is 10 when $I = 12$.

11. The marginal propensity to save is given by $\frac{dS}{dI} = \frac{1}{4} + \frac{1}{6\sqrt{I}}$, where S is savings and I is income. Find the total savings S if $S = 2$ when $I = 25$.

ANSWERS

1. $xp^k = c$ **2.** $xp^{1/2} = c$ **3.** $p = 4 - x$

4. $p = ce^{-kx}$ **5.** $x = \frac{5(p+3)}{6(p-2)}$ **6.** $x = \frac{2(p-1)}{p-2}$

7. $C = 10\sqrt{3x+4}$ **9.** $C = 0.5\,I + 99$ **10.** $C = \frac{3}{4}I - \frac{\sqrt{I}}{\sqrt{3}} + 3$

11. $S = \frac{1}{4}I + \frac{1}{3}\sqrt{I} - \frac{71}{12}$

10.3 CONSUMERS' AND PRODUCERS' SURPLUS

Determining the area of a region has applications in economics. Figure 10.2 shows the demand and supply curves for a product.

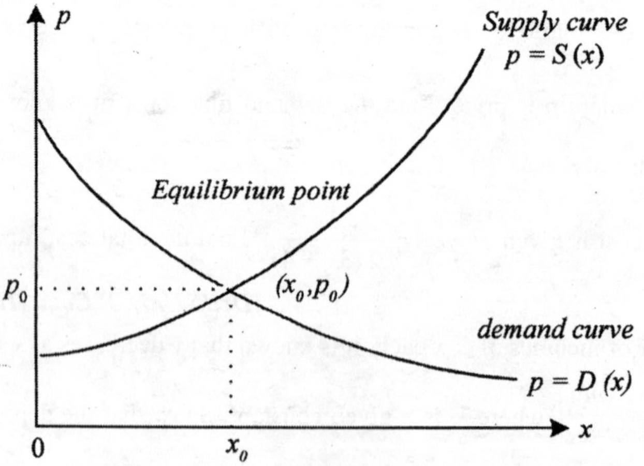

Fig. 10.2

The intersection of a demand and supply curve is known as the equilibrium point. It is the point where market equilibrium is attained and occurs at the price where the quantity demanded is equal to the quantity supplied.

Let us assume that the market is at equilibrium and the price per unit of the product is p_0. According to the demand curve, there are consumers who would be willing to pay more than p_0. These consumers will benefit by paying the lower equilibrium price. The total of their benefits is called the **consumers' surplus** (abbreviated *CS*) and is represented by the area between the line $p = p_0$ and the demand curve $p = D(x)$ from $x = 0$ to $x = x_0$ (see Fig. 10.3), where x_0 is the market demand. Thus

$$CS = \int_0^{x_0} D(x)\,dx - x_0 \times p_0$$

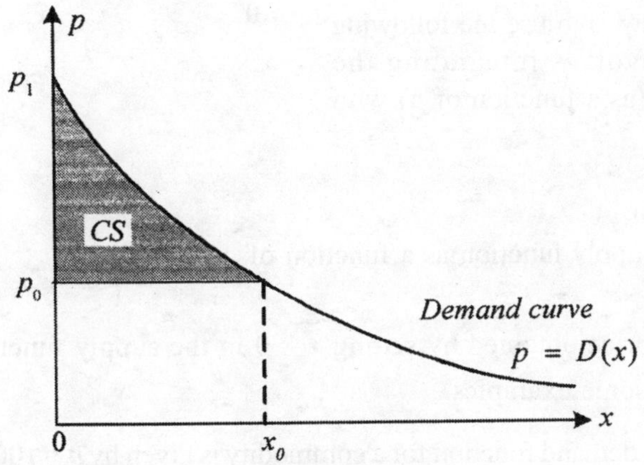

Fig. 10.3

There is an alternative method to find the consumers' surplus that involves integrating the demand function (as a function of p) with respect to p :

$$CS = \int_{p_0}^{p_1} (\text{demand function as a function of } p)\, dp,$$

where p_1 is the price at which quantity demanded is zero. It can be obtained by setting $x = 0$ in the demand function.

Consumers' Surplus under Pure Competition : *Under pure competition, the equilibrium demand x_0 is obtained by equating the demand and supply functions. Once x_0 is obtained, the corresponding market price p_0 is obtained by substituting the value of x_0 in the demand function or the supply function.*

Consumers' Surplus under Monopoly : *Under monopoly, the quantity x_0 is obtained by equating MR and MC (so as to maximize profit). The corresponding value of p_0 is obtained by substituting the value of x_0 in the demand function.*

According to the supply curve, some of the producers also benefit from the equilibrium price, since they are willing to supply the product at prices less than p_0. The total of their benefits is called the **producers' surplus** (abbreviated *PS*) and is represented by the area between the line $p = p_0$ and the supply curve $p = S(x)$ from $x = 0$ to $x = x_0$ (See Figure 11.4). Thus

$$PS = x_0 \times p_0 - \int_0^{x_0} S(x)\, dx$$

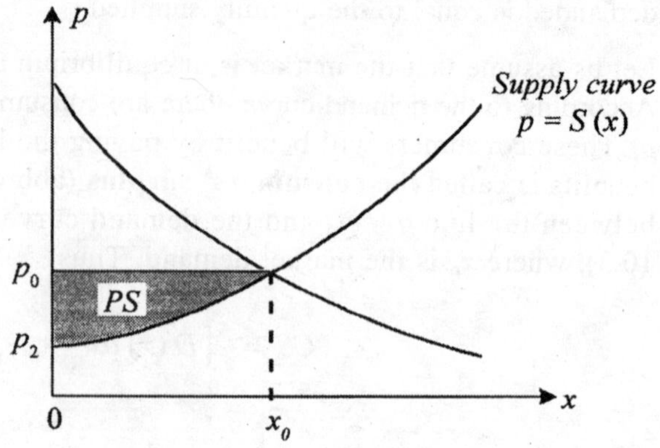

Fig. 10.4

An alternative way is to use the following formula that involves integrating the supply function (as a function of p) with respect to p :

$$PS = \int_{p_2}^{p_0} (\text{supply function as a function of } p)\, dp,$$

where p_2 is the price obtained by setting $x = 0$ in the supply function.

Let us consider some examples.

Example 30. The demand function for a commodity is given by $p = 100 - 8x$. Find the consumers' surplus corresponding to the market price $p = 4$.

Solution. The market price is given to be $p_0 = 4$. Substituting this value of p_0 in the demand function $p = 100 - 8x$, we get the market demand $x_0 = 12$.

$$\therefore \quad CS = \int_0^{x_0} D(x)\, dx - x_0 \times p_0$$

$$= \int_0^{12} (100 - 8x)\, dx - 12 \times 4$$

$$= (100x - 4x^2) \Big|_0^{12} - 48$$

$$= [100(12) - 4(12)^2] - 48 = 576$$

Hence consumers' surplus is 576.

Example 31. The demand law for a commodity is $p = 20 - 2x - x^2$. Find the consumers' surplus when the market demand is 3. **[*Delhi Univ. B.Com. (H) 1982*]**

Solution. The market demand is given to be $x_0 = 3$. Substituting this value of x_0 in the demand function $p = 20 - 2x - x^2$, we get the market price $p_0 = 5$.

$$\therefore \quad CS = \int_0^{x_0} D(x)\, dx - x_0 \times p_0$$

$$= \int_0^3 (20 - 2x - x^2)\, dx - 3 \times 5$$

$$= \left[20x - x^2 - \frac{x^3}{3} \right]_0^3 - 15$$

$$= (60 - 9 - 9) - 0 - 15 = 27$$

Hence consumers' surplus is 27.

Example 32. The demand equation of a product is $p - 10\, e^{-x} = 0$. Find the consumer surplus when the market price $p = 1$. **[*Delhi Univ. B.Com. (H) 1998*]**

Solution. The demand equation is : $p = 10\, e^{-x}$

and the market price $p = 1$. Substituting $p = 1$ in the demand equation, the quantity demanded x will be given by

$$1 = 10\, e^{-x} \quad \Rightarrow \quad e^{-x} = \frac{1}{10} \quad \text{or} \quad e^{x} = 10$$

$$\Rightarrow \quad x \log_{10} e = \log_{10} 10 = 1 \quad \Rightarrow \quad x(0.4343) = 1 \quad \Rightarrow \quad x = \frac{1}{0.4343} = 2.30$$

$$\therefore \quad CS = \int_0^{2.30} 10\, e^{-x}\, dx - 1 \times 2.30$$

$$= -10 \left| e^{-x} \right|_0^{2.30} - 2.30$$

$$= -10\,[e^{-2.30} - e^{0}] - 2.30 = -10 \left[\frac{1}{10} - 1 \right] - 2.30 = 6.70.$$

Example 33. The supply function of a producer is given by : $p = 0.4e^{2x}$, where x denotes thousand units. Find producer's surplus when sales are 2000 units .

[*Delhi Univ. B.Com. (H) 2008*]

Solution . The supply function is : $p = 0.4e^{2x}$

When sales are 2000 units (*i.e.*, $x_0 = 2$), $p_0 = 0.4e^4 = 0.4 \times 54.60 = 21.84$

$$\therefore \quad PS = p_0 \times x_0 - \int_0^{x_0} S(x)dx$$

$$= 21.84 \times 2 - \int_0^2 0.4\, e^{2x} dx = 43.68 - 0.2 \left| e^{2x} \right|_0^2$$

$$= 43.68 - 0.2\left[e^4 - e^0\right] = 43.68 - 0.2\,(54.60 - 1) = 43.68 - 10.72 = 32.96.$$

Example 34. If the supply function for a particular commodity is $100\,p = (x + 20)^2$ and the market price is ₹ 25, find the producer's surplus by two methods : (*a*) integrate with respect to x, (*b*) integrate with respect to p.

Solution. (*a*) Using Integration with respect to x, the producer's surplus is :

$$PS = x_0 \times p_0 - \int_0^{x_0} S(x)\, dx$$

where (x_0, p_0) is the equilibrium point and $p = S(x)$ is the supply function.

It is given that $p_0 = 25$. Substituting this value in the supply function : $100\,p = (x + 20)^2$, we obtain $x_0 = 30$. Thus

$$PS = 25 \times 30 - \int_0^{30} \frac{(x+20)^2}{100}\, dx$$

$$= 750 - \frac{1}{300}(x+20)^3 \Big|_0^{30}$$

$$= 750 - \frac{1}{300}\left[(50)^3 - (20)^3\right] = 360$$

(*b*) Using Integration with respect to p, the producer's surplus is :

$$PS = \int_{p_1}^{p_0} (\text{supply function as a function of } p)\, dp,$$

where p_0 is the market price and p_1 is obtained by setting $x = 0$ in the supply function (see Figure 10.5).

We are given $p_0 = 25$. To find p_1, we set $x = 0$ in the supply function: $100\,p = (x + 20)^2$ and obtain $p_1 = 4$. Now solving the supply function for x in terms of p, we get

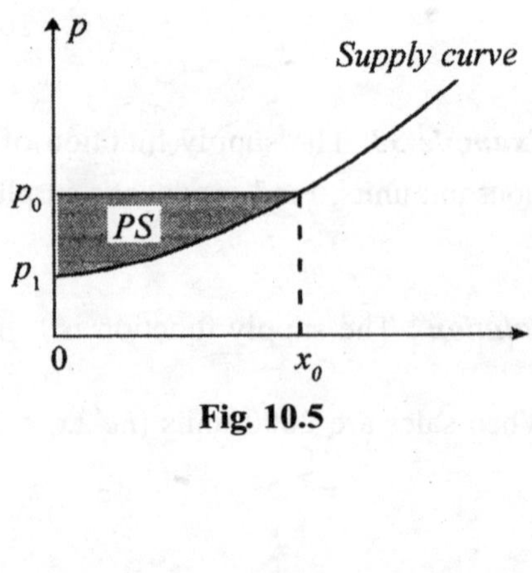

Fig. 10.5

$$x = 10\sqrt{p} - 20$$

Thus $$PS = \int_4^{25} (10\sqrt{p} - 20)\, dp$$

$$= \left(\frac{20p^{3/2}}{3} - 20p\right)\Big|_4^{25}$$

$$= \left[\frac{20}{3}(25)^{3/2} - 20(25)\right] - \left[\frac{20}{3}(4)^{3/2} - 20(4)\right] = 360,$$

which is same as before.

Example 35. The supply curve for a commodity is $p = \sqrt{9+x}$ and the quantity sold is 7 units. Find the producer's surplus. Can you find the consumer's surplus? If yes, find it, if not explain with the help of a diagram why not. **[*Delhi Univ. B.Com. (H) 2006*]**

Solution . We are given that the quantity sold is 7 units, *i.e.,* $x_0 = 7$. Substituting this value in the supply function $p = \sqrt{9+x}$, we get $p_0 = 4$.

$$\therefore \quad PS = 7 \times 4 - \int_0^7 \sqrt{9+x}\, dx$$

$$= 28 - \left[\frac{2}{3}(9+x)^{\frac{3}{2}}\right]_0^7$$

$$= 28 - \frac{2}{3}\left[16^{3/2} - 9^{3/2}\right] = 28 - \frac{2}{3}\left[4^3 - 3^3\right] = \frac{10}{3}.$$

We can not find consumer's surplus unless demand function is given, since consumers surplus is represented by the area between the line $p = p_0$ and the demand curve $p = D(x)$ from $x = 0$ to $x = x_0$ (See Figure 10.6), where (x_0, p_0) is the equilibrium point.

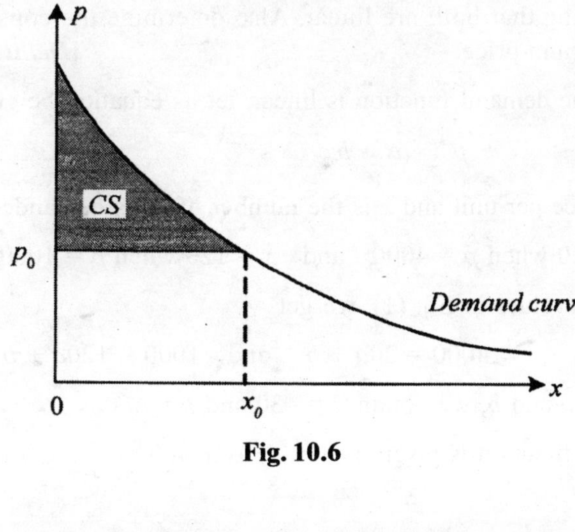

Fig. 10.6

Example 36. Find the consumer's surplus and producers' surplus under pure competition for demand function $p = \frac{8}{x+1} - 2$ and supply function $p = \frac{1}{2}(x+3)$, where p is price and x is quantity. **[*Delhi Univ. B.Com. (H) 1992, 2002, 2011, 2012, 2017, 2019*]**

Solution. Under pure competition, the market demand is obtained by equating the demand and supply functions. Thus

$$\frac{8}{x+1} - 2 = \frac{1}{2}(x+3) \quad \Rightarrow \quad 16 - 4(x+1) = (x+3)(x+1)$$

$$\Rightarrow \quad x^2 + 8x - 9 = 0 \quad \text{or} \quad (x-1)(x+9) = 0$$

$\Rightarrow$ $x = 1$, as the negative value $x = -9$ is meaningless.

Substituting $x = 1$ in the demand function, we obtain $p = \frac{8}{1+1} - 2 = 2$. Thus the equilibrium point is (1, 2).

The consumers' surplus is :

$$CS = \int_0^1 \left(\frac{8}{x+1} - 2\right) dx - 1 \times 2 = [8 \log (x+1) - 2x]\Big|_0^1 - 2 = 8 \log 2 - 4$$

The producers' surplus is :

$$PS = 1 \times 2 - \int_0^1 \frac{1}{2}(x+3)\, dx = 2 - \left[\frac{(x+3)^2}{4}\right]\Bigg|_0^1 = \frac{1}{4}.$$

Example 37. When the price of washing machine averaged ₹ 4000; *ABC* Co. Ltd. sold 20 every month. When the price dropped to an average ₹ 1000; 120 were sold every month by the same company. When the price was ₹ 4000; 200 machines were available per month for sale. When the price reached ₹ 1000, only 50 remained. Find the demand and supply functions, assuming that both are linear. Also determine the consumer and the producer's surplus at equilibrium price. **[*Delhi Univ. B.Com. (H) 2009*]**

Solution. Since the demand function is linear, let its equation be given by

$$p = ax + b, \qquad \text{..(1)}$$

where p is the price per unit and x is the number of units demanded at this price.

Given : $x = 20$ when $p = 4000$ and $x = 120$ when $p = 1000$

Substituting these values in Eq. (1), we get

$$4000 = 20a + b \quad \text{and} \quad 1000 = 120a + b$$

Solving these for a and b, we obtain $a = -30$ and $b = 4600$.

Thus the demand function is given by $p = -30x + 4600$.

Similarly, it can be verified that the equation of the supply function is given by $p = 20x$.

Under equilibrium, demand = supply

i.e., $-30x + 4600 = 20x \Rightarrow x = 92$

Substituting $x = 92$ in the demand function : $p = -30x + 4600$, we get $p = 1840$.

$$\therefore \quad CS = \int_0^{92} (4600 - 30x)\, dx - 92 \times 1840$$

$$= \left|4600x - 15x^2\right|_0^{92} - 1{,}69{,}280$$

$$= 4600(92) - 15(92)^2 - 1{,}69{,}280 = 1{,}26{,}960$$

and

$$PS = 92 \times 1840 - \int_0^{92} 20x\, dx$$

$$= 1{,}69{,}280 - \left|10x^2\right|_0^{92}$$

$$= 1{,}69{,}280 - 10(92)^2 = 1{,}69{,}280 - 84{,}640 = 84{,}640$$

Example 38. A monopolist's demand function is $x = 210 - 3p$, where x is the quantity demanded when price is ₹ p per unit. With the average cost function:

$$AC = x + 6 + \frac{10}{x},$$

find the consumer surplus at the price which monopolist will like to fix.

[Delhi Univ. B.Com. (H) 1999, 2018 (modified)]

Solution. Given $x = 210 - 3p$ and $AC(x) = x + 6 + \dfrac{10}{x}$

$$\therefore \quad p = 70 - \frac{x}{3} \text{ and } TC(x) = x^2 + 6x + 10$$

$$\Rightarrow \quad TR(x) = 70x - \frac{x^2}{3} \text{ and } MC(x) = 2x + 6$$

Under monopoly, $MR = MC \Rightarrow 70 - \dfrac{2x}{3} = 2x + 6 \Rightarrow x = 24$

Also, $\dfrac{dMR}{dx} = -\dfrac{2}{3}$ and $\dfrac{dMC}{dx} = 2$

$$\therefore \quad \frac{dMR}{dx} < \frac{dMC}{dx} \text{ at } x = 24$$

$\therefore$ Profit is maximum at $x = 24$. Thus $x_0 = 24$ and $p_0 = 70 - 8 = 62$. Hence

$$CS = \int_0^{24} \left(70 - \frac{x}{3}\right) dx - 62 \times 24$$

$$= \left|70x - \frac{x^2}{6}\right|_0^{24} - 1488$$

$$= (1680 - 96) - 1488 = 96.$$

Example 39. For a monopolist, the demand law is $p = 50 - x^2$ and marginal cost $MC = 1 + x^2$. Determine the consumer's surplus at the price which monopolist will like to fix. If there is pure competition in this case, what will be the consumer's surplus at equilibrium price ? **[*Delhi Univ. B.Com. (H) 2004*]**

Solution. The demand function is $p = 50 - x^2$. If R denotes the revenue function, then

$$R = px = (50 - x^2)x = 50x - x^3$$

The marginal revenue is : $MR = \dfrac{dR}{dx} = 50 - 3x^2$

The marginal cost is : $MC = 1 + x^2$

Under monopoly, the quantity x_0 is obtained by equating MR and MC. Thus

$$50 - 3x^2 = 1 + x^2 \quad \Rightarrow \quad 4x^2 = 49 \quad \Rightarrow \quad x = \frac{7}{2},$$

rejecting the negative value $x = -\dfrac{7}{2}$. Therefore, $x_0 = \dfrac{7}{2}$. Substituting this value in the demand law $p = 50 - x^2$, we obtain $p_0 = \dfrac{151}{4}$.

$$\therefore \quad CS = \int_0^{7/2} (50 - x^2)\,dx - \frac{7}{2} \times \frac{151}{4}$$

$$= \left(50x - \frac{x^3}{3}\right)\Big|_0^{7/2} - \frac{1057}{8}$$

$$= \left(175 - \frac{343}{24}\right) - \frac{1057}{8} = \frac{343}{12}.$$

However, if there is a case of pure competition, the equilibrium point is obtained by equating demand and supply functions. Identifying supply function with the marginal cost function, we have

$$50 - x^2 = 1 + x^2 \quad \Rightarrow \quad 2x^2 = 49 \quad \Rightarrow \quad x = \frac{7}{\sqrt{2}}$$

Substituting this value of x in the demand function, we have $p = 50 - \frac{49}{2} = \frac{51}{2}$.

Thus $x_0 = \frac{7}{\sqrt{2}}$ and $p_0 = \frac{51}{2}$.

$$\therefore \quad CS = \int_0^{7/\sqrt{2}} (50 - x^2)\, dx - \frac{7}{\sqrt{2}} \times \frac{51}{2}$$

$$= \left(50x - \frac{x^3}{3}\right)\Bigg|_0^{\frac{7}{\sqrt{2}}} - \frac{357}{2\sqrt{2}} = \frac{350}{\sqrt{2}} - \frac{343}{6\sqrt{2}} - \frac{357}{2\sqrt{2}} = \frac{343}{3\sqrt{2}}.$$

Example 40. Demand and supply functions are: $p = D(x) = (12 - 2x)^2$ and $p = S(x) = 56 + 4x$ respectively. Determine CS under monopoly (so at to maximize the profit) and the supply function is identified with the marginal cost function.

Solution. The demand function is:

$$p = D(x) = (12 - 2x)^2 = 144 + 4x^2 - 48x$$

If R denotes the revenue function, then

$$R = px = 144x + 4x^3 - 48x^2$$

$$\Rightarrow \quad MR = 144 + 12x^2 - 96x$$

The supply function is: $\quad p = S(x) = 56 + 4x$

Since the supply function is identified with the marginal cost function, therefore

$$MC = S(x) = 56 + 4x$$

Under monopoly, the quantity x_0 is obtained by equating MR and MC. Thus

$$144 + 12x^2 - 96x = 56 + 4x$$

$$\Rightarrow \quad 12x^2 - 100x + 88 = 0 \quad \Rightarrow \quad x = \frac{22}{3} \text{ or } 1 \qquad \textit{(on simplification)}$$

Now $\quad \frac{dMR}{dx} = 24x - 96$ and $\frac{dMC}{dx} = 4$

At $x = \frac{22}{3}$, $\quad \frac{dMR}{dx} = 80$ and $\frac{dMC}{dx} = 4$

$\therefore \quad \frac{dMR}{dx}$ is not less than $\frac{dMC}{dx}$ at $x = \frac{22}{3}$

Hence the value $x = \frac{22}{3}$ is rejected. However, at $x = 1$

$$\frac{dMR}{dx} = -72 \quad \text{and} \quad \frac{dMC}{dx} = 4$$

$$\therefore \quad \frac{dMR}{dx} < \frac{dMC}{dx} \text{ at } x = 1$$

Thus $x_0 = 1$. Substituting this value in the demand function $p = (12 - 2x)^2$, we obtain $p_0 = 100$.

$$\therefore \quad CS = \int_0^1 (12 - 2x)^2 \, dx - 1 \times 100 = \frac{64}{3} \quad \textit{(on simplification)}$$

Example 41. The demand and supply functions for a commodity are given by $p = 24e^{-x}$ and $p = 3e^{x/2}$ respectively. Determine the consumers' and producers' surplus at equilibrium price.

Solution. Under equilibrium, demand = supply

$$\therefore \quad 24e^{-x} = 3e^{x/2} \quad \Rightarrow \quad e^{3x/2} = 8$$

Taking natural logarithm on both sides, we get

$$\frac{3x}{2} = \log 8 = 3\log 2 \quad \Rightarrow \quad x = 2\log 2 = \log 4$$

Substituting this value of x in the demand function, we get $p = 24\, e^{-\log 4} = 6$.

$$\therefore \quad CS = \int_0^{\log 4} 24e^{-x} \, dx - 6\log 4$$

$$= (-24\, e^{-x}) \Big|_0^{\log 4} - 6 \log 4 = (-24)\left[e^{-\log 4} - e^0\right] - 6 \log 4$$

$$= (-24)\left[\frac{1}{4} - 1\right] - 6 \log 4 = 18 - 6 \log 4$$

and
$$PS = 6 \log 4 - \int_0^{\log 4} 3e^{x/2} \, dx = 6 \log 4 - 6\, e^{x/2} \Big|_0^{\log 4}$$

$$= 6\log 4 - 6[e^{\log 2} - e^0]$$

$$= 6\log 4 - 6(2 - 1) = 6\log 4 - 6.$$

EXERCISE 10.5

1. Explain the concept of consumer's surplus. What data would you require to determine this when there is (*i*) pure competition, (*ii*) monopoly ? Give reasons. **[*Delhi Univ. B.Com. (H) 1987*]**
2. The demand law of a commodity is $p = 4 - x^2$. Find the consumers' surplus when the market demand is 1.

3. The demand law for a commodity is $p = 35 - 2x - x^2$. Find the consumers' surplus at equilibrium price $p = 20$.
4. The demand function for a commodity is $p = 20e^{-x/10}$. Find the consumers' surplus at equilibrium price $p = 2$.
5. If the demand function for a commodity is $10\,p = \sqrt{900 - x}$ and the demand is fixed at 500, find the consumers' surplus by two methods : (*a*) integrate with respect to x, (*b*) integrate with respect to p.
6. The supply function for a commodity is $p = 4 + x$. Determine producers' surplus if 12 units of goods are sold. **[*Delhi Univ. B.Com. (H) 1985*]**
7. The supply function for a commodity is $100p = (x + 20)^2$. Find the producers' surplus when the market price is ₹ 25.
8. The demand and supply laws for a commodity are $p_d = 18 - 2x - x^2$ and $p_s = 2x - 3$. Find the consumer's surplus and producer's surplus at equilibrium price.
9. The demand and supply laws for a commodity are $p_d = 56 - x^2$ and $P_s = 8 + \dfrac{x^2}{3}$. Find the consumer's surplus and producer's surplus at equilibrium price.
10. For a particular commodity, the demand function is $p = D(x) = 16 - \dfrac{x^2}{100}$ and the supply function is $p = S(x) = \dfrac{x^2}{400} + 6$. Determine to the nearest rupee the consumer's surplus and producer's surplus if market equilibrium prevails. **[*Delhi Univ. B.Com. (H) 2005*]**
11. The demand and supply functions under pure competition are $p_d = 25 - x^2$ and $p_s = 2x + 1$ respectively. Find the consumers' surplus and producers' surplus. **[*Delhi Univ. B.Com. (H) 1995*]**
12. The demand and supply functions under perfect competition are $p = 16 - x^2$ and $p = 2x^2 + 4$ respectively. Find the consumers' and producers' surplus. **[*Delhi Univ. B.Com. (H) 1983*]**
13. The demand and supply functions under pure competition are $p = 1600 - x^2$ and $p = 2x^2 + 400$ respectively. Find the consumers' and producers' surplus. **[*Delhi Univ. B.Com. (H) 2000*]**
14. The demand and supply functions under perfect competition are $x = \dfrac{25}{4} - \dfrac{p}{8}$ and $p = 5 + x$ respectively. Find the consumers' surplus and producers' surplus. **[*Delhi Univ. B.Com. (H) 1996*]**
15. When the price of pocket calculators averaged ₹ 400, a company sold 20 every month. When the price dropped to an average of ₹ 100, 120 were sold every month by the same company. When the price was ₹ 400, 200 calculators were available per month for sale. When the price reached ₹ 100, only 50 remained. Find the demand and supply functions, assuming both are linear. Also determine consumers' and producers' surplus at equilibrium price.
16. A monopolist's demand function is $x = 240 - 10\,p$, where x is quantity demanded when price is ₹ p per unit. With the average cost function: $AC = 10 + \dfrac{x}{25}$, find the consumer surplus at the price which monopolist will like to fix. **[*Delhi Univ. B.Com. (H) 2018*]**

17. The demand function for a monopolist is $x = \dfrac{60 - 10p}{3}$ and his average cost function is $\dfrac{20}{x} + 1 + 0.2x$, where p and x refer to the price and quantity of the commodity respectively. Determine consumer surplus. **[*Delhi Univ. B.Com. (H) 2003, 2006, 2007, 2014*]**

18. For a monopolist, the demand law is $p = 600 - 8x$ and marginal cost $MC = 2x + 78$. Determine the consumer's surplus at the price which monopolist will like to fix.

19. For a monopolist, the demand function is $x = \sqrt{2200 - p}$ with $MC = 200 + 2x^2$. Determine the consumer's surplus at the price which the monopolist will like to fix.

[*Delhi Univ. B.Com. (H) 2022*]

20. Demand and supply functions are : $p = D(x) = (6 - x)^2$ and $p = S(x) = 14 + x$ respectively. Determine CS under monopoly (so as to maximize the profit) and the supply function is identified with the marginal cost function.

21. For a monopolist's product, the demand function is $p = 200 - \dfrac{x}{2}$ and the supply function is $p = 120 + x$. Determine the consumer's surplus at the price which monopolist will like to fix, the supply function being identified with the marginal cost function.

ANSWERS

2. $CS = \dfrac{2}{3}$ **3.** $CS = 27$ **4.** $CS = 133.96$ **5.** $CS = 266.66$

6. $PS = 72$ **7.** $PS = 360$ **8.** $CS = 27,\ PS = 9$

9. $CS = 144,\ PS = 48$ **10.** $CS = 150.40,\ PS = 37.60$ **11.** $CS = \dfrac{128}{3},\ PS = 16$

12. $CS = \dfrac{16}{3},\ PS = \dfrac{32}{3}$ **13.** $CS = \dfrac{16,000}{3},\ PS = \dfrac{32,000}{3}$ **14.** $CS = 100,\ PS = \dfrac{25}{2}$

15. $CS = 12,696,\ PS = 8464$ **16.** $CS = 125$ **17.** $CS = \dfrac{15}{4}$

18. $CS = 3364$ **19.** $\dfrac{16,000}{3}$ **20.** $CS = \dfrac{16}{3}$ **21.** $CS = 400$

CHAPTER 11

MATHEMATICS OF FINANCE

OBJECTIVES

After studying the material in this chapter, you should be able to:

- *Know the difference between simple interest and compound interest.*
- *Explain the concept of continuous compounding.*
- *Find the present value of money due in the future or the future value of money presently invested.*
- *Distinguish between nominal and effective rates of interest.*
- *Know various methods of computing the annual depreciation.*
- *Identify different types of annuities.*
- *Determine the amount and present value of an annuity.*
- *Solve problems on sinking funds.*
- *Use present value concept to solve problems on leasing and capital expenditure.*

11.1 INTEREST

Certain terms are used so often in business transactions that it is better to define them explicitly.

Capital is a stock of accumulated wealth—money or its equivalent.

Consideration is a fee.

Interest is consideration for the use of invested or loaned capital.

The *principal* is the capital originally invested.

The *time* is the number of years or fraction of a year for which the principal is borrowed or loaned.

The *amount* is the sum of the principal and the interest.

The *rate of interest* is the amount charged for the use of the principal for a given length of time, usually on a yearly or per annum basis.

Rates of interest are usually expressed as percentages : 5% per annum; 10% per annum and so on. However, when using rates of interest in calculations, we use the decimal equivalent : 0.05 for 5%; 0.10 for 10% and so on.

There are two kinds of interest. If interest is computed on the original principal, it is called **simple interest.** Thus, in transactions involving simple interest, the principal on which interest is computed remains unchanged throughout the term of the loan and the interest becomes due either at the end of the loan or at the end of stated intervals during the term of the loan. However, when the principal is increased by interest at the end of each period and interest is thus computed on a principal which grows periodically, we are dealing with **compound interest.**

11.2 SIMPLE INTEREST

The simple interest on a given principal at a given rate for a given time is the product of three factors: the principal, the rate of interest (per annum) and the time (in years). Thus the formula for simple interest expressed in symbols is :

$$\boldsymbol{I = Prt}$$

where

I = Interest in rupees (or other units of money)

P = Principal in rupees (or other units of money)

r = Rate of interest (in decimal form)

t = Time (in years)

If S is used to symbolize the amount, then the definition of amount leads to the formula:

$$\boldsymbol{S = P + I = P + Prt = P\,(1 + rt)}$$

These formulas can be used to calculate the interest on a stated principal for a given time at a given rate, and to find the amount.

Example 1. Find the simple interest on ₹ 500 for 3 years at 4%, and find the amount.

Solution. Here $P = 500$, $r = 0.04$ and $t = 3$. Since

$$I = Prt$$

$$\therefore \quad I = (500)\,(0.04)\,(3) = ₹\ 60$$

and $$S = P + I = 500 + 60 = ₹\ 560$$

Example 2. Find the time required for ₹ 2500 to yield ₹ 300 in simple interest at 8%.

Solution. From the formula $I = Prt$, we obtain

$$t = \frac{I}{Pr}$$

Now, $I = 300, P = 2500$ and $r = 0.08$.

$$\therefore \quad t = \frac{300}{2500\,(0.08)} = \frac{300}{200} = 1.5 \text{ (years)}.$$

Example 3. At what interest rate will ₹ 3000 yield ₹ 120 in simple interest in 6 months ?

Solution. From the formula $I = Prt$, we obtain

$$r = \frac{I}{Pt}$$

Substituting $I = 120$, $P = 3000$ and $t = \frac{1}{2}$, we have

$$r = \frac{120}{(3000)\left(\frac{1}{2}\right)} = 0.08 \text{ or } 8\%$$

Example 4. What principal will amount to ₹ 645 in $1\frac{1}{2}$ years at 5% simple interest ?

Solution. From the formula $S = P\,(1 + rt)$, we have

$$P = \frac{S}{1 + rt}$$

Substituting $S = 645$, $r = 0.05$ and $t = 1.5$, we obtain

$$P = \frac{645}{1 + (0.05)\,(1.5)} = ₹\ 600.$$

EXERCISE 11.1

1. Find the simple interest and amount of a principal of ₹ 600 for 2 years at 5%.
2. A person invests ₹ 1000 in a savings bank paying 6% simple interest. What is the balance or amount of the savings account after 6 months ?
3. Find the amount at 6% simple interest of ₹ 1200 due in 9 months.
4. Find the simple interest on ₹ 600 for 2 months at 7%.
5. A savings account opened 3 months ago now has a balance of ₹ 2040. If the bank pays 8% simple interest, how much money was deposited?
6. What sum of money placed at simple interest at 4% will yield ₹ 50 interest in 6 months?
7. At what rate will ₹ 1500 yield ₹ 25 simple interest in 8 months?
8. How long will it take for ₹ 3000 to amount to ₹ 3300 at 4% simple interest?
9. What time will be required for a sum of money to double itself at 5% simple interest?
10. The simple interest on ₹ 600 for 9 months is ₹ 27. Find the interest rate.

ANSWERS

1. ₹ 60; ₹ 660 **2.** ₹ 1030 **3.** ₹ 1254 **4.** ₹ 7 **5.** ₹ 2000

6. ₹ 2500 **7.** 2.5% **8.** 2.5 years **9.** 20 years **10.** 6%

11.3 COMPOUND INTEREST

In transactions involving compound interest, the interest earned by an invested amount of money (or principal) is reinvested so that it too earns interest. That is, interest is converted (or compounded) into principal and hence there is "interest on interest".

Interest may be converted into principal annually, semiannually, quarterly, monthly or at any other regular periods of time. The **frequency of conversion** is a number indicating how many times interest is compounded in one year. The time between two successive conversions of interest is called the **conversion period.** Thus, if interest is compounded quarterly, the frequency of conversion is 4 and the conversion period is 3 months. The total amount due at the end of the last period is called the **compound amount.** The difference between the compound amount and the original principal is called the **compound interest.**

Remark. It may be remarked that regardless of the frequency of conversion, the rate of interest is usually expressed as an annual rate. When the conversion period is other than a year, the rate per conversion period is found by dividing the stated annual rate by the number of conversion periods in a year. Thus, if the quoted rate is 8% compounded quarterly, the rate per conversion period is 2% or 0.02. Further, when the frequency of conversion is not stated, interest will be understood to be converted *annually.* Thus, the expression "interest at 4%" or "money worth 4%" will mean 4% converted annually.

11.4 FORMULA FOR COMPOUND AMOUNT

In this section we shall develop a formula for computing the amount, when interest is compounded. Let

P = the original principal

n = the number of conversion periods

i = the interest rate (in decimal form) per conversion period

S = the amount at the end of n periods

Then

the amount at the end of first conversion period $= P + Pi = P(1 + i)$

the amount at the end of second conversion period

$$= P(1 + i) + P(1 + i)\,i = P(1 + i)(1 + i) = P(1 + i)^2$$

the amount at the end of third conversion period

$$= P(1 + i)^2 + P(1 + i)^2\,i = P(1 + i)^3$$

$$\vdots$$

the amount at the end of the nth conversion period $= P(1 + i)^n$

Hence the compound amount S of a principal P at the end of n conversion periods at the interest rate of i per conversion period is given by

$$S = P(1 + i)^n \qquad ...(1)$$

This formula is usually referred to as the **compound interest formula.** When using the compound interest formula, remember that

$$i = \frac{\text{annual rate of interest}}{\text{number of conversion periods per year}}$$

For example, if the annual rate of interest is 8% and the compounding is semiannually, then there are 2 conversion periods per year and

$$i = 0.08 / 2 = 0.04$$

If the values of i and n are given, the value of $(1 + i)^n$ may be found directly from Table I. The table tells us what one rupee will amount to for various values of i and n.

***Note* 1.** It may be noted that if a principal of P is invested for 't' years at an annual interest rate of r and interest is compounded m times a year, then the interest rate per period is r/m and there are mt conversion periods throughout the term of the investment. Hence [using Formula (1)], the compound amount S at the end of t years becomes

$$S = P\left(1 + \frac{r}{m}\right)^{mt}$$

That is, the compound amount *S* at the end of '*t*' years at an annual interest of *r* compounded *m* times a year is given by

$$S = P\left(1 + \frac{r}{m}\right)^{mt} \qquad ...(2)$$

***Note* 2.** Once compound amount is obtained, compound interest may then be obtained by using the following formula:

$$\text{Compound Interest} = S - P$$

***Example* 5.** Find the compound amount of ₹ 2000 for 4 years at 6% converted :

(*i*) Annually (*ii*) Semiannually

(*iii*) Quarterly (*iv*) Monthly

Solution. (*i*) *When the interest is converted annually, we have*

$$P = 2000, \quad i = 0.06 \quad \text{and} \quad n = 4.$$

$$\therefore \quad S = P(1 + i)^n = 2000\ (1.06)^4$$
$$= 2000\ (1.262476) \qquad \textit{(using Table I)}$$
$$= ₹\ 2524.95$$

(ii) When the interest is converted semiannually, we have

$$i = 0.06/2 = 0.03 \quad \text{and} \quad n = 4(2) = 8$$

$$\therefore \quad S = P(1+i)^n = 2000\,(1.03)^8$$
$$= 2000\,(1.266770) \qquad \textit{(using Table I)}$$
$$= ₹\,2533.54$$

(iii) When the interest is converted quarterly, we have

$$i = 0.06/4 = 0.015 \quad \text{and} \quad n = 4(4) = 16$$

$$\therefore \quad S = P(1+i)^n = 2000\,(1.015)^{16}$$
$$= 2000\,(1.268985) \qquad \textit{(using Table I)}$$
$$= ₹\,2537.97$$

(iv) When the interest is converted monthly, we have

$$i = 0.06/12 = 0.005 \quad \text{and} \quad n = 4(12) = 48$$

$$\therefore \quad S = P(1+i)^n = 2000\,(1.005)^{48}$$
$$= 2000\,(1.270489) \qquad \textit{(using Table I)}$$
$$= ₹\,2540.97$$

Example 6. Find the compound amount and the compound interest of ₹ 700 invested for 15 years at 8% compounded semiannually.

Solution. Here $P = 700$. With interest compounded semiannually, we have $n = 15\,(2) = 30$ and the periodic rate i is 0.08/2 = 0.04. Substituting these values in Formula (1), we obtain

$$S = 700\,(1.04)^{30} = 700\,(3.243397) \qquad \textit{(using Table I)}$$
$$= ₹\,2270.37\,(app)$$

and compound interest $= S - P = 2270.37 - 700 = ₹1570.37$

Example 7. How long will it take for ₹ 1200 to amount to ₹ 2000 at 5% converted quarterly?

Solution. Substituting $S = 2000$, $P = 1200$, $i = 0.0125$ in the formula $S = P\,(1 + i)^n$, we obtain

$$2000 = 1200\,(1.0125)^n \quad \Rightarrow \quad (1.0125)^n = \frac{2000}{1200} = 1.6667$$

Taking logarithm of both sides, we get

$$n \log(1.0125) = \log(1.6667) \quad \Rightarrow \quad n = \frac{\log(1.6667)}{\log(1.0125)} = \frac{0.2217}{0.0052} = 42\,(app.)$$

Thus it takes about 42 quarterly periods or 10 years and 5 months.

Example 8. A man borrows ₹ 750 from a moneylender and the bill is renewed after every half-year at an increase of 21%. After how much time the amount of bill will become ₹ 7500? ***[Delhi Univ. B. Com. (H) 2007]***

Solution. Suppose it takes x years for the amount of bill to become ₹ 7500. Then

$$750(1+0.21)^{2x} = 7500 \quad \Rightarrow \quad (1.21)^{2x} = 10 \quad \Rightarrow \quad 2x\log(1.21) = \log 10 = 1$$

i.e. $2x(0.0828) = 1$ or $x(0.1656) = 1 \quad \Rightarrow \quad x = 1/0.1656 = 6.04$

Thus it takes about 6 years for the amount of bill to become ₹ 7500.

Example 9. (*i*) A certain sum of money is invested at 4% compounded annually. The interest for second year is ₹ 25. Find the interest for the third year.

(*ii*) A sum of money is put at compound interest for two years at 20% per annum. It would fetch ₹ 482 more, if the interest were payable half yearly than if it were payable yearly. Find the sum. ***Delhi Univ. B. Com. (H) 2003]***

Solution. (*i*) Let P be the original principal, which is being invested at 4% compounded annually. Then

The amount at the end of first year $= P\,(1.04)$

The amount at the end of second year $= P\,(1.04)^2$

The amount at the end of third year $= P\,(1.04)^3$

$\Rightarrow$ interest for the second year $= P\,(1.04)^2 - P\,(1.04)$

but interest for the second year $= 25$ (given)

$\therefore \quad P\,(1.04)^2 - P\,(1.04) = 25$

$\Rightarrow \quad P\,(1.04)\,(1.04 - 1) = 25$

$\Rightarrow \quad P\,(1.04)\,(0.04) = 25$ ...(1)

The interest for the third year $= P\,(1.04)^3 - P\,(1.04)^2$

$= P\,(1.04)^2\,(1.04 - 1)$

$= P\,(1.04)\,(1.04)\,(0.04)$

$= 25 \times 1.04$ [*using* (1)]

$=$ ₹ 26

(*ii*) Let the sum invested be ₹ P. Then the amount of ₹ P at the end of two years at 20% per annum compounded annually is :

$$P\,(1.20)^2 = P\,(1.44)$$

Also, the amount of ₹ P at the end of two years at 20% per annum compounded semiannually is :

$$P(1.10)^4 = P(1.4641)$$

Given $P\,(1.4641) - P\,(1.44) = 482 \quad \Rightarrow \quad P\,(0.0241) = 482 \quad \Rightarrow \quad P =$ ₹ 20,000.

Example 10. A sum of money is deposited in a bank which compounds interest semiannually. The amount at the end of 4 years is ₹ 6333.85 and the amount became ₹ 8023.53 at the end of 8 years Find the money deposited and the interest rate.

[*Delhi Univ. B. Com. (H) 2010, 2012*]

Solution. Let P (in ₹) be the money deposited and r be the rate of interest compounded semiannually. Then

$$P\left(1+\frac{r}{2}\right)^8 = 6333.85 \quad \text{and} \quad P\left(1+\frac{r}{2}\right)^{16} = 8023.53$$

On division, we get $\left(1+\frac{r}{2}\right)^8 = \frac{8023.53}{6333.85} = 1.2667698 \cong 1.267$

$$\Rightarrow \quad 8\log\left(1+\frac{r}{2}\right) = \log(1.267) = 0.1028 \Rightarrow \log\left(1+\frac{r}{2}\right) = \frac{0.1028}{8} = 0.01285$$

$$\Rightarrow \quad 1+\frac{r}{2} = AL(0.01285) = 1.030 \Rightarrow \frac{r}{2} = 0.03 \Rightarrow r = 0.06 \text{ or } 6\%$$

$$\therefore \quad P(1.03)^8 = 6333.85 \Rightarrow P(1.2667008) = 6333.85 \Rightarrow P = 5000$$

11.5 INTEREST COMPOUNDED CONTINUOUSLY

If a principal of P is invested for t years at an annual rate of r and interest is compounded m times a year, then the interest rate per conversion period is r/m and there are mt conversion periods. Thus if we make use of Formula (1) from the previous section, the compound amount S at the end of t years becomes

$$S = P\left(1+\frac{r}{m}\right)^{mt}$$

Now we ask what happens to the amount as the number of times m that the interest is compounded per year gets larger and larger. When interest is so compounded, we say that interest is **compounded continuously.** Thus in the case of continuous compounding, the amount S is given by

$$S = \lim_{m\to\infty}\left[P\left(1+\frac{r}{m}\right)^{mt}\right] = P\left[\lim_{m\to\infty}\left(1+\frac{r}{m}\right)^{mt}\right]$$

$$= P\lim_{m\to\infty}\left[\left(\left(1+\frac{r}{m}\right)^{m/r}\right)\right]^{rt} = P\left[\lim_{m\to\infty}\left(1+\frac{r}{m}\right)^{m/r}\right]^{rt}$$

If we let $\frac{r}{m} = x$, then as $m \to \infty$, we have $x \to 0$. Thus

$$S = P\left[\lim_{x\to 0}(1+x)^{1/x}\right]^{rt} = Pe^{rt}$$

> ***Hence the compound amount S of a principal P after t years at an annual interest rate of r compounded continuously is given by***
>
> $$S = Pe^{rt}$$

Example 11. A person deposits ₹ 5000 in a bank which pays an interest of 11% per annum compounded continuously. How much amount will be in his account after 10 years ?

[Delhi Univ. B.Com. (H) 1986]

Solution. Substituting $P = 5000$, $r = 0.11$ and $t = 10$ in the formula $S = Pe^{rt}$, the amount S after 10 years is given by

$$S = 5000\, e^{0.11\,(10)} = 5000e^{1.1}$$

Taking logarithm of both sides, we obtain

$$\log S = \log 5000 + (1.1) \log e$$

$$= 3.6990 + (1.1)(0.4343) = 4.1767$$

$$\Rightarrow \quad S = AL\,(4.1767) = ₹15{,}020\,(app.)$$

Example 12. A National Savings Certificate costs ₹ 15 and realises ₹ 20 after 10 years Find the rate of interest involved when it is added (*i*) yearly (*ii*) continuously.

[Delhi Univ. B.A. Econ. (H) 1991]

Solution. (*i*) *When interest is added yearly*

Substituting $P = 15$, $S = 20$ and $n = 10$ in the formula $S = P(1 + i)^n$, we obtain

$$20 = 15\,(1 + i)^{10} \quad \Rightarrow \quad \frac{4}{3} = (1 + i)^{10}$$

Taking logarithm of both sides, we obtain

$$\log 4 - \log 3 = 10 \log (1 + i)$$

$$\Rightarrow \quad \log (1 + i) = \frac{\log 4 - \log 3}{10} = \frac{0.6021 - 0.4771}{10} = 0.0125$$

$$\Rightarrow \quad 1 + i = AL\,(0.0125) = 1.0292 \quad \Rightarrow \quad i = 0.0292 \text{ or } 2.92\%$$

(*ii*) *When interest is added continuously*

Substituting $P = 15$, $S = 20$ and $t = 10$ in the continuous compounding amount formula $S = Pe^{rt}$, we obtain

$$20 = 15\,e^{10r} \quad \text{or} \quad e^{10r} = \frac{4}{3}$$

$$\Rightarrow \quad 10r \log e = \log 4 - \log 3 = 0.6021 - 0.4771 = 0.125$$

$$\Rightarrow \quad r = \frac{0.125}{10 \log e} = \frac{0.125}{10\,(0.4343)} = 0.0288 \text{ or } 2.88\%$$

Example 13. If interest is compounded continuously at an annual rate of 7%, how many years would it take for a principal P to triple ? Give your answer to the nearest year.

Solution. Suppose it takes t years for a principal P to triple at an annual rate of 7% compounded continuously. Then

$$3P = Pe^{0.07t} \quad \Rightarrow \quad e^{0.07t} = 3$$

Taking logarithm of both sides, we obtain

$$0.07t \log e = \log 3 \quad \Rightarrow \quad 0.07(0.4343)t = 0.4771$$

or
$$t = \frac{0.4771}{(0.07)\ (0.4343)} - 15.69$$

Thus the answer to the nearest year is 16.

Example 14. If interest is compounded continuously, at what annual rate will a principal of P double in 10 years ?

Solution. Let r be the annual interest rate compounded continuously. Then

$$2P = Pe^{10r} \quad \Rightarrow \quad e^{10r} = 2 \quad \Rightarrow \quad 10r \log e = \log 2$$

$$\Rightarrow \quad 10(0.4343)r = 0.3010 \quad \Rightarrow \quad r = \frac{0.3010}{4.343} = 0.069 \quad \text{or} \quad 6.9\%$$

Thus the annual rate of interest compounded continuously is 6.9%.

11.6 COMPOUND AMOUNT AT CHANGING RATES

Thus far we have assumed a constant rate of interest for the entire duration of an investment. However, interest rates may change from time to time. Thus, a bank, which pays interest at 8% when a deposit is made, may, after a number of years, raise the rate to 9% and later on perhaps reduce it to 7%. The final compound amount is the product of the original principal and two or more factors of the form $(1 + i)^n$ or e^{rt}, each with its proper value for i and n or r and t. This is best illustrated with the help of following examples.

Example 15. A man made a deposit of ₹ 2500 in a savings account. The deposit was left to accumulate at 6% compounded quarterly for the first 5 years and at 8% compounded semiannually for the next 8 years. Find the compound amount at the end of 13 years.

[*Delhi Univ. B.Com. (H) 2014*]

Solution. For the first 5 years, $i = 0.06/4 = 0.015$, $n = 5\ (4) = 20$. For the next 8 years, $i = 0.08/2 = 0.04$, $n = 8\ (2) = 16$. Hence the amount at the end of 13 years is:

$$S = 2500\ (1.015)^{20}\ (1.04)^{16}$$
$$= 2500\ (1.34685)\ (1.87298)$$
$$= ₹\ 6306.55\ (\textit{app.})$$

Example 16. Mr. X deposited ₹ 10,000 in a bank for 3 years offering interest at the rate of 6% compounded half yearly during first year, at the rate of 12% compounded quarterly during second year and at 10% compounded continuously during third year. Find his balance after 3 years. **[*Delhi Univ. B. Com. (H) 1999, 2018 (modified)*]**

Solution. Required balance after 3 years is :

$$S = 10{,}000\left(1+\frac{0.06}{2}\right)^2\left(1+\frac{0.12}{4}\right)^4 e^{0.10}$$

$$= 10{,}000\,(1.03)^2\;(1.03)^4\;e^{0.10}$$

$$= 10{,}000\,(1.03)^6\,e^{0.10}$$

$$= 10{,}000\,(1.19405)\,(1.1052) = ₹\,13196.64.$$

EXERCISE 11.2

1. Find the compound amount and the compound interest of ₹ 4000 invested for 7 years at 6% compounded annually.
2. Find the amount if ₹ 1800 is invested at 5% compounded semiannually for 8 years. Also determine the compound interest.
3. If ₹ 1000 is invested at an annual rate of interest of 15%, what is the amount after 5 years if the compounding takes place monthly ?
4. To what amount will ₹ 10,000 accumulate in 6 years if invested at 8% compounded quarterly?
5. If ₹ 1000 is invested at an annual rate of 9% compounded continuously, find the amount at the end of 6 years.
6. If ₹ 2000 is deposited in a savings account that earns interest at an annual rate of 6% compounded continuously, what is the value of the account at the end of 3 years ?
7. A person deposits ₹ 4000 in a bank which pays an interest of 5% per annum compounded continuously. How much amount will be in his account after 4 years ?
8. How long will it take for ₹ 5000 to amount to ₹ 7000 if it is invested at 8% compounded quarterly ?
9. How long will it take for a principal to double if money is worth 12% compounded monthly ?
10. Approximately how long will it take to triple an investment at 10% compounded annually?
11. At what annual rate of interest, compounded yearly, will money double in 8 years ?
12. How long will it take for ₹ 4000 to amount to ₹ 7000 if it is invested at 7% compounded continuously ?
13. How long will it take for a principal to double if money is worth 6% compounded continuously ?
14. A person deposited ₹ 1000 in a bank at 5% compounded annually. After 5 years, the rate of interest was increased to 6% and after 4 more years, the rate was further increased to 7%. The money was withdrawn at the end of 12 years. Find the amount.

15. Find the amount of ₹ 2000 after 10 years at 8% converted quarterly for the first 4 years and at 6% converted monthly thereafter.

16. Find the amount of ₹ 5000 after 8 years if invested at 6% compounded semiannually for the first 2 years, at 8% compounded quarterly for the next 3 years, and at 7% compounded continuously thereafter.

17. A person deposited ₹ 4000 in a bank at 6% compounded continuously. After 3 years, the rate of interest was increased to 7% and after 5 more years, the rate was further increased to 8%. The money was withdrawn at the end of 10 years. Find the amount.

[*Delhi Univ. B. Com. (H) 2005*]

18. Mr. X deposited ₹ 50,000 in a bank for five years offering interest at the rate of 8% per annum compounded half yearly during first two years, at the rate of 10% per annum compounded quarterly for the third year and at 9% per annum compounded continuously for the next two years. Find his balance after five years. **[*Delhi Univ. B. Com. (H) 2018*]**

ANSWERS

1. ₹ 6014.52; ₹ 2014.52	**2.** ₹ 2672.10; ₹ 872.10	**3.** ₹ 2107.18
4. ₹ 16,084.30	**5.** ₹ 1716	**6.** ₹ 2394.40
7. ₹ 4885.60	**8.** $4\frac{1}{4}$ years	**9.** 5.84 years
10. 11.5 years	**11.** 9%	**12.** 8 years (app.)
13. 11.5 years	**14.** ₹ 1973.86	**15.** ₹ 3931.78
16. ₹ 8804.94	**17.** ₹ 7974.85	**18.** ₹ 73,567 (app.)

11.7 NOMINAL AND EFFECTIVE RATES OF INTEREST

In transactions involving compound interest, the stated annual rate of interest in called the **nominal rate of interest.** Thus if an investment is made at 6% converted semiannually, the nominal rate of interest on this investment is 6%. It may be noted that the actual interest earned on the given investment will be more than 6% per year. For example, ₹ 100 invested at 6% converted semiannually amounts in one year to 100 $(1.03)^2$ = ₹ 106.09. Thus the interest actually earned on this principal of ₹ 100 is ₹ 6.09 which represents an annual return of 6.09%. We say that the **effective rate** in this case is 6.09%. When the conversion period is a year, the effective rate is the same as the stated annual rate.

Relationship Between the Effective Rate and the Nominal Rate. Let r_e denote the effective rate corresponding to the nominal rate r, converted m times a year. We shall use i exclusively for rate per conversion period. Thus $i = r / m$. At the rate i, the principal P amounts in one year to $P(1 + i)^m$. Since an effective rate is the actual rate compounded annually, therefore at the effective rate r_e, the principal P amounts in one year to $P(1 + r_e)$. Thus

$$P(1 + r_e) = P(1 + i)^m \quad \text{or} \quad 1 + r_e = (1 + i)^m \quad \Rightarrow \quad r_e = (1 + i)^m - 1$$

Thus the effective rate, r_e, equivalent to the nominal rate r converted m times a year is given by $r_e = (1 + i)^m - 1 = \left(1 + \dfrac{r}{m}\right)^m - 1$ ...(1)

By use of Formula (1), we can find the effective rate equivalent to the nominal rate r converted m times a year, that is, equivalent to the rate $i = r/m$ per conversion period. However, if the nominal rate r is compounded continuously, the effective rate r_e is given by

$$r_e = \lim_{m\to\infty}\left[\left(1 + \frac{r}{m}\right)^m - 1\right] = \lim_{m\to\infty}\left(1 + \frac{r}{m}\right)^m - 1$$

$$= \lim_{m\to\infty}\left[\left(1 + \frac{r}{m}\right)^{m/r}\right]^r - 1 = \left[\lim_{m\to\infty}\left(1 + \frac{r}{m}\right)^{m/r}\right]^r - 1$$

Let $\dfrac{r}{m} = x$, then as $m \to \infty$, $x \to 0$. Thus

$$r_e = \left[\lim_{x\to 0}(1 + x)^{1/x}\right]^r - 1 = e^r - 1$$

Hence in the case of nominal rate r compounded continuously, the effective rate is given by $r_e = e^r - 1$ ... (2)

Force of Interest. The nominal rate r compounded continuously and equivalent to a given effective rate r_e is called the **force of interest.**

Example 17. What effective rate is equivalent to a nominal rate of 8% converted quarterly ?

Solution. Using Formula (1), the effective rate r_e equivalent to a nominal rate of 8% converted quarterly is given by

$$r_e = \left(1 + \frac{r}{m}\right)^m - 1 = \left(1 + \frac{0.08}{4}\right)^4 - 1 = (1.02)^4 - 1$$

$$= 1.0824 - 1 = 0.0824 \quad \text{or} \quad 8.24\%$$

Thus the effective rate is 8.24%. This means that the rate 8.24% compounded *annually* yields the same interest as the nominal rate 8% compounded quarterly.

Example 18. Find the effective rate equivalent to the nominal rate 6% converted (*i*) monthly, (*ii*) continuously. **[*Delhi Univ. B. Com. (H) 2014*]**

Solution. (*i*) Here $r = 0.06$ and $m = 12$. Thus the equivalent effective rate r_e is given by

$$r_e = \left(1 + \frac{r}{m}\right)^m - 1 = \left(1 + \frac{0.06}{12}\right)^{12} - 1 = (1.005)^{12} - 1 = 1.0616 - 1 = 0.0616$$

i.e. $\quad r_e = 0.0616 \quad \text{or} \quad 6.16\%$

(*ii*) Using Formula (2), the effective rate equivalent to the nominal rate 6% converted continuously is given by

$$r_e = e^r - 1 = e^{0.06} - 1 = 1.0618 - 1 = 0.0618$$

i.e., $\quad r_e = 0.0618 \quad \text{or} \quad 6.18\%.$

Example 19. Find the force of interest corresponding to the effective rate 8%.

Solution. Substituting $r_e = 0.08$ in the formula $r_e = e^r - 1$, we get

$$0.08 = e^r - 1 \quad \Rightarrow \quad e^r = 1 + 0.08 = 1.08 \quad \Rightarrow \quad r \log e = \log(1.08)$$

i.e., $\quad r(0.4343) = 0.0334 \quad \text{or} \quad r = \dfrac{0.0334}{0.4343} = 0.0769 \quad \text{or} \quad 7.69\%$

Thus the force of interest is 7.69%. This means that 7.69% compounded continuously and 8% effective are equivalent rates.

Example 20. A money-lender charges 'interest' at the rate of 10 paise per rupee per month, payable in advance. What effective rate of interest does he charge per annum ?

[*Delhi Univ. B. Com. (H) 1996, 2017 (modified)*]

Solution. Since the money lender charges interest at the rate of 10 paise per rupee per month, payable in advance, therefore 10 paise may be treated as interest on 90 paise for one month. Consequently, interest rate per month is $i = 10/90 = 1/9$. Thus the equivalent effective rate r_e per annum is given by

$$r_e = (1 + i)^{12} - 1 = \left(1 + \frac{1}{9}\right)^{12} - 1 = \left(\frac{10}{9}\right)^{12} - 1 \quad \Rightarrow \quad 1 + r_e = \left(\frac{10}{9}\right)^{12}$$

$\Rightarrow \quad \log(1 + r_e) = 12(\log 10 - \log 9) = 12(1 - 0.9542) = 12(0.0458) = 0.5496$

Hence $\quad 1 + r_e = AL(0.5496) = 3.545 \quad \Rightarrow \quad r_e = 2.545 \quad \text{or} \quad 254.5\%$

Example 21. Mr. *X* took a loan of ₹ 2000 for 6 months. Lender deducts ₹ 200 as interest while lending. Find the effective rate of interest charged by lender.

[*Delhi Univ. B. Com. (H) 2005 (SOL)*]

Solution. Since the money lender deducts ₹ 200 as interest while lending a loan of ₹ 2000 for 6 months, therefore ₹ 200 may be treated as interest on ₹ 1800 for 6 months.

Consequently, interest rate per six month is $i = 200 / 1800 = 1/9$. Thus, the equivalent effective rate of interest, r_e, is given by

$$r_e = (1+i)^2 - 1 = \left(1+\frac{1}{9}\right)^2 - 1 = 0.23456 \quad \text{or} \quad 23.45\%$$

Example 22. To what amount will ₹ 12,000 accumulate in 12 years if invested at an effective rate of 5% ?

Solution. Since an effective rate is the actual rate compounded annually, we have

$$S = 12{,}000\ (1 + 0.05)^{12} = 12{,}000\ (1.05)^{12}$$

Taking logarithm of both sides, we obtain

$$\log S = \log 12{,}000 + 12 \log (1.05)$$

$$= 4.0792 + 12\ (0.0212) = 4.3336$$

Hence $\quad S = AL\ (4.3336) =$ ₹ 21,560 (*app.*)

Example 23. Find, for each of the following, the amount to which ₹ 100 will accumulate :

(*i*) At the rate of interest 12% per annum compounded quarterly for 10 years.

(*ii*) At the force of interest 3% per annum for 3.5 years.

(*iii*) At the effective rates of interest 3% per annum for 10 years, 4% per annum for 4 years and 5% per annum for 2 years.

(*iv*) At the rate of interest corresponding to 3% per annum effective rate of discount for 8 years.

(*v*) What constant force of interest would produce the same amount after 16 years as the rates in (*iii*) above ? **[*Delhi Univ. B. Com. (H)2002, 2004, 2011*]**

Solution. (*i*) Required amount $= 100\ (1.03)^{40} =$ ₹ $100(3.2620) =$ ₹326.20

(*ii*) Required amount $= 100 e^{.03(3.5)} = 100 e^{0.105}$

$$= 100\left[AL(0.105 \log e)\right]$$

$$= 100\left[AL\ (0.105 \times 0.4343)\right]$$

$$= 100\left[AL\ (0.04561)\right]$$

$$= 100 \times 1.111 = ₹\,111.1$$

(*iii*) Required amount $= 100\ (1.03)^{10}\ (1.04)^4\ (1.05)^2$

$$= 100\ (1.3439)\ (1.1698)\ (1.1025)$$

$$= 100\ (1.7331) = ₹\,173.31$$

(*iv*) Let i be the rate of interest corresponding to an effective rate of discount of $d = 0.03$ per annum. Then it is known that

$$1 - d = \frac{1}{1+i} \quad \Rightarrow \quad 1 + i = \frac{1}{1-d} = \frac{1}{0.97}$$

$\therefore$ Required amount

$$= 100\left(\frac{1}{0.97}\right)^8 = 100\ (1.0309)^8 = 100\ (1.2756) = ₹\,127.56$$

(*v*) Let δ be the constant force of interest. Then δ is given by

$$100e^{16\delta} = 173.31 \quad \Rightarrow \quad e^{16\delta} = 1.7331 \quad \Rightarrow \quad 16\delta \log e = \log 1.7331$$

$$\Rightarrow \quad \delta = \frac{\log 1.7333}{16 \log e} = \frac{0.2388}{16 \times 0.4343} = 0.0343 \quad \text{or} \quad 3.43\%$$

Remark. It may be remarked that effective rates are used to compare different interest rates, that is, which is "best", as is illustrated in the following example.

Example 24. Which is better from the standpoint of the investor, 9.1% converted semiannually or 9% converted monthly?

Solution. The effective rate corresponding to 9.1% converted semiannually is :

$$\left(1 + \frac{0.091}{2}\right)^2 - 1 = (1.0455)^2 - 1 = 0.09307025 \cong 9.30\%$$

The effective rate corresponding to 9% converted monthly is :

$$\left(1 + \frac{0.09}{12}\right)^{12} - 1 = (1.0075)^{12} - 1 = 0.09380609 \cong 9.38\%$$

Thus the *second* choice is *preferable*.

Example 25. Mr. Y has two investment options–either at 10% per annum compounded semi-annually or 9.5% per annum compounded continuously. Which option is preferable and why?

[*Delhi Univ. B. Com. (H) 2005 (SOL)*]

Solution. The effective rate corresponding to 10% per annum compounded semiannually is :

$$\left(1 + \frac{0.10}{2}\right)^2 - 1 = (1.05)^2 - 1 = 1.1025 - 1 = 0.1025 \quad \text{or} \quad 10.25\%$$

The effective rate corresponding to 9.5% per annum compounded continuously is :

$$e^{0.095} - 1 = 1.0997 - 1 = 0.0997 \quad \text{or} \quad 9.97\%$$

Thus the *first* investment is *preferable*.

Example 26. Mr X wants to make an investment of ₹ 50,000 for six years. He has two alternatives. First alternative fetches him a return of 8% compounded annually and the second alternative fetches him a return of 7.5% compounded semiannually. Which investment should he go for? **[*Delhi Univ. B. Com. (H) 2007 (SOL)*]**

Solution. The amount of ₹ 50,000 after six years @ 8% compounded annually is given by

$$50,000\,(1+0.08)^6 = 50,000\,(1.5868) = ₹79,340$$

The amount of ₹ 50,000 after six years @ 7.5% compounded semiannually is given by

$$50,000\left(1+\frac{0.075}{2}\right)^{12} = 50,000\,(1.0375)^{12} = 50,000(1.5555) = ₹77,775$$

Thus he should go for the first alternative.

EXERCISE 11.3

1. Find the effective rate that is equivalent to a nominal rate of 6% compounded (*i*) semiannually (*ii*) quarterly (*iii*) monthly (*iv*) continuously.

2. Find the effective rate that is equivalent to a nominal rate of 12% compounded monthly.

3. Find the effective rate equivalent to the nominal rate 5% compounded continuously.

4. Find the effective rate equivalent to the nominal rate 7% compounded continuously.

5. Find the force of interest corresponding to the effective rate 6%.

6. What annual rate compounded continuously is equivalent to an effective rate of 5% ?

7. Which yields more interest : 7.8% compounded semiannually or 8% effective?

8. Which is better from the standpoint of the investor, 6.2% compounded semiannually or 6% compounded monthly ?

9. A certain bank offers an interest rate of 6% per annum compounded annually. A competing bank compounds its interest continuously. What rate should the competing bank offer so that the effective rates of interest for the two banks are equal?

[*Delhi Univ. B. Com. (H) 2022*]

10. Suppose that you had the opportunity of investing ₹ 40,000 in a business such that the value of the investment after 5 years would be ₹ 53,000. On the other hand, you could instead put ₹ 40,000 in a savings account that pays 6% compounded semiannually. Which investment is better ?

11. If interest at 100 r% compounded annually is equivalent to interest at 100 s % compounded m times a year, show that $r = \left(1+\dfrac{s}{m}\right)^m - 1$. Hence deduce that r exceeds s by an amount equal to $\dfrac{m-1}{2m}s^2$ when s is small.

12. If 100 r % compounded annually and 100 s % compounded continuously are equivalent, show that $s = \log(1 + r)$.

Hint : $r = e^s - 1$.

13. A money-lender charges 'interest' at the rate of 5 rupees per 100 rupees per quarter, payable in advance. What effective rate does he charge per annum?

14. A money-lender charges 'interest' at the rate of 10 rupees per 100 rupees per half-year, payable in advance. What effective rate of interest does he charge per annum ?

15. To what sum will ₹ 6000 accumulate in 8 years if invested at an effective rate of 8%?

16. How many years will it take for money to double at the effective rate of 6%?

17. To what sum will ₹ 2000 amount in 8 years if invested at 6% effective rate for the first two years, at 6% compounded semi-annually for the next three years, and at 6% compounded continuously thereafter.

18. Distinguish between the nominal and effective rate of interest. Also establish the relationship between nominal and effective rate of interest when compounded n times a year and when compounded continuously. **[*Delhi Univ. B. Com. (H) 1997, 2003, 2012*]**

19. Explain the concepts of effective rate of interest, nominal rate of interest and force of interest. Bring out the realationship among them. **[*Delhi Univ. B. Com. (H) 2006, 2007*]**

ANSWERS

1. (*i*) 6.09% (*ii*) 6.13% (*iii*) 6.16% (*iv*) 6.18% **2.** 12.68% **3.** 5.13%
4. 7.25% **5.** 5.82% **6.** 4.88%
7. Second **8.** First **9.** 5.91%
10. Second **13.** 22.8% **14.** 23.5%
15. ₹ 11,105.40 **16.** 11.9 years **17.** ₹ 3212.27

11.8 PRESENT VALUE

When making future plans we would often like to know how much money we must invest now to receive a certain desired amount S at some later date. In other words, we are asking for the original invested principal, which is called the **present value** or **capital value** of the amount. Thus if money is worth i per period, the present value of S due in n periods is that principal which, invested *now* at the rate i per period, will amount to S in n periods.

Formula for Present Value

To obtain a formula for the present value, we solve the compound amount formula $S = P(1 + i)^n$ for P by dividing both sides by $(1 + i)^n$. This gives $P = S(1 + i)^{-n}$.

Thus the present value of S due n periods hence at the rate i per period is given by

$$P = S(1+i)^{-n} \qquad \ldots (1)$$

The quantity $(1 + i)^{-n}$ in the above formula is called the **discount factor.** It represents the present value of ₹ 1 due n periods hence at the rate i per period. Values of $(1 + i)^{-n}$ for various n and i are given in Table II.

Note. It should be noted that P and S represent the value of the same obligation *at different dates. P* is the present value of a given obligation, while S is the future value of the same obligation. P (now) is just as good as S (n periods hence).

To obtain a formula for the present value in the case of continuous compounding, we solve the equation $S = Pe^{rt}$ for P. This gives $P = Se^{-rt}$.

Thus the present value of S due at the end of t years at the annual rate of r compounded continuously is given by

$$P = Se^{-rt} \qquad \ldots (2)$$

Example 27. Find the present value of ₹ 500 due 10 years hence when interest of 10% is compounded (*i*) half yearly (*ii*) continuously. **[*Delhi Univ. B.A. Eco. (H) 1989*]**

Solution. In this problem we want to find the principal P when we know that the amount S after 10 years is going to be ₹ 500.

(*i*) *When the interest is compounded half yearly*

We use the formula

$$P = S(1+i)^{-n}$$

where $S = 500,\ i = 0.10/2 = 0.05,\ n = 10\,(2) = 20$

$$\therefore \quad P = 500\,(1.05)^{-20}$$
$$= 500\,(0.3768) \qquad \text{(using Table II)}$$
$$= ₹188.40$$

(*ii*) *When interest is compounded continuously*

In this case, we use the formula

$$P = Se^{-rt}$$

where $S = 500$, $r = 0.10$, and $t = 10$.

$$\therefore \quad P = 500e^{-0.10\,(10)} = 500e^{-1} = 500\,(0.36788) = ₹\,183.94$$

Example 28. How much should be invested at 6% per annum so that after 4 years the amount will be ₹ 2500 when the interest is compounded :

(*i*) annually ? (*ii*) quarterly ? (*ii*) monthly ? (*iv*) continuously ?

Solution. We want to find the present value of ₹ 2500 due in 4 years at 6% per annum when

(*i*) *interest is compounded annually*

We use the formula : $P = S(1+i)^{-n}$, where $S = 2500$, $i = 0.06$, $n = 4$

$\therefore$ $P = 2500\,(1 + 0.06)^{-4} = 2500\,(0.7920)$ (*using Table II*)

= ₹ 1980

(*ii*) *interest is compounded quarterly*

We use the formula :

$P = S(1+i)^{-n}$, where $S = 2500$, $i = 0.06/4 = 0.015$, $n = 4(4) = 16$

$\therefore$ $P = 2500\,(1 + 0.015)^{-16} = 2500\,(0.7880)$ (*using Table*

= ₹ 1970

II)

(*iii*) *interest is compounded monthly*

We use the formula :

$P = S(1+i)^{-n}$, where $S = 2500$, $i = 0.06/12 = 0.005$, $n = 4(12) = 48$

$\therefore$ $P = 2500\,(1 + 0.005)^{-48} = 2500\,(0.7870) =$ ₹1967.50

(*iv*) *interest is compounded continuously*

In this case, we use the formula : $P = Se^{-rt}$, where $S = 2500$, $r = 0.06$, and $t = 4$.

$\therefore$ $P = 2500\,e^{-0.06(4)} = 2500\,e^{-0.24} = 2500\,(0.7866) =$ ₹ 1966.5.

Example 29. Mr. *X* left ₹ 50,000 to be divided between his two daughters Divya and Neha. Divya's share was to amount to a certain sum of money at the end of 5 years and Neha's share was to amount to an equal amount at the end of 7 years. If the rate of interest is 6% compounded annually, find the amount.

Solution. Let Divya's share amount to *S* at the end of 5 years. Then Neha's share also amounts to *S* at the end of 7 years.

Since the rate of interest is 6% compounded annually, therefore

the present value of Divya's amount = $S\,(1 + 0.06)^{-5}$

the present value of Neha's amount = $S\,(1 + 0.06)^{-7}$

Thus

$$S\,(1 + 0.06)^{-5} + S\,(1 + 0.06)^{-7} = 50{,}000$$

i.e., $S[(1.06)^{-5} + (1.06)^{-7}] = 50{,}000$

$\Rightarrow$ $S(0.7472 + 0.6650) = 50{,}000$

$\Rightarrow$ $S(1.4122) = 50{,}000$

$\Rightarrow$ $S = 50{,}000/1.4122 = ₹\ 35{,}405.75$

EXERCISE 11.4

1. Find the present value of ₹ 6000 due in 20 years at 5% compounded annually.
2. Find the present value of ₹ 4000 due after 6 years if the interest rate is 7% compounded semiannually.
3. Find the present value of ₹ 1000 due after 3 years if the interest rate is 9% compounded monthly.
4. Find the present value of ₹ 2500 due after 15 months if the interest rate is 8% compounded quarterly.
5. A person invests money in a bank paying 6% compounded semiannually. If the person expects to receive ₹ 8000 in 6 years, what is the present value of the investment ?
6. Find the present value of ₹ 1000 due at end of 4 years if money is worth 7% effective.
7. What sum will amount in 6 years to ₹ 5000 at 8% converted quarterly ?
8. Find the present value of ₹ 2500 due 8 years from now if the interest rate is 8% compounded continuously.
9. What is the present value of ₹ 6000 due after 5 years from now if the interest is compounded continuously at the rate of interest of 6% ?
10. A trust fund is being set up by a single payment so that at the end of 20 years there will be ₹ 50,000 in the fund. If interest is compounded continuously at an annual rate of 5%, how much money should be paid into the fund initially ?

ANSWERS

1. ₹ 2261.33 **2.** ₹ 2647.13 **3.** ₹ 764.15 **4.** ₹ 2264.32 **5.** ₹ 5611.03

6. ₹ 762.89 **7.** ₹ 3108.60 **8.** ₹ 1318.22 **9.** ₹ 4444.92 **10.** ₹ 18,394

11.9 EQUATION OF VALUE

A sum of money has different values at different times. Let a non-interest-bearing debt of ₹ x be due at some specified time. Its value at the due date is of course x. If not paid at that time and if money is worth i per period, the value of the debt n periods *after* the due date is more than x and is found by multiplying x by $(1 + i)^n$. But its value n period before the due date is less than x and is found by multiplying x by $(1 + i)^{-n}$.

An equation of value is an equation which states that the sum of the values on a given date, of one set of obligations is the same as the sum of the values, on this same date, of another set of obligations. Any date chosen for comparing the values of two sets of obligations is called the *comparison date* or *focal date.*

Note. It is important to note that, if two sets of obligations have equal values on one date, they will have equal values on any other date, provided compound interest is used and rate of interest is same for each obligation.

Example 30. A debt of ₹ 200 due 2 years hence and another of ₹ 500 due 7 years hence are to be paid off by a single payment 3 years hence. If rate of interest is 5% per annum effective, how much is this payment ?

Solution. Let the unknown payment be x. To shorten computations, let the focal date be 3 years hence, that is, the date connected with the unknown payment. Prepare a table of values as follows :

Focal Date : 3 Years Hence

Old Obligations	*Value of Each at Focal Date*	*New Obligations*	*Value of Each at Focal Date*
₹ 200 due in 2 years	$200\ (1.05)^1$	₹ x due in 3 years	₹ x
₹ 500 due in 7 years	$500\ (1.05)^{-4}$		

The sum of values in column *II* must be equal to the value in column *IV*. The equation of value is then

$$200\ (1.05) + 500\ (1.05)^{-4} = x$$

$$\Rightarrow \quad x = 200\ (1.05) + 500\ (0.8227) = 210 + 411.35 = ₹\ 621.35.$$

Example 31. In return of a promise to pay ₹ 500 at the end of 10 years, a person agrees to pay ₹ 100 now, ₹ 200 at the end of 6 years and a final payment at the end of 12 years. If the rate of interest is 2% per annum effective, what should the final payment be ?

[*Delhi Univ. B. Com. (H) 1996*]

Solution. Let x be the final payment due at the end of 12 years. For computational convenience, we shall set up an equation of value to represent the situation at the end of 12 years. In other words, let the focal date be 12 years hence. Now the old obligations and the new obligations can be shown by means of the following table :

Focal Date : 12 Years Hence

Old Obligations	*Value of Each at Focal Date*	*New Obligations*	*Value of Each at Focal Date*
₹ 500 at the end of 10 years	$500\ (1.02)^2$	₹ 100 now	$100\ (1.02)^{12}$
		₹ 200 at the end of 6 years	$200\ (1.02)^6$
		₹ x at the end of 12 years	x

The equation of value is then

$$500(1.02)^2 = 100\,(1.02)^{12} + 200\,(1.02)^6 + x$$

$$\Rightarrow \quad x = 500\,(1.02)^2 - 100\,(1.02)^{12} - 200\,(1.02)^6$$

$$= 500\,(1.0404) - 100\,(1.2682) - 200\,(1.1261)$$

$$= 520.20 - 126.82 - 225.22 = ₹\,168.16$$

Example 32. A man owes ₹ 800 due in 2 years without interest and ₹ 300 due in 9 years with 6% annual effective rate of interest. He wishes to pay these debts in two equal instalments due 4 and 5 years hence, respectively. If money is worth 4% converted semiannually, how much should each instalment be?

Solution. Let the focal date be 5 years hence and the amount of each instalment be ₹ x. Prepare a table of values as shown below :

Focal Date : 5 Years Hence

Old Obligations	*Value of Each at Focal Date*	*New Obligations*	*Value of Each at Focal Date*
₹ 800 due in 2 years	$800\,(1.02)^6$	₹ x due in 4 years	$x\,(1.02)^2$
₹ $300\,(1.06)^9$ due in 9 years	$300\,(1.06)^9\,(1.02)^{-8}$	₹ x due in 5 years	x

The equation of value is :

$$x\,(1.02)^2 + x = 800\,(1.02)^6 + 300\,(1.06)^9\,(1.02)^{-8}$$

$$\Rightarrow \quad 1.0404x + x = 800\,(1.1261) + 300\,(1.6894)\,(0.8534)$$

or

$$2.0404x = 900.88 + 432.52 = 1333.40$$

$$\Rightarrow \quad x = \frac{1333.40}{2.0404} = ₹\,653.50$$

Example 33. A debt of ₹ 5000 due 5 years from now and ₹ 5000 due 10 years from now is to be repaid by a payment of ₹ 2000 in 2 years, a payment of ₹ 4000 in 4 years and a final payment at the end of 6 years. If the interest rate is 7% compounded annually, how much is the final payment? ***[Delhi Univ. B. Com. (H) 2012, 2019 (Modified)]***

Solution. Let ₹ x be the final payment and 6 years from now be the focal date. Prepare a table of values as follows :

Focal Date : 6 Years Hence

Old Obligations	*Value of Each at Focal Date*	*New Obligations*	*Value of Each at Focal Date*
₹ 5000 due in 5 years	$5000(1.07)^1$	₹ 2000 due in 2 years	$2000(1.07)^4$
₹ 5000 due in 10 years	$5000(1.07)^{-4}$	₹ 4000 due in 4 years	$4000(1.07)^2$
		₹ x due in 6 years	x

The equation of value is :

$$2000\,(1.07)^4 + 4000\,(1.07)^2 + x = 5000\,(1.07) + 5000\,(1.07)^{-4}$$

$$2000\,(1.3107) + 4000\,(1.1449) + x = 5000\,(1.07) + 5000\,(0.7628)$$

$$2621.4 + 4579.6 + x = 5350 + 3814$$

$$7201 + x = 9164 \quad \Rightarrow \quad x = 9164 - 7201 = ₹\,1963$$

Example 34. A person borrows ₹ 12,000. He pays ₹ 4000 at the end of 6 months and ₹ 5000 at the end of one year. What final payment should be made at the end of 2 years to settle the debt if interest rate is 12% compounded continuously. **[*Delhi Univ. B. Com. (H) 2010, 2014, 2017*]**

Solution. Let the size of final payment be ₹ x. Taking 0 or present date as the focal date, the equation of value is :

$$12{,}000 = 4000e^{-0.06} + 5000e^{-0.12} + xe^{-0.24}$$

$$= 4000(0.94176) + 5000(0.94176)^2 + x(0.94176)^4$$

$$= 3767.04 + 5000(0.88691) + x(0.78661)$$

$$= 3767.04 + 4434.55 + 0.78661x$$

$$\Rightarrow \quad 0.78661x = 12{,}000 - 3767.04 - 4434.55 = 3798.41$$

$$\Rightarrow \quad x = 3798.41/0.78661 = 4828.83$$

Hence the size of final payment is ₹ 4828.83.

Example 35. The sums of ₹ 2000, ₹ 3000 and ₹ 4000 are due at the end of 2, 4 and 8 years respectively. It is proposed to replace this series of payments by a single sum of ₹ 9000 payable at the end of n years. If the rate of interest is 10% per annum effective, find the value of n. **[*Delhi Univ. B. Com. (H) 1998, 2009, 2018 (Modified)*]**

Solution. To find value of n, we shall set up an equation of value to represent the situation at the current year. In other words, let the focal date be 0 year. The equation of value is then

$$9000\,(1 + 0.10)^{-n} = 2000\,(1 + 0.10)^{-2} + 3000\,(1 + 0.10)^{-4} + 4000\,(1 + 0.10)^{-8}$$

$$\Rightarrow \quad 9\,(1.10)^{-n} = 2\,(1.10)^{-2} + 3\,(1.10)^{-4} + 4\,(1.10)^{-8}$$

$$= 2\,(0.8264) + 3\,(0.6830) + 4\,(0.4665) = 5.5678$$

$$\Rightarrow \quad (1.10)^{-n} = 5.5678/9 = 0.6186$$

Taking logarithm of both sides, we obtain

$$-n \log(1.10) = \log(0.6186) = -0.2086$$

or $$n\,(0.0414) = 0.2086 \quad \Rightarrow \quad n = \frac{0.2086}{0.0414} = 5 \text{ years } (app.)$$

EXERCISE 11.5

1. A debt of ₹ 2000 due 2 years hence and another of ₹ 5000 due 5 years hence are to be paid off by a single payment 4 years hence. If rate of interest is 4% compounded semiannually, how much is this payment?

2. What single payment 5 years hence will discharge debts of ₹ 800 and ₹ 500 due in 3 and 9 years respectively, if money is worth 6% compounded quarterly?

3. A debt of ₹ 3000, which is due 6 years from now, is instead to be paid off by three payments: ₹ 500 now, ₹ 1500 in 3 years and a final payment at the end of 5 years. If the interest rate is 6% compounded annually, how much is the final payment?

4. A debt of ₹ 5000 due in 5 years is to be repaid by a payment of ₹ 2000 now and a second payment at the end of 6 years. How much should the second payment be if the interest rate is 6% compounded quarterly ?

5. *A* owes *B* two sums of money : ₹ 1000 due in 2 years and ₹ 600 due in 5 years. If *A* wishes to pay off the total debt now by a single payment, find how much should the payment be if money is worth 8% compounded quarterly?

6. A debt of ₹ 8,000 due 2 years from now and ₹ 10, 000 due 5 years from now is to be repaid by a payment of ₹ 2,000 in 3 years, a payment of ₹ 5,000 in 6 years, and a final payment at the end of 4 years. If the interest rate is 7% p.a. compounded semi-annually, how much is the final payment? **[*Delhi Univ. B. Com (H) 2022*]**

7. A debt of ₹ 2000 due in 3 years and ₹ 3000 due in 7 years is to be repaid by a single payment of ₹ 1000 now and two equal payments which are due 1 year from now and 4 years from now. If the interest rate is 6% compounded annually, how much are each of the equal payments ?

8. *A* owes *B* two sums of money : ₹ 1000 plus interest at 7% compounded annually, which is due in 5 years, and ₹ 2000 plus interest at 8% compounded semiannually, which is due in 7 years. If both debts are to be paid off by a single payment at the end of 6 years, find the amount of the payment if money is worth 6% compounded quarterly.

9. Mr. *X* took a loan of ₹ 50,000 payable with interest at 10% p.a. compounded semiannually. If he pays ₹ 10,000 each at the end of first year and second year, find the balance payable at the end of third year if the rate of interest remains same. **[*Delhi Univ. B. Com (H) 2013*]**

10. Mr. *X* borrows ₹ 20,000 and agrees to pay ₹ 5000 in 2 years, ₹ 7000 in 4 years and a final payment at the end of 5 years. If the interest rate is 6% compounded continuously, how much is the final payment ?

11. The sums of ₹ 2000 due 2 years hence and another of ₹ 5000 due 5 years hence are to be replaced by a single sum of ₹ 6970 at the end of n years. If the rate of interest is 4% compounded semiannually, find the value of n.

12. *A* debt of ₹ 3000 which is due 6 years from now, is instead to be paid off by three payments: ₹ 500 now, ₹ 1500 in 3 years and a final payment of ₹ 475 at the end of n years. If the rate of interest is 6% per annum effective, find the value of n.

[*Delhi Univ. B. Com (H) 2005, 2018*]

13. A money is worth 4% effective. What equal payments x at the end of 1 year and 3 years will replace the obligation ₹ 20,000 due in 3 years without interest and ₹ 40,000 plus interest at 4% compounded semiannually which is due in 6 years ?

ANSWERS

1. ₹ 6970.30	**2.** ₹ 1295.28	**3.** ₹ 475.68
4. ₹ 2447.81	**5.** ₹ 1257.27	**6.** ₹ 12,015.60
7. ₹ 1541.01	**8.** ₹ 4751.72	**9.** ₹ 43,820
10. ₹ 13,578.18	**11.** 4	**12.** 5
13. ₹ 31,273.42		

11.10 DEPRECIATION

The fixed assets of a business enterprise, such as buildings, machinery and equipment of all kinds diminish in value as time goes on, on account of wear and tear, exposure etc. That part of the decrease in value which can not be made by current repairs is called **depreciation.** The value of a depreciable asset at the end of its useful life is called the **scrap value.** The difference between the original cost and the scrap value is the **total depreciation** or **wearing value** of the asset. The **book value** of an asset on a given date is the original cost of that asset minus the accumulated depreciation at that time.

11.11 METHODS OF COMPUTING THE ANNUAL DEPRECIATION

The computation of annual depreciation depends on the following factors: the original cost, the useful life and the scrap value. The original cost of the asset is, of course, known definitely. However, the useful life and scrap value are determined by experts in engineering or in valuation.

Before discussing some of the methods used in computing annual depreciation, the following symbols will be defined :

C = the original cost of the asset

S = the estimated scrap value

W = the total depreciation

D = the annual depreciation

n = the useful life in years

Straight line method. By this method the annual depreciation of the asset is found by dividing the total depreciation by the number of years in its estimated useful life. Thus the straight line method of depreciation involves the following formula

$$D = \frac{C - S}{n} = \frac{W}{n}$$

Example 36 . A machine costing ₹ 50,000 has a useful life of 4 years. The estimated scrap value is ₹ 10,000. Using the straight line method, find the annual depreciation and construct a schedule for depreciation.

Solution . Here $C = 50{,}000$, $S = 10{,}000$, and $n = 4$. Hence

$$D = \frac{C - S}{n} = \frac{50{,}000 - 10{,}000}{4} = ₹\, 10{,}000$$

At the beginning of the first year, the book value of the machine is ₹ 50,000. At the end of the first year, the accumulated depreciation is ₹ 10,000; hence the depreciation charge for the first year is ₹ 10,000 and the book value is reduced to ₹ 50,000 – ₹ 10,000 = ₹ 40,000. At the end of the second year, the accumulated depreciation is ₹ 20,000; hence the depreciation charge is ₹ 20,000 – ₹ 10,000 = ₹ 10,000; and the book value is ₹ 40,000 – ₹ 10,000 = ₹ 30,000. It is evident that at the end of each year, the accumulated depreciation is increased by ₹ 10,000, and the book value is reduced by ₹ 10,000, until, at the end of the 4th year, the accumulated depreciation is ₹ 40,000, and the book value is reduced to ₹ 10,000. These charges are shown in the following table :

DEPRECIATION SCHEDULE

	Date of purchase	*Year 1*	*Year 2*	*Year 3*	*Year 4*
Annual Depreciation	₹ 0	₹ 10,000	₹ 10,000	₹ 10,000	₹ 10,000
Accumulated Depreciation	₹ 0	₹ 10,000	₹ 20,000	₹ 30,000	₹ 40,000
Book Value	₹ 50,000	₹ 40,000	₹ 30,000	₹ 20,000	₹ 10,000

Sum-of-the-years-digits method. By this method a greater fraction of the cost of the asset is depreciated in the earlier years of the life of the asset. The fraction of the asset to be depreciated each year is determined by putting the digit of the year in reverse order over the sum of the digits of the life periods. This method can be best illustrated with the help of the following example.

Example 37 . A machine which costs ₹ 15,000 is expected to have a useful life of 5 years and a scrap value of zero. Find the annual depreciation and form a depreciation schedule using the sum-of-the-years-digits method.

Solution. The annual depreciation charged each year is determined by putting the digit of the year in reverse order over the sum of the digits of the life periods. The appropriate depreciation schedule is shown in the following table :

DEPRECIATION SCHEDULE

Year	*Digit of year in Reverse Order*	*Fraction of Asset to be Depreciated*	*Annual Depreciation*	*Accumulated Depreciation*
1	5	5/15	₹ 5000	₹5000
2	4	4/15	₹ 4000	₹9000
3	3	3/15	₹ 3000	₹12,000
4	2	2/15	₹ 2000	₹14,000
5	1	1/15	₹ 1000	₹15,000
Sum of the years digits =	15		₹ 15,000	

Constant Percentage Method (or Diminishing Balance Method). By this method the annual depreciation is a constant percentage of the book value of the depreciated asset at the end of the preceding year. This constant percentage must be determined so that the book value of the asset at the end of its estimated life is reduced to the scrap value. This method is also referred to as "Diminishing Balance Method" because the annual depreciation calculated every year is based on the book value of the asset at the end of the preceding year which does not remain constant but keeps on diminishing from year to year.

Let r be the constant percentage also called the rate of depreciation. At the end of the first year, the depreciation charge is Cr and the book value is $C - Cr = C(1 - r)$. That is, the book value at the end of the first year is $(1 - r)$ times the book value at the beginning of that year. At the end of the second year, the depreciation charge is $Cr(1 - r)$ and the book value is $C(1 - r) - Cr(1 - r) = C(1 - r)^2$. That is, the book value at the end of the second year is $(1 - r)^2$ times the book value at the beginning of the first year. In general, the book value at the end of the nth year is $C(1 - r)^n$, which must be equal to S. Hence

$$S = C(1 - r)^n$$

Note . To find the depreciation charge for the kth year using constant percentage method, simply subtract the book value of the asset at the end of the kth year from the book value of the asset at the end of the $(k - 1)^{th}$ year, where the book value of the asset at the end of the kth year is $C(1 - r)^k$, as seen from the derivation of above formula.

Example 38. A machine costing ₹ 50,000 depreciates at a constant rate of 8%. What is the depreciation charge for the 8th year ? If the estimated useful life of the machine is 10 years, determine its scrap value.

Solution . It is given that $C = 50{,}000$ and $r = 0.08$. The depreciation charge for the 8th year is obtained by subtracting the book value at the end of the 8th year from the book value at the end of the 7th year.

The book value at the end of the 7th year $= C(1-r)^7 = 50{,}000\,(1-0.08)^7$

$= 50{,}000\,(0.92)^7$

$= 50{,}000\,(0.5578466) =$ ₹ 27892.33

The book value at the end of the 8th year $= C(1-r)^8 = 50{,}000\,(1-0.08)^8$

$= 50{,}000\,(0.92)^8$

$= 50{,}000\,(0.5132188) =$ ₹ 25660.94

$\therefore$ The depreciation charge for the 8th year = ₹ 27892.33 – ₹ 25660.94 = ₹ 2231.39

The scrap value of the machine is given by $S = C(1-r)^{10} = 50{,}000\,(1-0.08)^{10}$

$= 50{,}000\,(0.92)^{10}$

$= 50{,}000\,(0.4343884) =$ ₹ 21719.42

Example 39. A mainframe computer whose cost is ₹ 5,00,000 will depreciate to a scrap value of ₹ 50,000 in 5 years. What is the book value of the computer at the end of third year,

(*i*) if the reducing balance method of depreciation is used.

(*ii*) if the straight line method of depreciation is used. **[*Delhi Univ. B. Com. (H) 2009*]**

Solution. *(i)* First we'll obtain the rate of depreciation. Let the rate of depreciation be r. Then

$$5{,}00{,}000\,(1-r)^5 = 50{,}000 \Rightarrow (1-r)^5 = \frac{50{,}000}{5{,}00{,}000} = \frac{1}{10}$$

$$\Rightarrow \quad 5\log(1-r) = \log 1 - \log 10 = -1$$

$$\Rightarrow \quad \log(1-r) = -\frac{1}{5} = -0.2 = \bar{1}.8 \Rightarrow 1-r = AL\,(\bar{1}.8) = 0.631$$

$\therefore$ The book value at the end of 3rd year $= 5{,}00{,}000\,(1-r)^3$

$= 5{,}00{,}000\,(0.631)^3$

$=$ ₹1,25,620 (*app.*)

(*ii*) The annual depreciation, using the straight line method, is given by

$$\frac{5{,}00{,}000 - 50{,}000}{5} = ₹\ 90{,}000$$

$\therefore$ The book value at the end of 3rd year = 5,00,000 – 3 (90,000) = ₹ 2,30,000

***Example 40*.** A machine is purchased for ₹ 10,000. It is depreciated at a constant rate 6% for the first 4 years and after that at 10% for the next 6 years. Find the value of the machine after a period of 10 years.

Solution. The value of the machine at the end of the 4th year is :

$$10{,}000\,(1-0.06)^4 = 10{,}000\,(0.94)^4 = 10{,}000\,(0.78074) = ₹7807.40$$

After a period of 4 years, the machine is depreciated at a rate of 10% for the next 6 years. Hence the value of the machine after a period of 10 years is :

$$7807.40\,(1-0.10)^6 = 7807.40\,(0.90)^6 = 7807.40\,(0.53144) = ₹4149.16$$

Example 41. A machine depreciates at the rate of 8% for the first two years, at 10% for next three years and then at the rate of 15% p.a. Find the value of the machine at the end of 10 years, if the value of the machine is ₹ 1,00,000 initially. Find also the average rate of depreciation. **[*Delhi Univ. B. Com. (H) 2010, 2017 (modified)*]**

Solution . Depreciated value of the machine at the end of 10th year

$$\begin{aligned} &= 1,00,000\,(1-0.08)^2\,(1-0.10)^3\,(1-0.15)^5 \\ &= 1,00,000\,(0.8464)\,(0.729)\,(0.4437) \\ &= ₹27377.42 \end{aligned}$$

Let r be the average rate of depreciation. Then r is given by

$$\begin{aligned} & 1,00,000\,(1-r)^{10} = 27377.42 \\ \Rightarrow\ & (1-r)^{10} = 0.2737742 \cong 0.2738 \\ \Rightarrow\ & 10\log(1-r) = \log 2738 - \log 10,000 = 3.4375 - 4 = -0.5625 \\ \Rightarrow\ & \log(1-r) = -0.05625 = \bar{1}.94375 \\ \Rightarrow\ & 1-r = 0.8785 \quad \Rightarrow \quad r = 0.1215 \text{ or } 12.15\% \end{aligned}$$

Hence the average rate of depreciation is 12.15%.

Example 42. An asset costing ₹ 4500 will depreciate to a scrap value of ₹ 500 in 10 years. Find the rate of depreciation. **[*Delhi Univ. B. Com. (H) 2008, 2014*]**

Solution. Let r be the rate of depreciation. Then

$$\begin{aligned} & 4500\,(1-r)^{10} = 500 \\ \Rightarrow\ & (1-r)^{10} = 1/9 \\ \Rightarrow\ & 10\log(1-r) = \log(1/9) = \log 1 - \log 9 = -0.9542 \\ \Rightarrow\ & \log(1-r) = -0.09542 = \bar{1}.9046 \\ \Rightarrow\ & 1-r = AL\left(\bar{1}.9046\right) = 0.8028 \\ \Rightarrow\ & r = 1 - 0.8028 = 0.1972 \text{ or } 19.72\% \end{aligned}$$

Hence the rate of depreciation is 19.72%.

EXERCISE 11.6

1. A machine costing ₹ 30,000 is expected to have a useful life of 13 years and a final scrap value of ₹ 4000. Find the annual depreciation charge using the straight line method.

2. An asset costing ₹ 15,000 is expected to have a useful life of 5 years and a scrap value of ₹ 3000. Find the annual depreciation charge using the straight line method.

3. A piece of machinery costing ₹ 10,000 is expected to have a useful life of 4 years and a scrap value of zero. Find the annual depreciation charge using the sum-of-the-years-digits method.

4. A machine, the life of which is estimated to be 15 years, costs ₹ 40,000. Calculate the scrap value at the end of its life if it is depreciated at a constant rate of 10% per annum.
5. A machine costing ₹ 5000 depreciates at a constant rate of 5%. What is the depreciation charge for the 5th year?
6. If the estimated useful life of the machine in problem 5 is 15 years, determine its scrap value.
7. A machine is purchased for ₹ 10,600. Depreciation is calculated at 8% for the first 3 years and after that 10% per annum for the next seven years, depreciation being calculated on diminishing value. Find the value of the machine after a period of 10 years.
8. A machine depreciates at the rate of 10% per annum for the first two years and then 7% per annum for the next 3 years, depreciation being calculated on the diminishing value. If the value of the machine be ₹ 10,000 initially, find the depreciated value of the machine at the end of the fifth year. Also, find the average rate of depreciation. ***[Delhi Univ. B. Com. (H) 2006, 2017]***
9. An asset depreciates at the rate of 10% p.a. for the first two years and then 8% p.a. for the next three years, depreciation being calculated on the diminishing value. If the cost of the asset is ₹ 45,000, find the average rate of depreciation and the depreciated value of the asset at the end of the fifth year. ***[Delhi Univ. B. Com. (H) 2022]***
10. A machine costing ₹ 75,000 is depreciated at the rate of 10% p.a. for the first 5 years and then at 12% p.a. for next 3 years, both on diminishing balance basis. Find its book value at the end of 8th year. Using this, also find average rate of depreciation. ***[Delhi Univ. B. Com. (H) 2013]***
11. A machine depreciates at the rate of 8% of its value at the beginning of a year. The machine was purchased for ₹ 1,00,000 and the scrap value realised when sold was ₹ 43,440. Find the number of years the machine was used.
12. An item being depreciated in such a way that the value of the item at the end of any year is 90% of the value at the beginning of the year. The cost of the item is ₹ 4000 and it was sold eventually as waste metal for ₹ 1500. Obtain the number of years the item was in use.
13. A machine costing ₹ 5600 will depreciate to a scrap value of ₹ 1951 in 10 years. Find the rate of depreciation.
14. An asset costing ₹ 2000 will depreciate to a scrap value of ₹ 160 in 10 years. Find the rate of depreciation.
15. A machine costing ₹ 80,000 would reduce to ₹ 20,000 in 8 years. Find the rate of yearly depreciation, given that depreciation is calculated using diminishing balance method. ***[Delhi Univ. B.Com. (H) 2000]***

ANSWERS

1. ₹ 2000 **2.** ₹ 2400 **3.** ₹ 4000, ₹ 3000, ₹ 2000, ₹ 1000

4. ₹ 8224 **5.** ₹ 203.50 **6.** ₹ 2315 **7.** ₹ 3946

8. ₹ 6516; 8.21% **9.** ₹ 28383.61; 8.8% **10.** ₹ 30180.18; 10.755%

11. 10 years (app.) **12.** 9.3 years (app.) **13.** 10% (app.) **14.** 22.33%

15. 15.91%

11.12 ANNUITY

An *annuity* is a sequence of payments, usually equal in size, and made at equal intervals of time. Although the word "annuity" indicates that payments must be annual, its meaning in common use has been broadened to include payments at regular intervals which may be more or less than one year. Some examples are : premiums on life insurance, monthly deposits in a bank, instalment loan payments and the like.

The **periodic rent** or **periodic payment** of an annuity is the size of each payment. The **rent period** or **payment interval** is the period of time elapsing between two successive payment dates. The **term** of an annuity is the time from the beginning of the first rent period to the end of the last rent period. The **amount or future value** of an annuity is the sum of the compound amounts of all the payments accumulated at the end of the term. The **present value** or **capital value** of an annuity is the sum of the present values of all the payments. In other words, the present value of an annuity represents the amount of money that must be invested now to purchase the payments due in the future.

11.13 TYPES OF ANNUITIES

Ordinary Annuity. An *ordinary annuity* or *immediate annuity* is an annuity the first payment of which is made at the end of the first payment interval.

Annuity Due. An *annuity due* is an annuity the first payment of which is made at the beginning of the first payment interval.

Annuity Certain. An *annuity certain* is an annuity whose payments extend over a fixed period of time.

Contingent Annuity. A *contingent annuity* is one whose payments continue for a period of time which depends on events whose dates of occurrence cannot be accurately foretold. For example, the premiums on a life insurance policy constitute such an annuity.

Deferred Annuity. A *deferred annuity* is one the first payment of which is postponed until the expiration of a period of time equivalent to a certain number of payment intervals.

Perpetuity. A *perpetuity* is an annuity whose payments continue forever.

Note. In what is to follow, it is understood that the payment interval coincides with the interest period unless statement to the contrary is made.

11.14 AMOUNT OF AN ORDINARY ANNUITY

The amount (or future value) of an annuity is the value, at the end of the term, of all payments. That is, it is the sum of the compound amounts of all payments. In this section we shall derive the formula for the amount of an ordinary annuity in which each payment is made at the end of a payment period.

To develop a formula for the amount of an ordinary annuity, let us consider an annuity of n payment of ₹ R each, where the interest rate per period is i and the first payment is made one period from now (see Fig. 11.1).

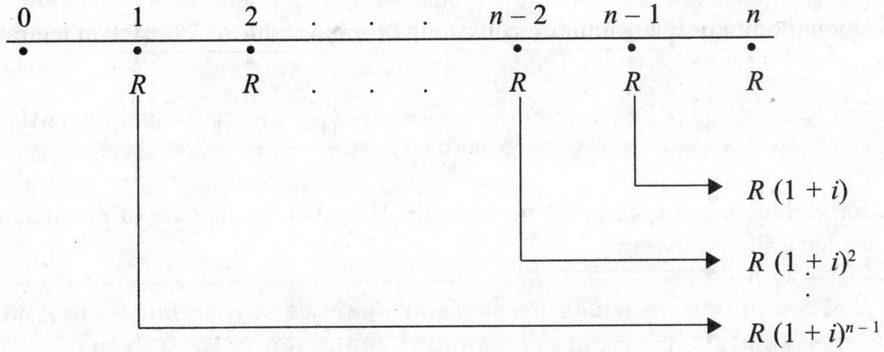

Fig. 11.1

The first payment, made at the end of the first period, accrues interest for $(n-1)$ periods, hence its compound amount would be $R(1+i)^{n-1}$. The second payment of ₹ R accrues interest for $(n-2)$ periods, the third for $(n-3)$ periods and so on. The last payment, made at the end of the term, does not accrue any interest and its value is simply R. We have, then

$$
\begin{aligned}
\text{Amount of Ist payment} &= R(1+i)^{n-1}\\
\text{Amount of IInd payment} &= R(1+i)^{n-2}\\
&\ \ \vdots\\
\text{Amount of } (n-1)\text{th payment} &= R(1+i)\\
\text{Amount of } n\text{th payment} &= R
\end{aligned}
$$

Thus the amount, S, of this annuity is given by

$$S = R + R(1+i) + \ldots + R(1+i)^{n-2} + R(1+i)^{n-1}$$

This is a geometric series consisting of n terms with first term R and common ratio $1+i$. Hence its sum is given by

$$S = R\left[\frac{(1+i)^n - 1}{(1+i)-1}\right] = R\left[\frac{(1+i)^n - 1}{i}\right]$$

Thus the formula $S = R\left[\frac{(1+i)^n - 1}{i}\right]$ gives the amount of an ordinary annuity consisting of n payments of ₹ R each at the rate i per period.

The expression $\left[\dfrac{(1+i)^n - 1}{i}\right]$ is denoted by $s_{\overline{n}|i}$ and is read "*s* sub *n* at the rate *i*". The symbol $s_{\overline{n}|i}$ means the amount of an ordinary annuity consisting of *n* payments of ₹ 1 each at the rate *i* per period. Thus

$$S = R s_{\overline{n}|i}, \quad \text{where } s_{\overline{n}|i} = \frac{(1+i)^n - 1}{i}$$

Values of $s_{\overline{n}|i}$, for various *n* and *i* are given in Table III. However, in the case of continuous compounding, we have the following :

The amount S of an annuity in which R rupees are paid each year for N years at the rate of interest of r% per annum compounded continuously is given by

$$S = \int_0^N R e^{rt} \, dt$$

Example 43. Find the amount of an annuity of ₹ 500 payable at the end of each year for 12 years if money is worth 6% effective.

Solution. The amount of the annuity is given by using the formula

$$S = R s_{\overline{n}|i}, \quad \text{where } R = 500, \; n = 12 \text{ and } i = 0.06$$

From Table III, we have $s_{\overline{12}|0.06} = 16.869941$.

$\therefore \quad S = 500\,(16.869941) =$ ₹ 8434.97.

Example 44. Find the amount of an annuity consisting of payments of ₹ 600 at the end of every 3 months for 3 years at the rate of 8% compounded quarterly.

Solution. The amount of the annuity is given by using the formula

$$S = R s_{\overline{n}|i}, \quad \text{where } R = 600, \; n = 3\,(4) = 12 \text{ and } i = 0.08/4 = 0.02$$

From Table III, we have $s_{\overline{12}|0.02} = 13.412089$.

$\therefore \quad S = 600\,(13.412089) =$ ₹ 8047.25.

Example 45. Find the amount of an annuity if payment of ₹ 200 is made at the end of every month for 3 years at the rate of 15% compounded monthly.

Solution. The amount of the annuity is given by using the formula

$$S = R s_{\overline{n}|i}, \quad \text{where } R = 200, \; n = 3\,(12) = 36 \text{ and } i = 0.15/12 = 0.0125$$

From Table III, we have $s_{\overline{36}|0.0125} = 45.115505$.

$\therefore \quad S = 200\,(45.115505) =$ ₹ 9023.10.

Example 46. At six-month intervals, A deposited ₹ 100 in a savings account which credit interest at 10% per annum compounded semiannually. The first deposit was made when A's son was six month old and the last deposit was made when his son was 8 years old. The money remained in the account and was presented to the son on his 10th birthday. How much did he receive ? ***[Delhi Univ B. Com. (H) 1998, 2011]***

Solution. First we find the amount accumulated in the account when A's son was 8 years old. Since A deposited ₹ 100 every six months for 8 years and the first deposit was made when A's son was six month old, the required amount accumulated in the account when A's son was 8 years old is the amount of an ordinary annuity consisting of 16 semi-annual payments of ₹ 100 each at the rate 0.05 per period. Hence it is given by

$$100 s_{\overline{16}|0.05} = 100\,(23.6574) = ₹2365.74$$

Next, we find the amount received by the son on his 10th birthday. Since the amount ₹ 2365.74 remained in the account for 2 more years at interest rate of 10% per annum compounded semiannually, the required amount received by the son is :

$$2365.74\left(1+\frac{0.10}{2}\right)^4 = 2365.74\,(1.05)^4 = 2365.74\,(1.215506) = ₹\ 2875.57$$

Example 47. Mr. X deposits in his son's account ₹ 500 times his son's age at the end of each birthday. Find the balance accumulated at the tenth birthday, if the rate of interest is 10% per annum compounded annually. ***[Delhi Univ B. Com. (H) 2004]***

Solution. Let the balance accumulated on the tenth birthday be S. Then S is given by

$$S = 500\,(1.10)^9 + 1000\,(1.10)^8 + 1500\,(1.10)^7 + \ldots + 5000$$

$$\Rightarrow \quad \frac{S}{1.10} = 500\,(1.10)^8 + 1000\,(1.10)^7 + \ldots + 4500 + \frac{5000}{1.10}$$

(*dividing throughout by* 1.10)

Subtracting, we obtain

$$S\left(1-\frac{1}{1.10}\right) = [500\,(1.10)^9 + 500\,(1.10)^8 + 500\,(1.10)^7 + \ldots + 500] - \frac{5000}{1.10}$$

$$= [500 + 500\,(1.10) + 500\,(1.10)^2 + \ldots + 500\,(1.10)^9] - \frac{5000}{1.10}$$

$$= 500\left[\frac{(1.10)^{10}-1}{1.10-1}\right] - \frac{5000}{1.10}$$

$$= 5000\,[(1.10)^{10}-1] - \frac{5000}{1.10}$$

$$\Rightarrow \quad \frac{S}{11} = 5000\,[2.59374246-1] - \frac{5000}{1.10} = 7968.71 - 4545.45 = 3423.26$$

$$\Rightarrow \quad S = 3423.26 \times 11 = ₹37{,}655.86$$

Example 48. **If you want to accumulate ₹ 50,000 by making equal payments at the end of each quarter for the next five years, what will be the size of these investments if money is worth 6% converted quarterly ?**

Solution. Let size of each quarterly payment be ₹ *R*. Then *R* is given by

$$R\, s_{\overline{n}|i} = 50{,}000, \quad \text{where } n = 5\,(4) = 20 \text{ and } i = 0.06/4 = 0.015$$

From Table III, we have $s_{\overline{20}|0.015} = 23.123667$.

$$\therefore \quad R\,(23.123667) = 50{,}000 \quad \Rightarrow \quad R = \frac{50{,}000}{23.123667} = ₹2162.29$$

Example 49. **A bank pays interest @ 6% per annum compounded continuously. If a person places ₹ 1000 in the bank each year, how much will be in his account after 3 years ?**

[Delhi Univ. B. Com. (H) 1989]

Solution. The amount *S* in the account after 3 years is given by

$$S = \int_0^N R e^{rt}\, dt = \int_0^3 1000 e^{0.06t}\, dt = \frac{1000}{0.06} \left| e^{0.06t} \right|_0^3 = \frac{50{,}000}{3}(e^{0.18} - e^0)$$

$$= \frac{50{,}000}{3}(1.1972 - 1) = ₹3286.67$$

Example 50. **A bank pays interest at the rate of 6% per annum compounded continuously. Find how much should be deposited in the bank each year in order to accumulate ₹ 6000 in 3 years ?** ***[Delhi Univ. B. Com. (H) 1992, 2006]***

Solution. Let ₹ *R* be deposited in the bank each year for 3 years in order to accumulate ₹ 6000 at the rate of interest of 6% per annum compounded continuously, then

$$6000 = \int_0^3 R e^{0.06t}\, dt = \frac{R}{0.06} \left| e^{0.06t} \right|_0^3 = \frac{R}{0.06}(e^{0.18} - e^0)$$

$$= \frac{R}{0.06}(1.1972 - 1) = \frac{R}{0.06}(0.1972)$$

$$\Rightarrow \quad R = \frac{6000 \times 0.06}{0.1972} = \frac{36{,}00{,}000}{1972} = ₹1825.56$$

EXERCISE 11.7

1. Find the amount of an annuity of ₹ 2000 payable at the end of each year for 5 years if money is worth 8% effective.
2. Find the amount of an annuity of ₹ 500 payable at the end of each year for 14 years if money is worth 5% effective.
3. Find the amount of an annuity of ₹ 200 payable at the end of each 6 months for 4 years if money is worth 6% compounded semiannually.
4. Find the amount of an annuity of ₹ 800 payable at the end of each 3 months for 6 years if money is worth 6% compounded quarterly.

5. A person decides to put aside ₹ 100 at the end of every month in a money market fund that pays 8% compounded monthly. After making 12 deposits, how much money does he have ?

6. Find the amount of an annuity consisting of payments of ₹ 500 made at the end of every 3 months for 4 years at the rate of 6% compounded quarterly.

7. A man requires ₹ 20,00,000 to purchase a house after 5 years. He has an opportunity to invest the funds in an account which can earn 6% per annum compounded quarterly. Find how much must be deposited in each quarter so as to have the required amount at the end of 5 years.

[*Delhi Univ. B. Com. (H) 2006*]

8. At six-month intervals, A deposited ₹ 2000 in a savings account which credit interest at 10% per annum compounded semi-annually. The first deposit was made when A's son was six month old and the last deposit was made when his son was 8 years old. The money remained in the account and was presented to the son on his 10th birthday. How much did he receive ?

(*Given*: $(1.05)^4 = 1.21550625$, $(1.05)^{16} = 2.18287459$, $s_{\overline{15}|0.05} = 21.57585636$, $s_{\overline{16}|0.05} = 23.6574918$)

[*Delhi Univ. B. Com. (H) 2008*]

9. If a bank pays 6% interest, compounded quarterly, what equal deposits have to be made at the end of each quarter for 3 years if you want to have ₹ 1500 at the end of 3 years ?

10. An annuity consisting of equal payments at the end of each month for 2 years is to be purchased for ₹ 2000. If the interest rate is 6% compounded monthly, how much is each payment ?

11. What sum must be invested at the end of each year to provide for the replacement at the end of 10 years of a machine costing ₹ 10,000 if money is worth 5% effective ?

12. A bank pays interest @ 10% per annum compounded continuously. If a person deposits ₹ 10,000 in the bank each year, how much will be in his account after 5 years ?

13. If ₹ 500 is deposited each year in a savings account paying 5.5% per annum compounded continuously, how much will be in the account after 4 years ?

14. An account fetches interest at 5% per annum compounded continuously. An individual deposits ₹ 1000 each year in the account. Find how much will be the amount in the account after 5 years ?

15. An annuity of ₹ 5000 per annum is flowing continuously for 10 years. Find its future value if the rate of interest is 10 % compounded continuously. **[*Delhi Univ. B. Com. (H), 2013*]**

16. A bank pays interest at the rate of 8% per annum compounded continuously. Find how much should be deposited in the bank each year in order to accumulate ₹ 10,000 in 5 years.

ANSWERS

1. ₹ 11,733.20	**2.** ₹ 9799.32	**3.** ₹ 1778.47	**4.** ₹ 22,906.82
5. ₹ 1244.99	**6.** ₹ 8966.18	**7.** ₹ 86,491.47	**8.** ₹ 57,511.65
9. ₹ 115.01	**10.** ₹ 78.64	**11.** ₹ 795.05	**12.** ₹ 64,870
13. ₹ 2237.27	**14.** ₹ 5680	**15.** ₹ 85,915	**16.** ₹ 1626.67

11.15 PRESENT VALUE OF AN ORDINARY ANNUITY

The *present value or capital value* of an annuity is the sum of the present values of all payments. In other words, the present value of an annuity represents the amount of money that must be invested now to purchase the payments due in the future.

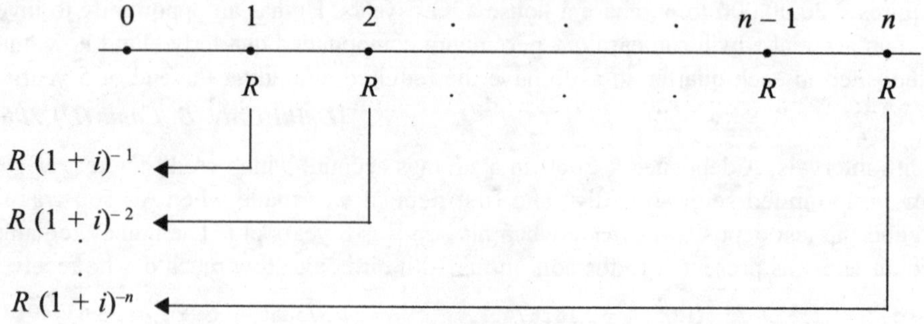

Fig. 11.2

To develop a formula for the present value of an ordinary annuity, let us consider an annuity of n payments of ₹ R each, where the interest rate per period is i and the first payment is due one period from now (see Fig. 11.2).

The present value of the first payment $= R\,(1+i)^{-1}$

The present value of the second payment $= R\,(1+i)^{-2}$

$\vdots$

The present value of the nth payment $= R\,(1+i)^{-n}$

Thus the present value P of the annuity is given by

$$P = R\,(1+i)^{-1} + R\,(1+i)^{-2} + \dots + R\,(1+i)^{-n}$$

This a geometric series of n terms with first term $R\,(1+i)^{-1}$ and common ratio $(1+i)^{-1}$. Hence its sum is given by

$$P = R\,(1+i)^{-1}\left[\frac{1-(1+i)^{-n}}{1-(1+i)^{-1}}\right] = R\left[\frac{1-(1+i)^{-n}}{i}\right]$$

Thus the formula $P = R\left[\dfrac{1-(1+i)^{-n}}{i}\right]$ gives the present value of an ordinary annuity consisting of n payments of ₹ R each at the rate i per period.

The expression $\left[\dfrac{1-(1+i)^{-n}}{i}\right]$ is denoted by $a_{\overline{n}|i}$ and is read "a sub n at the rate i". The symbol $a_{\overline{n}|i}$ means the present value of an ordinary annuity consisting of n payments of ₹ 1 each at the rate i per period. Thus

$$P = R a_{\overline{n}|i}, \quad \text{where} \quad a_{\overline{n}|i} = \frac{1-(1+i)^{-n}}{i}$$

Values of $a_{\overline{n}|i}$ for various n and i are given in Table IV.

However, in the case of continuous compounding, we have the following :

The present value P of an annuity consisting of an income stream of ₹ R per year for N years at the rate of interest of r% per annum compounded continuously is given by

$$P = \int_0^N R e^{-rt}\, dt$$

Example 51. Prove that $i = \frac{1}{a_{\overline{n}|i}} - \frac{1}{s_{\overline{n}|i}}$.

Solution. $RHS = \frac{1}{a_{\overline{n}|i}} - \frac{1}{s_{\overline{n}|i}} = \frac{i}{1-(1+i)^{-n}} - \frac{i}{(1+i)^n - 1}$

$$= i\left[\frac{(1+i)^n}{(1+i)^n - 1} - \frac{1}{(1+i)^n - 1}\right]$$

$$= i\left[\frac{(1+i)^n - 1}{(1+i)^n - 1}\right] = i = LHS$$

This completes the proof.

Example 52. If the present value and amount of an ordinary annuity of ₹1 per annum for n years are ₹ 8.1109 and ₹ 12.0061 respectively, find the rate of interest and the value of n without consulting the compound interest table.

[*Delhi Univ. B. Com. (H) 2002, 2004, 2008, 2017*]

Solution. We are given : $a_{\overline{n}|i} = 8.1109$ and $s_{\overline{n}|i} = 12.0061$

We know that $i = \frac{1}{a_{\overline{n}|i}} - \frac{1}{s_{\overline{n}|i}}$. (*Refer Example* 53)

$$\therefore \quad i = \frac{1}{8.1109} - \frac{1}{12.0061} = 0.1232 - 0.0832 = 0.04 \text{ or } 4\%$$

We also know that $s_{\overline{n}|i} = \frac{(1+i)^n - 1}{i}$. Substituting $i = 0.04$ in this formula, we obtain

$$12.0061 = \frac{(1.04)^n - 1}{0.04} \quad \Rightarrow \quad (1.04)^n = 12.0061 \times 0.04 + 1 = 1.480$$

$$\Rightarrow \quad n \log(1.04) = \log(1.480) \quad i.e. \quad (0.0170)n = 0.1703$$

$$\Rightarrow \quad n = \frac{0.1703}{0.0170} = 10.$$

Example 53. Find the present value of an annuity of ₹ 500 payable at the end of each year for 8 years, if money is worth 5% effective.

Solution. The present value of the annuity is given by

$$P = Ra_{\overline{n}|i}, \text{ where } R = 500, \ n = 8 \text{ and } i = 0.05.$$

From Table IV, $a_{\overline{8}|0.05} = 6.463212$.

$$\therefore \quad P = 500a_{\overline{8}|0.05} = 500(6.463212) = ₹3231.61$$

This means that ₹ 3231.61 *now* is just as good as 8 annual payments of ₹ 500 each or just as good as the amount of the annuity 8 years hence.

Check. The single sum ₹ 3231.61, invested at 5% effective, will amount in 8 years to

$$3231.61(1.05)^8 = 3231.61(1.477455) = ₹4774.56$$

On the other hand, the amount of this annuity is given by

$$S = 500s_{\overline{8}|0.05} = 500(9.549108) = ₹4774.56$$

Thus the determined present value of the annuity is correct.

Example 54. Find the present value of an annuity of ₹ 400 payable at the end of each month for 7 years, if money is worth 6% converted monthly.

Solution. The present value of the annuity is given by

$$P = Ra_{\overline{n}|i}, \text{ where } R = 400, \ n = 7(12) = 84 \text{ and } i = 0.06/12 = 0.005.$$

From Table IV, $a_{\overline{84}|0.005} = 68.453042$.

$$\therefore \quad P = 400a_{\overline{84}|0.005} = 400(68.453042) = ₹27,381.22.$$

Example 55. *X* buys a piece of land priced at ₹ 5,00,000 for which he agrees to make equal payments at the end of each year for 10 years. If money is worth 5% effective, what is the size of these annual payments ?

Solution. Let the size of each annual payment be ₹ *R*. Then *R* is given by

$$5,00,000 = Ra_{\overline{10}|0.05} \quad \Rightarrow \quad R = \frac{5,00,000}{a_{\overline{10}|0.05}} = \frac{5,00,000}{7.721734} = ₹64,752.30.$$

Example 56. Find the present value of an annuity of ₹ 100 per annum, assumed to be payable continuously for 10 years, at the rate of interest of 4% per annum compounded continuously. **[*Delhi Univ. B. Com. (H) 1996*]**

Solution . The required present value is given by the formula

$$P = \int_0^N R\, e^{-rt}\, dt,$$

where $R = 100$, $r = 0.04$ and $N = 10$.

$$P = \int_0^{10} 100\, e^{-0.04t}\, dt$$

$$\therefore \quad = 100 \left| \frac{e^{-0.04t}}{-0.04} \right|_0^{10} = \frac{100}{0.04}(1 - e^{-0.4}) = 2500(1 - 0.67032) = ₹824.20.$$

Example 57. An equipment is purchased on an instalment basis such that ₹ 5000 is to be paid on the signing of the contract and four yearly instalments of ₹ 3000 each payable at the end of first, second, third and the fourth year. If interest is charged at 5% p.a., what would be the cash down price ?

Solution. The cash down price of the equipment is given by :

5000 + present value of four yearly instalments of ₹ 3000 each payable at the end of each year

$$= 5000 + 3000\, a_{\overline{4}|0.05}$$

$$= 5000 + 3000(3.54595) \qquad \textit{(using Table IV)}$$

$$= 5000 + 10{,}637.85 = ₹15{,}637.85.$$

Example 58. Mr. X paid a loan in 8 quarterly instalments of ₹ 5000 each at the end of each quarter and a final payment of ₹ 4000 at the end of next quarter. Find the initial loan taken if the rate of interest is 12% p.a. compounded quarterly and the total interest on loan.

[Delhi Univ. B. Com. (H) 2013]

Solution. The loan amount = present value of an ordinary annuity consisting of 8 quarterly instalments of ₹ 5000 each at the rate 0.03 per quarter + present value of ₹ 4000 due at the end of 9th quarter

$$= 5000\, a_{\overline{8}|0.03} + 4000(1.03)^{-9}$$

$$= 5000 \times 7.01969 + 4000 \times 0.76641$$

$$= 35{,}098.45 + 3065.64 = ₹\ 38{,}164.09.$$

Total interest $= 5000 \times 8 + 4000 - 38{,}164.09 = ₹5835.91.$

Example 59. A purchased a television paying ₹ 5000 down and promising to pay ₹ 200 every three months for next 4 years. The seller charges interest at 8% per annum compounded quarterly.

(*i*) What is the cash price of television ?

(*ii*) If *A* missed the first three payments, what must he pay at the time the fourth payment is due to bring him upto date ? **[*Delhi Univ. B. Com. (H) 1998*]**

Solution.

(*i*) The cash price of television = 5000 + present value of ordinary annuity consisting of 16 quarterly payments of ₹ 200 each at the rate 0.02 per quarter

$$= 5000 + 200\, a_{\overline{16}|0.02}$$

$$= 5000 + 200(13.5777) \quad \text{(using Table IV)}$$

$$= 5000 + 2715.54 = ₹7715.54.$$

(*ii*) If *A* missed the first three payments, then the amount that he must pay at the time the fourth is due is simply the amount of an ordinary annuity consisting of 4 quarterly payments of ₹ 200 each at the rate 0.02 per quarter. Thus it is given by

$$200 s_{\overline{4}|0.02} = 200(4.1216) \quad \text{(using Table III)}$$

$$= ₹824.32$$

Example 60. *M* purchased a house paying ₹ 5,00,000 down and promising to pay ₹ 20,000 every 3 months for next 10 years. The seller charges interest at the rate of 6% per annum compounded quarterly.

(*i*) What was the cash price of house ?

(*ii*) If *M* missed the first 12 payments, what must he pay at the time the 13th is due to bring him upto date ?

(*iii*) After making 8 payments, *M* wishes to discharge his remaining indebtness by a single payment at the time when 9th regular payment was due. What must he pay in addition to regular payment ?

(*iv*) If *M* missed the first 10 payments, what must he pay when the 11th payment is due to discharge his entire indebtness ?

Solution. (*i*) The cash price of house= 5,00,000 + present value of ordinary annuity consisting of 40 quarterly payments of ₹ 20,000 each at the rate 0.015 per quarter

$$= 5{,}00{,}000 + 20{,}000\, a_{\overline{40}|0.015}$$

$$= 5{,}00{,}000 + 20{,}000(29.91584520)$$

$$= 5{,}00{,}000 + 5{,}98{,}316.90 = ₹10{,}98{,}316.90.$$

(*ii*) If *M* missed the first 12 payments, then the amount he must pay at the time when 13th payment is due is given by

$$20{,}000 s_{\overline{13}|0.015} = 20{,}000(14.23682960) = ₹2{,}84{,}736.59$$

(*iii*) After making 8 payments, the amount he must pay to discharge his remaining indebtness by a single payment at the time when 9th regular payment was due is given by

$$20{,}000\, a_{\overline{31}|0.015} = 20{,}000(24.64614582) = ₹4{,}92{,}922.92$$

This is in addition to 9th regular payment of ₹ 20,000. Thus, the total amount paid at the time when 9th regular payment was due

$$= 20{,}000 + 4{,}92{,}922.92 = ₹5{,}12{,}922.92$$

(*iv*) If *M* missed the first 10 payments, the amount he must pay when the 11th payment is due to discharge his entire indebtness is given by

$$20{,}000 s_{\overline{11}|0.015} + 20{,}000\, a_{\overline{29}|0.015} = 20{,}000(11.86326249) + 20{,}000(23.37607558)$$
$$= 2{,}37{,}265.25 + 4{,}67{,}521.51 = ₹7{,}04{,}786.51.$$

Example 61. An automobile advertisement reads :

"No money down. ₹ 5000 per month for 30 months. Interest rate equals 12% per year compounded montlhy on unpaid balance."

How much of ₹ 1,50,000 to be paid will go for interest and how much for the vehicle itself? (Given $a_{\overline{30}|0.01} = 25.807708$) **[*Delhi Univ. B. Com. (H) 2007 (SOL)*]**

Solution. We are given :

$$R = \text{value of the instalment} = ₹5000$$
$$n = \text{number of instalments} = 30$$
$$i = \text{rate of interest per month} = 0.01$$

$$\therefore \quad \text{Cost of vehicle} = R a_{\overline{n}|i} = 5000\, a_{\overline{30}|0.01} = 5000(25.807708) = ₹1{,}29{,}038.54$$
$$\text{Interest} = ₹1{,}50{,}000 - ₹1{,}29{,}038.54 = ₹20{,}961.46.$$

Example 62. What should be the monthly sales volume of a company if it desires to earn a 12% annual return convertible monthly on its investment of ₹ 2,00,000 ? Monthly costs are ₹ 3000. The investment will have eight year life with no scrap value.

[*Delhi Univ. B. Com. (H) 2003, 2007 (SOL)*]

Solution. Let monthly sales volume be ₹ *x*. Then *x* is given by

$$2{,}00{,}000 = (x - 3000)\, a_{\overline{96}|0.01} = (x - 3000)\,(61.52770299)$$

$$\Rightarrow \quad x - 3000 = \frac{2{,}00{,}000}{61.52770299} = 3250.57$$

$$\Rightarrow \quad x = 3000 + 3250.57 = ₹\ 6250.57.$$

Example 63. Earning after taxes (cost savings or profits) from a new machine are expected to be ₹ 45,000 per year. The machine costs ₹ 2,00,000 and after five years has no resale or scrap value. A loan can be made for this amount payable in five equal annual instalments at 6% per annum on the unpaid balance of the loan. Should management buy the machine?

(Given : $a_{\overline{5}|0.06} = 4.212364$) **[*Delhi Univ. B. Com. (H) 2008*]**

Solution. Let size of each annual loan instalment be ₹ R. Then the cost of machine may be considered as the present value of an annuity of ₹ R for 5 years at 6% per annum. Hence R is given by :

$$2{,}00{,}000 = Ra_{\overline{5}|0.06} = R(4.212364)$$

$$\Rightarrow \qquad R = 2{,}00{,}000/4.212364 = ₹\ 47{,}479.28,$$

which exceeds the earning after taxes (*i.e.*, ₹ 45,000 per year).

Hence it is not advisable to buy the machine.

Example 64. You have just won a million rupees, the prize is awarded in 10 annual payments of ₹ 1,00,000 each. Annual payments are received at the end of each year. You are given the option of receiving a single lump-sum payment of ₹ 6,00,000 instead of the million rupees annuity. You want to find out which option is more in today's rupees. Money is worth 10% compounded annually. **[*Delhi Univ. B. Com. (H) 2009*]**

Solution. The present value of an annuity consisting of 10 annual payments of ₹1,00,000 each at the rate 0.10 per annum is given by :

$$1{,}00{,}000\ a_{\overline{10}|0.10} = 1{,}00{,}000 \times 6.145 = ₹\ 6{,}14{,}500,$$

which exceeds a single lump-sum payment of ₹ 6,00,000.

∴ Option of 10 annual payments of ₹ 1,00,000 each is better than a single lump-sum payment of ₹ 6,00,000.

Example 65. Mr. X sells his old car for ₹ 1,00,000 to buy a new one costing ₹ 2,58,000. He pays ₹ x cash and balance by payments of ₹ 7000 at the end of each month for 18 months. If rate of interest is 9% compounded monthly, find x. (Given : $(1.0075)^{-18} = 0.8741566$).

[*Delhi Univ. B. Com. (H) 2010*]

Solution. The present value of annuity consisting of 18 monthly payments of ₹ 7000 each at the rate of 9% compounded monthly is given by

$$7000a_{\overline{18}|0.0075} = 7000\left[\frac{1-(1.0075)^{-18}}{0.0075}\right] = 7000\left[\frac{1-0.8741566}{0.0075}\right] = ₹\ 1{,}17{,}454$$

$$\therefore \qquad x = 2{,}58{,}000 - 1{,}00{,}000 - 1{,}17{,}454 = ₹\ 40{,}546.$$

Example 66. Shares in a mining company are expected to produce dividends of ₹ 0, 30, 24, 16 and 8 in the present and in the four following years and to be worth nothing thereafter. If the interest is added once yearly at 5%, find the present or the capital value of the holdings.

[*Delhi Univ B.A. Eco (Hons.) 1992*]

Solution. The present value P of the holding is :

$$\begin{aligned} & 0 + 30\,(1.05)^{-1} + 24\,(1.05)^{-2} + 16\,(1.05)^{-3} + 8\,(1.05)^{-4} \\ &= 30\,(0.95238) + 24\,(0.90702) + 16\,(0.86383) + 8\,(0.82270) \\ &= 28.571 + 21.768 + 13.821 + 6.581 = ₹\ 70.74 \end{aligned}$$

Example 67. What is the present value of a continuous income stream of ₹ 3000 per year for two years if it is discounted continuously at the rate of 0.06 per year ?

[*Delhi Univ. B. Com. (H) 2008*]

Solution. The present value of a continuous income stream of ₹ 3000 per year for 2 years discounted continuously at the rate 0.06 per year is given by

$$P = \int_0^2 3000 e^{-0.06t}\, dt = 3000 \left| \frac{e^{-0.06t}}{-0.06} \right|_0^2$$

$$= \frac{3000}{0.06}(1 - e^{-0.12})$$

$$= 50{,}000(1 - 0.8869) = ₹5655.$$

Example 68. Find out the capital value of a uniform income stream of ₹ R per year for m years, reckoning interest continuously at 100 r% per year. What will be the result if income is for ever ?

[*Delhi Univ. B. Com. (H) 2006*]

Solution. The capital value of a uniform income stream of ₹ R per year for m years, reckoning interest continuously at 100 r% per year, is given by

$$P = \int_0^m Re^{-rt}\, dt = R \left| \frac{e^{-rt}}{-r} \right|_0^m = \frac{R}{r}(1 - e^{-rm})$$

However, if the income is forever, then its present value is :

$$P = \lim_{m \to \infty} \frac{R}{r}(1 - e^{-rm}) = \frac{R}{r}(1 - e^{-\infty}) = \frac{R}{r} \qquad (\because e^{-\infty} = 0)$$

Example 69. A person desires to create an endowment fund to provide for a prize of ₹ 5000 per year. If the fund can be invested at 12% per annum compounded continuously, find the amount of the endowment.

[*Delhi Univ. B. Com. (H) 2011*]

Solution. The required amount of endowment is the present value of perpetuity of ₹ 5000, reckoning interest at 12% per annum compounded continuously. Hence it is given by

$$\int_0^\infty 5000\, e^{-0.12t}\, dt = 5000 \left| \frac{e^{-0.12t}}{-0.12} \right|_0^\infty = -\frac{5000}{0.12}(e^{-\infty} - 1) = ₹41{,}666.67 \qquad (\because e^{-\infty} = 0)$$

Example 70. A company's dividend on equity share is according to $D_t = D_0\, e^{gt}$, where D_0 is a constant, g is the growth rate growing continuously. Determine the total present value of dividend of first t years if rate of interest is 100 r% compounded continuously. Also determine the total present value if $t \to \infty$ (assuming $r > g$).

[*Delhi Univ. B. Com. (H) 1997*]

Solution. The present value of dividend of first t years is given by

$$P_t = \int_0^t D_0\, e^{gx}\, e^{-rx}\, dx = D_0 \int_0^t e^{(g-r)x}\, dx = D_0 \left| \frac{e^{(g-r)x}}{g-r} \right|_0^t = \frac{D_0}{g-r}[e^{(g-r)t} - 1]$$

The present value, if $t \to \infty$, is given by

$$\lim_{t \to \infty} P_t = \frac{D_0}{g - r} \lim_{t \to \infty} [e^{(g-r)t} - 1] = \frac{D_0}{g - r} \lim_{t \to \infty} [e^{-(r-g)t} - 1]$$

$$= \frac{D_0}{g-r}(e^{-\infty} - 1) = \frac{D_0}{g-r}(-1) = \frac{D_0}{r-g}.$$

Example 71. An income stream decreases continuously over time for x years, the income rate at t years from now being ₹ ae^{-bt} per year. What is its present value if interest be reckoned at $100\,r\%$ compounded continuously ? Show that this equals the capital value of a uniform income stream of ₹ a per year for x years if the rate of interest is raised to $100\,(r + b)$ per cent per year.

[Delhi Univ. B. Com. (H) 1999, 2006 (SOL)]

Solution. The capital value of an income stream of ₹ ae^{-bt} per year for x years at $100\,r\%$ compounded continuously is given by

$$P = \int_0^x ae^{-bt}\, e^{-rt}\, dt = a\int_0^x e^{-(b+r)t}\, dt = \frac{a}{-(b+r)} \left| e^{-(b+r)t} \right|_0^x = -\frac{a}{b+r}(e^{-(b+r)x} - 1)$$

$$= \frac{a}{b+r}(1 - e^{-(r+b)x})$$

The capital value of a uniform income stream of ₹ a per year for x years, if the rate of interest is $100\,(r + b)\%$ per year, is given by

$$\int_0^x ae^{-(r+b)t}\, dt = \frac{a}{-(r+b)} \left| e^{-(r+b)t} \right|_0^x = -\frac{a}{r+b}(e^{-(r+b)x} - 1) = \frac{a}{r+b}(1 - e^{-(r+b)x}),$$

which is same as before.

Example 72. According to an investment proposal, an initial investment of ₹ 1,00,000 is expected to yield a uniform income stream of ₹ 10,000 per annum. If money is worth 8% per annum compounded continuously, what is the expected payback period, *i.e.* after what time, the initial investment will be recovered? **[Delhi Univ. B. Com. (H) 2009]**

Solution. Suppose it takes N years to recover the initial investment. Then N is given by

$$1{,}00{,}000 = \int_0^N 10{,}000\, e^{-0.08t}\, dt = \frac{10{,}000}{-0.08} \left| e^{-0.08t} \right|_0^N = 1{,}25{,}000\,(1 - e^{-0.08N})$$

$$\Rightarrow 1 - e^{-0.08N} = \frac{1{,}00{,}000}{1{,}25{,}000} = 0.8 \Rightarrow e^{-0.08N} = 1 - 0.8 = 0.2 \Rightarrow e^{0.08N} = \frac{1}{0.2} = 5$$

$$\Rightarrow 0.08N \log e = \log 5 \quad \text{or} \quad (0.08)\,(0.4343)\,N = 0.6990$$

$$\text{i.e.,} \quad 0.034744\,N = 0.6990 \Rightarrow N = \frac{0.6990}{0.034744} = 20.12 \text{ years.}$$

Example 73. According to the will of a man, his life insurance of ₹ 50,000 should be deposited in a bank soon after his death. The bank has to pay ₹ 8,000 annually to his heir. If the interest is 5% p.a. compounded continuously, how many annual payments could be made?

[*Delhi Univ. B. Com. (H) 2019*]

Solution. If N annual payments of ₹ 8000 each are paid to the heir @ 5% per annum compounded continuously, then N is given by

$$50{,}000 = \int_0^N 8000\, e^{-0.05t}\, dt$$

$$= 8000\left|\frac{e^{-0.05t}}{-0.05}\right|_0^N = 1{,}60{,}000\,(1 - e^{-0.05N})$$

$$\Rightarrow \quad 0.3125 = 1 - e^{-0.05N} \quad \text{or} \quad e^{-0.05N} = 0.6875$$

$$\Rightarrow \quad -0.05\, N \log e = \log (0.6875) = \log 6875 - \log 10{,}000 = 3.8373 - 4 = -0.1627$$

i.e., $(0.05)\,(0.4343)\, N = 0.1627$

or, $(0.021715)\, N = 0.1627 \quad \Rightarrow \quad N = 7.5$ (app.)

EXERCISE 11.8

1. Find the present value of an annuity of ₹ 5000 payable at the end of each year for 6 years, if money is worth 7% effective.
2. Find the present value of an annuity of ₹ 2000 payable at the end of each 6 months for 10 years, if money is worth 4% effective.
3. Find the present value of an annuity of ₹ 800 payable at the end of each 6 months for 5 years, if money is worth 6% converted semiannually.
4. Find the present value of an annuity which pays ₹ 200 at the end of each 3 months for 10 years, assuming money to be worth 5% converted quarterly.
5. A person buys a house for which he agrees to pay ₹ 5000 at the end of each month for 8 years. If money is worth 12% converted monthly, what is the cash price of the house ?
6. A person buys a house paying 50,000 in cash and the balance in 20 instalments of ₹ 8000 each at the end of each year. If interest be reckoned at 16% p.a., how much he should have paid if he had purchased it cash down?
7. M purchased a television paying ₹ 5000 down and promising to pay ₹ 200 every quarter for next 10 years. The seller charges interest at the rate of 12% per annum compounded quarterly. If M missed the first 10 payments, what must he pay when the 11th payment is due to discharge his entire loan ? **[*Delhi Univ. B. Com. (H) 1999*]**
8. Mohan has first purchased a house for ₹ 7,00,000 and has made a down payment of ₹ 1,50,000. He proposes to repay the balance in 25 years by monthly instalments at 9% per annum compounded monthly.

(*i*) What are the monthly payments ?

(*ii*) What is the total interest payment ? **[*Delhi Univ. B. Com. (H) 2001*]**

9. Mr. X purchases a house for ₹ 2,00,000. He agrees to pay for the house in 5 equal instalments at the end of each year. If money is worth 5% per annum effective, what would be the size of each instalment? In case X makes a down payment of ₹ 50,000 what would be size of each instalment? **[*Delhi Univ. B. Com. (H) 2012*]**

10. A man borrows ₹ 30,000 at 6% p.a. and promises to pay off the debt in 20 annual payments beginning at the end of the first year. What is the annual payment necessary ?

11. A man wishes to make lump sum deposit of an amount which will provide payments of ₹ 1000 a year. Five withdrawls of ₹ 1000 will be made, first taking place one year after the account was opened. What should be the amount invested if the interest is 6% compounded annually and the account has to be exhausted in 5 withdrawls ?

12. A machine is purchased for ₹ 30,000 down and payments of ₹ 2500 at the end of every six months for 6 years. If the interest be reckoned at 8% compounded semiannually, find the corresponding cash price of the machine.

13. What is the present value of a continuous income stream of ₹ 3500 per year for four years at the rate of interest of 5% per annum compounded continuously ?

14. What is the present value of a countinous income stream of ₹ 5000 per year for three years if it is discounted countinously at the rate of 6% per year?

[*Delhi Univ. B. Com. (H) 2006 (C.C)*]

15. Find the present value of a continuous income stream of ₹ 1000 per year for three years at the rate of 4% per annum compounded continuously.

16. Calculate the present value of an annuity of ₹ 30,000 per annum, assumed to be payable continuously for 10 years, at the rate of interest of 8 per cent per annum compounded continuously. **[*Delhi Univ. B. Com. (H) 2000*]**

17. A two-wheeler is purchased on instalment basis such that ₹ 1800 is to be paid each year for 5 years. If interest is charged @ 11% per annum compounded continuously, what would be the cash price or present value of the two-wheeler ?

18. A person sells his old car for ₹ 1,00,000 to buy a new car for ₹ 2,50,000. He pays ₹ x in cash and the balance in 18 instalments of ₹ 7000 at the end of each month. If interest be reckoned at 6% per annum compounded monthly, find the value of x.

19. A man buys a car worth ₹ 2,00,000 by paying 40% in cash and the balance in 20 equal annual instalments. Calculate the value of each instalment if money is worth $7\frac{1}{2}\%$ per annum compounded continuously.

20. Derive, by using calculus, the present value of an annuity of ₹ 1 per annum assumed to be payable continuously for n years at the rate of interest i per rupee per annum convertible continuously. **[*Delhi Univ. B. Com. (H) 2003, 2006, 2007*]**

21. Deduce that a/r is the present value of an income stream of ₹ a per year for ever, interest being 100 r% per annum compounded continuously. **[*Delhi Univ. B. Com. (H) 1984*]**

22. According to the will of a man, his Life Insurance of ₹ 40,000 should be deposited in a bank soon after his death. The bank has to pay ₹ 5000 annually to his heir and if the interest is 5% per annum compounded continuously how many annual payments could be made? **[Delhi Univ. B. Com. (H) 2012]**

23. According to an investment proposal an initial investment of ₹ 50,000 is expected to yield a uniform income stream of ₹ 5000 per annum. If money is worth 5% per annum compounded continuously, what is the expected payback period, *i.e.*, after what time, the initial investment will be recovered ?

ANSWERS

1. ₹ 23,832.70	**2.** ₹ 16,221.79	**3.** ₹ 6824.16	**4.** ₹ 6265.38	**5.** ₹ 3,07,638.50
6. ₹ 97,440	**7.** ₹ 6399.26	**8.** (*i*) ₹ 4615.58	(*ii*) ₹ 8,34,674	**9.** ₹ 46,195.78;
₹ 34,646.83	**10.** ₹ 2615.54	**11.** ₹ 4212.36	**12.** ₹ 53,462.68	**13.** ₹ 12,688.90
14. ₹ 13,727.50	**15.** ₹ 2827	**16.** ₹ 2,06,501.54	**17.** ₹ 6922.64	**18.** ₹ 29,790.62
19. ₹ 11,584	**20.** $\dfrac{1-e^{-in}}{i}$	**22.** 10	**23.** 13.87 years	

11.16 LEASING AND CAPITAL EXPENDITURE

The concept of present value of an annuity plays an important role in leasing, capital expenditure and bonds.

Leasing. A company may have to decide whether to obtain a machine either by leasing it for n years (the useful life) at an annual rent of, say, ₹ R or by purchasing the machine for ₹ P. If company can borrow money at r% per annum, the question is, "Which alternative is preferable ?" In such situations, we find the present value of an ordinary annuity of ₹ R for n years at the interest rate r% per annum. We compare this value with the purchase price and then decide accordingly.

Example 74. A company may obtain a particular machine either by leasing it for 5 years (the useful life) at an annual rent of ₹ 1200 or by purchasing the machine for ₹ 4500.

(*i*) Which alternative is preferable if the company can borrow money at 10% per annum ?

(*ii*) What if the cost of borrowing is 14% per annum ?

Solution. (*i*) The present value of an annuity of ₹ 1200 for 5 years at 10% per annum is given by

$$1200\, a_{\overline{5}|0.10} = 1200\left[\frac{1-(1.10)^{-5}}{0.10}\right] = 1200\times 3.7908 = ₹\,4548.96,$$

which exceeds the purchase price. Hence, purchase is preferable.

(*ii*) If company can borrow money at 14% per annum, the present value of the annuity is given by

$$1200\, a_{\overline{5}|0.14} = 1200\left[\frac{1-(1.14)^{-5}}{0.14}\right] = 1200\times 3.4331 = ₹\,4119.77,$$

which is less than the purchase price. Hence, leasing is preferable.

Capital Expenditure. A corporation may be faced with a choice between two machines, both of which are designed to improve operations by saving on labour costs. Knowing the time value of money, we can find the net annual savings of each machine and then decide accordingly. To illustrate, let us consider the following example.

Example 75. Machine *A* costs ₹ 10,000 and has a useful life of 8 years. Machine *B* costs ₹ 8000 and has a useful life of 6 years. Suppose machine A generates an annual labour savings of ₹ 2000 while machine *B* generates an annual labour savings of ₹ 1800. Assuming the time value of money is 10% per annum, find which machine is preferable ?

[Delhi Univ B. Com. (H) 1997, 2006, 2011, 2014, 2018 (modified)]

Solution. Machine *A* costs ₹ 10,000 and has a life of 8 years. If ₹ C_1 is the equivalent annual cost of machine *A*, then the cost of machine *A* may be considered as the present value of an annuity of ₹ C_1 for 8 years at 10% per annum. Hence C_1 is given by

$$10{,}000 = C_1\, a_{\overline{8}|0.10} \quad \Rightarrow \quad C_1 = \frac{10{,}000}{a_{\overline{8}|0.10}} = \frac{10{,}000}{5.335} = ₹\,1874.41$$

Machine *B* costs ₹ 8000 and has a life of 6 years. If ₹ C_2 is the equivalent annual cost of machine *B*, then the cost of machine *B* may be considered as the present value of an annuity of ₹ C_2 for 6 years at 10% per annum. Hence C_2 is given by

$$8000 = C_2\, a_{\overline{6}|0.10} \quad \Rightarrow \quad C_2 = \frac{8000}{a_{\overline{6}|0.10}} = \frac{8000}{4.356} = ₹\,1836.55$$

Thus, the net annual savings of each machine can be obtained as follows :

	Machine A	*Machine B*
Labour Savings	₹ 2000.00	₹ 1800.00
Equivalent annual cost	₹ 1874.41	₹ 1836.55
Net Savings	₹ 125.59	– ₹ 36.55

Hence, machine *A* is preferable.

ALITER : Since machine *A* generates an annual labour savings of ₹ 2,000 for 8 years, therefore present value of a sequence of annual savings of ₹ 2,000 for 8 years @ 10% per annum is given by

$$2000\, a_{\overline{8}|0.10} = 2000\left[\frac{1-(1+0.10)^{-8}}{0.10}\right] = 2000 \times 5.335 = ₹10{,}670$$

Cost of Machine *A* = ₹ 10,000

∴ Net savings from Machine *A* = ₹ 10,670 – ₹ 10,000 = ₹ 670

Further, since machine *B* generates an annual laboursavings of ₹ 1800 for 6 years, therefore present value of a sequence of annual savings of ₹ 1800 for 6 years @ 10% per annum is given by

$$1800\,a_{\overline{6}|0.10} = 1800\left[\frac{1-(1+0.10)^{-6}}{0.10}\right] = 1800\times 4.356 = ₹\,7840.80$$

$$\text{Cost of Machine } B = ₹\,8000$$

$$\therefore \quad \text{Net savings from Machine } B = ₹\,7840.80 - ₹\,8000$$

$$= -₹\,159.20$$

Thus, Machine *B* costs ₹ 159.20 more than it will save in labour costs.

Hence Machine *A* is preferable.

EXERCISE 11.9

1. A company may obtain a particular machine either by leasing it for 5 years (the useful life) at an annual rent of ₹ 2000 or by purchasing the machine for ₹ 8100. If the company can borrow money at 10% per annum, which alternative is preferable ?

   [*Delhi Univ B. Com. (H) 1997, 2011*]

2. A company may obtain a particular machine either by leasing it for 4 years (the useful life) at an annual rent of ₹ 1000 or by purchasing the machine for ₹ 3000. If the company can borrow money at 10% per annum, which alternative is preferable ?

3. If the company in Problem 2 can borrow money at 14% per annum, which alternative is preferable ?

4. A company is considering proposal of purchasing a machine either by making a full payment of ₹ 4000 or by leasing it for 4 years at an annual rent of ₹ 1250. Which course of action is preferable if the company can borrow money at 14% per annum ?

5. Mr. Ram, a businessman, plans to shift his business from Delhi to Faridabad because of closure of certain type of industries in Delhi. He has to make a choice between two machines, both of which are designed to improve operations by saving on labour costs. Machine *A* costs ₹ 9750 and will generate an annual labour savings of ₹ 2400. Machine *B* costs ₹ 7750 and will save ₹ 2300 annually. Machine *A* has a useful life of 8 years while Machine *B* has a useful life of 6 years only. If time value of money is 10% per annum, which machine is preferable ? (Assume annual compounding and that the savings is realised at the end of each year.)

   [*Delhi Univ B. Com. (H) 2001*]

6. Machine A costs ₹ 8000 and has a useful life of 7 years. Machine *B* costs ₹ 6000 and has a useful life of 5 years. Suppose Machine *A* generates an annual labour savings of ₹ 2000 while Machine *B* generates an annual labour savings of ₹ 1800. Assuming that the time value of money of the corporation is 10% per annum, which machine is preferable ?

7. A machine with useful life of 7 years costs ₹ 10,000 while another machine with useful life of 5 years costs ₹ 8000. The first machine saves labour expenses of ₹ 1900 annually and the second one saves labour expenses by ₹ 2200 annually. Assuming that the time value of the corporation is 10% per annum, which machine is preferable ?

   [*Delhi Univ B. Com. (H) 2005*]

8. A company wishes to invest 10,000 in a machine that will yield an income stream of 1200 per year for the next 8 years. If money is worth 8% per annum effective, find whether the investment is worthwhile. **[*Delhi Univ B. Com. (H) 2012*]**

ANSWER

1. Leasing is preferable
2. Purchase is preferable
3. Leasing is preferable
4. Leasing is preferable
5. Machine *A* is preferable
6. Machine *A* is preferable
7. Second machine is preferable
8. No

11.17 ANNUITY DUE

As already defined, an annuity due is an annuity in which the payments are made at the *beginning* of each rent period, the first payment being due now. The term of an annuity due begins at the time the first payment is made and ends one period after the last payment is made.

Amount of an Annuity Due

The amount of an annuity due is the value of the annuity at the end of its term. To develop a formula for the amount of an annuity due, let us consider an annuity of n payments of ₹ R each, where the interest rate per period is i and the first payment is due at the beginning of the term (See Figure 11.3).

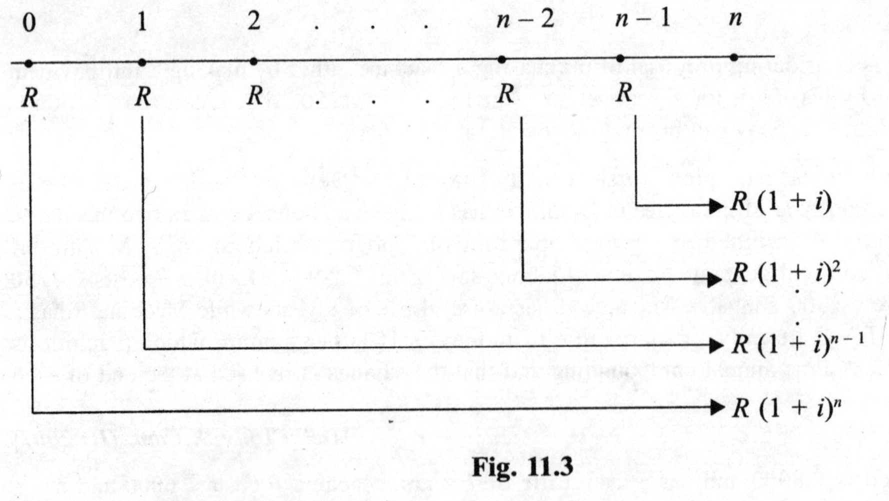

Fig. 11.3

As in Article 11.14, we have, then

$$\text{amount of Ist payment} = R(1+i)^n$$

$$\text{amount of IInd payment} = R(1+i)^{n-1}$$

$$\vdots \qquad \vdots$$

$$\text{amount of last payment} = R(1+i)$$

Adding these amounts in the reverse order, and denoting the amount of this annuity by S, we have

$$S = R(1+i) + R(1+i)^2 + \ldots + R(1+i)^n$$

This is a geometric series consisting of n terms with first term $R\,(1+i)$ and common ratio $1+i$. Hence its sum is

$$S = R(1+i)\left[\frac{(1+i)^n - 1}{1+i-1}\right] = R\left[\frac{(1+i)^{n+1} - (1+i)}{i}\right]$$

$$= R\left[\frac{(1+i)^{n+1} - 1}{i} - 1\right] = R\left(s_{\overline{n+1}|i} - 1\right).$$

Hence the formula for the amount of an annuity due consisting of n payments of ₹ R each at the rate i per period is :

$$S = R\left(s_{\overline{n+1}|i} - 1\right)$$

Example 76. Find the amount at the end of 10 years of an annuity of ₹ 400 payable at the beginning of each year until 10 payments have been made if money is worth 6% effective.

Solution. This represents an annuity due of 10 annual payments of ₹ 400 each, the rate of interest being 6% effective. Hence the amount of this annuity is given by

$$S = R\left(s_{\overline{n+1}|i} - 1\right),$$

where $R = 400$, $n = 10$, $i = 0.06$.

$$\therefore \quad S = 400\left(s_{\overline{11}|0.06} - 1\right)$$

$$= 400(14.97164 - 1) = 400(13.97164) = ₹5588.66.$$

Example 77. At the beginning of each quarter, ₹ 200 is deposited into a saving accounts that pays 8% compounded quarterly. Find the balance in the account at the end of 3 years.

Solution. The given deposits represent annuity due consisting of 12 quarterly payments of ₹ 200 each at the rate of 0.02 per quarter. Hence the balance in the account at the end of 3 years is given by

$$S = 200\left(s_{\overline{13}|0.02} - 1\right)$$

$$= 200(14.68033 - 1) = 200(13.68033) = ₹2736.07.$$

Example 78. What sum of money deposited in a fund at the beginning of each 6 months until 20 deposits have been made will accumulate to ₹ 10,000 at the end of 10 years, if the fund earns interest at 6% compounded semiannually.

Solution. Let ₹ R be deposited in the fund at the beginning of each 6 months for 10 years in order to accumulate ₹ 10,000 at the rate of interest of 6% compounded semiannually. Then R is given by

$$10{,}000 = R\left(s_{\overline{21}|0.03} - 1\right) = R(28.67648 - 1) = R(27.67648)$$

$$\Rightarrow \qquad R = \frac{10{,}000}{27.67648} = ₹361.32.$$

Example 79. Mr. X wants ₹ 5,00,000 at the end of 7 years. If the rate of interest is 8%, what amount shall be deposited at the beginning of each quarter so as to get the above amount? **[*Delhi Univ. B. Com. (H) 2007*]**

Solution. Let ₹ R be deposited at the beginning of each quarter for 7 years in order to get ₹ 5,00,000 at the end of 7 years. Then R is given by

$$R\left(s_{\overline{n+1}|i} - 1\right) = 5{,}00{,}000,$$

where $n = 4 \times 7 = 28$ and $i = 0.08/4 = 0.02$.

$$\therefore \quad R\left(s_{\overline{29}|0.02} - 1\right) = 5{,}00{,}000 \quad \Rightarrow \quad R(38.7922 - 1) = 5{,}00{,}000$$

$$\Rightarrow \quad R(37.7922) = 5{,}00{,}000 \quad \Rightarrow \quad R = \frac{5{,}00{,}000}{37.7922} = ₹13{,}230.24.$$

Present Value of an Annuity Due

The present value of an annuity due is the sum of the present values of all the payments. To develop a formula for the present value of an annuity due, let us consider an annuity consisting of n payments of ₹ R each, where the interest rate is i per period and each payment is due at the beginning of a payment period. (See Fig. 11.4).

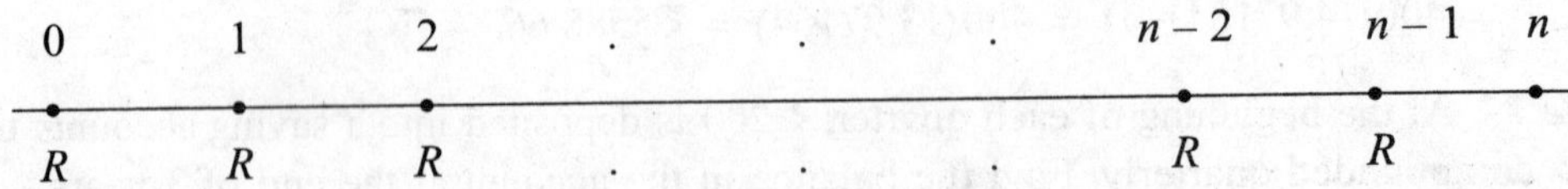

Fig. 11.4

The given annuity can be thought of as an initial payment of ₹ R followed by an ordinary annuity of ₹ R for $(n - 1)$ periods. Thus the required present value is given by

$$R + Ra_{\overline{n-1}|i} = R(1 + a_{\overline{n-1}|i})$$

Hence the formula for the present value of an annuity due consisting of *n* payments of ₹ *R* each at the rate *i* per period is :

$$P = R(a_{\overline{n-1}|i} + 1)$$

In particular, when each periodic payment is ₹ 1, the present value of annuity due, denoted by $\ddot{a}_{\overline{n}|i}$, is given by

$$\ddot{a}_{\overline{n}|i} = a_{\overline{n-1}|i} + 1$$

Example 80. What is the present value, at 6% compounded semiannually, of an annuity due of ₹ 400 payable semiannually for 12 years ?

Solution. The present value of the given annuity is obtained by using the formula

$$P = R(a_{\overline{n-1}|i} + 1),$$

where $R = 400$, $n = 24$ and $i = 0.03$.

$$\therefore \quad P = 400(a_{\overline{23}|0.03} + 1)$$
$$= 400(16.44360 + 1)$$
$$= 400(17.44360) = ₹6977.44.$$

Example 81. *A* purchases a house from *B* and agrees to pay *B* twenty annual instalments of ₹ 20,000 each, the first to be paid immediately. If money is worth 8% effective, what is the equivalent cash price of this house ?

Solution. The cash price of the house is given by :

$$P = R(a_{\overline{n-1}|i} + 1), \quad \text{where } R = 20,000, \ n = 20 \text{ and } i = 0.08$$

$$P = 20,000(a_{\overline{19}|0.08} + 1)$$
$$\therefore \quad = 20,000(9.60359 + 1)$$
$$= 20,000(10.60359) = ₹2,12,071.80.$$

Example 82. What equal payments made at the beginning of each year for 10 years will pay for a piece of property priced at ₹ 6,00,000, if money is worth 5% effective ?

[*Delhi Univ. B. Com. (H) 2019 (modified)*]

Solution. It is given that $P = 6,00,000$, $n = 10$ and $i = 0.05$. It is required to find *R*. Substituting these quantities in the formula $P = R(a_{\overline{n-1}|i} + 1)$, we obtain

$$6,00,000 = R(a_{\overline{9}|0.05} + 1) = R(7.10782 + 1) = R(8.10782)$$

$$\Rightarrow \quad R = \frac{6,00,000}{8.10782} = ₹74,002.63.$$

Example 83. Mr. *X* purchased an asset for ₹ 1,00,000 on instalment basis. Each instalment is to be paid at the beginning of each quarter. Find the size of each instalment if the money is to be repaid in three years and rate of interest is 6% compounded quarterly.

[*Delhi Univ. B. Com. (H) 2007*]

Solution. Let R be the size of each quarterly instalment. Then R is given by

$$1,00,000 = R(a_{\overline{n-1}|i} + 1), \quad \text{where } n = 12 \text{ and } i = 0.06/4 = 0.015$$

$$\Rightarrow \qquad R(a_{\overline{11}|0.05} + 1) = 1,00,000 \quad \Rightarrow \quad R(10.0711 + 1) = 1,00,000$$

$$\Rightarrow \qquad R(11.0711) = 1,00,000 \quad \Rightarrow \quad R = \frac{1,00,000}{11.0711} = ₹9032.53$$

EXERCISE 11.10

1. Find the amount at the end of 12 years of an annuity of ₹ 1200 payable at the beginning of each year until 12 payments have been made if money is worth 8% effective.
2. At the beginning of each quarter, ₹ 600 is deposited into a savings account that pays 6% compounded quarterly. Find the balance in the account at the end of 5 years.
3. Find the amount of an annuity of ₹ 2000 payable at the beginning of each month for 3 years at the rate of 15% compounded monthly.
4. A bank pays interest @ 8% per annum compounded quarterly. Find how much should be deposited in the bank at the biginning of each quarter in order to accumulate ₹ 7000 in 3 years?
5. What is the present value, at 8% compounded quarterly, of an annuity due of ₹ 1000 payable quarterly for 5 years ?
6. A man purchases a house for which he agrees to pay ₹ 5000 at the beginning of each 3 months until he has made 10 payments. If money is worth 6% compounded quarterly, what is the equivalent cash price of the house ?
7. The premiums on an insurance policy are ₹ 700 a quarter, payable at the beginning of each quarter. If the policyholder wishes to pay 1 year's premium in advance, how much should be paid provided that the interest rate is 10% compounded quarterly ?
8. What equal payments made at the beginning of each month for 3 years will pay for a house priced at ₹ 4,00,000, if money is worth 15% compounded monthly ?
9. What equal payments made at the beginning of each year for 10 years will pay for a piece of property priced at ₹ 80,000, if money is worth 7% effective ?
10. The amount of an annuity due consisting of 15 annual payments invested at 8% effective is ₹ 10,000. Find the size of each payment.

11. A debt of ₹ 5000 with interest at the rate of 8% compounded quarterly is to be discharged by 8 equal quarterly payments, the first being due today. Find the size of each payment.

12. Mr. *X* purchased a house for ₹ 30,00,000. He agrees to pay for the house in 6 equal instalments in the beginning of each year. If money is worth 6% p.a. effective, what would be the size of each instalment? **[*Delhi Univ. B. Com. (H) 2019*]**

ANSWERS

1. ₹ 24,594.35 **2.** ₹ 14,082.31 **3.** ₹ 91,358.90 **4.** ₹ 511.68

5. ₹ 16,678.46 **6.** ₹ 46,802.58 **7.** ₹ 2699.22 **8.** ₹ 13,694.94

9. ₹ 10,645.05 **10.** ₹ 341.01 **11.** ₹ 669.17 **12.** ₹ 5,75,561.65

11.18 DEFERRED ANNUITY

Recall that a deferred annuity is an ordinary annuity in which the first payment is postponed until the expiration of a certain number of payment intervals. When the term of an ordinary annuity of n periods begins m periods hence, the annuity is said to be an annuity of n periods, **deferred** m periods (see Fig. 11.5).

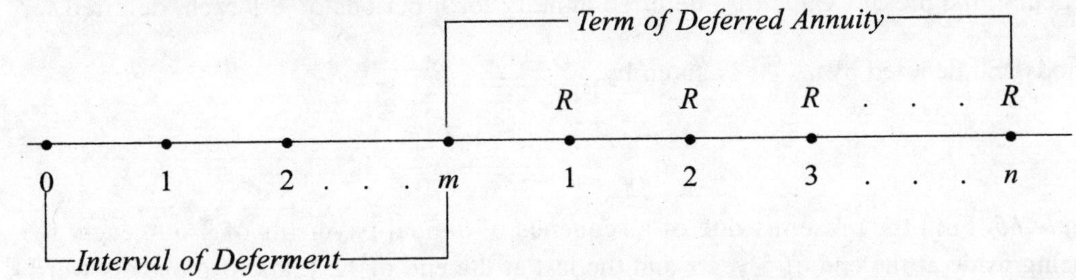

Fig. 11.5

The length of time between now and the beginning of the term of the deferred annuity is called the **interval of deferment**, which ends one period before the first payment is due.

Amount of a Deferred Annuity

The amount of a deferred annuity for n periods, deferred m periods, is the value of the annuity at the end of its term. Thus if a deferred annuity consists of n payments of ₹ R each, then the amount of this annuity is the amount of an ordinary annuity consisting of n payments of ₹ R each. Hence

$$S = R s_{\overline{n}|i}$$

Example 84. An annuity consists of 5 annual payments of ₹ 1500 each, the first being made at the end of 4 years. Find the amount of this annuity if money is worth 7% effective.

Solution. This is a deferred annuity consisting of 5 annual payments, the first being deferred for 3 years. Hence the amount of this annuity is :

$$S = 1500 s_{\overline{5}|0.07} = 1500(5.75073) = ₹8626.10$$

Present Value of a Deferred Annuity

The present value of a deferred annuity for *n* periods, deferred *m* periods, is the value of the annuity at the beginning of the interval of deferment. This is not the same as the value of an ordinary annuity for *n* periods at the beginning of its term. However, the present value of this annuity may be found by the following procedure :

First, find the present value of an ordinary annuity consisting of payments of ₹ *R* each made at the end of each of *m* payment intervals, together with payments of ₹ R each made at the end of each of *n* additional payment intervals. The present value of this annuity is :

$$P_1 = Ra_{\overline{m+n}|i}$$

Next, find the present value of the ordinary annuity consisting of payments of ₹ *R* each made at the end of each of *m* payment intervals. The present value of this annuity is :

$$P_2 = Ra_{\overline{m}|i}$$

Hence the present value of the deferred annuity is :

$$P = P_1 - P_2 = Ra_{\overline{m+n}|i} - Ra_{\overline{m}|i} = R(a_{\overline{m+n}|i} - a_{\overline{m}|i})$$

In particular, the present value of a deferred annuity for *n* periods of ₹ 1 each, deferred for *m* periods and denoted by $\overset{(m)}{a}_{\overline{n}|i}$, is given by

$$\overset{(m)}{a}_{\overline{n}|i} = a_{\overline{m+n}|i} - a_{\overline{m}|i}$$

Example 85. Find the present value of a sequence of annual payments of ₹ 600 each, the first being made at the end of 5 years and the last at the end of 12 years, if money is worth 6% effective.

Solution. This is a deferred annuity consisting of 8 annual payments, the first being deferred for 4 years. Consequently, the present value of this annuity is given by the formula

$$P = R(a_{\overline{m+n}|i} - a_{\overline{m}|i}),$$

where $R = 600,\ m = 4,\ n = 8$ and $i = 0.06$. Hence

$$\begin{aligned} P &= 600(a_{\overline{12}|0.06} - a_{\overline{4}|0.06}) \\ &= 600(8.38384 - 3.46510) \\ &= 600(4.91874) = ₹2951.24 \end{aligned}$$

Example 86. *X* buys a piece of land for which he agrees to make 10 annual payments of ₹ 20,000 each, the first being made at the end of 3 years. Find the equivalent cash price of this property if money is worth 5% effective.

Solution. The cash price of the property is nothing but the present value of a deferred annuity consisting of 10 annual payments of ₹ 20,000 each, the first being deferred for 2 years. Hence the cash price is given by

$$P = R(a_{\overline{m+n}|i} - a_{\overline{m}|i}),$$

where $R = 20{,}000$, $m = 2$, $n = 10$ and $i = 0.05$. Hence

$$\begin{aligned} P &= 20{,}000(a_{\overline{12}|0.05} - a_{\overline{2}|0.05}) \\ &= 20{,}000(8.86325 - 1.85941) \\ &= 20{,}000(7.00384) = ₹1{,}40{,}076.80 \end{aligned}$$

Example 87. Find the present value of an annuity consisting of 41 monthly payments of ₹ 100 each, the first being made at the end of 2 years and money is worth 6% compounded monthly.

Solution. Since payments are made monthly and the first payment is made at the end of 2 years, the annuity is deferred for 23 periods. Hence its present value is given by

$$P = R(a_{\overline{m+n}|i} - a_{\overline{m}|i}),$$

where $R = 100$, $m = 23$, $n = 41$ and $i = 0.005$. Hence

$$\begin{aligned} P &= 100(a_{\overline{64}|0.005} - a_{\overline{23}|0.005}) \\ &= 100(54.65434 - 21.67568) \\ &= 100(32.97866) = ₹3297.87 \end{aligned}$$

Example 88. A house is sold for ₹ 50,000 down and 10 semi-annual payments of ₹ 5000 each, the first due 3 years hence. Find the cash price of the house if money is worth 20% compounded semi-annually. **[*Delhi Univ. B. Com. (H) 2009, 2017*]**

Solution. The cash price of the house = ₹ (50,000 + P), where P is the present value of a deferred annuity consisting of 10 semi-annual payments of ₹ 5000 each, the first being deferred for 5 periods. Hence P is given by

$$P = R(a_{\overline{m+n}|i} - a_{\overline{m}|i}),$$

where $R = 5000$, $m = 5$, $n = 10$ and $i = 0.10$. Hence

$$\begin{aligned} P &= 5000(a_{\overline{15}|0.10} - a_{\overline{5}|0.10}) \\ &= 5000(7.606 - 3.791) \\ &= 5000(3.815) = ₹19{,}075 \end{aligned}$$

Hence, required cash price of the house = 50,000 + 19,075 = ₹ 69,075.

Example 89. An orchard will yield its first full crop at the end of 5 years and is expected to maintain an annual income of ₹ 5,000 for 20 years in all. Find the cash price of the orchard if money is worth 3% per annum. **[*Delhi Univ. B. Com. (H) 2018*]**

Solution. The cash price, P, of the orchard is the present value of a deferred annuity consisting of 20 annual payments of ₹ 5000 each, the first being deferred for 4 years. Hence P is given by

$$P = R\left(a_{\overline{m+n}|i} - a_{\overline{m}|i}\right),$$

where $R = 5000$, $m = 4$, $n = 20$, and $i = 0.03$. Hence

$$P = 5000\left(a_{\overline{24}|0.03} - a_{\overline{4}|0.03}\right)$$

$$= 5000\ (16.9355 - 3.7170)$$

$$= 5000\ (13.2185) = ₹\ 66092.50$$

11.19 RELATIONSHIP AMONG $a_{\overline{n}|i}$, $\ddot{a}_{\overline{n}|i}$ AND $a^{(m)}_{\overline{n}|i}$

By definition, $a_{\overline{n}|i}$ represents the present value of an ordinary annuity of n periodic payments of ₹ 1 each at the rate i per period and is given by the formula

$$a_{\overline{n}|i} = \frac{1-(1+i)^{-n}}{i}$$

Also, $\ddot{a}_{\overline{n}|i}$ represents the present value of an annuity due of n periodic payments of ₹ 1 each at the rate i per period and is given by the formula

$$\ddot{a}_{\overline{n}|i} = a_{\overline{n-1}|i} + 1$$

$$= \frac{1-(1+i)^{-n+1}}{i} + 1$$

$$= \frac{1-(1+i)^{-n+1} + i}{i}$$

$$= \frac{(1+i)-(1+i)^{-n+1}}{i}$$

$$= (1+i)\left[\frac{1-(1+i)^{-n}}{i}\right] = (1+i)a_{\overline{n}|i}$$

Hence we have, $\ddot{a}_{\overline{n}|i} = a_{\overline{n-1}|i} + 1$

Further, $\overset{(m)}{a}_{\overline{n}|i}$ represents the present value of a deferred annuity consisting of n periodic payments of ₹ 1 each and deferred for m periods at the rate i per period and is given by the formula

$$\begin{aligned}\overset{(m)}{a}_{\overline{n}|i} &= a_{\overline{m+n}|i} - a_{\overline{m}|i}\\ &= \frac{1-(1+i)^{-(m+n)}}{i} - \frac{1-(1+i)^{-m}}{i}\\ &= \frac{1-(1+i)^{-(m+n)} - 1 + (1+i)^{-m}}{i}\\ &= (1+i)^{-m}\left[\frac{1-(1+i)^{-n}}{i}\right] = (1+i)^{-m}\, a_{\overline{n}|i}\end{aligned}$$

Hence we have, $\overset{(m)}{a}_{\overline{n}|i} = (1+i)^{-m}\, a_{\overline{n}|i}$

EXERCISE 11.11

1. Find the amount and the present value of a sequence of annual payments of ₹ 6000 each, the first being made at the end of 6 years and the last at the end of 10 years, if money is worth 5% effective.
2. An annuity consists of 12 semiannual payments of ₹ 200 each, the first being payable at the end of 2 years. Find the amount and the present value of this annuity if money is worth 6% compounded semiannually.
3. At 6 months interval, Mr. *A* deposited ₹ 100 in a savings account which credit interest at the rate of 8% per annum compounded semiannually. The first deposit was made when *A*'s son was 6 years old and the last deposit was made when his son was 21 years old. The money remained in the account and was presented to the son on his 25th birthday. How much did he receive ?
4. Mr. *A* wishes to provide for his son's education by investing a certain sum of money three months after his son's 12th birthday, and an equal sum quarterly thereafter. If this investment yields interest at the rate of 6% compounded quarterly, find the quarterly investment if ₹ 40, 000 is to be available for the son on his 17th birthday.
5. Find the present value of a sequence of annual payments of ₹ 1500 each, the first being made at the end of 7 years and the last at the end of 16 years, if money is worth 7% effective. Also find the amount of this annuity.
6. Find the present value of 20 semiannual payments of ₹ 500 each, the first being made at the end of $7\frac{1}{2}$ years, if money is worth 10% compounded semiannually.
7. Find the present value of an annuity of ₹ 200 per quarter for 8 years, deferred $3\frac{1}{2}$ years, if money is worth 8% converted quarterly.

8. X buys a piece of property for which he agrees to make 5 annual payments of ₹ 20,000 each, the first being made at the end of 2 years. Find the equivalent cash price of this property if money is worth 7% effective.

9. A house sells for ₹ 50,000 down and 10 semiannual payments of ₹ 5000 each, the first due 3 years hence. Find the cash price of the house if money is worth 6% compounded semiannually.

[*Delhi Univ. B. Com. (H) 2005*]

10. An orchard will yield its first full crop at the end of 5 years and is expected to maintain an annual income of ₹ 5000 for 20 years in all. Find the cash price of the orchard if money is worth 3% per annum. **[*Delhi Univ. B. Com. (H) 1999, 2018*]**

ANSWERS

1. ₹ 33,153.79, ₹ 20,353.92	**2.** ₹ 2838.40, ₹ 1821.86	**3.** ₹ 8119.07
4. ₹ 1729.83	**5.** ₹ 7020.16; ₹ 20,724.67	**6.** ₹ 3147.13
7. ₹ 3557.21	**8.** ₹ 76,639.20	**9.** ₹ 86,791.14
10. ₹ 66,092.55		

11.20 PERPETUITY

A **perpetuity** is an annuity whose payments continue for ever. The "amount" of a perpetuity is undefined, since it increases beyond all bounds as time goes on. In defining present value, we shall distinguish between two types of perpetuities.

TYPE I. In the first type, we consider a perpetuity of ₹ R payable at the end of each period, the first payment due one period hence. Let money be worth i per period. The present value of this perpetuity is defined to be that sum of money which, invested now at the rate i per period, will yield ₹ R at the end of each period forever. Thus the present value, in this type, is given by

$$R(1+i)^{-1} + R(1+i)^{-2} + R(1+i)^{-3} + \ldots = \frac{R(1+i)^{-1}}{1-(1+i)^{-1}} = \frac{R}{i}$$

Thus $\frac{R}{i}$ is the cash equivalent of an unlimited number of periodic payments of ₹ R each, the first being due one period hence.

TYPE II. In this type, we consider a perpetuity of ₹ R payable at the beginning of each period, the first payment due now. Let money be worth i per period. To find the present value in this case, we observe that the given perpetuity can be thought of as an initial payment of ₹ R followed by a perpetuity of ₹ R of Type *I*. Thus the present value, in this type, is given by

$$R + \frac{R}{i}$$

Thus $R + \frac{R}{i}$ is the cash equivalent of an unlimited number of periodic payments of ₹ R each, the first being due now.

Example 90. Find the present value of a sequence of payments of ₹ 80 made at the end of each 6 months and continuing for ever, if money is worth 4% compounded semiannually.

Solution. This is a perpetuity of Type *I*, since payments are made at the end of each period. Denoting the present value of this perpetuity by P_∞, we have

$$P_\infty = \frac{R}{i},$$

where $R = 80$ and $i = 0.04/2 = 0.02$.

$$\therefore \quad P_\infty = \frac{80}{0.02} = ₹\ 4000$$

Example 91. How much money is needed to endure a series of lectures costing ₹ 2500 at the beginning of each year indefinitely, if money is worth 3% compounded annually ?

[*Delhi Univ. B. Com. (H) 1999, 2014*]

Solution. It is required to find how much to be invested now that would provide for an unlimited number of payments of ₹ 2500 each year, the first due now. Thus this is a perpetuity of Type *II* and hence required money needed is given by

$$P_\infty = R + \frac{R}{i},$$

where $R = 2500$ and $i = 0.03$. Substituting in (1), we obtain

$$P_\infty = 2500 + \frac{2500}{0.03} = ₹85{,}833.33$$

Example 92. At what rate converted semiannually will the present value of a perpetuity of ₹ 450 payable at the end of each 6 months be ₹ 20,000 ?

Solution. Let r be the interest rate converted semiannually. Then i, the interest rate per period, is $r/2$. Since

$$P_\infty = \frac{R}{i},$$

where $P_\infty = 20{,}000$ and $R = 450$, we get

$$i = \frac{R}{P_\infty} = \frac{450}{20{,}000} = 0.0225$$

i.e.,

$$\frac{r}{2} = 0.0225 \quad \text{or} \quad r = 0.045 \quad \text{or} \quad 4.5\%$$

EXERCISE 11.12

1. If money is worth 5%, compare the present value of a perpetuity of ₹ 200 payable at the end of each year with that of an ordinary annuity of ₹ 200 per year for 100 years.

2. At 6% converted quarterly, find the present value of a perpetuity of ₹ 300 payable at the end of each quarter.

3. Find the present value of a sequence of payments of ₹ 140 made at the end of each 6 months and continuing forever, if money is worth 7% converted semiannually.

4. What sum invested at 8% compounded semiannually will provide payments of ₹ 100 each at the end of each 6 months forever ?

5. Find the present value of a perpetuity of ₹ 780 payable at the beginning of each year, if money is worth 6% effective.

6. What sum of money invested now would establish a scholarship of ₹ 600 to be awarded at the end of each year to a deserving candidate, if money is worth 4% compounded annually ?

7. If the cash equivalent of a perpetuity of ₹ 300 payable at the end of each quarter is ₹ 24,000. Find the rate of interest convertible quarterly.

8. The present value of a perpetual income of ₹ x at the end of each 6 months is ₹ 36,000. Find the value of x if money is worth 6% compounded semiannually.

ANSWERS

1. ₹ 4000; ₹ 3969.58	2. ₹ 20,000	3. ₹ 4000	4. ₹ 2500
5. ₹ 13,780	6. ₹ 15,000	7. 5%	8. ₹ 1080

11.21 SINKING FUND

A *sinking fund* is a fund that is set up to accumulate a specific sum of money by a definite date, through regular, equal payments. The use of sinking funds is more common in business where they are used to plan ahead for anticipated expenses. For example a company may create a sinking fund for replacing a machine by a new model at the end of its life time by setting aside equal amount at the end of each year. Similarly, a person may create a sinking fund by saving equal amount regularly to meet the cost of his son's higher education abroad in the near future.

The problems relating to sinking fund are solved by using the known formulas for the *amount of an ordinary annuity or annuity due* as the case may be depending upon whether the payments are set aside at the *end* or the *beginning* of each payment interval.

Example 93. Suppose a machine costing ₹ 50,000 is to be replaced at the end of 10 years, at which time it will have a salvage value of ₹ 5000. In order to provide money at that time for a machine costing the same amount, a sinking fund is set up. The amount in the fund at that time is to be difference between the replacement cost and salvage value. If equal payments are placed in the fund at the end of each quarter and the fund earns 8% compounded quarterly, what should each payment be ?

Solution. The amount needed after 10 years = replacement cost – scrap value

= 50,000 – 5000 = ₹ 45,000

The payments into the sinking fund form an ordinary annuity consisting of 40 quarterly payments at the rate 0.02 per quarter. Let each quarterly payment be ₹ *R*. Then *R* is given by

$$45,000 = R s_{\overline{40}|0.02} = R(60.40198)$$

$$\Rightarrow \qquad R = \frac{45,000}{60.40198} = ₹745$$

Example 94. A machine costs a company ₹ 52,000 and its effective life is estimated to be 12 years. A sinking fund is created for replacing the machine by a new model at the end of its life time, when its scrap realizes a sum of ₹ 5000 only. The price of new model is estimated to be 25% higher than the price of the present one. Find what amount should be set aside at the end of each year, out of the profits, for the sinking fund, if it accumulates at 10% effective. ***[Delhi Univ. B. Com. (H) 2009, 2019 (modified)]***

Solution. Cost of the new machine = 52,000 + 0.25 (52,000) = ₹ 65,000

Scrap value of the machine = ₹ 5000

∴ Net cash required at the end of 12 years to purchase the new machine

= 65000 – 5000 = ₹ 60,000

Let ₹ *R* be the amount set aside each year. Then *R* is given by

$$S = R s_{\overline{n}|i}, \text{ where } S = 60,000, \; i = 0.10 \text{ and } n = 12$$

$$\therefore \qquad 60,000 = R s_{\overline{12}|0.10} = R(21.384)$$

$$\Rightarrow \qquad R = \frac{60,000}{21.384} = ₹2805.84$$

Example 95. A machinery purchased for ₹ 50,000 is depreciated at 10% per annum on diminishing values. In order to make up the loss due to depreciation during the period of 8 years, a sinking fund is created by setting aside a sum of money every year. What is the value of each payment if the accumulated amount in the fund is just equal to the total depreciation and if rate of interest is 5% compounded annually ? ***[Delhi Univ. B. Com. (H) 2018]***

Solution. Value of the machinery after 8 years is :

$$V = 50,000(1 - 0.10)^8 = 40,000(0.90)^8 = 50,000(0.4305) = ₹21,525$$

∴ Total depreciation = 50,000 – 21,525 = ₹ 28,475

Let ₹ *R* be set aside each year in the sinking fund in order to accumulate ₹ 28,475. Then *R* is given by

$$28,475 = R s_{\overline{8}|0.05} = R(9.5491) \quad \Rightarrow \quad R = \frac{28,475}{9.5491} = ₹\,2981 \; (app.)$$

Example 96. A company establishes a sinking fund to provide for the payment of a ₹ 1,00,000 debt, maturing in 4 years. Contributions to the fund are to be made at the end of every year. Find the amount of each annual deposit if interest is 18% per annum.

Solution. Let each annual deposit to the sinking fund be ₹ R. Then R is given by

$$1,00,000 = R s_{\overline{4}|0.18} = R\left[\frac{(1+0.18)^4 - 1}{0.18}\right]$$

$$= R\left[\frac{1.9388 - 1}{0.18}\right] = R\left[\frac{0.9388}{0.18}\right] = R(5.2156)$$

$$\Rightarrow \quad R = \frac{1,00,000}{5.2156} = ₹\ 19,173.25.$$

Example 97. Mr X plans to send his son for higher studies abroad after 10 years. He expects the cost of these studies to be ₹ 1,00,000. How much should he save at the beginning of each year to accumulate this amount at the end of 10 years, if the interest rate is 12% compounded annually?

Solution. Let the size of each annual payment be ₹ R. These payments represent annuity due consiting of 10 annual payments at the rate 0.12 per annum. Thus using the following formula for the amount of annuity due :

$$S = R(s_{\overline{n+1}|i} - 1),$$

where $S = 1,00,000$, $n = 10$ and $i = 0.12$, we obtain

$$1,00,000 = R(s_{\overline{11}|0.12} - 1) = R\left[\frac{(1.12)^{11} - 1}{0.12} - 1\right] = R(19.65458)$$

$$\Rightarrow \quad R = \frac{1,00,000}{19.65458} = ₹ 5087.87.$$

EXERCISE 11.13

1. In 10 years a machine costing ₹ 40,000 will have a salvage value of ₹ 4000. A new machine at that time is expected to sell for ₹ 52,000. In order to provide funds for the difference between the replacement cost and the salvage cost, a sinking fund is set up into which equal payments are placed at the end of each year. If the fund earns interest @7% compounded annually, how much should each payment be ?

2. Suppose a machine costing ₹ 70,000 is to be replaced at the end of 5 years, at which time it will have a salvage value of ₹ 10,000. In order to provide money at that time for a new machine costing the same amount, a sinking fund is set up. The amount in the fund at that time is to be the difference between the replacement cost and salvage value. If equal payments are placed in the fund at the end of each quarter and the fund earns 8% compounded quarterly, what should each payment be ?

3. Mr. X plans to send his son for higher studies abroad after 10 years. He expects the cost of these studies to be ₹ 1,00,000. How much must he set aside at the end of each year for 10 years to accumulate this amount, if interest rate is 12% effective ?

4. If you know you will need ₹ 2,00,000 for your children's education, how much must you set aside at the end of each quarter for 10 years to accumulate this amount, if money is worth 6% compounded quarterly ?

5. A machine costs a company ₹ 80,000 and its effective life is estimated to be 15 years. A sinking fund is created for replacing the machine by a new model at the end of its life, when its scrap realises a sum of ₹ 8000 only. The price of the new model is estimated to be 20% higher than the price of the present one. Find what amount should be set aside at the end of each year, out of the profits, for the sinking fund, if the fund earns interest @ 7% per annum compounded annually ?

6. A machine purchased for ₹ 50,000 is depreciated at 10% per annum on diminishing values. In order to make up the loss due to depreciation during the period of 8 years, a sinking fund is created by setting aside a sum of money at the end of each year. What is the value of each payment if the accumulated amount in the fund is just equal to the total depreciation and if the fund earns interest @ 5% compounded annually ? ***[Delhi Univ. B. Com. (H) 2018]***

7. A company establishes a sinking fund to provide for the payment of a ₹ 2,00,000 debt maturing in 10 years. Contributions to the fund are to be made at the end of each quarter. Find the amount of each quarterly deposit if rate of interest is 9% compounded quarterly.

8. A sinking fund is created for the redemption of debentures of ₹ 1,00,000 at the end of 25 years. How much money should be provided out of profits at the end of each year for the sinking fund, if the investment can earn interest at 4% effective ?

9. A company sets aside a sum of ₹ 5000 at the end of each year for 10 years to pay off a debenture issue of ₹ 60,000. If the money earns interest at 5% effective, find the surplus after full redemption of the debenture issue.

10. A machine costs a company ₹ 50,000 and its effective life is estimated to be 10 years. A sinking fund is created for replacing the machine at the end of its effective life when its scrap realises a sum of rupees five thousand only. Find what amount should be set aside at the end of each year, out of the profits, for the sinking fund, if the fund earns interest @ 5% per annum compounded continuously ?

11. Suppose a machine costing ₹ 5,25,000 is to be replaced at the end of 20 years, at which time it will have a scrap value of ₹ 25,000. In order to provide money at that time for a new machine costing the same amount, a sinking fund is set up. The amount in the fund at that time is to be the difference between the replacement cost and scrap value. If equal payments are placed in the fund at the beginning of each year and the fund earns interest @ 5% compounded annually, what should each payment be?

12. Suppose a machine costing ₹ 90,000 is to be replaced at the end of 8 years, at which time it will have a salvage value of ₹ 12,000. In order to provide money at that time for a new machine costing 20% higher than price of the present one, a sinking fund is set up. Find what amount should be set aside at the beginning of each quarter, out of profits, for the sinking fund, if the fund earns interest at the rate of 8% p.a. compounded quarterly.

ANSWERS

1. ₹ 3474.12	**2.** ₹ 2469.40	**3.** ₹ 5701.25	**4.** ₹ 3685.42
5. ₹ 33,541.29	**6.** ₹ 2983.52	**7.** ₹ 3135.48	**8.** ₹ 2401.20
9. ₹ 2889.46	**10.** ₹ 3468.47	**11.** ₹ 14,371.43	**12.** ₹ 2128.06

APPENDIX

TABLES

TABLE I : AMOUNT AT COMPOUND INTEREST

$$s = (1 + i)^n$$

Periods n	Rate i .0025 $\left(\frac{1}{4}\%\right)$	.004167 $\left(\frac{5}{12}\%\right)$	.005 $\left(\frac{1}{2}\%\right)$	.005833 $\left(\frac{7}{12}\%\right)$	.0075 $\left(\frac{3}{4}\%\right)$
1	1.0025 0000	1.0041 6667	1.0050 0000	1.0058 3333	1.0075 0000
2	1.0050 0625	1.0083 5069	1.0100 2500	1.0117 0069	1.0150 5625
3	1.0075 1877	1.0215 5216	1.0150 7513	1.0176 0228	1.0226 6917
4	1.0100 3756	1.0167 7112	1.0201 5050	1.0235 3830	1.0303 3919
5	1.0125 6266	1.0210 0767	1.0252 5125	1.0295 0894	1.0380 6673
6	1.0150 9406	1.0252 6187	1.0303 7751	1.0355 1440	1.0458 5224
7	1.0176 3180	1.0295 3379	1.0355 2940	1.0415 5490	1.0536 9613
8	1.0201 7588	1.0338 2352	1.0407 0704	1.0476 3064	1.0615 9885
9	1.0227 2632	1.0381 3111	1.0459 1058	1.0537 4182	1.0695 6084
10	1.0252 8313	1.0424 5666	1.0511 4013	1.0598 8865	1.0775 8255
11	1.0278 4634	1.0468 0023	1.0563 9583	1.0660 7133	1.0856 6441
12	1.0304 1596	1.0511 6190	1.0616 7781	1.0722 9008	1.0938 0690
13	1.0329 9200	1.0555 4174	1.0669 8620	1.0785 4511	1.1020 1045
14	1.0355 7448	1.0599 3983	1.0723 2113	1.0848 3662	1.1102 7553
15	1.0381 6341	1.0643 5625	1.0776 8274	1.0911 6483	1.1186 0259
16	1.0407 5882	1.0687 9106	1.0830 7115	1.0975 2996	1.1269 9211
17	1.0433 6072	1.0732 4436	1.0884 8651	1.1039 3222	1.1354 4455
18	1.0459 6912	1.0777 1621	1.0939 2894	1.1103 7182	1.1439 6039
19	1.0485 8404	1.0822 0670	1.0993 9858	1.1168 4899	1.1525 4009
20	1.0512 0550	1.0867 1589	1.1048 9558	1.1233 6395	1.1611 8414
21	1.0538 3352	1.0912 4387	1.1104 2006	1.1299 1590	1.1698 9302
22	1.0564 6810	1.0957 9072	1.1159 7216	1.1365 0808	1.1786 6722
23	1.0591 0927	1.1003 5652	1.1215 5202	1.1431 3771	1.1875 0723
24	1.0617 5704	1.1049 4134	1.1271 5978	1.1498 0602	1.1964 1353
25	1.0644 1144	1.1095 4526	1.1327 9558	1.1565 1322	1.2053 8663
26	1.0670 7247	1.1141 6836	1.1384 5955	1.1632 5955	1.2144 2703
27	1.0697 4015	1.1188 1073	1.1441 5185	1.1700 4523	1.2235 3523
28	1.0724 1450	1.1234 7244	1.1498 7261	1.1768 7049	1.2327 1175
29	1.0750 9553	1.1281 5358	1.1556 2197	1.1837 3557	1.2419 5709
30	1.0777 8327	1.1328 5422	1.1614 0008	1.1906 4069	1.2512 7176
31	1.0804 7773	1.1375 7444	1.1672 0708	1.1975 8610	1.2606 5630
32	1.0831 7892	1.1423 1434	1.1730 4312	1.2045 7202	1.2701 1122
33	1.0858 8687	1.1470 7398	1.1789 0833	1.2115 9869	1.2796 3706
34	1.0886 0159	1.1518 5346	1.1848 0288	1.2186 6634	1.2892 3434
35	1.0913 2309	1.1566 5284	1.1907 2689	1.2257 7523	1.2989 0359
36	1.0940 5140	1.1614 7223	1.1966 8052	1.2329 2559	1.3086 4537
37	1.0967 8653	1.1663 1170	1.2026 6393	1.2401 1765	1.3184 6021
38	1.0995 2850	1.1711 7133	1.2086 7725	1.2473 5167	1.3283 4866
39	1.1022 7732	1.1760 5121	1.2147 2063	1.2546 2789	1.3383 1128
40	1.1050 3301	1.1809 5142	1.2207 9424	1.2619 4655	1.3483 4861
41	1.1077 9559	1.1858 7206	1.2268 9821	1.2693 0791	1.3584 6123
42	1.1105 6508	1.1908 1319	1.2330 3270	1.2767 1220	1.3686 4969
43	1.1133 4149	1.1957 7491	1.2391 9786	1.2841 5969	1.3789 1456
44	1.1161 2485	1.2007 5731	1.2453 9385	1.2916 5062	1.3892 5642
45	1.1189 1516	1.2057 6046	1.2516 2082	1.2991 8525	1.3996 7584
46	1.1217 1245	1.2107 8446	1.2578 7892	1.3067 6383	1.4101 7341
47	1.1245 1673	1.2158 2940	1.2641 6832	1.3143 8662	1.4207 4971
48	1.1273 2802	1.2208 9536	1.2704 8916	1.3220 5388	1.4314 0533
49	1.1301 4634	1.2259 8242	1.2768 4161	1.3297 6586	1.4421 4087
50	1.1329 7171	1.2310 9068	1.2832 2581	1.3375 2283	1.4529 5693

TABLE I : AMOUNT AT COMPOUND INTEREST

$$s = (1 + i)^n$$

Periods n	Rate i .0025 $\left(\frac{1}{4}\%\right)$	.004167 $\left(\frac{5}{12}\%\right)$	.005 $\left(\frac{1}{2}\%\right)$	.005833 $\left(\frac{7}{12}\%\right)$	.0075 $\left(\frac{3}{4}\%\right)$
51	1.1358 0414	1.2362 2022	1.2896 4194	1.3453 2504	1.4638 5411
52	1.1386 4365	1.2413 7114	1.2960 9015	1.3531 7277	1.4748 3301
53	1.1414 9026	1.2465 4352	1.3025 7060	1.3610 6628	1.4858 9426
54	1.1443 4398	1.2517 3745	1.3090 8346	1.3690 0583	1.4970 3847
55	1.1472 0484	1.2569 5302	1.3156 2887	1.3769 9170	1.5082 6626
56	1.1500 7285	1.2621 9033	1.3222 0702	1.3850 2415	1.5195 7825
57	1.1529 4804	1.2674 4946	1.3288 1805	1.3931 0346	1.5309 7509
58	1.1558 3041	1.2727 3050	1.3354 6214	1.4012 2990	1.5424 5740
59	1.1587 1998	1.2780 3354	1.3421 3946	1.4094 0374	1.5540 2583
60	1.1616 1678	1.2833 5868	1.3488 5015	1.4176 2526	1.5656 8103
61	1.1645 2082	1.2887 0601	1.3555 9440	1.4258 9474	1.5774 2363
62	1.1674 3213	1.2940 7561	1.3623 7238	1.4342 1246	1.5892 5431
63	1.1703 5071	1.2994 6700	1.3691 8424	1.4425 7870	1.6011 7372
64	1.1732 7658	1.3048 8204	1.3760 3016	1.4509 9374	1.6131 8252
65	1.1762 0977	1.3103 1905	1.3829 1031	1.4594 5787	1.6252 8139
66	1.1791 5030	1.3157 7872	1.3898 2486	1.4679 7138	1.6374 7100
67	1.1820 9817	1.3212 6113	1.3967 7399	1.4765 3454	1.6497 5203
68	1.1850 5342	1.3267 6638	1.4037 5785	1.4851 4766	1.6621 2517
69	1.1880 1605	1.3322 9458	1.4107 7664	1.4938 1102	1.6745 9111
70	1.1909 8609	1.3378 4580	1.4178 3053	1.5025 2492	1.6871 5055
71	1.1939 6356	1.3434 2016	1.4249 1968	1.5112 8965	1.6998 0418
72	1.1969 4847	1.3490 1774	1.4320 4428	1.5201 0550	1.7125 5271
73	1.1999 4084	1.3546 3865	1.4392 0450	1.5289 7279	1.7253 9685
74	1.2029 4069	1.3602 8298	1.4464 0052	1.5378 9179	1.7383 3733
75	1.2059 4804	1.3659 5082	1.4536 3252	1.5468 6283	1.7513 7486
76	1.2089 6291	1.3716 4229	1.4609 0069	1.5558 8620	1.7645 1017
77	1.2119 8532	1.3773 5746	1.4682 0519	1.5649 6220	1.7777 4400
78	1.2150 1528	1.3830 9645	1.4755 4622	1.5740 9115	1.7910 7708
79	1.2180 5282	1.3888 5935	1.4829 2395	1.5832 7334	1.8045 1015
80	1.2210 9795	1.3946 4627	1.4903 3857	1.5925 0910	1.8180 4398
81	1.2241 5070	1.4004 5789	1.4977 9026	1.6017 9874	1.8316 7931
82	1.2272 1108	1.4062 9253	1.5052 7921	1.6111 4257	1.8454 1691
83	1.2302 7910	1.4121 5209	1.5128 0561	1.6205 4090	1.8592 5753
84	1.2333 5480	1.4180 3605	1.5203 6964	1.6299 9405	1.8732 0196
85	1.2364 3819	1.4239 4454	1.5279 7148	1.6395 0235	1.8872 5098
86	1.2395 2928	1.4298 7764	1.5356 1134	1.6490 6612	1.9014 0536
87	1.2426 2811	1.4358 3546	1.5432 8940	1.6586 8567	1.9156 6590
88	1.2457 3468	1.4418 1811	1.5510 0585	1.6683 6134	1.9300 3339
89	1.2488 4901	1.4478 2568	1.5587 6087	1.6780 9344	1.9445 0865
90	1.2519 7114	1.4538 5829	1.5665 5468	1.6878 8232	1.9590 9246
91	1.2551 0106	1.4599 1603	1.5743 8745	1.6977 2830	1.9737 8565
92	1.2582 3882	1.4659 9902	1.5822 5939	1.7076 3172	1.9885 8905
93	1.2613 8441	1.4721 0735	1.5901 7069	1.7175 9290	2.0035 0346
94	1.2645 3787	1.4782 4113	1.5981 2154	1.7276 1219	2.0185 2974
95	1.2676 9922	1.4844 0047	1.6061 1215	1.7376 8993	2.0336 6871
96	1.2708 6847	1.4905 8547	1.6141 4271	1.7478 2646	2.0489 2123
97	1.2740 4564	1.4967 9624	1.6222 1342	1.7580 2211	2.0642 8814
98	1.2772 3075	1.5030 3289	1.6303 2449	1.7682 7724	2.0797 7030
99	1.2804 2383	1.5092 9553	1.6384 7611	1.7785 9219	2.0953 6858
100	1.2836 2489	1.5155 8496	1.6466 6849	1.7889 6731	2.1110 8384

TABLE I : AMOUNT AT COMPOUND INTEREST

$$s = (1 + i)^n$$

Periods n	Rate i .01 (1%)	.01125 $\left(1\frac{1}{8}\%\right)$	.0125 $\left(1\frac{1}{4}\%\right)$	.015 $\left(1\frac{1}{2}\%\right)$	.0175 $\left(1\frac{3}{4}\%\right)$
1	1.0100 0000	1.0112 5000	1.0125 0000	1.0150 0000	1.0175 0000
2	1.0201 0000	1.0226 2656	1.0251 5625	1.0302 2500	1.0353 0625
3	1.0303 0100	1.0341 3111	1.0379 7070	1.0456 7838	1.0534 2411
4	1.0406 0401	1.0457 6509	1.0509 4534	1.0613 6355	1.0718 5903
5	1.0510 1005	1.0575 2994	1.0640 8215	1.0772 8400	1.0906 1656
6	1.0615 2015	1.0694 2716	1.0773 8318	1.0934 4326	1.1097 0235
7	1.0721 3535	1.0814 5821	1.0908 5047	1.1098 4491	1.1291 2215
8	1.0828 5671	1.0936 2462	1.1044 8610	1.1264 9259	1.1488 8178
9	1.0936 8527	1.1059 2789	1.1182 9218	1.1433 8998	1.1689 8721
10	1.1046 2213	1.1183 6958	1.1322 7083	1.1605 4083	1.1894 4449
11	1.1156 6835	1.1309 5124	1.1464 2422	1.1779 4894	1.2102 5977
12	1.1268 2503	1.1436 7444	1.1607 5452	1.1956 1817	1.2314 3931
13	1.1380 9328	1.1565 4078	1.1752 6395	1.2135 5244	1.2529 8950
14	1.1494 7421	1.1695 5186	1.1899 5475	1.2317 5573	1.2749 1682
15	1.1609 6896	1.1827 0932	1.2048 2918	1.2502 3207	1.2972 2786
16	1.1725 7864	1.1960 1480	1.2198 8955	1.2689 8555	1.3199 2935
17	1.1843 0443	1.2094 6997	1.2351 3817	1.2880 2033	1.3430 2811
18	1.1961 4748	1.2230 7650	1.2505 7739	1.3073 4064	1.3665 3111
19	1.2081 0895	1.2368 3611	1.2662 0961	1.3269 5075	1.3904 4540
20	1.2201 9004	1.2507 5052	1.2820 3723	1.3468 5501	1.4147 7820
21	1.2323 9194	1.2648 2146	1.2980 6270	1.3670 5783	1.4395 3681
22	1.2447 1586	1.2790 5071	1.3142 8848	1.3875 6370	1.4647 2871
23	1.2571 6302	1.2934 4003	1.3307 1709	1.4083 7715	1.4903 6146
24	1.2697 3465	1.3079 9123	1.3473 5105	1.4295 0281	1.5164 4279
25	1.2824 3200	1.3227 0613	1.3641 9294	1.4509 4535	1.5429 8054
26	1.2952 5631	1.3375 8657	1.3812 4535	1.4727 0953	1.5699 8269
27	1.3082 0888	1.3526 3442	1.3985 1092	1.4948 0018	1.5974 5739
28	1.3212 9097	1.3678 5156	1.4159 9230	1.5172 2218	1.6254 1290
29	1.3345 0388	1.3832 3989	1.4336 9221	1.5399 8051	1.6538 5762
30	1.3478 4892	1.3988 0134	1.4516 1336	1.5630 8022	1.6828 0013
31	1.3613 2740	1.4145 3785	1.4697 5853	1.5865 2642	1.7122 4913
32	1.3749 4068	1.4304 5140	1.4881 3051	1.6103 2432	1.7422 1349
33	1.3886 9009	1.4465 4398	1.5067 3214	1.6344 7918	1.7727 0223
34	1.4025 7699	1.4628 1760	1.5255 6629	1.6589 9637	1.8037 2452
35	1.4166 0276	1.4792 7430	1.5446 3587	1.6838 8132	1.8352 8970
36	1.4307 6878	1.4959 1613	1.5639 4382	1.7091 3954	1.8674 0727
37	1.4450 7647	1.5127 4519	1.5834 9312	1.7347 7663	1.9000 8689
38	1.4595 2724	1.5297 6357	1.6032 8678	1.7607 9828	1.9333 3841
39	1.4741 2251	1.5469 7341	1.6233 2787	1.7872 1025	1.9671 7184
40	1.4888 6373	1.5643 7687	1.6436 1946	1.8140 1841	2.0015 9734
41	1.5037 5237	1.5819 7611	1.6641 6471	1.8412 2868	2.0366 2530
42	1.5187 8989	1.5997 7334	1.6849 6677	1.8688 4712	2.0722 6624
43	1.5339 7779	1.6177 7079	1.7060 2885	1.8968 7982	2.1085 3090
44	1.5493 1757	1.6359 7071	1.7273 5421	1.9253 3302	2.1454 3019
45	1.5648 1075	1.6543 7538	1.7489 4614	1.9542 1301	2.1829 7522
46	1.5804 5885	1.6729 8710	1.7708 0797	1.9835 2621	2.2211 7728
47	1.5962 6344	1.6918 0821	1.7929 4306	2.0132 7910	2.2600 4789
48	1.6122 2608	1.7108 4105	1.8153 5485	2.0434 7829	2.2995 9872
49	1.6283 4834	1.7300 8801	1.8380 4679	2.0741 3046	2.3398 4170
50	1.6446 3182	1.7495 5150	1.8610 2237	2.1052 4242	2.3807 8893

TABLE I : AMOUNT AT COMPOUND INTEREST

$$s = (1 + i)^n$$

Periods n	Rate i .01 (1%)	.01125 $\left(1\frac{1}{8}\%\right)$	.0125 $\left(1\frac{1}{4}\%\right)$	.015 $\left(1\frac{1}{2}\%\right)$	.0175 $\left(1\frac{3}{4}\%\right)$
51	1.6610 7814	1.7692 3395	1.8842 8515	2.1368 2106	2.4224 5274
52	1.6776 8892	1.7891 3784	1.9078 3872	2.1688 7337	2.4648 4566
53	1.6944 6581	1.8092 6564	1.9316 8670	2.2014 0647	2.5079 8046
54	1.7114 1047	1.8296 1988	1.9558 3279	2.2344 2757	2.5518 7012
55	1.7285 2457	1.8502 0310	1.9802 8070	2.2679 4398	2.5965 2785
56	1.7458 0982	1.8710 1788	2.0050 3420	2.3019 6314	2.6419 6708
57	1.7632 6792	1.8920 6684	2.0300 9713	2.3364 9259	2.6882 0151
58	1.7809 0060	1.9133 5259	2.0554 7335	2.3715 3998	2.7352 4503
59	1.7987 0960	1.9348 7780	2.0811 6676	2.4071 1308	2.7831 1182
60	1.8166 9670	1.9566 4518	2.1071 8135	2.4432 1978	2.8318 1628
61	1.8348 6367	1.9786 5744	2.1335 2111	2.4798 6807	2.8813 7306
62	1.8532 1230	2.0009 1733	2.1601 9013	2.5170 6609	2.9317 9709
63	1.8717 4443	2.0234 2765	2.1871 9250	2.5548 2208	2.9831 0354
64	1.8904 6187	2.0461 9121	2.2145 3241	2.5931 4442	3.0353 0785
65	1.9093 6649	2.0692 1087	2.2422 1407	2.6320 4158	3.0884 2574
66	1.9284 6015	2.0924 8949	2.2702 4174	2.6715 2221	3.1424 7319
67	1.9477 4475	2.1160 2999	2.2986 1976	2.7115 9504	3.1974 6647
68	1.9672 2220	2.1398 3533	2.3273 5251	2.7522 6896	3.2534 2213
69	1.9868 9442	2.1639 0848	2.3564 4442	2.7935 5300	3.3103 5702
70	2.0067 6337	2.1882 5245	2.3858 9997	2.8354 5629	3.3682 8827
71	2.0268 3100	2.2128 7029	2.4157 2372	2.8779 8814	3.4272 3331
72	2.0470 9931	2.2377 6508	2.4459 2027	2.9211 5796	3.4872 0990
73	2.0675 7031	2.2629 3994	2.4764 9427	2.9649 7533	3.5482 3607
74	2.0882 4601	2.2883 9801	2.5074 5045	3.0094 4996	3.6103 3020
75	2.1091 2847	2.3141 4249	2.5387 9358	3.0545 9171	3.6735 1098
76	2.1302 1975	2.3401 7659	2.5705 2850	3.1004 1059	3.7377 9742
77	2.1515 2195	2.3665 0358	2.6026 6011	3.1469 1674	3.8032 0888
78	2.1730 3717	2.3931 2675	2.6351 9336	3.1941 2050	3.8697 6503
79	2.1947 6754	2.4200 4942	2.6681 3327	3.2420 3230	3.9374 8592
80	2.2167 1522	2.4472 7498	2.7014 8494	3.2906 6279	4.0063 9192
81	2.2388 8237	2.4748 0682	2.7352 5350	3.3400 2273	4.0765 0378
82	2.2612 7119	2.5026 4840	2.7694 4417	3.3901 2307	4.1478 4260
83	2.2838 8390	2.5308 0319	2.8040 6222	3.4409 7492	4.2204 2984
84	2.3067 2274	2.5592 7473	2.8391 1300	3.4925 8954	4.2942 8737
85	2.3297 8997	2.5880 6657	2.8746 0191	3.5449 7838	4.3694 3740
86	2.3530 8787	2.6171 8232	2.9105 3444	3.5981 5306	4.4459 0255
87	2.3766 1875	2.6466 2562	2.9469 1612	3.6521 2535	4.5237 0584
88	2 4003 8494	2.6764 0016	2.9837 5257	3.7069 0723	4.6028 7070
89	2.4243 8879	2.7065 0966	3.0210 4948	3.7625 1084	4.6834 2093
90	2.4486 3267	2.7369 5789	3.0588 1260	3.8189 4851	4.7653 8080
91	2.4731 1900	2.7677 4867	3.0970 4775	3.8762 3273	4.8487 7496
92	2.4978 5019	2.7988 8584	3.1357 6085	3.9343 7622	4.9336 2853
93	2.5228 2869	2.8303 7331	3.1749 5786	3.9933 9187	5.0199 6703
94	2.5480 5698	2.8622 1501	3.2146 4483	4.0532 9275	5.1078 1645
95	2.5735 3755	2.8944 1492	3.2548 2789	4.1140 9214	5.1972 0324
96	2,5992 7293	2.9269 7709	3.2955 1324	4.1758 0352	5.2881 5429
97	2.6252 6565	2.9599.0559	3.3367 0716	4.2384 4057	5.3806 9699
98	2.6515 1831	2.9932 0452	3.3784 1600	4.3020 1718	5.4748 5919
99	2.6780 3349	3.0268 7807	3.4206 4620	4.3665 4744	5.5706 6923
100	2.7048 1383	3.0609 3045	3.4634 0427	4.4320 4565	5.6681 5594

TABLE I : AMOUNT AT COMPOUND INTEREST

$$s = (1 + i)^n$$

Periods n	Rate i .02 (2%)	.0225 $\left(2\frac{1}{4}\%\right)$	.025 $\left(2\frac{1}{2}\%\right)$	.0275 $\left(2\frac{3}{4}\%\right)$	.03 (3%)
1	1.0200 0000	1.0225 0000	1.0250 0000	1.0275 0000	1.0300 0000
2	1.0404 0000	1.0455 0625	1.0506 2500	1.0557 5625	1.0609 0000
3	1.0612 0800	1.0690 3014	1.0768 9063	1.0847 8955	1.0927 2700
4	1.0824 3216	1.0930 8332	1.1038 1289	1.1146 2126	1.1255 0881
5	1.1040 8080	1.1176 7769	1.1314 0821	1.1452 7334	1.1592 7407
6	1.1261 6242	1.1428 2544	1.1596 9342	1.1767 6836	1.1940 5230
7	1.1486 8567	1.1685 3901	1.1886 8575	1.2091 2949	1.2298 7387
8	1.1716 5938	1.1948 3114	1.2184 0290	1.2423 8055	1.2667 7008
9	1.1950 9257	1.2217 1484	1.2488 6297	1.2765 4602	1.3047 7318
10	1.2189 9442	1.2492 0343	1.2800 8454	1.3116 5103	1.3439 1638
11	1.2433 7431	1.2773 1050	1.3120 8666	1.3477 2144	1.3842 3387
12	1.2682 4179	1.3060 4999	1.3448 8882	1.3847 8378	1.4257 6089
13	1.2936 0663	1.3354 3611	1.3785 1104	1.4228 6533	1.4685 3371
14	1.3194 7876	1.3654 8343	1.4129 7382	1.4619 9413	1.5125 8972
15	1.3458 6834	1.3962 0680	1.4482 9817	1.5021 9896	1.5579 6742
16	1.3727 8571	1.4276 2146	1.4845 0562	1.5435 0944	1.6047 0644
17	1.4002 4142	1.4597 4294	1.5216 1826	1.5859 5595	1.6528 4763
18	1.4282 4625	1.4925 8716	1.5596 5872	1.6295 6973	1.7024 3306
19	1.4568 1117	1.5261 7037	1.5986 5019	1.6743 8290	1.7535 0605
20	1.4859 4740	1.5605 0920	1.6386 1644	1.7204 2843	1.8061 1123
21	1.5156 6634	1.5956 2066	1.6795 8185	1.7677 4021	1.8602 9457
22	1.5459 7967	1.6315 2212	1.7215 7140	1.8163 5307	1.9161 0341
23	1.5768 9926	1.6682 3137	1.7646 1068	1.8663 0278	1.9735 8651
24	1.6084 3725	1.7057 6658	1.8087 2595	1.9176 2610	2.0327 9411
25	1.6406 0599	1.7441 4632	1.8539 4410	1.9703 6082	2.0937 7793
26	1.6734 1811	1.7833 8962	1.9002 9270	2.0245 4575	2.1565 9127
27	1.7068 8648	1.8235 1588	1.9478 0002	2.0802 2075	2.2212 8901
28	1.7410 2421	1.8645 4499	1.9964 9502	2.1374 2682	2.2879 2768
29	1.7758 4469	1.9064 9725	2.0464 0739	2.1962 0606	2.3565 6551
30	1.8113 6158	1.9493 9344	2.0975 6758	2.2566 0173	2.4272 6247
31	1.8475 8882	1.9932 5479	2.1500 0677	2.3186 5828	2.5000 8035
32	1.8845 4059	2.0381 0303	2.2037 5694	2.3824 2138	2.5750 8276
33	1.9222 3140	2.0839 6034	2.2588 5086	2.4479 3797	2.6523 3524
34	1.9606 7603	2.1308 4945	2.3153 2213	2.5152 5626	2.7319 0530
35	1.9998 8955	2.1787 9356	2.3732 0519	2.5844 2581	2.8138 6245
36	2.0398 8734	2.2278 1642	2.4325 3532	2.6554 9752	2.8982 7833
37	2.0806 8509	2.2779 4229	2.4933 4870	2.7285 2370	2.9852 2668
38	2.1222 9879	2.3291 9599	2.5556 8242	2.8035 5810	3.0747 8348
39	2.1647 4477	2.3816 0290	2.6195 7448	2.8806 5595	3.1670 2698
40	2.2080 3966	2.4351 8897	2.6850 6384	2.9598 7399	3.2620 3779
41	2.2522 0046	2.4899 8072	2.7521 9043	3.0412 7052	3.3598 9893
42	2.2972 4447	2.5460 0528	2.8209 9520	3.1249 0546	3.4606 9589
43	2.3431 8936	2.6032 9040	2.8915 2008	3.2108 4036	3.5645 1677
44	2.3900 5314	2.6618 6444	2.9638 0808	3.2991 3847	3.6714 5227
45	2.4378 5421	2.7217 5639	3.0379 0328	3.3898 6478	3.7815 9584
46	2.4866 1129	2.7829 9590	3.1138 5086	3.4830 8606	3.8950 4372
47	2.5363 4352	2.8456 1331	3.1916 9713	3.5788 7093	4.0118 9503
48	2.5870 7039	2.9096 3961	3.2714 8956	3.6772 8988	4.1322 5188
49	2.6388 1179	2.9751 0650	3.3532 7680	3.7784 1535	4.2562 1944
50	2.6915 8803	3.0420 4640	3.4371 0872	3.8823 2177	4.3839 0602

TABLE I : AMOUNT AT COMPOUND INTEREST

$$s = (1 + i)^n$$

Periods n	Rate i .02 (2%)	$.0225\left(2\frac{1}{4}\%\right)$	$.025\left(2\frac{1}{2}\%\right)$	$.0275\left(2\frac{3}{4}\%\right)$	.03 (3%)
51	2.7454 1979	3.1104 9244	3.5230 3644	3.9890 8562	4.5154 2320
52	2.8003 2819	3.1804 7852	3.6111 1235	4.0987 8547	4.6508 8590
53	2.8563 3475	3.2520 3929	3.7013 9016	4.2115 0208	4.7904 1247
54	2.9134 6144	3.3252 1017	3.7939 2491	4.3273 1838	4.9341 2485
55	2.9717 3067	3.4000 2740	3.8887 7303	4.4463 1964	5.0821 4859
56	3.0311 6529	3.4765 2802	3.9859 9236	4.5685 9343	5.2346 1305
57	3.0917 8859	3.5547 4990	4.0856 4217	4.6942 2975	5.3916 5144
58	3.1536 2436	3.6347 3177	4.1877 8322	4.8233 2107	5.5534 0098
59	3.2166 9685	3.7165 1324	4.2924 7780	4.9559 6239	5.7200 0301
60	3.2810 3079	3.8001 3479	4.3997 8975	5.0922 5136	5.8916 0310
61	3.3466 5140	3.8856 3782	4.5097 8449	5.2322 8827	6.0683 5120
62	3.4135 8443	3.9730 6467	4.6225 2910	5.3761 7620	6.2504 0173
63	3.4818 5612	4.0624 5862	4.7380 9233	5.5240 2105	6.4379 1379
64	3.5514 9324	4.1538 6394	4.8565 4464	5.6759 3162	6.6310 5120
65	3.6225 2311	4.2473 2588	4.9779 5826	5.8320 1974	6.8299 8273
66	3.6949 7357	4.3428 9071	5.1024 0721	5.9924 0029	7.0348 8222
67	3.7688 7304	4.4406 0576	5.2299 6739	6.1571 9130	7.2459 2868
68	3.8442 5050	4.5405 1939	5.3607 1658	6.3265 1406	7.4633 0654
69	3.9211 3551	4.6426 8107	5.4947 3449	6.5004 9319	7.6872 0574
70	3.9995 5822	4.7471 4140	5.6321 0286	6.6792 5676	7.9178 2191
71	4.0795 4939	4.8539 5208	5.7729 0543	6.8629 3632	8.1553 5657
72	4.1611 4038	4.9631 6600	5.9172 2806	7.0516 6706	8.4000 1727
73	4.2443 6318	5.0748 3723	6.0651 5876	7.2455 8791	8.6520 1778
74	4.3292 5045	5.1890 2107	6.2167 8773	7.4448 4158	8.9115 7832
75	4.4158 3546	5.3057 7405	6.3722 0743	7.6495 7472	9.1789 2567
76	4.5041 5216	5.4251 5396	6.5315 1261	7.8599 3802	9.4542 9344
77	4.5942 3521	5.5472 1993	6.6948 0043	8.0760 8632	9.7379 2224
78	4.6861 1991	5.6720 3237	6.8621 7044	8.2981 7869	10.0300 5991
79	4.7798 4231	5.7996 5310	7.0337 2470	8.5263 7861	10.3309 6171
80	4.8754 3916	5.9301 4530	7.2095 6782	8.7608 5402	10.6408 9056
81	4.9729 4794	6.0635 7357	7.3898 0701	9.0017 7751	10.9601 1727
82	5.0724 0690	6.2000 0397	7.5745 5219	9.2493 2639	11.2889 2079
83	5.1738 5504	6.3395 0406	7.7639 1599	9.5036 8286	11.6275 8842
84	5.2773 3214	6.4821 4290	7.9580 1389	9.7650 3414	11.9764 1607
85	5.3828 7878	6.6279 9112	8.1569 6424	10.0335 7258	12.3357 0855
86	5.4905 3636	6.7771 2092	8.3608 8834	10.3094 9583	12.7057 7981
87	5.6003 4708	6.9296 0614	8.5699 1055	10.5930 0696	13.0869 5320
88	5.7123 5402	7.0855 2228	8.7841 5832	10.8843 1465	13.4795 6180
89	5.8266 0110	7.2449 4653	9.0037 6228	11.1836 3331	13.8839 4865
90	5.9431 3313	7.4079 5782	9.2288 5633	11.4911 8322	14.3004 6711
91	6.0619 9579	7.5746 3688	9.4595 7774	11.8071 9076	14.7294 8112
92	6.1832 3570	7.7450 6621	9.6960 6718	12.1318 8851	15.1713 6556
93	6.3069 0042	7.9193 3020	9.9384 6886	12.4655 1544	15.6265 0652
94	6.4330 3843	8.0975 1512	10.1869 3058	12.8083 1711	16.0953 0172
95	6.5616 9920	8.2797 0921	10.4416 0385	13.1605 4584	16.5781 6077
96	6.6929 3318	8.4660 0267	10.7026 4395	13.5224 6085	17.0755 0559
97	6.8267 9184	8.6564 8773	10.9702 1004	13.8943 2852	17.5877 7076
98	6.9633 2768	8.8512 5871	11.2444 6530	14.2764 2255	18.1154 0388
99	7.1025 9423	9.0504 1203	11.5255 7693	14.6690 2417	18.6588 660
100	7.2446 4612	9.2540 4630	11.8137 1635	15.0724 2234	19.2186 3198

TABLE I : AMOUNT AT COMPOUND INTEREST

$$s = (1 + i)^n$$

Periods n	Rate i .035 $\left(3\frac{1}{2}\%\right)$	.04(4%)	.045 $\left(4\frac{1}{2}\%\right)$	.05 (5%)	.055 $\left(5\frac{1}{2}\%\right)$
1	1.0350 0000	1.0400 0000	1.0450 0000	1.0500 0000	1.0550 0000
2	1.0712 2500	1.0816 0000	1.0920 2500	1.1025 0000	1.1130 2500
3	1.1087 1788	1.1248 6400	1.1411 6613	1.1576 2500	1.1742 4138
4	1,1475 2300	1.1698 5856	1.1925 1860	1.2155 0625	1.2388 2465
5	1.1876 8631	1.2166 5290	1.2461 8194	1.2762 8156	1.3069 6001
6	1.2292 5533	1.2653 1902	1.3022 6012	1.3400 9564	1.3788 4281
7	1.2722 7926	1.3159 3178	1.3608 6183	1.4071 0042	1.4546 7916
8	1.3168 0904	1.3685 6905	1.4221 0061	1.4774 5544	1.5346 8651
9	1.3628 9735	1.4233 1181	1.4860 9514	1.5513 2822	1.6190 9427
10	1.4105 9876	1.4802 4428	1.5529 6942	1.6288 9463	1.7081 4446
11	1.4599 6972	1.5394 5406	1.6228 5305	1.7103 3936	1.8020 9240
12	1.5110 6866	1.6010 3222	1.6958 8143	1.7958 5633	1.9012 0749
13	1.5639 5606	1.6650 7351	1.7721 9610	1.8856 4914	2.0057 7390
14	1.6186 9452	1.7316 7645	1.8519 4492	1.9799 3160	2.1160 9146
15	1.6753 4883	1.8009 4351	1.9352 8244	2.0789 2818	2.2324 7649
16	1.7339 8604	1.8729 8125	2.0223 7015	2.1828 7459	2.3552 6270
17	1.7946 7555	1.9479 0050	2.1133 7681	2.2920 1832	2.4848 0215
18	1.8574 8920	2.0258 1652	2.2084 7877	2.4066 1923	2.6214 6627
19	1.9225 0132	2.1068 4918	2.3078 6031	2.5269 5020	2.7656 4691
20	1.9897 8886	2.1911 2314	2.4117 1402	2.6532 9771	2.9177 5749
21	2.0594 3147	2.2787 6807	2.5202 4116	2.7859 6259	3.0782 3415
22	2.1315 1158	2.3699 1879	2.6336 5201	2.9252 6072	3.2475 3703
23	2.2061 1448	2.4647 1554	2.7521 6635	3.0715 2376	3.4261 5157
24	2.2833 2849	2.5633 0416	2.8760 1383	3.2250 9994	3.6145 8990
25	2.3632 4498	2.6658 3633	3.0054 3446	3.3863 5494	3.8133 9235
26	2.4459 5856	2.7724 6978	3.1406 7901	3.5556 7269	4.0231 2893
27	2.5315 6711	2.8833 6858	3.2820 0956	3.7334 5632	4.2444 0102
28	2.6201 7196	2.9987 0332	3.4296 9999	3.9201 2914	4.4778 4307
29	2.7118 7798	3.1186 5145	3.5840 3649	4.1161 3560	4.7241 2444
30	2.8067 9370	3.2433 9751	3.7453 1813	4.3219 4238	4.9839 5129
31	2.9050 3148	3.3731 3341	3.9138 5745	4.5380 3949	5.2580 6861
32	3.0067 0759	3.5080 5875	4.0899 8104	4.7649 4147	5.5472 6238
33	3.1119 4235	3.6483 8110	4.2740 3018	5.0031 8854	5.8523 6181
34	3.2208 6033	3.7943 1634	4.4663 6154	5.2533 4797	6.1742 4171
35	3.3335 9045	3.9460 8899	4.6673 4781	5.5160 1537	6.5138 2501
36	3.4502 6611	4.1039 3255	4.8773 7846	5.7918 1614	6.8720 8538
37	3.5710 2543	4.2680 8986	5.0968 6049	6.0814 0694	7.2500 5008
38	3.6960 1132	4.4388 1345	5.3262 1921	6.3854 7729	7.6488 0283
39	3.8253 7171	4.6163 6599	5.5658 9908	6.7047 5115	8.0694 8699
40	3.9592 5972	4.8010 2063	5.8163 6454	7.0399 8871	8.5133 0877
41	4.0978 3381	4.9930 6145	6.0781 0094	7.3919 8815	8.9815 4076
42	4.2412 5799	5.1927 8391	6.3516 1548	7.7615 8756	9.4755 2550
43	4.3897 0202	5.4004 9527	6.6374 3818	8.1496 6693	9.9966 7940
44	4.5433 4160	5.6165 1508	6.9361 2290	8.5571 5028	10.5464 9677
45	4.7023 5855	5.8411 7568	7.2482 4843	8.9850 0779	11.1265 5409
46	4.8669 4110	6.0748 2271	7.5744 1961	9.4342 5818	11.7385 1456
47	5.0372 8404	6.3178 1562	7.9152 6849	9.9059 7109	12.3841 3287
48	5.2135 8898	6.5705 2824	8.2714 5557	10.4012 6965	13.0652 6017
49	5.3960 6459	6.8333 4937	8.6436 7107	10.9213 3313	13.7838 4948
50	5.5849 2686	7.1066 8335	9.0326 3627	11.4673 9979	14.5419 6120

TABLE I : AMOUNT AT COMPOUND INTEREST

$$s = (1 + i)^n$$

Periods n	Rate i .06 (6%)	.065 $\left(6\frac{1}{2}\%\right)$	.07 (7%)	.075 $\left(7\frac{1}{2}\%\right)$	.08 (8%)
1	1.0600 0000	1.0650 0000	1.0700 0000	1.0750 0000	1.0800 0000
2	1.1236 0000	1.1342 2500	1.1449 0000	1.1556 2500	1.1664 0000
3	1.1910 1600	1.2079 4963	1.2250 4300	1.2422 9688	1.2597 1200
4	1.2624 7696	1.2864 6635	1.3107 9601	1.3354 6914	1.3604 8896
5	1.3382 2558	1.3700 8666	1.4025 5173	1.4356 2933	1.4693 2808
6	1.4185 1911	1.4591 4230	1.5007 3035	1.5433 0153	1.5868 7432
7	1.5036 3026	1.5539 8655	1.6057 8148	1.6590 4914	1.7138 2427
8	1.5938 4807	1.6549 9567	1.7181 8618	1.7834 7783	1.8509 3021
9	1.6894 7896	1.7625 7039	1.8384 5921	1.9172 3866	1.9990 0463
10	1.7908 4770	1.8771 3747	1.9671 5136	2.0610 3156	2.1589 2500
11	1.8982 9856	1.9991 5140	2.1048 5195	2.2156 0893	2.3316 3900
12	2.0121 9647	2.1290 9624	2.2521 9159	2.3817 7960	2.5181 7012
13	2.1329 2826	2.2674 8750	2.4098 4500	2.5604 1307	2.7196 2373
14	2.2609 0396	2.4148 7418	2.5785 3415	2.7524 4405	2.9371 9362
15	2.3965 5819	2.5718 4101	2.7590 3154	2.9588 7735	3.1721 6911
16	2.5403 5168	2.7390 1067	2.9521 6375	3.1807 9315	3.4259 4264
17	2.6927 7279	2.9170 4637	3.1588 1521	3.4193 5264	3.7000 1805
18	2.8543 3915	3.1066 5438	3.3799 3228	3.6758 0409	3.9960 1950
19	3.0255 9950	3.3085 8691	3.6165 2754	3.9514 8940	4.3157 0106
20	3.2071 3547	3.5236 4506	3.8696 8446	4.2478 5110	4.6609 5714
21	3.3995 6360	3.7526 8199	4.1405 6237	4.5664 3993	5.0338 3372
22	3.6035 3742	3.9966 0632	4.4304 0174	4.9089 2293	5.4365 4041
23	3.8197 4966	4.2563 8573	4.7405 2986	5.2770 9215	5.8714 6365
24	4.0489 3464	4.5330 5081	5.0723 6695	5.6728 7406	6.3411 8074
25	4.2918 7072	4.8276 9911	5.4274 3264	6.0983 3961	6.8484 7520
26	4.5493 8296	5.1414 9955	5.8073 5292	6.5557 1508	7.3963 5321
27	4.8223 4594	5.4756 9702	6.2138 6763	7.0473 9371	7.9880 6147
28	5.1116 8670	5.8316 1733	6.6488 3836	7.5759 4824	8.6271 0639
29	5.4183 8790	6.2106 7245	7.1142 5705	8.1441 4436	9.3172 7490
30	5.7434 9117	6.6143 6616	7.6122 5504	8.7549 5519	10.0626 5689
31	6.0881 0064	7.0442 9996	8.1451 1290	9.4115 7683	10.8676 6944
32	6.4533 8668	7.5021 7946	8.7152 7080	10.1174 4509	11.7370 8300
33	6.8405 8988	7.9898 2113	9.3253 3975	10.8762 5347	12.6760 4964
34	7.2510 2528	8.5091 5950	9.9781 1354	11.6919 7248	13.6901 3361
35	7.6860 8679	9.0622 5487	10.6765 8148	12.5688 7042	14.7853 4429
36	8.1472 5200	9.6513 0143	11.4239 4219	13.5115 3570	15.9681 7184
37	8.6360 8712	10.2786 3603	12.2236 1814	14.5249 0088	17.2456 2558
38	9.1542 5235	10.9467 4737	13.0792 7141	15.6142 6844	18.6252 7563
39	9.7035 0749	11.6582 8595	13.9948 2041	16.7853 3858	20.1152 9768
40	10.2857 1794	12.4160 7453	14.9744 5784	18.0442 3897	21.7245 2150
41	10.9028 6101	13.2231 1938	16.0226 6989	19.3975 5689	23.4624 8322
42	11.5570 3267	14.0826 2214	17.1442 5678	20.8523 7366	25.3394 8187
43	12.2504 5463	14.9979 9258	18.3443 5475	22.4163 0168	27.3666 4042
44	12.9854 8191	15.9728 6209	19.6284 5959	24.0975 2431	29.5559 7166
45	13.7646 1083	17.0110 9813	21.0024 5176	25.9048 3863	31.9204 4939
46	14.5904 8748	18.1168 1951	22.4726 2338	27.8477 0153	34.4740 8534
47	15.4659 1673	19.2944 1278	24.0457 0702	29.9362 7915	37.2320 1217
48	16.3938 7173	20.5485 4961	25.7289 0651	32.1815 0008	40.2105 7314
49	17.3775 0403	21.8842 0533	27.5299 2997	34.5951 1259	43.4274 1899
50	18.4201 5427	23.3066 7868	29.4570 2506	37.1897 4603	46.9016 1251

TABLE II : PRESENT VALUE AT COMPOUND INTEREST

$$v^n = (1+i)^{-n}$$

Periods n	Rate i .0025 $\left(\frac{1}{4}\%\right)$	.004167 $\left(\frac{5}{12}\%\right)$	.005 $\left(\frac{1}{2}\%\right)$	.005833 $\left(\frac{7}{12}\%\right)$	.0075 $\left(\frac{3}{4}\%\right)$
1	.9975 0623	.9958 5062	.9950 2488	.9942 0050	.9925 5583
2	.9950 1869	.9917 1846	.9900 7450	.9884 3463	.9851 6708
3	.9925 3734	.9876 0345	.9851 4876	.9827 0220	.9778 3333
4	.9900 6219	.9835 0551	.9802 4752	.9770 0301	.9705 5417
5	.9875 9321	.9794 2457	.9753 7067	.9713 3688	.9633 2920
6	.9851 3038	.9753 6057	.9705 1808	.9657 0361	.9561 5802
7	.9826 7370	.9713 1343	.9656 8963	.9601 0301	.9490 4022
8	.9802 2314	.9672 8308	.9608 8520	.9545 3489	.9419 7540
9	.9777 7869	.9632 6946	.9561 0468	.9489 9906	.9349 6318
10	.9753 4034	.9592 7249	.9513 4794	.9434 9534	.9280 0315
11	.9729 0807	.9552 9211	.9466 1487	.9380 2354	.9210 9494
12	.9704 8187	.9513 2824	.9419 0534	.9325 8347	.9142 3815
13	.9680 6171	.9473 8082	.9372 1924	.9271 7495	.9074 3241
14	.9656 4759	.9434 4978	.9325 5646	.9217 9779	.9006 7733
15	.9632 3949	.9395 3505	.9279 1688	.9164 5182	.8939 7254
16	.9608 3740	.9356 3657	.9233 0037	.9111 3686	.8873 1766
17	.9584 4130	.9317 5426	.9187 0684	.9058 5272	.8807 1231
18	.9560 5117	.9278 8806	.9141 3616	.9005 9922	.8741 5614
19	.9536 6700	.9240 3790	.9095 8822	.8953 7619	.8676 4878
20	.9512 8878	.9202 0372	.9050 6290	.8901 8346	.8611 8985
21	.9489 1649	.9163 8544	.9005 6010	.8850 2084	.8547 7901
22	.9465 5011	.9125 8301	.8960 7971	.8798 8815	.8484 1589
23	.9441 8964	.9087 9636	.8916 2160	.8747 8524	.8421 0014
24	.9418 3505	.9050 2542	.8871 8567	.8697 1192	.8358 3140
25	.9394 8634	.9012 7013	.8827 7181	.8646 6802	.8296 0933
26	.9371 4348	.8975 3042	.8783 7991	.8596 5338	.8234 3358
27	.9348 0646	.8938 0623	.8740 0986	.8546 6782	.8173 0380
28	.9324 7527	.8900 9749	.8696 6155	.8497 1117	.8112 1966
29	.9301 4990	.8864 0414	.8653 3488	.8447 8327	.8051 8080
30	.9278 3032	.8827 2611	.8610 2973	.8398 8394	.7991 8690
31	.9255 1653	.8790 6335	.8567 4600	.8350 1303	.7932 3762
32	.9232 0851	.8754 1578	.8524 8358	.8301 7037	.7873 3262
33	.9209 0624	.8717 8335	.8482 4237	.8253 5580	.7814 7158
34	.9186 0972	.8681 6599	.8440 2226	.8205 6914	.7756 5418
35	.9163 1892	.8645 6365	.8398 2314	.8158 1025	.7698 8008
36	.9140 3384	.8609 7624	.8356 4492	.8110 7896	.7641 4896
37	.9117 5445	.8574 0373	.8314 8748	.8063 7510	.7584 6051
38	.9094 8075	.8538 4604	.8273 5073	.8016 9853	.7528 1440
39	.9072 1272	.8503 0311	.8232 3455	.7970 4907	.7472 1032
40	.9049 5034	.8467 7488	.8191 3886	.7924 2659	.7416 4796
41	.9026 9361	.8432 6129	.8150 6354	.7878 3091	.7361 2701
42	.9004 4250	.8397 6228	.8110 0850	.7832 6188	.7306 4716
43	.8981 9701	.8362 7779	.8069 7363	.7787 1935	.7252 0809
44	.8959 5712	.8328 0776	.8029 5884	.7742 0316	.7198 0952
45	.8937 2281	.8293 5212	.7989 6402	.7697 1317	.7144 5114
46	.8914 9407	.8259 1083	.7949 8907	.7652 4922	.7091 3264
47	.8892 7090	.8224 8381	.7910 3390	.7608 1115	.7038 5374
48	.8870 5326	.8190 7102	.7870 9841	.7563 9883	.6986 1414
49	.8848 4116	.8156 7238	.7831 8250	.7520 1209	.6934 1353
50	.8826 3457	.8122 8785	.7792 8607	.7476 5079	.6882 5165

TABLE II : PRESENT VALUE AT COMPOUND INTEREST

$$v^n = (1 + i)^{-n}$$

Periods *n*	Rate *i* .0025 $\left(\frac{1}{4}\%\right)$	.004167 $\left(\frac{5}{12}\%\right)$	.005 $\left(\frac{1}{2}\%\right)$	.005833 $\left(\frac{7}{12}\%\right)$	.0075 $\left(\frac{3}{4}\%\right)$
51	.8804 3349	.8089 1736	.7754 0902	.7433 1479	.6831 2819
52	.8782 3790	.8055 6086	.7715 5127	.7390 0393	.6780 4286
53	.8760 4778	.8022 1828	.7677 1270	.7347 1808	.6729 9540
54	.8738 6312	.7988 8957	.7638 9324	.7304 5708	.6679 8551
55	.8716 8391	.7955 7468	.7600 9277	.7262 2079	.6630 1291
56	.8695 1013	.7922 7354	.7563 1122	.7220 0907	.6580 7733
57	.8673 4178	.7889 8610	.7525 4847	.7178 2178	.6531 7849
58	.8651 7883	.7857 1230	.7488 0445	.7136 5877	.6483 1612
59	.8630 2128	.7824 5208	.7450 7906	.7095 1990	.6434 8995
60	.8608 6911	.7792 0539	.7413 7220	.7054 0504	.6386 9970
61	.8587 2230	.7759 7217	.7376 8378	.7013 1404	.6339 4511
62	.8565 8085	.7727 5237	.7340 1371	.6972 4677	.6292 2592
63	.8544 4474	.7695 4593	.7303 6190	.6932 0308	.6245 4185
64	.8523 1395	.7663 5279	.7267 2826	.6891 8285	.6198 9266
65	.8501 8848	.7631 7291	.7231 1269	.6851 8593	.6152 7807
66	.8480 6831	.7600 0621	.7195 1512	.6812 1219	.6106 9784
67	.8459 5343	.7568 5266	.7159 3544	.6772 6150	.6061 5170
68	.8438 4382	.7537 1219	.7123 7357	.6733 3372	.6016 3940
69	.8417 3947	.7505 8476	.7088 2943	.6694 2872	.5971 6070
70	.8396 4037	.7474 7030	.7053 0291	.6655 4637	.5927 1533
71	.8375 4650	.7443 6876	.7017 9394	.6616 8653	.5883 0306
72	.8354 5786	.7412 8009	.6983 0243	.6578 4908	.5839 2363
73	.8333 7142	.7382 0424	.6948 2829	.6540 3388	.5795 7681
74	.8312 9618	.7351 4115	.6913 7143	.6502 4081	.5752 6234
75	.8292 2312	.7320 9078	.6870 3177	.6464 6973	.5709 7999
76	.8271 5523	.7290 5306	.6845 0923	.6427 2053	.5667 2952
77	.8250 9250	.7260 2794	.6811 0371	.6389 9306	.5625 1069
78	.8230 3491	.7230 1537	.6777 1513	.6352 8723	.5583 2326
79	.8209 8246	.7200 1531	.6743 4342	.6316 0288	.5541 6701
80	.8189 3512	.7170 2770	.6709 8847	.6279 3989	.5500 4170
81	.8168 9289	.7140 5248	.6676 5022	.6242 9816	.5459 4710
82	.8148 5575	.7110 8960	.6643 2858	.6206 7754	.5418 8297
83	.8128 2369	.7081 3902	.6610 2346	.6170 7792	.5378 4911
84	.8107 9670	.7052 0069	.6577 3479	.6134 9917	.5338 4527
85	.8087 7476	.7022 7454	.6544 6248	.6099 4118	.5298 7123
86	.8067 5787	.6993 6054	.6512 0644	.6064 0382	.5259 2678
87	.8047 4600	.6964 5863	.6479 6661	.6028 8698	.5220 1169
88	.8027 3915	.6935 6876	.6447 4290	.5993 9054	.5181 2575
89	.8007 3731	.6906 9088	.6415 3522	.5959 1437	.5142 6873
90	.7987 4046	.6878 2495	.6383 4350	.5924 5836	.5104 4043
91	.7967 4859	.6849 7090	.6351 5666	.5890 2240	.5066 4063
92	.7947 6168	.6821 2870	.6320 0763	.5856 0636	.5028 6911
93	.7927 7973	.6792 9829	.6288 6331	.5822 1014	.4991 2567
94	.7908 0273	.6764 7962	.6257 3464	.5788 3361	.4954 1009
95	.7888 3065	.6736 7265	.6226 2153	.5754 7666	.4917 2217
96	.7868 6349	.6708 7733	.6195 2391	.5721 3918	.4880 6171
97	.7849 0124	.6680 9361	.6164 4170	.5688 2106	.4844 2850
98	.7829 4388	.6653 2143	.6133 7483	.5655 2218	.4808 2233
99	.7809 9140	.6625 6076	.6103 2321	.5622 4243	.4772 4301
100	.7790 4379	.6598 1155	.6072 8678	.5589 8171	.4736 9033

TABLE II : PRESENT VALUE AT COMPOUND INTEREST

$$v^n = (1+i)^{-n}$$

Periods n	Rate i .01 (1%)	.01125 $\left(1\frac{1}{8}\%\right)$	.0125 $\left(1\frac{1}{4}\%\right)$	.015 $\left(1\frac{1}{2}\%\right)$	.0175 $\left(1\frac{3}{4}\%\right)$
1	.9900 9901	.9888 7515	.9876 5432	.9852 2167	.9828 0098
2	.9802 9605	.9778 7407	.9754 6106	.9706 6175	.9658 9777
3	.9705 9015	.9669 9537	.9634 1833	.9563 1699	.9492 8528
4	.9609 8034	.9562 3770	.9515 2428	.9421 8423	8429 5851
5	.9514 6569	.9455 9970	.9397 7706	.9282 6033	.9169 1254
6	.9420 4524	.9350 8005	.9281 7488	.9145 4219	.9011 4254
7	.9327 1805	.9246 7743	.9167 1593	.9010 2679	.8856 4378
8	.9234 8322	.9143 9054	.9053 9845	.8877 1112	.8704 1157
9	.9143 3982	.9042 1808	.8942 2069	.8745 9224	.8554 4135
10	.9052 8695	.8941 5880	.8831 8093	.8616 6723	.8407 2860
11	.8963 2372	.8842 1142	.8722 7746	.8489 3323	.8262 6889
12	.8874 4923	.8743 7470	.8615 0860	.8363 8742	.8120 5788
13	.8786 6260	.8646 4742	.8508 7269	.8240 2702	.7980 9128
14	.8699 6297	.8550 2835	.8403 6809	.8118 4928	.7843 6490
15	.8613 4947	.8455 1629	.8299 9318	.7998 5150	.7708 7459
16	.8528 2126	.8361 1005	.8197 4635	.7880 3104	.7576 1631
17	.8443 7749	.8268 0846	.8096 26.02	.7763 8526	.7445 8605
18	.8360 1731	.8176 1034	.7996 3064	.7649 1159	.7317 7990
19	.8277 3992	.8085 1455	.7897 5866	.7536 0747	.7191 9401
20	.8195 4447	.7995 1995	.7800 0855	.7424 7042	.7068 2458
21	.8114 3017	.7906 2542	.7703 7881	.7314 9795	.6946 6789
22	.8033 9621	.7818 2983	.7608 6796	.7206 8763	.6827 2028
23	.7954 4179	.7731 3210	.7514 7453	.7100 3708	.6709 7817
24	.7875 6613	.7645 3112	.7421 9707	.6995 4392	.6594 3800
25	.7797 6844	.7560 2583	.7330 3414	.6892 0583	.6480 9632
26	.7720 4796	.7476 1516	.7239 8434	.6790 2052	.6369 4970
27	.7644 0392	.7392 9806	.7150 4626	.6689 8574	.6259 9479
28	.7568 3557	.7310 7348	.7062 1853	.6590 9925	.6152 2829
29	.7493 4215	.7229 4040	.6974 9978	.6493 5887	.6046 4697
30	.7419 2292	.7148 9780	.6888 8867	.6397 6243	.5942 4764
31	.7345 7715	.7069 4467	.6803 8387	.6303 0781	.5840 2716
32	.7273 0411	.6990 8002	.6719 8407	.6209 9292	.5739 8247
33	.7201 0307	.6913 0287	.6636 8797	.6118 1568	.5641 1053
34	.7129 7334	.6836 1223	6554 9429	.6027 7407	.5544 0839
35	.7059 1420	.6760 0715	.6474 0177	.5938 6608	.5448 7311
36	.6989 2495	.6684 8667	.6394 0916	.5850 8974	.5355 0183
37	.6920 0490	.6610 4986	.6315 1522	.5764 4309	.5262 9172
38	.6851 5337	.6536 9578	.6237 1873	.5679 2423	.5172 4002
39	.6783 6967	.6464 2352	.6160 1850	.5595 3126	.5083 4400
40	.6716 5314	.6392 3216	.6084 1334	.5512 6232	.4996 0098
41	.6650 0311	.6321 2080	.6009 0206	.5431 1559	.4910 0834
42	.6584 1892	.6250 8855	.5934 8352	.5350 8925	.4825 6348
43	.6518 9992	.6181 3454	.5861 5656	.5271 8153	.4742 6386
44	.6454 4546	.6112 5789	.5789 2006	.5193 9067	.4661 0699
45	.6390 5492	.6044 5774	.5717 7290	.5117 1494	.4580 9040
46	.6327 2764	.5977 3324	.5647 1397	.5041 5265	.4502 1170
47	.6264 6301	.5910 8355	.5577 4219	.4967 0212	.4424 6850
48	.6202 6041	.5845 0784	.5508 5649	.4893 6170	.4348 5848
49	.6141 1921	.5780 0528	.5440 5579	.4821 2975	.4273 7934
50	.6080 3882	.5715 7506	.5373 3905	.4750 0468	.4200 2883

TABLE II: PRESENT VALUE AT COMPOUND INTEREST

$$v^n = (1 + i)^{-n}$$

Periods *n*	Rate *i* .01 (1%)	.01125 $\left(1\frac{1}{8}\%\right)$	.0125 $\left(1\frac{1}{4}\%\right)$	.015 $\left(1\frac{1}{2}\%\right)$	.0175 $\left(1\frac{3}{4}\%\right)$
51	.6020 1864	.5652 1637	.5307 0524	.4679 8491	.4128 0475
52	.5960 5806	.5589 2843	.5241 5332	.4610 6887	.4057 0492
53	.5901 5649	.5527 1044	.5176 8229	.4542 5505	.3987 2719
54	.5843 1336	.5465 6162	.5112 9115	.4475 4192	.3918 6947
55	.5785 2808	.5404 8120	.5049 7892	.4409 2800	.3851 2970
56	.5728 0008	.5344 6843	.4387 4461	.4344 1182	.3785 0585
57	.5671 2879	.5285 2256	.4925 8727	.4279 9194	.3719 9592
58	.5615 1365	.5226 4282	.4865 0594	.4216 6694	.3655 9796
59	.5559 5411	.5168 2850	.4804 9970	.4154 3541	.3593 1003
60	.5504 4962	.5110 7887	.4745 6760	.4092 9597	.3531 3025
61	.5449 9962	5053 9319	.4687 0874	.4032 4726	.3470 5676
62	.5396 0358	.4997 7077	.4629 2222	.3972 8794	.3410 8772
63	.5342 6097	.4942 1090	.4572 0713	.3914 1669	.3352 2135
64	.5289 7126	.4887 1288	.4515 6259	.3856 3221	.3294 5587
65	.5237 3392	.4832 7602	.4459 8775	.3799 3321	.3237 8956
66	.5185 4844	.4778 9965	.4404 8173	.3743 1843	.3182 2069
67	.5134 1429	.4725 8309	.4350 4368	.3687 8663	.3127 4761
68	.5083 3099	.4673 2568	.4296 7277	.3633 3658	.3073 6866
69	.5032 9801	.4621 2675	.4243 6817	.3579 6708	.3020 8222
70	.4983 1486	.4569 8566	.4191 2905	.3526 7692	.2968 8670
71	.4933 8105	.4519 0177	.4139 5462	.3474 6495	.2917 8054
72	.4884 9609	.4468 7443	.4088 4407	.3423 3000	.2867 6221
73	.4836 5949	.4419 0302	.4037 9661	.3372 7093	.2818 3018
74	.4788 7078	.4369 8692	.3988 1147	.3322 8663	.2769 8298
75	.4741 2949	.4321 2551	.3938 8787	.3273 7599	.2722 1914
76	.4694 3514	.4273 1818	.3890 2506	.3225 3793	.2675 3724
77	.4647 8726	.4225 6433	.3842 2228	.3177 7136	.2629 3586
78	.4601 8541	.4178 6337	.3794 7879	.3130 7523	.2584 1362
79	.4556 2912	.4132 1470	.3747 9387	.3084 4850	.2539 6916
80	.4511 1794	.4086 1775	.3701 6679	.3038 9015	.2496 0114
81	.4466 5142	.4040 7194	.3655 9683	.2993 9916	.2453 0825
82	.4422 2913	.3995 7670	.3610 8329	.2949 7454	.2410 8919
83	.4378 5063	.3951 3148	.3566 2547	.2906 1531	.2369 4269
84	.4335 1547	.3907 3570	.3522 2268	.2863 2050	.2328 6751
85	.4292 2324	.3863 8882	.3478 7426	.2820 8917	.2288 6242
86	.4249 7350	.3820 9031	.3435 7951	.2779 2036	.2249 2621
87	.4207 6585	.3778 3961	.3393 3779	.2738 1316	.2210 5770
88	.4165 9985	.3736 3621	.3351 4843	.2697 6666	.2172 5572
89	.4124 7510	.3694 7956	.3310 1080	.2657 7996	.2135 1914
90	.4083 9119	.3653 6916	.3269 2425	.2618 5218	.2098 4682
91	.4043 4771	.3613 0448	.3228 8814	.2579 8245	2062 3766
92	.4003 4427	.3572 8503	.3189 0187	.2541 6990	.2026 9057
93	.3963 8046	.3533 1029	.3149 6481	.2504 1369	.1992 0450
94	.3924 5590	.3493 7976	.3110 7636	.2467 1300	.1957 7837
95	.3885 7020	.3454 9297	.3072 3591	.2430 6699	.1924 1118
96	.3847 2297	.3416 4941	.3034 4287	.2394 7487	.1891 0190
97	.3809 1383	.3378 4861	.2996 9666	.2359 3583	.1858 4953
98	.3771 4241	.3340 9010	.2959 9670	.2324 4909	.1826 5310
99	.3734 0832	.3303 7340	.2923 4242	.2290 1389	1795 1165
100	.3697 1121	.3266 9805	.2887 3326	.2256 2944	.1764 2422

TABLE II : PRESENT VALUE AT COMPOUND INTEREST

$$v^n = (1 + i)^{-n}$$

Periods n	Rate i .02 (2%)	.0225 $\left(2\frac{1}{4}\%\right)$	.025 $\left(2\frac{1}{2}\%\right)$	.0275 $\left(2\frac{3}{4}\%\right)$	.03 (3%)
1	.9803 9216	.9779 9511	.9756 0976	.9732 3601	.9708 7379
2	.9611 6878	.9564 7444	.9518 1440	.9471 8833	.9425 9591
3	.9423 2233	.9354 2732	.9285 9941	.9218 3779	.9151 4166
4	.9238 4543	.9148 4335	.9059 5064	.8971 6573	.8884 8705
5	.9057 3081	.8947 1232	.8838 5429	.8731 5400	.8626 0878
6	.8879 7148	.8750 2427	.8622 9687	.8497 8491	.8374 8426
7	.8705 6018	.8557 6946	.8412 6524	.8270 4128	.8130 9151
8	.8534 9037	.8369 3835	.8207 4657	.8049 0635	.7894 0923
9	.8367 5527	.8185 2161	.8007 2836	.7833 6385	.7664 1673
10	.8203 4830	.8005 1013	.7811 9840	.7623 9791	.7440 9391
11	.8042 6304	.7828 9499	.7621 4478	.7419 9310	.7224 2128
12	.7884 9318	.7656 6748	.7435 5589	.7221 3440	.7013 7988
13	.7730 3253	.7488 1905	.7254 2038	.7028 0720	.6809 5134
14	.7578 7502	.7323 4137	.7077 2720	.6839 9728	.6611 1781
15	.7430 1473	.7162 2628	.6904 6556	.6656 9078	.6418 6195
16	.7284 4581	.7004 6580	.6736 2493	.6478 7424	.6231 6694
17	.7141 6256	.6850 5212	.6571 9506	.6305 3454	.6050 1645
18	.7001 5937	.6699 7763	.6411 6591	.6136 5892	.5873 9461
19	.6864 3076	.6532 3484	.6255 2772	.5972 3496	.5702 8603
20	.6729 7133	.6408 1647	.6102 7094	.5812 5057	.5336 7575
21	.6597 7582	.6267 1538	.5953 8629	.5656 9398	.5375 4928
22	.6468 3904	.6129 2457	.5808 6467	.5505 6375	.5218 9250
23	.6341 5592	.5994 3724	.5666 9724	.5358 1874	.5066 9175
24	.6217 2149	.5862 4668	.5528 7535	.5214 7809	.4919 3374
25	.6095 3087	.5733 4639	.5393 9059	.5075 2126	.4776 0557
26	.5975 7928	.5607 2997	.5262 3472	.4939 3796	.4636 9473
27	.5858 6204	.5483 9117	.5133 9973	.4807 1821	.4501 8906
28	.5743 7455	.5363 2388	.5008 7778	.4678 5227	.4370 7675
29	.5631 1231	.5245 2213	.4886 6125	.4553 3068	.4243 4636
30	.5520 7089	.5129 8008	.4767 4269	.4431 4421	.4119 8676
31	.5412 4597	.5016 9201	.4651 1481	.4312 8391	.3999 8715
32	.5306 3330	.4906 5233	.4537 7055	.4097 4103	.3883 3703
33	.5202 2873	.4798 5558	.4427 0298	.4085 0708	.3770 2625
34	.5100 2817	.4692 9641	.4319 0534	.3975 7380	.3660 4490
35	.5000 2761	.4589 6960	.4213 7107	.3869 3314	.3553 8340
36	.4902 2315	.4488 7002	.4110 9372	.3765 7727	.3450 3243
37	.4806 1093	.4389 9268	.4010 6705	.3664 9856	.3349 8294
38	.4711 8719	.4293 3270	.3912 8492	.3566 8959	.3252 2615
39	.4619 4822	.4198 8528	.3817 4139	.3471 4316	.3157 5355
40	.4528 9042	.4106 4575	.3724 3062	.3378 5222	.3065 5684
41	.4440 1021	.4016 0954	.3633 4695	.3288 0995	.2976 2800
42	.4353 0413	.3927 7216	.3544 8483	.3200 0968	.2889 5922
43	.4267 6875	.3841 2925	.3458 3886	.3114 4495	.2805 4294
44	.4184 0074	.3756 7653	.3374 0376	.3031 0944	.2723 7178
45	.4101 9680	.3674 0981	.3291 7440	.2949 9702	.2644 3862
46	.4021 5373	.3593 2500	.3211 4576	.2871 0172	.2567 3653
47	.3942 6836	.3514 1809	.3133 1294	.2794 1773	.2492 5876
48	.3865 3761	.3436 8518	.3056 7116	.2719 3940	.2419 9880
49	.3789 5844	.3361 2242	.2982 1576	.2646 6122	.2349 5029
50	.3715 2788	.3287 2608	.2909 4221	.2575 7783	.2281 0708

TABLE II : PRESENT VALUE AT COMPOUND INTEREST

$$v^n = (1+i)^{-n}$$

Periods n	Rate i .02 (2%)	.0225 $\left(2\frac{1}{4}\%\right)$	.025 $\left(2\frac{1}{2}\%\right)$	.0275 $\left(2\frac{3}{4}\%\right)$	.03 (3%)
51	.3642 4302	.3214 9250	.2838 4606	.2506 8402	.2214 6318
52	.3571 0100	.3144 1810	.2769 2298	.2439 7471	.2150 1280
53	.3500 9902	.3074 9936	.2701 6876	.2374 4497	.2087 5029
54	.3432 3433	.3007 3287	.2635 7928	.2310 9000	.2026 7019
55	.3365 0425	.2941 1528	.2571 5052	.2249 0511	.1967 6717
56	.3299 0613	.2876 4330	.2508 7855	.2188 8575	.1910 3609
57	.3234 3738	.2813 1374	.2447 5056	.2130 2749	.1854 7193
58	.3170 9547	.2751 2347	.2387 8982	.2073 2603	.1800 6984
59	.3108 7791	.2690 6940	.2329 6568	.2017 7716	.1748 2508
60	.3047 8227	.2631 4856	.2272 8359	.1963 7679	.1697 3309
61	.2988 0614	.2573 5801	.2217 4009	.1911 2097	.1647 8941
62	.2929 4720	.2516 9487	.2163 3179	.1860 0581	.1599 8972
63	.2872 0314	.2461 5635	.2110 5541	.1810 2755	.1553 2982
64	.2815 7170	.2407 3971	.2059 0771	.1761 8253	.1508 0565
65	.2760 5069	.2354 4226	.2008 8557	.1714 6718	.1464 1325
66	.2706 3793	.2302 6138	.1959 8593	.1668 7804	.1421 4879
67	.2653 3130	.2251 9450	.1912 0578	.1624 1172	.1380 0853
68	.2601 2873	.2202 3912	.1865 4223	.1580 6493	.1339 8887
69	.2550 2817	.2153 9278	.1819 9241	.1538 3448	.1300 8628
70	.2500 2761	.2106 5309	.1775 5358	.1497 1726	.1262 9736
71	.2451 2511	.2060 1769	.1732 2300	.1457 1023	.1226 1880
72	.2403 1874	.2014 8429	.1689 9805	.1418 1044	.1190 4737
73	.2356 0661	.1970 5065	.1648 7615	.1380 1503	.1155 7998
74	.2309 8687	.1927 1458	.1608 5478	.1343 2119	.1122 1357
75	.2264 5771	.1884 7391	.1569 3149	.1307 2622	.1089 4521
76	.2220 1737	.1843 2657	.1531 0389	.1272 2747	.1057 7205
77	.2176 6408	.1802 7048	.1493 6965	.1238 2235	.1026 9131
78	.2133 9616	.1763 0365	.1457 2649	.1205 0837	.0997 0030
79	.2092 1192	.1724 2411	.1421 7218	.1172 8309	.0967 9641
80	.2051 0973	.1686 2993	.1387 0457	.1141 4412	.0939 7710
81	.2010 8797	.1649 1925	.1353 2153	.1110 8917	.0912 3990
82	.1971 4507	.1612 9022	.1320 2101	.1081 1598	.0885 8243
83	.1932 7948	.1577 4105	.1288 0098	.1052 2237	.0860 0236
84	.1894 8968	.1542 6997	.1256 5949	.1024 0620	.0834 9743
85	.1857 7420	.1508 7528	.1225 9463	.0996 6540	.0810 6547
86	.1821 3157	.1475 5528	.1196 0452	.0969 9795	.0787 0434
87	.1785 6036	.1443 0835	.1166 8733	.0944 0190	.0764 1198
88	.1750 5918	.1411 3286	.1138 4130	.0918 7533	.0741 8639
89	.1716 2665	.1380 2724	.1110 6468	.0894 1638	.0720 2562
90	.1682 6142	.1349 8997	.1083 5579	.0870 2324	.0699 2779
91	.1649 6217	.1320 1953	.1057 1296	.0846 9415	.0678 9105
92	.1617 2762	.1291 1445	.1031 3460	.0824 2740	.0659 1364
93	.1585 5649	.1262 7331	.1006 1912	.0802 2131	.0639 9383
94	.1554 4754	.1234 9468	.0981 6500	.0780 7427	.0621 2993
95	.1523 9955	.1207 7719	.0957 7073	.0759 8469	.0603 2032
96	.1494 1132	.1181 1950	.0934 3486	.0739 5104	.0585 6342
97	.1464 8169	.1155 2029	.0911 5596	.0719 7181	.0568 5769
98	.1436 0950	.1129 7828	.0889 3264	.0700 4556	.0552 0164
99	.1407 9363	.1104 9221	.0867 6355	.0681 7086	.0535 9383
100	.1380 3297	.1080 6084	.0864 4737	.0663 4634	.0520 3284

TABLE II : PRESENT VALUE AT COMPOUND INTEREST

$$v^n = (1 + i)^{-n}$$

Periods n	Rate i .035 $\left(3\frac{1}{2}\%\right)$	.04(4%)	.045 $\left(4\frac{1}{2}\%\right)$	.05 (5%)	.055 $\left(5\frac{1}{2}\%\right)$
1	.9661 8357	.9615 3846	.9569 3780	.9523 8095	.9478 6730
2	.9335 1070	.9245 5621	.9157 2995	.9070 2948	.8984 5242
3	.9019 4271	.8889 9636	.8762 9660	.8638 3760	.8516 1366
4	.8714 4223	.8548 0419	.8385 6134	.8227 0247	.8072 1674
5	.8419 7317	.8219 2711	.8024 5105	.7835 2617	.7651 3435
6	.8135 0064	.7903 1453	.7678 9574	.7462 1540	.7252 4583
7	.7859 9096	.7599 1781	.7348 2846	.7106 8133	.6874 3681
8	.7594 1156	.7306 9021	.7031 8513	.6768 3936	.6515 9887
9	.7337 3097	.7025 8674	.6729 0443	.6446 0892	.6176 2926
10	.7089 1881	.6755 6417	.6439 2768	.6139 1325	.5854 3058
11	.6849 4571	.6495 8093	.6161 9874	.5846 7929	.5549 1050
12	.6617 8330	.6245 9705	.5896 6386	.5568 3742	.5259 8152
13	.6394 0415	.6005 7409	.5642 7164	.5303 2135	.4985 6068
14	.6177 8179	.5774 7508	.5399 7286	.5050 6795	.4725 6937
15	.5968 9062	.5552 6450	.5167 2044	.4810 1710	.4479 3305
16	.5767 0591	.5339 0818	.4944 6932	.4581 1152	.4245 8109
17	.5572 0378	.5133 7325	.4731 7639	.4362 9669	.4024 4653
18	.5383 6114	.4936 2812	.4528 0037	.4155 2065	.3814 6590
19	.5201 5569	.4746 4242	.4333 0179	.3957 3396	.3615 7906
20	.5025 6588	.4563 8695	.4146 4286	.3768 8948	.3427 2896
21	.4855 7090	.4388 3360	.3967 8743	.3589 4236	.3248 6158
22	.4691 5063	.4219 5539	.3797 0089	.3418 4987	.3079 2567
23	.4532 8563	.4057 2633	.3633 5013	.3255 7131	.2918 7267
24	.4379 5713	.3901 2147	.3477 0347	.3100 6791	.2766 5656
25	.4231 4699	.3751 1680	.3327 3060	.2953 0277	.2622 3370
26	.4088 3767	.3606 8923	.3184 0248	.2812 4073	.2485 6275
27	.3950 1224	.3468 1657	.3046 9137	.2678 4832	.2365 0450
28	.3816 5434	.3334 7747	.2915 7069	.2550 9364	.2233 2181
29	.3687 4815	.3206 5141	.2790 1502	.2429 4632	.2116 7944
30	.3562 7841	.3083 1867	.2670 0002	.2313 7745	.2006 4402
31	.3442 3035	.2964 6026	.2555 0241	.2203 5947	.1901 8390
32	.3325 8971	.2850 5794	.2444 9991	.2098 6617	.1802 6910
33	.3213 4271	.2740 9417	.2339 7121	.1998 7254	.1708 7119
34	.3104 7605	.2635 5209	.2238 9589	.1903 5480	.1619 6321
35	.2999 7686	.2534 1547	.2142 5444	.1812 9029	.1535 1963
36	.2898 3272	.2436 6872	.2050 2817	.1726 5741	.1455 1624
37	.2800 3161	.2342 9685	.1961 9921	.1644 3563	.1379 3008
38	.2705 6194	.2252 8543	.1877 5044	.1566 0536	.1307 3941
39	.2614 1250	.2166 2061	.1796 6549	.1491 4797	.1239 2362
40	.2525 7247	.2082 8904	.1719 2870	.1420 4568	.1174 6314
41	.2440 3137	.2002 7793	.1645 2507	.1352 8160	.1113 3947
42	.2357 7910	.1925 7493	.1574 4026	.1288 3962	.1055 3504
43	.2278 0590	.1851 6820	.1506 6054	.1227 0440	.1000 3322
44	.2201 0231	.1780 4635	.1441 7276	.1168 6133	.0948 1822
45	.2126 5924	.1711 9841	.1370 6437	.1112 9651	.0898 7509
46	.2054 6787	.1646 1386	.1320 2332	.1059 9668	.0851 8965
47	.1985 1968	.1582 8256	.1263 3810	.1009 4921	.0807 4849
48	.1918 0645	.1521 9476	.1208 9771	.0961 4211	.0765 3885
49	.1853 2024	.1463 4112	.1156 9158	.0915 6391	.0725 4867
50	.1790 5337	.1407 1262	1107 0965	.0872 0373	.0687 6652

TABLE II : PRESENT VALUE AT COMPOUND INTEREST

$$v^n = (1+i)^{-n}$$

Periods n	Rate i: .06 (6%)	.065 $\left(6\frac{1}{2}\%\right)$	.07 (7%)	.075 $\left(7\frac{1}{2}\%\right)$	.08 (8%)
1	.9433 9623	.9389 6714	.9345 7944	.9302 3256	.9259 2593
2	.8899 9644	.8816 5928	.8734 3873	.8653 3261	.8573 3882
3	.8396 1928	.8278 4909	.8162 9788	.8049 6057	.7938 3224
4	.7920 9366	.7773 2309	.7628 9521	.7488 0053	.7350 2985
5	.7472 5817	.7298 8084	.7129 8618	.6965 5863	.6805 8320
6	.7049 6054	.6853 3412	.6663 4222	.6479 6152	.6301 6963
7	.6650 5711	.6435 0621	.6227 4974	.6027 5490	.5834 9040
8	.6274 1237	.6042 3119	.5820 0910	.5607 0223	.5402 6888
9	.5918 9846	.5673 5323	.5439 3374	.5215 8347	.5002 4897
10	.5583 9478	.5327 2604	.5083 4929	.4851 9393	.4631 9349
11	.5267 8753	.5002 1224	.4750 9280	.4513 4319	.4288 8286
12	.4969 6936	.4696 8285	.4440 1196	.4198 5413	.3971 1376
13	.4688 3902	.4410 1676	.4149 6445	.3905 6198	.3676 9792
14	.4423 0096	.4141 0025	.3878 1724	.3633 1347	.3404 6104
15	.4172 6506	.3888 2652	.3624 4602	.3379 6602	.3152 4170
16	.3936 4628	.3650 9533	.3387 3460	.3143 8699	.2918 9047
17	.3713 6442	.3428 1251	.3165 7439	.2924 5302	.2702 6895
18	.3503 4379	.3218 8969	.2958 6392	.2720 4932	.2502 4903
19	.3305 1301	.3022 4384	.2765 0833	.2530 6913	.2317 1206
20	.3118 0473	.2837 9703	.2584 1900	.2354 1315	.2145 4821
21	.2941 5540	.2664 7608	.2415 1309	.2189 8897	.1986 5575
22	.2775 0510	.2502 1228	.2257 1317	.2037 1067	.1839 4051
23	.2617 9726	.2349 4111	.2109 4688	.1894 9830	.1703 1528
24	.2469 7855	.2206 0198	.1971 4662	.1762 7749	.1576 9934
25	.2329 9863	.2071 3801	.1842 4918	.1639 7906	.1460 1790
26	.2198 1003	.1944 9579	.1721 9549	.1525 3866	.1352 0176
27	.2073 6795	.1826 2515	.1609 3037	.1418 9643	.1251 8682
28	.1956 3014	.1714 7902	.1504 0221	.1319 9668	.1159 1372
29	.1845 5674	.1610 1316	.1405 6282	.1227 8761	.1073 2752
30	.1741 1013	.1511 8607	.1313 6712	.1142 2103	.0993 7733
31	.1642 5484	.1419 5875	.1227 7301	.1062 5212	.0920 1605
32	.1549 5740	.1332 9460	.1147 4113	.0988 3918	.0852 0005
33	.1461 8622	.1251 5925	.1072 3470	.0919 4343	.0788 8893
34	.1379 1153	.1175 2042	.1002 1934	.0855 2877	.0730 4531
35	.1301 0522	.1103 4781	.0936 6294	.0795 6164	.0676 3454
36	.1227 4077	.1036 1297	.0875 3546	.0740 1083	.0626 2458
37	.1157 9318	.0972 8917	.0818 0884	.0688 4729	.0579 8572
38	.1092 3885	.0913 5134	.0764 5686	.0640 4399	.0536 9048
39	.1030 5552	.0857 7590	.0714 5501	.0595 7580	.0497 1341
40	.0972 2219	.0805 4075	.0667 8038	.0554 1935	.0460 3093
41	.0917 1905	.0756 2512	.0624 1157	.0515 5288	.0426 2123
42	.0865 2740	.0710 0950	.0583 2857	.0479 5617	.0394 6411
43	.0816 2962	.0666 7559	.0545 1268	.0446 1039	.0365 4084
44	.0770 0908	.0626 0619	.0509 4643	.0414 9804	.0338 3411
45	.0726 5007	.0587 8515	.0476 1349	.0386 0283	.0313 2788
46	.0685 3781	.0551 9733	.0444 9859	.0359 0961	.0290 0730
47	.0646 5831	.0518 2848	.0415 8747	.0334 0428	.0268 5861
48	.0609 9840	.0486 6524	.0388 6679	.0310 7375	.0248 6908
49	.0575 4566	.0456 9506	.0363 2410	.0289 0582	.0230 2693
50	.0542 8836	.0429 0616	.0339 4776	.0268 8913	.0213 2123

TABLE III : AMOUNT OF AN ANNUITY

$$s_{\overline{n}|i} = \frac{(1+i)^n - 1}{i}$$

Periods n	Rate i .0025 $\left(\frac{1}{4}\%\right)$	.004167 $\left(\frac{5}{12}\%\right)$	.005 $\left(\frac{1}{2}\%\right)$	.005833 $\left(\frac{7}{12}\%\right)$	.0075 $\left(\frac{3}{4}\%\right)$
1	1.0000 0000	1.0000 0000	1.0000 0000	1.0000 0000	1.0000 0000
2	2.0025 0000	2.0041 6667	2.0050 0000	2.0058 3333	2.0075 0000
3	3.0075 0625	3.0125 1736	3.0150 2500	3.0175 3403	3.0225 5625
4	4.0150 2502	4.0250 6952	4.0301 0013	4.0351 3631	4.0452 2542
5	5.0250 6258	5.0418 4064	5.0502 5063	5.0586 7460	5.0755 6461
6	6.0376 2523	6.0628 4831	6.0755 0188	6.0881 8354	6.1136 3135
7	7.0527 1930	7.0881 1018	7.1058 7939	7.1236 9794	7.1594 8358
8	8.0703 5110	8.1176 4397	8.1414 0879	8.1652 5285	8.2131 7971
9	9.0905 2697	9.1514 6749	9.1821 1583	9.2128 8349	9.2747 7856
10	10.1132 5329	10.1895 9860	10.2280 2641	10.2666 2531	10.3443 3940
11	11.1385 3642	11.2320 5526	11.2791 6654	11.3265 1396	11.4219 2194
12	12.1663 8277	12.2788 5549	12.3355 6237	12.3925 8529	12.5075 8636
13	13.1967 9872	13.3300 1739	13.3972 4018	13.4648 7537	13.6013 9325
14	14.2297 9072	14.3855 5913	14.4642 2639	14.5434 2048	14.7034 0370
15	15.2653 6520	15.4454 9896	15.5365 4752	15.6282 5710	15.8136 7923
16	16.3035 2861	16.5098 5520	16.6142 3026	16.7194 2193	16.9322 8183
17	17.3442 8743	17.5786 4627	17.6973 0141	17.8169 5189	18.0592 7394
18	18.3876 4815	18.6518 9063	18.7857 8791	18.9208 8411	19.1947 1849
19	19.4336 1727	19.7296 0684	19.8797 1685	20.0312 5593	20.3386 7888
20	20.4822 0131	20.8118 1353	20.9791 1544	21.1481 0493	21.4912 1897
21	21.5334 0682	21.8985 2942	22.0840 1101	22.2714 6887	22.6524 0312
22	22.5872 4033	22.9897 7330	23.1944 3107	23.4013 8577	23.8222 9614
23	23.6437 0843	24.0855 6402	24.3104 0323	24.5378 9386	25.0009 6336
24	24.7028 1770	25.1859 2053	25.4319 5524	25.6810 3157	26.1884 7059
25	25.7645 7475	26.2908 6187	26.5591 1502	26.8308 3759	27.3848 8411
26	26.8289 8619	27.4004 0713	27.6919 1059	27.9873 5081	28.5902 7075
27	27.8960 5865	28.5145 7549	28.8303 7015	29.1506 1036	29.8046 9778
28	28.9657 9880	29.6333 8622	29.9745 2200	30.3206 5558	31.0282 3301
29	30.0382 1330	30.7568 5866	31.1243 9461	31.4975 2607	32.2609 4476
30	31.1133 0883	31.8850 1224	32.2800 1658	32.6812 6164	33.5029 0184
31	32.1910 9210	33.0178 6646	33.4414 1666	33.8719 0233	34.7541 7361
32	33.2715 6983	34.1554 4090	34.6086 2375	35.0694 8843	36.0148 2991
33	34.3547 4876	35.2977 5524	35.7816 6686	36.2740 6045	37.2849 4113
34	35.4406 3563	36.4448 2922	36.9605 7520	37.4856 5913	38.5645 7819
35	36.5292 3722	37.5966 8268	38.1453 7807	38.7043 2548	39.8538 1253
36	37.6205 6031	38.7533 3552	39.3361 0497	39.9301 0071	41.1527 1612
37	38.7146 1171	39.9148 0775	40.5327 8549	41.1630 2630	42.4613 6149
38	39.8113 9824	41.0811 1945	41.7354 4942	42.4031 4395	43.7798 2170
39	40.9109 2674	42.2522 9078	42.9441 2666	43.6504 9562	45.1081 7037
40	42.0132 0405	43.4283 4199	44.1588 4730	44.9051 2352	46.4464 8164
41	43.1182 3706	44.6092 9342	45.3796 4153	46.1670 7007	47.7948 3026
42	44.2260 3265	45.7951 6547	46.6065 3974	47.4363 7798	49.1532 9148
43	45.3365 9774	46.9859 7866	47.8395 7244	48.7130 9018	50.5219 4117
44	46.4499 3923	48.1817 5357	49.0787 7030	49.9972 4988	51.9008 5573
45	47.5660 6408	49.3825 1088	50.3241 6415	51.2889 0050	53.2901 1215
46	48.6849 7924	50.5882 7134	51.5757 8498	52.5880 8575	54.6897 8799
47	49.8066 9169	51.7990 5581	52.8336 6390	53.8948 4959	56.0999 6140
48	50.9312 0842	53.0148 8521	54.0978 3222	55.2092 3621	57.5207 1111
49	52.0585 3644	54.2357 8056	55.3683 2138	56.5312 9009	58.9521 1644
50	53.1886 8278	55.4617 6298	56.6451 6299	57.8610 5595	60.3942 5732

TABLE III : AMOUNT OF AN ANNUITY

$$s_{\overline{n}|i} = \frac{(1+i)^n - 1}{i}$$

Periods n	Rate i: .0025 $\left(\frac{1}{4}\%\right)$	.004167 $\left(\frac{5}{12}\%\right)$	.005 $\left(\frac{1}{2}\%\right)$	.005833 $\left(\frac{7}{12}\%\right)$	.0075 $\left(\frac{3}{4}\%\right)$
51	54.3216 545	56.6928 537	57.9283 888	59.1985 788	61.8472 142
52	55.4574 586	57.9290 739	59.2180 307	60.5439 038	63.3110 684
53	56.5961 023	59.1704 450	60.5141 209	61.8970 766	64.7859 014
54	57.7375 925	60.4169 885	61.8166 915	63.2581 429	66.2717 956
55	58.8819 365	61.6687 260	63.1257 750	64.6271 487	67.7688 341
56	60.0291 413	62.9256 790	64.4414 038	66.0041 404	69.2771 003
57	61.1792 142	64.1878 694	65.7636 109	67.3891 646	70.7966 786
58	62.3321 622	65.4553 188	67.0924 289	68.7822 680	72.3276 537
59	63.4879 926	66.7280 493	68.4278 911	70.1834 979	73.8701 111
60	64.6467 126	68.0060 828	69.7700 305	71 5929 016	75.4241 369
61	65.8083 294	69.2894 415	71.1188 807	73.0105 269	76.9898 180
62	66.9728 502	70.5781 475	72.4744 751	74.4364 216	78.5672 416
63	68.1402 824	71.8722 231	73.8368 474	75.8706 341	80.1564 959
64	69.3106 331	73.1716 907	75.2060 317	77.3132 128	81.7576 696
65	70.4839 096	74.4765 728	76.5820 618	78.7642 065	83.3708 521
66	71.6601 194	75.7868 918	77.9649 721	80.2236 644	84.9961 335
67	72.8392 697	77.1026 706	79.3547 970	81.6916 358	86.6336 045
68	74.0213 679	78.4239 317	80.7515 710	83.1681 703	88.2833 566
69	75.2064 213	79.7506 981	82.1553 288	84.6533 180	89.9454 817
70	76.3944 374	81.0829 926	83.5661 055	86.1471 290	91.6200 729
71	77.5854 235	82.4208 384	84.9839 360	87.6496 539	93.3072 234
72	78.7793 870	83.7642 586	86.4088 557	89.1609 436	95.0070 276
73	79.9763 355	85.1132 763	87.8409 000	90.6810 491	96.7195 803
74	81.1762 763	86.4679 150	89.2801 045	92.2100 219	98.4449 771
75	82.3792 170	87.8281 980	90.7265 050	93.7479 137	100.1833 145
76	83.5851 651	89.1941 488	92.1801 375	95.2947 765	101.9346 893
77	84.7941 280	90.5657 911	93.6410 382	96.8506 627	103.6991 995
78	86.0061 133	91.9431 485	95.1092 434	98.4156 249	105.4769 435
79	87.2211 286	93.3262 450	96.5847 896	99.9897 160	107.2680 206
80	88.4391 814	94.7151 044	98.0677 136	101.5729 894	109.0725 307
81	89.6602 793	96.1097 506	99.5580 521	103.1654 985	110.8905 747
82	90.8844 300	97.5102 079	101.0558 424	104.7672 972	112.7222 540
83	92.1116 411	98.9165 004	102.5611 216	106.3784 398	114.5676 709
84	93.3419 202	100.3286 525	104.0739 272	107.9989 807	116.4269 284
85	94.5752 750	101.7466 886	105.5942 969	109.6289 748	118.3001 304
86	95.8117 132	103.1706 331	107.1222 683	111.2684 771	120.1873 814
87	97.0512 425	104.6005 108	108.6578 797	112.9175 432	122.0887 867
88	98.2938 706	106.0363 462	110.2011 691	114.5762 289	124.0044 526
89	99.5396 053	107.4781 643	111.7521 749	116.2445 902	125.9344 860
90	100.7884 543	108.9259 900	113.3109 358	117.9226 837	127.8789 947
91	102.0404 254	110.3798 483	114.8774 905	119.6105 660	129.8380 871
92	103.2955 265	111.8397 643	116.4518 779	121.3082 943	131.8118 728
93	104.5537 653	113.3057 634	118.0341 373	123.0159 260	133.8004 618
94	105.8151 497	114.7778 707	119.6243 080	124.7335 189	135.8039 653
95	107.0796 876	116.2561 118	121.2224 295	126.4611 311	137.8224 951
96	108.3473 868	117.7405 123	122.8285 417	128.1988 210	139.8561 638
97	109.6182 553	119.2310 978	124.4426 844	129.9466 475	141.9050 850
98	110.8923 009	120.7278 940	126.0648 978	131.7046 696	143.9693 731
99	112.1695 317	122.2309 269	127.6952 223	133.4729 468	146.0491 434
100	113.4499 555	123.7402 224	129.3336 984	135.2515 390	148.1445 120

TABLE III : AMOUNT OF AN ANNUITY

$$s_{\overline{n}|i} = \frac{(1+i)^n - 1}{i}$$

Periods *n*	Rate *i*: .01 (1%)	.01125 $\left(1\frac{1}{8}\%\right)$	.0125 $\left(1\frac{1}{4}\%\right)$	.015 $\left(1\frac{1}{2}\%\right)$	.0175 $\left(1\frac{3}{4}\%\right)$
1	1.0000 0000	1.0000 0000	1.0000 0000	1.0000 0000	1.0000 0000
2	2.0100 0000	2.0112 5000	2.0125 0000	2.0150 0000	2.0175 0000
3	3.0301 0000	3.0338 7656	3.0376 5625	3.0452 2500	3.0528 0625
4	4.0604 0100	4.0680 0767	4.0756 2695	4.0909 0338	4.1062 3036
5	5.1010 0501	5.1137 7276	5.1265 7229	5.1522 6693	5.1780 8939
6	6.1520 1506	6.1713 0270	6.1906 5444	6.2295 5093	6.2687 0596
7	7.2135 3521	7.2407 2986	7.2680 3762	7.3229 9419	7.3784 0831
8	8.2856 7056	8.3221 8807	8.3588 8809	8.4328 3911	8.5075 3045
9	9.3685 2727	9.4158 1269	9.4633 7420	9.5593 3169	9.6564 1224
10	10.4622 1254	10.5217 4058	10.5816 6637	10.7027 2167	10.8253 9945
11	11.5668 3467	11.6401 1016	11.7139 3720	11.8632 6249	12.0148 4394
12	12.6825 0301	12.7710 6140	12.8603 6142	13.0412 1143	13.2251 0371
13	13.8093 2804	13.9147 3584	14.0211 1594	14.2368 2960	14.4565 4303
14	14.9474 2132	15.0712 7662	15.1963 7988	15.4503 8205	15.7095 3253
15	16.0968 9554	16.2408 2848	16.3863 3463	16.6821 3778	16.9844 4935
16	17.2578 6449	17.4235 3780	17.5911 6382	17.9323 6984	18.2816 7721
17	18.4304 4314	18.6195 5260	18.8110 5336	19.2013 5539	19.6016 0656
18	19.6147 4757	19.8290 2257	20.0461 9153	20.4893 7572	20.9446 3468
19	20.8108 9504	21.0520 9907	21.2967 6893	21.7967 1636	22.3111 6578
20	22.0190 0399	22.2889 3519	22.5629 7854	23.1236 6710	23.7016 1119
21	23.2391 9403	23.5396 8571	23.8450 1577	24.4705 2211	25.1163 8938
22	24.4715 8598	24.8045 0717	25.1430 7847	25.8375 7994	26.5559 2620
23	25.7163 0183	26.0835 5788	26.4573 6695	27.2251 4364	28.0206 5490
24	26.9734 6485	27.3769 9790	27.7880 8403	28.6335 2080	29.5110 1637
25	28.2431 9950	28.6849 8913	29.1354 3508	30.0630 2361	31.0274 5915
26	29.5256 3150	30.0076 9526	30.4996 2802	31.5139 6896	32.5704 3969
27	30.8208 8781	31.3452 8183	31.8808 7337	32.9866 7850	34.1404 2238
28	32.1290 9669	32.6979 1625	33.2793 8429	34.4814 7867	35.7378 7977
29	33.4503 8766	34.0657 6781	34.6953 7659	35.9987 0085	37.3632 9267
30	34.7848 9153	35.4490 0769	36.1290 6880	37.5386 8137	39.0171 5029
31	36.1327 4045	36.8478 0903	37.5806 8216	39.1017 6159	40.6999 5042
32	37.4940 6785	38.2623 4688	39.0504 4069	40.6882 8801	42.4121 9955
33	38.8690 0853	39.6927 9829	40.5385 7120	42.2986 1233	44.1544 1305
34	40.2576 9862	41.1393 4227	42.0453 0334	43.9330 9152	45.9271 1527
35	41.6602 7560	42.6021 5987	43.5708 6963	45.5920 8789	47.7308 3979
36	43.0768 7836	44.0814 3417	45.1155 0550	47.2759 6921	49.5661 2949
37	44.5076 4714	45.5773 5030	46.6794 4932	48.9851 0874	51.4335 3675
38	45.9527 2361	47.0900 9549	48.2629 4243	50.7198 8538	53.3336 2365
39	47.4122 5085	48.6198 5906	49.8662 2921	52.4806 8366	55.2669 6206
40	48.8863 7336	50.1668 3248	51.4895 5708	54.2678 9391	57.2341 3390
41	50.3752 3709	51.7312 0934	53.1331 7654	56.0819 1232	59.2357 3124
42	51.8789 8946	53.3131 8545	54.7973 4125	57.9231 4100	61.2723 5654
43	53.3977 7936	54.9129 5879	56.4823 0801	59.7919 8812	63.3446 2278
44	54.9317 5715	56.5307 2957	58.1883 3686	61.6888 6794	65.4531 5367
45	56.4810 7472	58.1667 0028	59.9156 9108	63.6142 0096	67.5985 8386
46	58.0458 8547	59.8210 7566	61.6646 3721	65.5684 1398	69.7815 5908
47	59.6263 4432	61.4940 6276	63.4354 4518	67.5519 4018	72.0027 3637
48	61.2226 0777	63.1858 7097	65.2283 8824	69.5652 1929	74.2627 8425
49	62.8348 3385	64.8967 1201	67.0437 4310	71.6086 9758	76.5623 8298
50	64.4631 8218	66.6268 0002	68.8817 8989	73.6828 2804	78.9022 2468

TABLE III : AMOUNT OF AN ANNUITY

$$s_{\overline{n}|i} = \frac{(1+i)^n - 1}{i}$$

Periods	Rate i				
n	.01 (1%)	.01125 $\left(1\frac{1}{8}\%\right)$	.0125 $\left(1\frac{1}{4}\%\right)$	.015 $\left(1\frac{1}{2}\%\right)$	.0175 $\left(1\frac{3}{4}\%\right)$
51	66.1078 140	68.3763 515	70.7428 123	75.7880 705	81.2830 136
52	67.7688 921	70.1455 855	72.6270 974	77.9248 915	83.7054 663
53	69.4465 811	71.9347 233	74.5349 361	80.0937 649	86.1703 120
54	71.1410 469	73.7439 890	76.4666 228	82.2951 714	88.6782 925
55	72.8524 573	75.5736 088	78.4224 556	84.5295 989	91.2301 626
56	74.5809 819	77.4238 119	80.4027 363	86.7975 429	93.8266 904
57	76.3267 917	79.2948 298	82.4077 705	89.0995 061	96.4686 575
58	78.0900 597	81.1868 966	84.4378 676	91.4359 987	99.1568 590
59	79.8709 603	83.1002 492	86.4933 410	93.8075 386	101.8921 041
60	81.6696 699	85.0351 270	88.5745 078	96.2146 517	104.6752 159
61	83.4863 666	86.9917 722	90.6816 891	98.6578 715	107.5070 322
62	85.3212 302	88.9704 297	92.8152 102	101.1377 396	110.3884 052
63	87.1744 425	90.9713 470	94.9754 003	103.6548 057	113.3202 023
64	89.0461 869	92.9947 746	97.1625 928	106.2096 277	116.3033 058
65	90.9366 488	95.0409 659	99.3771 253	108.8027 722	119.3386 137
66	92.8460 153	97.1101 767	101.6193 393	111.4348 137	122.4270 394
67	94.7744 755	99.2026 662	103.8895 811	114.1063 359	125.5695 126
68	96.7222 202	101.3186 962	106.1882 008	116.8179 310	128.7669 791
69	98.6894 424	103.4585 315	108.5155 533	119.5701 999	132.0204 012
70	100.6763 368	105.6224 400	110.8719 978	122.3637 529	135.3307 583
71	102.6831 002	107.8106 925	113.2578 977	125.1992 092	138.6990 465
72	104.7099 312	110.0235 628	115.6736 215	128.0771 974	142.1262 798
73	106.7570 305	112.2613 278	118.1195 417	130.9983 553	145.6134 897
74	108.8246 008	114.5242 678	120.5960 360	133.9633 307	149.1617 258
75	110.9128 468	116.8126 658	123.1034 864	136.9727 806	152.7720 560
76	113.0219 753	119.1268 083	125.6422 800	140.0273 723	156.4455 670
77	115.1521 951	121.4669 849	128.2128 085	143.1277 829	160.1833 644
78	117.3037 170	123.8334 885	130.8154 686	146.2746 997	163.9865 733
79	119.4767 542	126.2266 152	133.4506 620	149.4688 202	167.8563 383
80	121.6715 217	128.6466 646	136.1187 953	152.7108 525	171.7938 242
81	123.8882 369	131.0939 396	138.8202 802	156.0015 153	175.8002 162
82	126.1271 193	133.5687 464	141.5555 337	159.3415 380	179.8767 200
83	128.3883 905	136.0713 948	144.3249 779	162.7316 611	184.0245 625
84	130.6722 744	138.6021 980	147.1290 401	166.1726 360	188.2449 924
85	132.9789 971	141.1614 727	149.9681 531	169.6652 255	192.5392 798
86	135.3087 871	143.7495 393	152.8427 550	173.2102 039	196.9087 172
87	137.6618 750	146.3667 216	155.7532 895	176.8083 569	201.3546 197
88	140.0384 937	149.0133 472	158.7002 056	180.4604 823	205.8783 256
89	142.4388 787	151.6897 474	161.6839 581	184.1673 895	210.4811 962
90	144.8632 675	154.3962 571	164.7050 076	187.9299 004	215.1646 172
91	147.3119 001	157.1332 149	167.7638 202	191.7488 489	219.9299 980
92	149.7850 191	159.9009 636	170.8608 680	195.6250 816	224.7787 729
93	152.2828 693	162.6998 495	173.9966 288	199.5594 578	229.7124 015
94	154.8056 980	165.5302 228	177.1715 867	203.5528 497	234.7323 685
95	157.3537 550	168.3924 378	180.3862 315	207.6061 425	239.8401 850
96	159.9272 926	171.2868 527	183.6410 594	211.7202 346	245.0373 882
97	162.5265 655	174.2138 298	186.9365 726	215.8960 381	250.3255 425
98	165.1518 311	177.1737 354	190.2732 798	220.1344 787	255.7062 395
99	167.8033 494	180.1669 399	193.6516 958	224.4364 959	261.1810 987
100	170.4813 829	183.1938 180	197.0723 420	228.8030 433	266.7517 679

TABLE III : AMOUNT OF AN ANNUITY

$$s_{\overline{n}|i} = \frac{(1+i)^n - 1}{i}$$

Periods n	Rate i .02 (2%)	.0225 $\left(2\frac{1}{4}\%\right)$	.025 $\left(2\frac{1}{2}\%\right)$	.0275 $\left(2\frac{3}{4}\%\right)$	.03 (3%)
1	1.0000 0000	1.0000 0000	1.0000 0000	1.0000 0000	1.0000 0000
2	2.0200 0000	2.0225 0000	2.0250 0000	2.0275 0000	2.0300 0000
3	3.0604 0000	3.0680 0625	3.0756 2500	3.0832 5625	3.0909 0000
4	4.1216 0800	4.1370 3639	4.1525 1563	4.1680 4580	4.1836 2700
5	5.2040 4016	5.2301 1971	5.2563 2852	5.2826 6706	5.3091 3581
6	6.3081 2096	6.3477 9740	6.3877 3673	6.4279 4040	6.4684 0988
7	7.4342 8338	7.4906 2284	7.5474 3015	7.6047 0876	7.6624 6218
8	8.5829 6905	8.6591 6186	8.7361 1590	8.8138 3825	8.8923 3605
9	9.7546 2843	9.8539 9300	9.9545 1880	10.0562 1880	10.1591 0613
10	10.9497 2100	11.0757 0784	11.2033 8177	11.3327 6482	11.4638 7931
11	12.1687 1542	12.3249 1127	12.4834 6631	12.6444 1585	12.8077 9569
12	13.4120 8973	13.6022 2177	13.7955 5297	13.9921 3729	14.1920 2956
13	14.6803 3152	14.9082 7176	15.1404 4179	15.3769 2107	15.6177 9045
14	15.9739 3815	16.2437 0788	16.5189 5284	16.7997 8639	17.0863 2416
15	17.2934 1692	17.6091 9130	17.9319 2666	18.2617 8052	18.5989 1389
16	18.6392 8525	19.0053 9811	19.3802 2483	19.7639 7948	20.1568 8130
17	20.0120 7096	20.4330 1957	20.8647 3045	21.3074 8892	21.7615 8774
18	21.4123 1238	21.8927 6251	22.3863 4871	22.8934 4487	23.4144 3537
19	22.8405 5863	23.3853 4966	23.9460 0743	24.5230 1460	25.1168 6844
20	24.2973 6980	24.9115 2003	25.5446 5761	26.1973 9750	26.8703 7449
21	25.7833 1719	26.4720 2923	27.1832 7405	27.9178 2593	28.6764 8572
22	27.2989 8354	28.0676 4989	28.8628 5590	29.6855 6615	30.5367 8030
23	28.8449 6321	29.6991 7201	30.5844 2730	31.5019 1921	32.4528 8370
24	30.4218 6247	31.3674 0338	32.3490 3798	33.3682 2199	34.4264 7022
25	32.0302 9972	33.0731 6996	34.1577 6393	35.2858 4810	36.4592 6432
26	33.6709 0572	34.8173 1628	36.0117 0803	37.2562 0892	38.5530 4225
27	35.3443 2383	36.6007 0590	37.9120 0073	39.2807 5467	40.7096 3352
28	37.0512 1031	38.4242 2178	39.8598 0075	41.3609 7542	42.9309 2252
29	38.7922 3451	40.2887 6677	41.8562 9577	43.4984 0224	45.2188 5020
30	40.5680 7921	42.1952 6402	43.9027 0316	45.6946 0831	47.5754 1571
31	42.3794 4079	44.1446 5746	46.0002 7074	47.9512 1003	50.0026 7818
32	44.2270 2961	46.1379 1226	48.1502 7751	50.2698 6831	52.5027 5852
33	46.1115 7020	48.1760 1528	50.3540 3445	52.6522 8969	55.0778 4128
34	48.0338 0160	50.2599 7563	52.6128 8531	55.1002 2765	57.7301 7652
35	49.9944 7763	52.3908 2508	54.9282 0744	57.6154 8391	60.4620 8181
36	51.9943 6719	54.5696 1864	57.3014 1263	60.1999 0972	63.2759 4427
37	54.0342 5453	56.7974 3506	59.7339 4794	62.8554 0724	66.1742 2259
38	56.1149 3962	59.0753 7735	62.2272 9664	65.5839 3094	69.1594 4927
39	58.2372 3841	61.4045 7334	64.7829 7906	68.3874 8904	72.2342 3275
40	60.4019 8318	63.7861 7624	67.4025 5354	71.2681 4499	75.4012 5973
41	62.6100 2284	66.2213 6521	70.0876 1737	74.2280 1898	78.6632 9753
42	64.8622.2330	68.7113 4592	72.8398 0781	77.2692 8950	82.0231 9645
43	67.1594 6777	71.2573 5121	75.6608 0300	80.3941 9496	85.4838 9234
44	69.5026 5712	73.8606 4161	78.5523 2308	83.6050 3532	89.0484 0911
45	71.8927 1027	76.5225 0605	81.5161 3116	86.9041 7379	92.7198 6139
46	74.3305 6447	79.2442 6243	84.5540 3443	90.2940 3857	96.5014 5723
47	76.8171 7576	82.0272 5834	87.6678 8530	93.7771 2463	100.3965 0095
48	79.3535 1928	84.8728 7165	90.8595 8243	97.3559 9556	104.4083 9598
49	81.9405 8966	87.7825 1126	94.1310 7199	101.0332 8544	108.5406 4785
50	84.5794 0145	90.7576 1776	97.4843 4879	104.8117 0079	112.7968 6729

TABLE III : AMOUNT OF AN ANNUITY

$$s_{\overline{n}|i} = \frac{(1+i)^n - 1}{i}$$

Periods n	Rate i .02 (2%)	.0225 $\left(2\frac{1}{4}\%\right)$	.025 $\left(2\frac{1}{2}\%\right)$	.0275 $\left(2\frac{3}{4}\%\right)$	.03 (3%)
51	87.2709 895	93.7996 642	100.9214 575	108.6940 226	117.1807 733
52	90.0164 093	96.9101 566	104.4444 939	112.6831 082	121.6961 965
53	92.8167 375	100.0906 351	108.0556 063	116.7818 937	126.3470 824
54	95.6730 722	103.3426 744	111.7569 965	120.9933 957	131.1374 949
55	98.5865 337	106.6678 846	115.5509 214	125.3207 141	136.0716 197
56	101.5582 643	110.0679 120	119.4396 944	129.7670 337	141.1537 683
57	104.5894 296	113.5444 400	123.4256 868	134.3356 272	146.3883 814
58	107.6812 182	117.0991 899	127.5113 289	139.0298 569	151.7800 328
59	110.8348 426	120.7339 217	131.6991 121	143.8531 780	157.3334 338
60	114.0515 394	124.4504 349	135.9915 900	148.8091 404	163.0534 368
61	117.3325 702	128.2505 697	140.3913 797	153.9013 917	168.9450 399
62	120.6792 216	132.1362 075	144.9011 642	159.1336 800	175.0133 911
63	124.0928 060	136.1092 722	149.5236 933	164.5098 562	181.2637 928
64	127.5746 622	140.1717 308	154.2617 856	170.0338 773	187.7017 066
65	131.1261 154	144.3255 948	159.1183 303	175.7098 089	194.3327 578
66	134.7486 785	148.5729 207	164.0962 885	181.5418 286	201.1627 406
67	138.4436 521	152.9158 114	169.1986 957	187.5342 289	208.1976 228
68	142.2125 251	157.3564 171	174.4286 631	193.6914 202	215.4435 515
69	146.0567 756	161.8969 365	179.7893 797	200.0179 343	222.9068 580
70	149.9779 111	166.5396 176	185.2841 142	206.5184 275	230.5940 637
71	153.9774 694	171.2867 590	190.9162 171	213.1976 842	238.5118 856
72	158.0570 183	176.1407 111	196.6891 225	220.0606 205	246.6672 422
73	162.2181 591	181.1038 771	202.6063 506	227.1122 876	255.0672 595
74	166.4625 223	186.1787 143	208.6715 093	234.3578 755	263.7192 773
75	170.7917 728	191.3677 354	214.8882 970	241.8027 171	272.6308 556
76	175.2076 082	196.6735 094	221.2605 045	249.4522 918	281.8097 813
77	179.7117 604	202.0986 634	227.7920 171	257.3122 298	291.2640 747
78	184.3059 956	207.6458 833	234.4868 175	265.3883 162	301.0019 969
79	188.9921 155	213.3179 157	241.3489 880	273.6864 948	311.0320 568
80	193.7719 578	219.1175 688	248.3827 126	282.2128 735	321.3630 185
81	198.6473 970	225.0477 141	255.5922 805	290.9737 275	332.0039 091
82	203.6203 449	231.1112 876	262.9820 875	299.9755 050	342.9640 264
83	208.6927 518	237.3112 916	270.5566 397	309.2248 314	354.2529 472
84	213.8666 068	243.6507 957	278.3205 557	318.7285 142	365.8805 356
85	219.1439 390	250.1329 386	286.2785 695	328.4935 484	377.8569 517
86	224.5268 178	256.7609 297	294.4355 338	338.5271 209	390.1926 602
87	230.0173 541	263.5380 506	302.7964 221	348.8366 168	402.8984 400
88	235.6177 012	270.4676 567	311.3663 327	359.4296 237	415.9853 932
89	241.3300 552	277.5531 790	320.1504 910	370.3139 384	429.4649 550
90	247.1566 563	284.7981 255	329.1542 533	381.4975 717	443.3489 037
91	253.0997 894	292.2060 834	338.3831 096	392.9887 549	457.6493 708
92	259.1617 852	299.7807 202	347.8426 873	404.7959 457	472.3788 519
93	265.3450 209	307.5257 865	357.5387 545	416.9278 342	487.5502 174
94	271.6519 214	315.4451 166	367.4772 234	429.3933 496	503.1767 240
95	278.0849 598	323.5426 318	377.6641 540	442.2016 667	519.2720 257
96	284.6466 590	331.8223 410	388.1057 578	455.3622 126	535.8501 865
97	291.3395 922	340.2883 437	398.8084 018	468.8846 734	552.9256 920
98	298.1663 840	348.9448 314	409.7786 118	482.7790 019	570.5134 628
99	305.1297 117	357.7960 901	421.0230 771	497.0554 245	588.6288 667
100	312.2323 059	366.8465 021	432.5486 540	511.7244 487	607.2877 327

TABLE III : AMOUNT OF AN ANNUITY

$$s_{\overline{n}|i} = \frac{(1+i)^n - 1}{i}$$

Periods n	Rate i: .035 $\left(3\frac{1}{2}\%\right)$	.04(4%)	.045 $\left(4\frac{1}{2}\%\right)$	.05 (5%)	.055 $\left(5\frac{1}{2}\%\right)$
1	1.0000.000	1.0000 000	1.0000 000	1.0000 000	1.0000 000
2	2.0350 000	2.0400 000	2.0450 000	2.0500 000	2.0550 000
3	3.1062 250	3.1216 000	3.1370 250	3.1525 000	3.1680 250
4	4.2149 429	4.2464 640	4.2781 911	4.3101 250	4.3422 664
5	5.3624 659	5.4163 226	5.4707 097	5.5256 313	5.5810 910
6	6.5501 522	6.6329 755	6.7168 917	6.8019 128	6.8880 510
7	7.7794 075	7.8982 945	8.0191 518	8.1420 085	8.2668 938
8	9.0516 868	9.2142 263	9.3800 136	9.5491 089	9.7215 730
9	10.3684 958	10.5827 953	10.8021 142	11.0265 643	11.2562 595
10	11.7313 932	12.0061 071	12.2882 094	12.5778 925	12.8753 538
11	13.1419 919	13.4863 514	13.8411 788	14.2067 872	14.5834 982
12	14.6019 616	15.0258 055	15.4640 318	15.9171 265	16.3855 907
13	16.1130 303	16.6268 377	17.1599 133	17.7129 828	18.2867 981
14	17.6769 864	18.2919 112	18.9321 094	19.5986 320	20.2925 720
15	19.2956 809	20.0235 876	20.7840 543	21.5785 636	22.4086 635
16	20.9710 297	21.8245 311	22.7193.367	23.6574 918	24.6411 400
17	22.7050 157	23.6975 124	24.7417 069	25.8403 664	26.9964 027
18	24.4996 913	25.6454 129	26.8550 837	28.1323 847	29.4812 048
19	26.3571 805	27.6712 294	29.0635 625	30.5390 039	32.1026 711
20	28.2796 818	29.7780 786	31.3714 228	33.0659 541	34.8683 180
21	30.2694 707	31.9692 017	33.7831 368	35.7192 518	37.7860 755
22	32.3289 022	34.2479 698	36.3033 780	38.5052 144	40.8643 097
23	34.4604 137	36.6178 886	38.9370 300	41.4304 751	44.1118 467
24	36.6665 282	39.0826 041	41.6891 963	44.5019 989	47.5379 983
25	38.9498 567	41.6459 083	44.5652 101	47.7270 988	51.1525 882
26	41.3131 017	44.3117.446	47.5706 446	51.1134 538	54.9659 805
27	43.7590 602	47.0842 144	50.7113 236	54.6691 264	58.9891 094
28	46.2906 273	49.9675 830	53.9933 332	58.4025 828	63.2335 105
29	48.9107 993	52.9662 863	57.4230 332	62.3227 119	67.7113 535
30	51.6226 773	56.0849 378	61.0070 697	66.4388 475	72.4354 780
31	54.4294 710	59.3283 353	64.7523 878	70.7607 899	77.4194 293
32	57.3345 025	62.7014 687	68.6662 452	75.2988 294	82.6774 979
33	60.3412 101	66.2095 274	72.7562 263	80.0637 708	83.2247 603
34	63.4531 524	69.8579 085	77.0302 565	85.0669 594	94.0771 221
35	66.6740 127	73.6522 249	81.4966 180	90.3203 074	100.2513 638
36	70.0076 032	77.5983 138	86.1639 658	95.8363 227	106.7651 888
37	73.4578 693	81.7022 464	91.0413 443	101.6281 389	113.6372 742
38	77.0288 947	85.9703 363	96.1382 048	107.7095 458	120.8873 242
39	80.7249 060	90.4091 497	101.4644 240	114.0950 231	128.5361 271
40	84.5502 777	95.0255 157	107.0303 231	120.7997 742	136.6056 141
41	88.5095 375	99.8265 363	112.8466 876	127.8397 630	145.1189 228
42	92.6073 713	104.8195 978	118.9247 885	135.2317 511	154.1004 636
43	96.8486 293	110.0123 817	125.2764 040	142.9933 387	163.5759 891
44	101.2383 313	115.4128 770	131.9138 422	151.1430 056	173.5726 685
45	105.7816 729	121.0293 920	138.8499 651	159.7001 559	184.1191 653
46	110.4840 314	126.8705 677	146.0982 135	168.6851 637	195.2457 194
47	115.3509 725	132.9453 904	153.6726 331	178.1194 218	206.9842 339
48	120.3882 566	139.2632 060	161.5879 016	188.0253 929	219.3683 668
49	125.6018 456	145.8337 343	169.8593 572	198.4266 626	232.4336 270
50	130.9979 102	152.6670 837	178.5030 283	209.3479 957	246.2174 764

TABLE III : AMOUNT OF AN ANNUITY

$$s_{\overline{n}|i} = \frac{(1+i)^n - 1}{i}$$

Periods n	Rate i .06 (6%)	.065 $\left(6\frac{1}{2}\%\right)$	.07 (7%)	.075 $\left(7\frac{1}{2}\%\right)$	.08 (8%)
1	1.0000.000	1.0000 000	1.0000 000	1.0000 000	1.0000 000
2	2.0600 000	2.0650 000	2.0700 000	2.0750 000	2.0800 000
3	3.1835 000	3.1992 250	3.2149 000	3.2306 250	3.2464 000
4	4.3746 160	4.4071 746	4.4399 430	4.4729 219	4.5061 120
5	5.6370 930	5.6936 410	5.7507 390	5.8083 910	5.8666 010
6	6.9753 185	7.0637 276	7.1532 907	7.2440 203	7.3359 290
7	8.3938 376	8.5228 699	8.6540 211	8.7873 219	8.9228 034
8	9.8974 679	10.0768 565	10.2598 026	10.4463 710	10.6366 276
9	11.4913 160	11.7318 522	11.9779 887	12.2298 488	12.4875 578
10	13.1807 949	13.4944 225	13.8164 480	14.1470 875	14.4865 625
11	14.9716 426	15.3715 600	15.7835 993	16.2081 191	16.6454 875
12	16.8699 412	17.3707 114	17.8884 513	18.4237 280	18.9771 265
13	18.8821 377	19.4998 076	20.1406 429	20.8055 076	21.4952 966
14	21.0150 659	21.7672 951	22.5504 879	23.3659 207	24.2149 203
15	23.2759 699	24.1821 693	25.1290 220	26.1183 647	27.1521 139
16	25.6725 281	26.7540 103	27.8880 536	29.0772 421	30.3242 830
17	28.2128 798	29.4930 210	30.8402 173	32.2580 352	33.7502 257
18	30.9056 525	32.4100 674	33.9990 325	35.6773 879	37.4502 437
19	33.7599 917	35.5167 218	37.3789 648	39.3531 919	41.4462 632
20	36.7855 912	38.8253 087	40.9954 923	43.3046 813	45.7619 643
21	39.9927 267	42.3489 537	44.8651 768	47.5525 324	50.4229 214
22	43.3922 903	46.1016 357	49.0057 392	52.1189 724	55.4567 552
23	46.9958 277	50.0982 420	53.4361 409	57.0278 953	60.8932 956
24	50.8155 774	54.3546 278	58.1766 708	62.3049 874	66.7647 592
25	54.8645 120	58.8876 786	63.2490 377	67.9778 615	73.1059 400
26	59.1563 827	63.7153 777	68.6764 704	74.0762 011	79.9544 151
27	63.7057 657	68.8568 772	74.4838 233	80.6319 162	87.3507 684
28	68.5281 116	74.3325 743	80.6976 909	87.6793 099	95.3388 298
29	73.6397 983	80.1641 916	87.3465 293	95.2552 582	103.9659 362
30	79.0581 862	86.3748 640	94.4607 863	103.3994 025	113.2832 111
31	84.8016 774	92.9892 302	102.0730 414	112.1543 577	123.3458 680
32	90.8897 780	100.0335 302	110.2181 543	121.5659 345	134.2135 374
33	97.3431 647	107.5357 096	118.9334 251	131.6833 796	145.9506 204
34	104.1837 546	115.5255 308	128.2587 648	142.5596 331	158.6266 701
35	111.4347 799	124.0346 903	138.2368 784	154 2516 056	172.3168 037
36	119.1208 667	133.0969 451	148.9134 598	166.8204 760	187.1021 480
37	127.2681 187	142.7482 466	160.3374 020	180.3320 117	203.0703 198
38	135.9042 058	153.0268 826	172.5610 202	194.8569 126	220.3159 454
39	145.0584 581	163.9736 300	185.6402 916	210.4711 810	238.9412 210
40	154.7619 656	175.6319 159	199.6351 120	227.2565 196	259.0565 187
41	165.0476 836	188.0479 904	214.6095 698	245.3007 586	280.7810 402
42	175.9505 446	201.2711 098	230.6322 397	264.6983 155	304.2435 234
43	187.5075 772	215.3537 320	247.7764 965	285.5506 891	329.5830 053
44	199.7580 319	230.3517 245	266.1208 513	307.9669 908	356.9496 457
45	212.7435 138	246.3245 866	285.7493 108	332.0645 151	386.5056 174
46	226.5081 246	263.3356 848	306.7517 626	357.9693 537	418.4260 668
47	241.0986 121	281.4525 043	329.2243 860	385.8170 553	452.9001 521
48	256.5645 288	300.7469 170	353.2700 930	415.7533 344	490.1321 643
49	272.9584 006	321.2954 666	378.9989 995	447.9348 345	530.3427 374
50	290.3359 046	343.1796 720	406.5289 295	482.5299 471	573.7701 564

TABLE IV : PRESENT VALUE OF AN ANNUITY

$$a_{\overline{n}|i} = \frac{1-(1+i)^{-n}}{i}$$

Periods n	Rate i .0025 $\left(\frac{1}{4}\%\right)$	.004167 $\left(\frac{5}{12}\%\right)$	.005 $\left(\frac{1}{2}\%\right)$	.005833 $\left(\frac{7}{12}\%\right)$	.0075 $\left(\frac{3}{4}\%\right)$
1	0.9975 0623	0.9958 5062	0.9950 2488	0.9942 0050	0.9925 5583
2	1.9925 2492	1.9875 6908	1.9850 9938	1.9826 3513	1.9777 2291
3	2.9850 6227	2.9751 7253	2.9402 4814	2.9653 3732	2.9555 5624
4	3.9751 2446	3.9586 7804	3.9504 9566	3.9423 4034	3.9261 1041
5	4.9627 1766	4.9381 0261	4.9258 6633	4.9136 7722	4.8894 3961
6	5.9478 4804	5.9134 6318	5.8963 8441	5.8793 8083	5.8455 9763
7	6.9305 2174	6.8847 7661	6.8620 7404	6.8394 8384	6.7946 3785
8	7.9107 4487	7.8520 5970	7.8229 5924	7.7940 1874	7.7366 1325
9	8.8885 2357	8.8153 2916	8.7790 6392	8.7430 1780	8.6715 7642
10	9.8638 6391	9.7746 0165	9.7304 1186	9.6865 1314	9.5995 7958
11	10.8367 7198	10.7298 9376	10.6770 2673	10.6245 3667	10.5206 7452
12	11.8072 5384	11.6812 2200	11.6189 3207	11.5571 2014	11.4349 1267
13	12.7753 1555	12.6286 0283	12.5561 5131	12.4842 9509	12.3423 4508
14	13.7409 6314	13.5720 5261	13.4887 0777	13.4060 9288	13.2430 2242
15	14.7042 0264	14.5115 8766	14.4166 2465	14.3225 4470	14.1369 9495
16	15.6650 4004	15.4472 2422	15.3399 2502	15.2336 8156	15.0243 1261
17	16.6234 8133	16.3789 7848	16.2586 3186	16.1395 3427	15.9050 2492
18	17.5795 3250	17.3068 6654	17.1727 6802	17.0401 3350	16.7791 8107
19	18.5331 9950	18.2309 0443	18.0823 5624	17.9355 0969	17.6468 2984
20	19.4844 8828	19.1511 0815	18.9874 1915	18.8256 9315	18.5080 1969
21	20.4334 0477	20.0674 9359	19.8879 7925	19.7107 1398	19.3627 9870
22	21.3799 5488	20.9800 7661	20.7840 5896	20.5906 0213	20.2112 1459
23	22.3241 4452	21.8888 7297	21.6756 8055	21.4653 8738	21.0533 1473
24	23.2659 7957	22.7938 9839	22.5628 6622	22.3350 9930	21.8891 4614
25	24.2054 6591	23.6951 6853	23.4456 3803	23.1997 6732	22.7187 5547
26	25.1426 0939	24.5926 9895	24.3240 1794	24.0594 2070	23.5421 8905
27	26.0774 1585	25.4865 0517	25.1980 2780	24.9140 8852	24.3594 9286
28	27.0098 9112	26.3766 0266	26.0676 8936	25.7637 9968	25.1707 1251
29	27.9400 4102	27.2630 0680	26.9330 2424	26.6085 8295	25.9758 9331
30	28.8678 7134	28.1457 3291	27.7940 5397	27.4484 6689	26.7750 8021
31	29.7933 8787	29.0247 9626	28.6507 9997	28.2834 7993	27.5683 1783
32	30.7165 9638	29.9002 1205	29.5032 8355	29.1136 5030	28.3556 5045
33	31.6375 0262	30.7719 9540	30.3515 2592	29.9390 0610	29.1371 2203
34	32.5561 1234	31.6401 6139	31.1955 4818	30.7595 7524	29.9127 7621
35	33.4724 3126	32.5047 2504	32.0353 7132	31.5753 8549	30.6826 5629
36	34.3864 6510	33.3657 0128	32.8710 1624	32.3864 6445	31.4468 0525
37	35.2982 1955	34.2231 0501	33.7025 0372	33.1928 3955	32.2052 6576
38	36.2077 0030	35.0769 5105	34.5298 5445	33.9945 3808	32.9580 8016
39	37.1149 1302	35.9272 5416	35.3530 8900	34.7915 8716	33.7052 9048
40	38.0198 6336	36.7740 2904	36.1722 2786	35.5840 1374	34.4469 3844
41	38.9225 5697	37.6172 9033	36.9872 9141	36.3718 4465	35.1830 6545
42	39.8229 9947	38.4570 5261	37.7982 9991	37.1551 0653	35.9137 1260
43	40.7211 9648	39.2933 3040	38.6052 7354	37.9338 2588	36.6389 2070
44	41.6171 5359	40.1261 3816	39.4082 3238	38.7080 2904	37.3587 3022
45	42.5108 7640	40.9554 9028	40.2071 9640	39.4777 4221	38.0731 8136
46	43.4023 7048	41.7814 0111	41.0021 8547	40.2429 9143	38.7823 1401
47	44.2916 4137	42.6038 8492	41.7932 1937	41.0038 0258	39.4861 6775
48	45.1786 9464	43.4229 5594	42.5803 1778	41.7602 0141	40.1847 8189
49	46.0635 3580	44.2386 2832	43.3635 0028	42.5122 1349	40.8781 9542
50	46.9461 7037	45.0509 1617	44.1427 8635	43.2598 6428	41.5664 4707

TABLE IV : PRESENT VALUE OF AN ANNUITY

$$a_{\overline{n}|i} = \frac{1-(1+i)^{-n}}{i}$$

Periods n	Rate i .0025 $\left(\frac{1}{4}\%\right)$	.004167 $\left(\frac{5}{12}\%\right)$	.005 $\left(\frac{1}{2}\%\right)$	.005833 $\left(\frac{7}{12}\%\right)$	.0075 $\left(\frac{3}{4}\%\right)$
51	47.8266 0386	45.8598 3353	44.9181 9537	44.0031 7907	42.2495 7525
52	48.7048 4176	46.6653 6439	45.6897 4664	44.7421 8301	42.9276 1812
53	49.5808 8953	47.4676 1267	46.4574 5934	45.4769 0108	43.6006 1351
54	50.4547 5265	48.2665 0224	47.2213 5258	46.2073 5816	44.2685 9902
55	51.3264 3656	49.0620 7692	47.9814 4535	46.9335 7895	44.9316 1193
56	52.1959 4669	49.8543 5046	48.7377 5657	47.6555 8802	45.5896 8926
57	53.0632 8847	50.6433 3656	49.4903 0505	48.3734 0080	46.2428 6776
58	53.9284 6730	51.4290 4885	50.2391 0950	49.0870 6856	46.8911 8388
59	54.7914 8858	52.2115 0093	50.9841 8856	49.7965 8846	47.5346 7382
60	55.6523 5769	52.9907 0632	51.7255 6075	50.5019 9350	48.1733 7352
61	56.5110 7999	53.7666 7850	52.4632 4453	51.2033 0754	48.8073 1863
62	57.3676 6083	54.6394 3087	53.1972 5824	51.9005 5431	49.4365 4455
63	58.2221 0557	55.3089 7680	53.9276 2014	52.5937 5739	50.0610 8640
64	59.0744 1952	56.0753 2959	54.6543 4840	53.2829 4024	50.6809 7906
65	59.9246 0800	56.8385 0250	55.3774 6109	53.9681 2617	51.2962 5713
66	60.7726 7631	57.5985 0871	56.0969 7621	54.6493 3836	51.9069 5497
67	61.6186 2974	58.3553 6137	56.8129 1165	55.3265 9986	52.5131 0667
68	62.4624 7355	59.1090 7357	57.5252 8522	55.9999 3358	53.1147 4607
69	63.3042 1302	59.8596 5832	58.2341 1465	56.6693 6230	53.7119 0677
70	64.1438 5339	60.6071 2862	58.9394 1756	57.3349 0867	54.3046 2210
71	64.9813 9989	61.3514 9738	59.6412 1151	57.9965 9520	54.8929 2516
72	65.8168 5774	62.0927 7748	60.3395 1394	58.6544 4427	55.4768 4880
73	66.6502 3216	62.8309 8172	61.0343 4222	59.3084 7815	56.0564 2561
74	67.4815 2834	63.5661 2287	61.7257 1366	59.9587 1896	56.6316 8795
75	68.3107 5146	64.2982 1365	62.4136 4543	60.6051 8869	57.2026 6794
76	69.1379 0670	65.0272 6670	63.0981 5466	61.2479 0922	57.7693 9746
77	69.9629 9920	65.7532 9464	63.7792 5836	61.8869 0229	58.3319 0815
78	70.7860 3411	66.4763 1002	64.4569 7350	62.5221 8952	58.8902 3141
79	71.6070 1657	67.1963 2533	65.1313 1691	63.1537 9239	59.4443 9842
80	72.2459 5169	67.9133 5303	65.8023 0539	63.7817 3229	59.9944 4012
81	73.2428 4458	68.6274 0550	66.4699 5561	64.4060 3044	60.5403 8722
82	74.0577 0033	69.3384 9511	67.1342 8419	65.0267 0798	61.0822 7019
83	74.8705 2402	70.0466 3413	67.7953 0765	65.6437 8590	61.6201 1930
84	75.6813 2072	70.7518 3482	68.4530 4244	66.2572 8507	62.1539 6456
85	76.4900 9548	71.4541 0936	69.1075 0491	66.8672 2625	62.6838 3579
86	77.2968 5335	72.1534 6991	69.7587 1135	67.4736 3007	63.2097 6257
87	78.1015 9935	72.8499 2854	70.4066 7796	68.0765 1706	63.7317 7427
88	78.9043 3850	73.5434 9730	71.0514 2086	68.6759 0759	64.2499 0002
89	79.7050 7581	74.2341 8818	71.6929 5608	69.2718 2197	64.7641 6875
90	80.5038 1627	74.9220 1313	72.3312 9958	69.8642 8033	65.2746 0918
91	81.3005 6486	75.6069 8403	72.9664 6725	70.4533 0273	65.7812 4981
92	82.0953 2654	76.2891 1272	73.5984 7487	71.0389 0910	66.2841 1892
93	82.8881 0628	76.9684 1101	74.2273 3818	71.6211 1923	66.7832 4458
94	83.6789 0901	77.6448 9063	74.8530 7282	72.1999 5284	67.2786 5467
95	84.4677 3966	78.3185 6329	75.4756 9434	72.7754 2950	67.7703 7685
96	85.2546 0315	78.9894 4062	76.0952 1825	73.3475 6869	68.2584 3856
97	86.0395 0439	79.6575 3422	76.7116 5995	73.9163 8975	68.7428 6705
98	86.8224 4827	80.3228 5566	77.3250 3478	74.4819 1193	69.2236 8938
99	87.6034 3967	80.9854 1642	77.9353 5799	75.0441 5436	69.7009 3239
100	88.3824 8346	81.6452 2797	78.5426 4477	75.6031 3607	70.1746 2272

TABLE IV : PRESENT VALUE OF AN ANNUITY

$$a_{\overline{n}|i} = \frac{1-(1+i)^{-n}}{i}$$

Periods *n*	Rate *i* .01 (1%)	.01125 $\left(1\frac{1}{8}\%\right)$	.0125 $\left(1\frac{1}{4}\%\right)$	.015 $\left(1\frac{1}{2}\%\right)$	.0175 $\left(1\frac{3}{4}\%\right)$
1	0.9900 9901	0.9888 7515	0.9876 5432	0.9852 2167	0.9828 0098
2	1.9703 9506	1.9667 4923	1.9631 1538	1.9558 8342	1.9486 9875
3	2.9409 8521	2.9337 4460	2.9265 3371	2.9122 0042	2.8979 8403
4	3.9019 6555	3.8899 8230	3.8780 5798	3.8543 8465	3.8309 4254
5	4.8534 3124	4.8355 8200	4.8178 3504	4.7826 4497	4.7478 5508
6	5.7954 7647	5.7706 6205	5.7460 0992	5.6971 8717	5.6489 9762
7	6.7281 9453	6.6953 3948	6.6627 2585	6.5982 1396	6.5346 4139
8	7.6516 7775	7.6097 3002	7.5681 2429	7.4859 2508	7.4050 5297
9	8.5660 1758	8.5139 4810	8.4623 4498	8.3605 1732	8.2604 9432
10	9.4713 0453	9.4081 0690	9.3455 2591	9.2221 8455	9.1012 2291
11	10.3676 2825	10.2923 1832	10.2178 0337	10.0711 1779	9.9274 9181
12	11.2550 7747	11.1666 9302	11.0793 1197	10.9075 0521	10.7395 4969
13	12.1337 4007	12.0313 4044	11.9301 8466	11.7315 3222	11.5376 4097
14	13.0037 0304	12.8863 6880	12.7705 5275	12.5433 8150	12.3220 0587
15	13.8650 5252	13.7318 8509	13.6005 4592	13.3432 3301	13.0928 8046
16	14.7178 7378	14.5679 9514	14.4202 9227	14.1312 6405	13.8504 9677
17	15.5622 5127	15.3948 0360	15.2299 1829	14.9076 4931	14.5950 8282
18	16.3982 6858	16.2124 1395	16.0295 4893	15.6725 6089	15.3268 6272
19	17.2260 0850	17.0209 2850	16.8193 0759	16.4261 6837	16.0460 5673
20	18.0455 5297	17.8204 4845	17.5993 1613	17.1686 3879	16.7528 8130
21	18.8569 8313	18.6110 7387	18.3696 9495	17.9001 3673	17.4475 4919
22	19.6603 7934	19.3929 0371	19.1305 6291	18.6208 2437	18.1302 6948
23	20.4558 2113	20.1660 3580	19.8820 3744	19.3308 6145	18.8012 4764
24	21.2433 8726	20.9305 6693	20.6242 3451	20.0304 0537	19.4606 8566
25	22.0231 5570	21.6865 9276	21.3572 6865	20.7196 1120	20.1087 8196
26	22.7952 0366	22.4342 0792	22.0812 5299	21.3986 3172	20.7457 3166
27	23.5596 0759	23.1735 0598	22.7962 9925	22.0676 1746	21.3717 2644
28	24.3164 4316	23.9045 7946	23.5025 1778	22.7267 1671	21.9869 5474
29	25.0657 8530	24.6275 1986	24.2000 1756	23.3760 7558	22.5916 0171
30	25.8077 0822	25.3424 1766	24.8889 0623	24.0158 3801	23.1858 4934
31	26.5422 8537	26.0493 6233	25.5692 9010	24.6461 4582	23.7698 7650
32	27.2695 8947	26.7484 4236	26.2412 7418	25.2671 3874	24.3438 5897
33	27.9896 9255	27.4397 4522	26.9049 6215	25.8789 5442	24.9079 6951
34	28.7026 6589	28.1233 5745	27.5604 5644	26.4817 2849	25.4623 7789
35	29.4085 8009	28.7993 6460	28.2078 5822	27.0755 9458	26.0072 5100
36	30.1075 0504	29.4678 5127	28.8472 6737	27.6606 8431	26.5427 5283
37	30.7995 0994	30.1289 0114	29.4787 8259	28.2371 2740	27.0690 4455
38	31.4846 6330	30.7825 9692	30.1025 0133	28.8050 5163	27.5862 8457
39	32.1630 3298	31.4290 2044	30.7185 1983	29.3645 8288	28.0946 2857
40	32.8346 8611	32.0682 5260	31.3269 3316	29.9158 4520	28.5942 2955
41	33.4996 8922	32.7003 7340	31.9278 3522	30.4589 6079	29.0852 3789
42	34.1581 0814	33.3254 6195	32.5213 1874	30.9940 5004	29.5678 0136
43	34.8100 0806	33.9435 9649	33.1074 7530	31.5212 3157	30.0420 6522
44	35.4554 5352	34.5548 5438	33.6863 9536	32.0406 2223	30.5081 7221
45	36.0945 0844	35.1593 1212	34.2581 6825	32.5523 3718	30.9662 6261
46	36.7272 3608	35.7570 4536	34.8228 8222	33.0564 8983	31.4164 7431
47	37.3536 9909	36.3481 2891	35.3806 2442	33.5531 9195	31.8589 4281
48	37.9739 5949	36.9326 3674	35.9314 8091	34.0425 5365	32.2938 0129
49	38.5880 7871	37.5106 4202	36.4755 3670	34.5246 8339	32.7211 8063
50	39.1961 1753	38.0822 1708	37.0128 7575	34.9996 8807	33.1412 0946

TABLE IV : PRESENT VALUE OF AN ANNUITY

$$a_{\overline{n}|i} = \frac{1-(1+i)^{-n}}{i}$$

Periods n	Rate i .01 (1%)	.01125 $\left(1\frac{1}{8}\%\right)$	.0125 $\left(1\frac{1}{4}\%\right)$	.015 $\left(1\frac{1}{2}\%\right)$	.0175 $\left(1\frac{3}{4}\%\right)$
51	39.7981 3617	38.6474 3345	37.5435 8099	35.4676 7298	33.5540 1421
52	40.3941 9423	39.2063 6188	38.0677 3431	35.9287 4185	33.9597 1913
53	40.9843 5072	39.7590 7232	38.5854 1660	36.3829 9690	34.3584 4632
54	41.5686 6408	40.3056 3394	39.0967 0776	36.8305 3882	34.7503 1579
55	42.1471 9216	40.8461 1514	39.6016 8667	37.2714 6681	35.1354 4550
56	42.7199 9224	41.3805 8358	40.1004 3128	37.7058 7863	35.5139 5135
57	43.2871 2102	41.9091 0613	40.5930 1855	38.1338 7058	35.8859 4727
58	43.8486 3468	42.4317 4896	41.0795 2449	38.5555 3751	36.2515 4523
59	44.4045 8879	42.9485 7746	41.5600 2419	38.9709 7292	36.6108 5526
60	44.9550 3841	43.4596 5633	42.0345 9179	39.3802 6889	36.9639 8552
61	45.5000 3803	43.9650 4952	42.5033 0054	39.7835 1614	37.3110 4228
62	46.0396 4161	44.4648 2029	42.9662 2275	40.1808 0408	37.6521 3000
63	46.5739 0258	44.9590 3119	43.4234 2988	40.5722 2077	37.9873 5135
64	47.1028 7385	45.4477 4407	43.8749 9247	40.9578 5298	38.3168 0723
65	47.6266 0777	45.9310 2009	44.3209 8022	41.3377 8618	38.6405 9678
66	48.1451 5621	46.4089 1975	44.7614 6195	41.7121 0461	38.9588 1748
67	48.6585 7050	46.8815 0284	45.1965 0563	42.0808 9125	39.2715 6509
68	49.1669 0149	47.3488 2852	45.6261 7840	42.4442 2783	39.5789 3375
69	49.6701 9949	47.8109 5527	46.0505 4656	42.8021 9490	39.8810 1597
70	50.1685 1435	48.2679 4094	46.4696 7562	43.1548 7183	40.1779 0267
71	50.6618 9539	48.7198 4270	46.8836 3024	43.5023 3678	40.4696 8321
72	51.1503 9148	49.1667 1714	47.2924 7431	43.8446 6677	40.7564 4542
73	51.6340 5097	49.6086 2016	47.6962 7093	44.1819 3771	41.0382 7560
74	52.1129 2175	50.0456 0708	48.0950 8240	44.5142 2434	41.3152 5857
75	52.5870 5124	50.4777 3259	48.4889 7027	44.8416 0034	41.5874 7771
76	53.0564 8638	50.9050 5077	48.8779 9533	45.1641 3826	41.8550 1495
77	53.5212 7364	51.3276 1510	49.2622 1761	45.4819 0962	42.1179 5081
78	53.9814 5905	51.7454 7847	49.6416 9640	45.7949 8485	42.3763 6443
79	54.4370 8817	52.1586 9317	50.0164 9027	46.1034 3335	42.6303 3359
80	54.8882 0611	52.5673 1092	50.3866 5706	46.4073 2349	42.8799 3474
81	55.3348 5753	52.9713 8286	50.7522 5389	46.7067 2265	43.1252 4298
82	55.7770 8666	53.3709 5957	51.1133 3717	47.0016 9720	43.3663 3217
83	56.2149 3729	53.7660 9104	51.4699 6264	47.2923 1251	43.6032 7486
84	56.6484 5276	54.1568 2674	51.8221 8532	47.5786 3301	43.8361 4237
85	57.0776 7600	54.5432 1557	52.1700 5958	47.8607 2218	44.0650 0479
86	57.5026 4951	54.9253 0588	52.5136 3909	48.1386 4254	44.2899 3099
87	57.9234 1535	55.3031 4549	52.8529 7688	48.4124 5571	44.5109 8869
88	58.3400 1520	55.6767 8169	53.1881 2531	48.6822 2237	44.7282 4441
89	58.7524 9030	56.0462 6126	53.5191 3611	48.9480 0234	44.9417 6355
90	59.1608 8148	56.4116 3041	53.8460 6035	49.2098 5452	45.1516 1037
91	59.5652 2919	56.7729 3490	54.1689 4850	49.4678 3696	45.3578 4803
92	59.9655 7346	57.1302 1992	54.4878 5037	49.7220 0686	45.5605 3860
93	60.3619 5392	57.4835 3021	54.8028 1518	49.9724 2055	45.7597 4310
94	60.7544 0982	57.8329 0997	55.1138 9154	50.2191 3355	45.9555 2147
95	61.1429 8002	58.1784 0294	55.4211 2744	50.4622 0054	46.1479 3265
96	61.5277 0299	58.5200 5235	55.7245 7031	50.7016 7541	46.3370 3455
97	61.9086 1682	58.8579 0096	56.0242 6698	50.9376 1124	46.5228 8408
98	62.2857 5923	59.1919 9106	56.3202 6368	51.1700 6034	46.7055 3718
99	62.6591 6755	59.5223 6446	56.6126 0610	51.3990 7422	46.8850 4882
100	63.0288 7877	59.8490 6251	56.9013 3936	51.6247 0367	47.0614 7304

TABLE IV : PRESENT VALUE OF AN ANNUITY

$$a_{\overline{n}|i} = \frac{1-(1+i)^{-n}}{i}$$

Periods *n*	Rate *i* .02 (2%)	.0225 $\left(2\frac{1}{4}\%\right)$	.025 $\left(2\frac{1}{2}\%\right)$	.0275 $\left(2\frac{3}{4}\%\right)$	.03 (3%)
1	0.9803 9216	0.9779 9511	0.9756 0976	0.9732 3601	0.9708 7379
2	1.9415 6094	1.9344 6955	1.9274 2415	1.9204 2434	1.9134 6970
3	2.8838 8327	2.8698 9687	2.8560 2356	2.8422 6213	2.8286 1135
4	3.8077 2870	3.7847 4021	3.7619 7421	3.7394 2787	3.7170 9840
5	4.7134 5951	4.6794 5253	4.6458 2850	4.6125 8186	4.5797 0719
6	5.6014 3089	5.5544 7680	5.5081 2536	5.4623 6678	5.4171 9144
7	6.4719 9107	6.4102 4626	6.3493 9060	6.2894 0806	6.2302 8296
8	7.3254 8144	7.2471 8461	7.1701 3717	7.0943 1441	7.0196 9219
9	8.1622 3671	8.0657 0622	7.9708 6553	7.8776 7826	7.7861 0892
10	8.9825 8501	8.8662 1635	8.7520 6393	8.6400 7616	8.5302 0284
11	9.7868 4805	9.6491 1134	9.5142 0871	9.3820 6926	9.2526 2411
12	10.5753 4122	10.4147 7882	10.2577 6460	10.1042 0366	9.9540 0399
13	11.3483 7375	11.1635 9787	10.9831 8497	10.8070 1086	10.6349 5533
14	12.1062 4877	11.8959 3924	11.6909 1217	11.4910 0814	11.2960 7314
15	12.8492 6350	12.6121 6551	12.3813 7773	12.1566 9892	11.9379 3509
16	13.5777 0931	13.3126 3131	13.0550 0266	12.8045 7315	12.5611 0203
17	14.2918 7188	13.9976 8343	13.7121 9772	13.4351 0769	13.1661 1847
18	14.9920 3125	14.6676 6106	14.3533 6363	14.0487 6661	13.7535 1308
19	15.6784 6201	15.3228 9590	14.9788 9134	14.6460 0157	14.3237 9911
20	16.3514 3334	15.9637 1237	15.5891 6229	15.2272 5213	14.8774 7486
21	17.0112 0916	16.5904 2775	16.1845 4857	15.7929 4612	15.4150 2414
22	17.6580 4820	17.2033 5232	16.7654 1324	16.3434 9987	15.9369 1664
23	18.2922 0412	17.8027 8955	17.3321 1018	16.8793 1861	16.4436 0839
24	18.9139 2560	18.3890 3624	17.8849 8583	17.4007 9670	16.9355 4212
25	19.5234 5647	18.9623 8263	18.4243 7642	17.9083 1795	17.4131 4769
26	20.1210 3576	19.5231 1260	18.9506 1114	18.4022 5592	17.8768 4242
27	20.7068 9780	20.0715 0376	19.4640 1087	18.8829 7413	18.3270 3147
28	21.2812 7236	20.6078 2764	19.9648 8866	19.3508 2640	18.7641 0823
29	21.8443 8466	21.1323 4977	20.4535 4991	19.8061 5708	19.1884 5459
30	22.3964 5555	21.6453 2985	20.9302 9250	20.2493 0130	19.6004 4135
31	22.9377 0152	22.1470 2186	21.3954 0741	20.6805 8520	20.0004 2849
32	23.4683 3482	22.6376 7419	21.8491 7796	21.1003 2623	20.3887 6553
33	23.9885 6355	23.1175 2977	22.2918 8094	21.5088 3332	20.7657 9178
34	24.4985 9172	23.5868 2618	22.7237 8028	21.9064 0712	21.1318 3668
35	24.9986 1933	24.0457 9577	23.1451 5734	22.2933 4026	21.4872 2007
36	25.4888 4248	24.4946 6579	23.5562 5107	22.6699 1753	21.8322 5250
37	25.9694 5341	24.9336 5848	23.9573 1812	23.0364 1609	22.1672 3544
38	26.4406 4060	25.3629 9118	24.3486 0304	23.3931 0568	22.4924 6159
39	26.9025 8883	25.7828 7646	24.7303 4443	23.7402 4884	22.8082 1513
40	27.3554 7924	26.1935 2221	25.1027 7505	24.0781 0106	23.1147 7197
41	27.7994 8945	26.5951 3174	25.4661 2200	24.4069 1101	23.4123 9997
42	28.2347 9358	26.9879 0390	25.8206 0683	24.7269 2069	23.7013 5920
43	28.6615 6233	27.3720 3316	26.1664 4569	25.0383 6563	23.9819 0213
44	29.0799 6307	27.7477 0969	26.5038.4945	25.3414 7507	24.2542 7392
45	29.4901 5987	28.1151 1950	26.8330 2386	25.6364 7209	24.5187 1254
46	29.8923 1360	28.4744 4450	27.1541 6962	25.9235 7381	24.7754 4907
47	30.2865 8196	28.8258 6259	27.4674 8255	26.2029 9154	25.0247 0783
48	30.6731 1957	29.1695 4777	27.7731 5371	26.4749 3094	25.2667 0664
49	31.0520 7801	29.5056 7019	28.0713 6947	26.7395 9215	25.5016 5693
50	31.4236 0589	29.8343 9627	28.3623 1168	26.9971 6998	26.7297 6401

TABLE IV : PRESENT VALUE OF AN ANNUITY

$$a_{\overline{n}|i} = \frac{1-(1+i)^{-n}}{i}$$

Periods n	Rate i .02 (2%)	.0225 $\left(2\frac{1}{4}\%\right)$	.025 $\left(2\frac{1}{2}\%\right)$	.0275 $\left(2\frac{3}{4}\%\right)$	.03 (3%)
51	31.7878 4892	30.1558 8877	28.6461 5774	27.2478 5400	25.9512 2719
52	32.1449 4992	30.4703 0687	28.9230 8072	27.4918 2871	26.1662 3999
53	32.4950 4849	30.7778 0623	29.1932 4948	27.7292 7368	26.3749 9028
54	32.8382 8327	31.0785 3910	29.4568 2876	27.9603 6368	26.5776 6047
55	33.1747 8752	31.3726 5438	29.7139 7928	28.1852 6879	26.7744 2764
56	33.5046 9365	31.6602 9768	29.9648 5784	28.4041 5454	26.9654 6373
57	33.8281 3103	31.9416 1142	30.2096 1740	28.6171 8203	27.1509 3566
58	34.1452 2650	32.2167 3489	30.4484 0722	28.8245 0806	27.3310 0549
59	34 4561 0441	32.4858 0429	30.6813 7290	29.0262 8522	27.5058 3058
60	34.7608 8668	32.7489 5285	30.9086 5649	29.2226 6201	27.6755 6367
61	35.0596 9282	33.0063 1086	31.1303 9657	29.4137 8293	27.8403 5307
62	35.3526 4002	33.2580 0573	31.3467 2836	29.5997 8879	28.0003 4279
63	35.6398 4316	33.5041 6208	31.5577 8377	29.7808 1634	28.1556 7261
64	35.9214 1486	33.7449 0179	31.7636 9148	29.9569 9887	28.3064 7826
65	36.1974 6555	33.9803 4405	31.9645 7705	30.1284 6605	28.4528 9152
66	36.4681 0348	34.2106 0543	32.1605 6298	30.2053 4409	28.5950 4031
67	36.7334 3478	34.4357 9993	32.3517 6876	30.4577 5581	28.7330 4884
68	36.9935 6351	34.6560 3905	32.5383 1099	30.6158 2074	28.8670 3771
69	37.2485 9168	34.8714 3183	32.7203 0340	30.7696 5522	28.9971 2399
70	37.4986 1929	35.0820 8492	32.8978 5698	30.9193 7247	29.1234 2135
71	37.7437 4441	35.2881 0261	33.0710 7998	31.0650 8270	29.2460 4015
72	37.9840 6314	35.4895 8691	33.2400 7803	31.2068 9314	29.3650 8752
73	38 2196 6975	35.6866 3756	33.4049 5417	31.3449 0816	29.4806 6750
74	38.4506 5662	35.8793 5214	33.5658 0895	31.4792 2936	29.5928 8107
75	38.6771 1433	36.0678 2605	33.7227 4044	31.6099 5558	29.7018 2628
76	38.8991 3170	36.2521 5262	33.8758 4433	31.7371 8304	29.8075 9833
77	39.1167 9578	36.4324 2310	34.0252 1398	31.8610 0540	29.9102 8964
78	39.3301 9194	36.6087 2675	34.1709 4047	31.9815 1377	30.0099 8994
79	39.5394 0386	36.7811 5085	34.3131 1265	32.0987 9685	30.1067 8635
80	39.7445 1359	36.9497 8079	34.4518 1722	32.2129 4098	30.2007 6345
81	39.9456 0156	37.1147 0004	34.5871 3875	32.3240 3015	30.2920 0335
82	40.1427 4663	37.2759 9026	34.7191 5976	32.4321 4613	30.3805 8577
83	40.3360 2611	37.4337 3130	34.8479 6074	32.5373 6850	30.4665 8813
84	40.5255 1579	37.5880 0127	34.9736 2023	32.6397 7469	30.5500 8556
85	40.7112 8999	37.7388 7655	35.0962 1486	32.7394 4009	30.6311 5103
86	40.8934 2156	37.8864 3183	35.2158 1938	32.8364 3804	30.7098 5537
87	41.0719 8192	38.0307 4018	35.3325 0671	32.9308 3994	30.7862 6735
88	41.2470 4110	38.1718 7304	35.4463 4801	33.0227 1527	30.8604 5374
89	41.4186 6774	38.3099 0028	35.5574 1260	33.1121 3165	30.9324 7936
90	41.5869 2916	38.4448 9025	35.6657 6848	33.1991 5489	31.0024 0714
91	41.7518 9133	38.5769 0978	35.7714 8144	33.2838 4905	31.0702 9820
92	41.9136 1895	38.7060 2423	35.8746 1604	33.3662 7644	31.1362 1184
93	42.0721 7545	38.8322 9754	35.9752 3516	33.4464 9776	31.2002 0567
94	42.2276 2299	38.9557 9221	36.0734 0016	33.5245 7202	31.2623 3560
95	42.3800 2254	39.0765 6940	36.1691 7089	33.6005 5671	31.3226 5592
96	42.5294 3386	39.1946 8890	36.2626 0574	33.6745 0775	31.3812 1934
97	42.6759 1555	39.3102 0920	36.3537 6170	33.7464 7956	31.4380 7703
98	42.8195 2505	39.4231 8748	36.4426 9434	33.8165 2512	31.4932 7867
99	42.9603.1867	39.5336 7968	36.5294 5790	33.8846 9598	31.5468 7250
100	43.0983 5164	39.6417 4052	36.6141 0526	33.9510 4232	31.5989 0534

TABLE IV : PRESENT VALUE OF AN ANNUITY

$$a_{\overline{n}|i} = \frac{1-(1+i)^{-n}}{i}$$

Periods	Rate i				
n	.035 $\left(3\frac{1}{2}\%\right)$	.04(4%)	.045 $\left(4\frac{1}{2}\%\right)$	.05 (5%)	.055 $\left(5\frac{1}{2}\%\right)$
1	0.9661 8357	0.9615 3846	0.9569 3780	0.9523 8095	0.9478 6730
2	1.8996 9428	1.8860 9467	1.8726 6775	1.8594 1043	1.8463 1971
3	2.8016 3698	2.7750 9103	2.7489 6435	2.7232 4803	2.6979 3338
4	3.6730 7921	3.6298 9522	3.5875 2570	3.5459 5050	3.5051 5012
5	4.5150 5238	4.4518 2233	4.3899 7674	4.3294 7667	4.2702 8448
6	5.3285 5302	5.2421 3686	5.1578 7248	5.0756 9207	4.9955 3031
7	6.1145 4398	6.0020 5467	5.8927 0094	5.7863 7340	5.6829 6712
8	6.8739 5554	6.7327 4487	6.5958 8607	6.4632 1276	6.3345 6599
9	7.6076 8651	7.4353 3161	7.2687 9050	7.1078 2168	6.9521 9525
10	8.3166 0532	8.1108 9578	7.9127 1818	7.7217 3493	7.5376 2583
11	9.0015 5104	8.7604 7671	8.5289 1692	8.3064 1422	8.0925 3633
12	9.6633 3433	9.3850 7376	9.1185 8078	8.8632 5164	8.6185 1785
13	10.3027 3849	9.9856 4785	9.6828 5242	9.3935 7299	9.1170 7853
14	10.9205 2028	10.5631 2293	10.2228 2528	9.8986 4094	9.5896 4790
15	11.5174 1090	11.1183 8743	10.7395 4573	10.3796 5804	10.0375 8094
16	12.0941 1681	11.6522 9561	11.2340 1505	10.8377 6956	10.4621 6203
17	12.6513 2059	12.1656 6885	11.7071 9143	11.2740 6625	10.8646 0856
18	13.1896 8173	12.6592 9697	12.1599 9180	11.6895 8690	11.2460 7447
19	13.7098 3742	13.1339 3940	12.5932 9359	12.0853 2086	11.6076 5352
20	14.2124 0330	13.5903 2634	13.0079 3645	12.4622 1034	11.9503 8248
21	14.6979 7420	14.0291 5995	13.4047 2388	12.8211 5271	12.2752 4406
22	15.1671 2484	14.4511 1533	13.7844 2476	13.1630 0258	12.5831 6973
23	15.6204 1047	14.8568 4167	14.1477 7489	13.4885 7388	12.8750 4239
24	16.0583 6760	15.2469 6314	14.4954 7837	13.7986 4179	13.1516 9895
25	16.4815 1459	15.6220 7994	14.8282 0896	14.0939 4457	13.4139 3266
26	16.8903 5226	15.9827 6918	15.1466 1145	14.3751 8530	13.6624 9541
27	17.2853 6451	16.3295 8575	15.4513 0282	14.6430 3362	13.8980 9991
28	17.6670 1885	16.6630 6322	15.7428 7351	14.8981 2726	14.1214 2172
29	18.0357 6700	16.9837 1463	16.0218 8853	15.1410 7358	14.3331 0116
30	18.3920 4541	17.2920 3330	16.2888 8854	15.3724 5103	14.5337 4517
31	18.7362 7576	17.5884 9356	16.5443 9095	15.5928 1050	14.7239 2907
32	19.0688 6547	17.8735 5150	16.7888 9086	15.8026 7667	14.9041 9817
33	19.3902 0818	18.1476 4567	17.0228 6207	16.0025 4921	15.0750 6936
34	19.7006 8423	18.4111 9776	17.2467 5796	16.1929 0401	15.2370 3257
35	20.0006 6110	18.6646 1323	17.4610 1240	16.3741 9429	15.3905 5220
36	20.2904 9381	18.9082 8195	17.6660 4058	16.5468 5171	15.5360 6843
37	20.5705 2542	19.1425 7880	17.8622 3979	16.7112 8734	15.6739 9851
38	20.8410 8736	19.3678 6423	18.0499 9023	16.8678 9271	15.8047 3793
39	21.1024 9987	19.5844 8484	18.2296 5572	17.0170 4067	15.9286 6154
40	21.3550 7234	19.7927 7388	18.4015 8442	17.1590 8635	16.0461 2469
41	21.5991 0371	19.9930 5181	18.5661 0949	17.2943 6796	16.1574 6416
42	21.8348 8281	20.1856 2674	18.7235 4975	17.4232 0758	16.2629 9920
43	22.0626 8870	20.3707 9494	18.8742 1029	17.5459 1198	16.3630 3242
44	22.2827 9102	20.5488 4129	19.0183 8305	17.6627 7331	16.4578 5063
45	22.4954 5026	20.7200 3970	19.1563 4742	17.7740 6982	16.5477 2572
46	22.7009 1813	20.8846 5356	19.2883 7074	17.8800 6650	16.6329 1537
47	22.8994 3780	21.0429 3612	19.4147 0884	17.9810 1571	16.7136 6386
48	23.0912 4425	21.1951 3088	19.5356 0654	18.0771 5782	16.7902 0271
49	23.2765 6450	21.3414 7200	19.6512 9813	18.1687 2173	16.8627 5139
50	23.4556 1787	21.4821 8462	19.7620 0778	18.2559 2546	16.9315 1790

TABLE IV : PRESENT VALUE OF AN ANNUITY

$$a_{\overline{n}|i} = \frac{1-(1+i)^{-n}}{i}$$

Periods n	Rate i .06 (6%)	.065 $\left(6\frac{1}{2}\%\right)$	.07 (7%)	.075 $\left(7\frac{1}{2}\%\right)$	.08 (8%)
1	0.9433 9623	0.9389 6714	0.9345 7944	0.9302 3256	0.9259 2593
2	1.8333 9267	1.8206 2642	1.8080 1817	1.7955 6517	1.7832 6475
3	2.6730 1195	2.6484 7551	2.6243 1604	2.6005 2574	2.5770 9699
4	3.4651 0561	3.4257 9860	3.3872 1126	3.3493 2627	3.3121 2684
5	4.2123 6379	4.1556 7944	4.1001 9744	4.0458 8490	3.9927 1004
6	4.9173 2433	4.8410 1356	4.7665 3966	4.6938 4642	4.6228 7080
7	5.5823 8144	5.4845 1977	5.3892 8940	5.2966 0132	5.2063 7000
8	6.2097 9381	6.0887 5096	5.9712 9851	5.8573 0355	5.7466 3894
9	6.8016 9227	6.6561 0419	6.5152 3225	6.3788 8703	6.2468 8791
10	7.3600 8705	7.1888 3022	7.0235 8154	6.8640 8096	6.7100 8140
11	7.8868 7458	7.6890 4246	7.4986 7434	7.3154 2415	7.1389 6426
12	8.3838 4394	8.1587 2532	7.9426 8630	7.7352 7827	7.5360 7802
13	8.8526 8296	8.5997 4208	8.3576 5074	8.1258 4026	7.9037 7594
14	9.2949 8393	9.0138 4233	8.7454 6799	8.4891 5373	8.2442 3698
15	9.7122 4899	9.4026 6885	9.1079 1401	8.8271 1975	8.5594 7869
16	10.1058 9527	9.7677 6418	9.4466 4860	9.1415 0674	8.8513 6916
17	10.4772 5969	10.1105 7670	9.7632 2299	9.4339 5976	9.1216 3811
18	10.8276 0348	10.4324 6638	10.0590 8691	9.7060 0908	9.3718 8714
19	11.1581 1649	10.7347 1022	10.3355 9524	9.9590 7821	9.6035 9920
20	11.4699 2122	11.0185 0725	10.5940 1425	10.1944 9136	9.8181 4741
21	11.7640 7662	11.2849 8333	10.8355 2733	10.4134 8033	10.0168 0316
22	12.0415 8172	11.5351 9562	11.0612 4050	10.6171 9101	10.2007 4366
23	12.3033 7898	11.7701 3673	11.2721 8738	10.8066 8931	10.3710 5895
24	12.5503 5753	11.9907 3871	11.4693 3400	10.9829 6680	10.5287 5828
25	12.7833 5616	12.1978 7673	11.6535 8318	11.1469 4586	10.6747 7619
26	13.0031 6619	12.3923 7251	11.8257 7867	11.2994 8452	10.8099 7795
27	13.2105 3414	12.5749 9766	11.9867 0904	11.4413 8095	10.9351 6477
28	13.4061 6428	12.7464 7668	12.1371 1125	11.5733 7763	11.0510 7849
29	13.5907 2102	12.9074 8984	12.2776 7407	11.6961 6524	11.1584 0601
30	13.7648 3115	13.0586 7591	12.4090 4118	11.8103 8627	11.2577 8334
31	13.9290 8599	13.2006 3465	12.5318 1419	11.9166 3839	11.3497 9939
32	14.0840 4339	13.3339 2925	12.6465 5532	12.0154 7757	11.4349 9944
33	14.2302 2961	13.4590 8850	12.7537 9002	12.1074 2099	11.5138 8837
34	14.3681 4114	13.5766 0892	12.8540 0936	12.1929 4976	11.5869 3367
35	14.4982 4636	13.6869 5673	12.9476 7230	12.2725 1141	11.6545 6822
36	14.6209 8713	13.7905 6970	13.0352 0776	12.3465 2224	11.7171 9279
37	14.7367 8031	13.8878 5887	13.1170 1660	12.4153 6952	11.7751 7851
38	14.8460 1916	13.9792 1021	13.1934 7345	12.4794 1351	11.8288 6899
39	14.9490 7468	14.0649 8611	13.2649 2846	12.5389 8931	11.8785 8240
40	15.0462 9687	14.1455 2687	13.3317 0884	12.5944 0866	11.9246 1333
41	15.1380 1592	14.2211 5199	13.3941 2041	12.6459 6155	11.9672 3457
42	15.2245 4332	14.2921 6149	13.4524 4898	12.6939 1772	12.0066 9867
43	15.3061 7294	14.3588 3708	13.5069 6167	12.7385 2811	12.0432 3951
44	15.3831 8202	14.4214 4327	13.5579 0810	12.7800 2615	12.0770 7362
45	15.4558 3209	14.4802 2842	13.6055 2159	12.8186 2898	12.1084 0150
46	15.5243 6990	14.5354 2575	13.6500 2018	12.8545 3858	12.1374 0880
47	15.5890 2821	14.5872 5422	13.6916 0764	12.8879 4287	12.1642 6741
48	15.6500 2661	14.6359 1946	13.7304 7443	12.9190 1662	12.1891 3649
49	15.7075 7227	14.6818 1451	13.7667 9853	12.9479 2244	12.2121 6341
50	15.7618 6064	14.7245 2067	13.8007 4629	12.9748 1157	12.2334 8464

TABLE V : TABLE OF e^x and e^{-x}

x	e^x	e^{-x}
0.00	1.0000	1.00000
0.01	1.0101	0.99005
0.02	1.0202	.98020
0.03	1.0305	.97045
0.04	1.0408	.96079
0.05	1.0513	.95123
0.06	1.0618	.94176
0.07	1.0725	.93239
0.08	1.0833	.92312
0.09	1.0942	.91393
0.10	1.1052	.90484
0.11	1.1163	.89583
0.12	1.1275	.88692
0.13	1.1388	.87809
0.14	1.1503	.86936
0.15	1.1618	.86071
0.16	1.1735	.85214
0.17	1.1853	.84366
0.18	1.1972	.83527
0.19	1.2092	.82696
0.20	1.2214	.81873
0.21	1.2337	.81058
0.22	1.2461	.80252
0.23	1.2586	.79453
0.24	1.2712	.78663
0.25	1.2840	.77880
0.26	1.2969	.77105
0.27	1.3100	.76338
0.28	1.3231	.75578
0.29	1.3364	.74826
0.30	1.3499	.74082
0.31	1.3634	.73345
0.32	1.3771	.72615
0.33	1.3910	.71892
0.34	1.4049	.71177
0.35	1.4191	.70469
0.36	1.4333	.69768
0.37	1.4477	.69073
0.38	1.4623	.68386
0.39	1.4770	.67706
0.40	1.4918	.67032
0.41	1.5068	.66365
0.42	1.5220	.65705
0.43	1.5373	.65051
0.44	1.5527	.64404
0.45	1.5683	.63763
0.46	1.5841	.63128
0.47	1.6000	.62500
0.48	1.6161	.61878
0.49	1.6323	.61263
0.50	1.6487	.60653
0.51	1.6653	.60050
0.52	1.6820	.59452
0.53	1.6989	.58860

TABLE V : TABLE OF e^x and e^{-x}

x	e^x	e^{-x}
0.54	1.7160	.58275
0.55	1.7333	.57695
0.56	1.7507	.57121
0.57	1.7683	.56553
0.58	1.7860	.55990
0.59	1.8040	.55433
0.60	1.8221	.54881
0.61	1.8404	.54335
0.62	1.8589	.53794
0.63	1.8776	.53259
0.64	1.8965	.52729
0.65	1.9155	.52205
0.66	1.9348	.51685
0.67	1.9542	.51171
0.68	1.9739	.50662
0.69	1.9937	.50158
0.70	2.0138	.49659
0.71	2.0340	.49164
0.72	2.0544	.48675
0.73	2.0751	.48191
0.74	2.0959	.47711
0.75	2.1170	.47237
0.76	2.1383	.46767
0.77	2.1598	.46301
0.78	2.1815	.45841
0.79	2.2034	.45384
0.80	2.2255	.44933
0.81	2.2479	.44486
0.82	2.2705	.44043
0.83	2.2933	.43605
0.84	2.3164	.43171
0.85	2.3396	.42741
0.86	2.3632	.42316
0.87	2.3869	.41895
0.88	2.4109	.41478
0.89	2.4351	.41066
0.90	2.4596	.40657
0.91	2.4843	.40252
0.92	2.5093	.39852
0.93	2.5345	.39455
0.94	2.5600	.39063
0.95	2.5857	.38674
0.96	2.6117	.38289
0.97	2.6379	.37908
0.98	2.6645	.37531
0.99	2.6912	.37158
1.00	2.7183	.36788
1.10	3.0042	.33287
1.20	3.3201	.30119
1.30	3.6693	.27253
1.40	4.0552	.24660
1.50	4.4817	.22313
1.60	4.9530	.20190
1.70	5.4739	.18268

TABLE V : TABLE OF e^x and e^{-x}

x	e^x	e^{-x}
1.80	6.0496	.16530
1.90	6.6859	.14957
2.00	7.3891	.13534
2.10	8.1662	.12246
2.20	9.2050	.11080
2.30	9.9742	.10026
2.40	11.023	.09072
2.50	12.182	.08208
2.60	13.464	.07427
2.70	14.880	.06721
2.80	16.445	.06081
2.90	18.174	.05502
3.00	20.086	.04979
3.10	22.198	.04505
3.20	24.533	.04076
3.30	27.113	.03688
3.40	29.964	.03337
3.50	33.115	.03020
3.60	36.598	.02732
3.70	40.447	.02472
3.80	44.701	.02237
3.90	49.402	.02024
4.00	54.598	.01832
4.10	60.340	.01657
4.20	66.686	.01500
4.30	73.700	.01357
4.40	81.451	.01227
4.50	90.107	.01111
4.60	99.484	.01005
4.70	109.95	.00910
4.80	121.51	.00823
4.90	134.29	.00745
5.00	148.41	.00674
5.10	164.02	.00610
5.20	181.27	.00552
5.30	200.34	.00499
5.40	221.41	.00452
5.50	244.69	.00409
5.60	270.43	.00370
5.70	298.87	.00335
5.80	330.30	.00303
5.90	365.04	.00274
6.00	403.43	.00248
6.25	518.01	.00193
6.50	665.14	.00150
6.75	854.06	.00117
7.00	1096.6	.00091
7.50	1808.0	.00055
8.00	2981.0	.00034
8.50	4914.8	.00020
9.00	8103.1	.00012
9.50	13360	.00007
10.00	22026	.000050

TABLE VI : LOGARITHMS

	0	1	2	3	4	5	6	7	8	9	1	2	3	4	5	6	7	8	9
10	.0000	0043	0086	0120	0170	0212	0253	0294	0334	0374	4	8	12	17	21	25	29	33	37
11	.0414	0453	0492	0531	0569	0607	0645	0682	0719	0755	4	8	11	15	19	23	26	30	34
12	.0792	0828	0864	0899	0934	0969	1004	1038	1072	1106	3	7	10	14	17	21	24	28	31
13	.1139	1173	1206	1239	1271	1303	1335	1367	1399	1430	3	6	10	13	16	19	23	26	29
14	.1461	1492	1523	1553	1584	1614	1644	1673	1703	1732	3	6	9	12	15	18	21	24	27
15	1761	1790	1818	1847	1875	1903	1931	1959	1987	2014	3	6	8	11	14	17	20	22	25
16	.2041	2068	2095	2122	2148	2175	2201	2227	2253	2279	3	5	8	11	13	16	18	21	24
17	.2304	2330	2355	2380	2405	2430	2455	2480	2504	2529	2	5	7	10	12	15	17	20	22
18	.2553	2577	2601	2625	2648	2672	2695	2718	2742	2765	2	5	7	9	12	14	16	19	21
19	.2788	2810	2833	2856	2878	2900	2923	2945	2967	2989	2	4	7	9	11	13	16	18	20
20	.3010	3032	3054	3075	3096	3118	3139	3160	3181	3201	2	4	6	8	11	13	15	17	19
21	.3222	3243	3263	3284	3304	3324	3345	3365	3385	3404	2	4	6	8	10	12	14	16	18
22	.3424	3444	3464	3483	3502	3522	3541	3562	3579	3598	2	4	6	8	10	12	14	15	17
23	.3617	3636	3655	3674	3692	3711	3729	3747	3766	3784	2	4	6	7	9	11	13	15	17
24	.3802	3820	3838	3856	3874	3892	3909	3927	3945	3962	2	4	5	7	9	11	12	14	16
25	.3979	3997	4014	4031	4048	4065	4082	4099	4116	4133	2	3	5	7	9	10	12	14	15
26	.4150	4166	4183	4200	4216	4232	4249	4265	4281	4298	2	3	5	7	8	10	11	13	15
27	.4314	4330	4346	4362	4378	4393	4409	4425	4440	4456	2	3	5	6	8	9	11	13	14
28	.4472	4487	4502	4518	4533	4548	4564	4579	4594	4609	2	3	5	6	8	9	11	12	14
29	.4624	4639	4654	4669	4683	4698	4713	4728	4742	4757	1	3	4	6	7	9	10	12	13
30	.4771	4786	4800	4814	4829	4843	4857	4871	4886	4900	1	3	4	6	7	9	10	11	13
31	.4914	4928	4942	4955	4969	4983	4997	5011	5024	5038	1	3	4	6	7	8	10	11	12
32	.5051	5065	5079	5092	5105	5119	5132	5145	5159	5172	1	3	4	5	7	8	9	11	12
33	.5185	5198	5211	5224	5237	5250	5263	5276	5289	5302	1	3	4	5	6	8	9	10	12
34	.5315	5315	5340	5353	5366	5378	5391	5403	5416	5428	1	3	4	5	6	8	9	10	11
35	.5441	5453	5465	5478	5490	5502	5514	5527	5539	5551	1	2	4	5	6	7	9	10	11
36	.5563	5575	5587	5599	5611	5623	5635	5647	5658	5670	1	2	4	5	6	7	8	10	11
37	.5682	5694	5705	5717	5729	5740	5752	5763	5775	5786	1	2	3	5	6	7	8	9	10
38	.5798	5809	5821	5832	5843	5855	5866	5877	5888	5899	1	2	3	5	6	7	8	9	10
39	.5911	5922	5933	5944	5955	5966	5977	5988	5999	6010	1	2	3	4	5	7	8	9	10
40	.6021	6031	6042	6053	6065	6075	6085	6096	6107	6117	1	2	3	4	5	6	8	9	10
41	.6128	6138	6149	6160	6170	6180	6191	6201	6212	6222	1	2	3	4	5	6	7	8	9
42	.6232	6243	6253	6263	6274	6284	6294	6304	6314	6325	1	2	3	4	5	6	7	8	9
43	.6335	6345	6355	6365	6375	6385	6395	6405	6415	6425	1	2	3	4	5	6	7	8	9
44	.6435	6444	6454	6465	6474	6484	6493	6503	6513	6522	1	2	3	4	5	6	7	8	9
45	.6532	6542	6551	6561	6571	6580	6590	6599	6609	6618	1	2	3	4	5	6	7	8	9
46	.6628	6637	6646	6656	6665	6675	6684	6693	6702	6712	1	2	3	4	5	6	7	7	8
47	.6721	6730	6739	6749	6758	6767	6776	6785	6794	6803	1	2	3	4	5	5	6	7	8
48	.6812	6821	6830	6839	6848	6857	6866	6875	6884	6893	1	2	3	4	4	5	6	7	8
49	.6902	6911	6920	6928	6937	6946	6955	6964	6972	6981	1	2	3	4	4	5	6	7	8
50	.6990	6993	7007	7016	7024	7033	7042	7050	7059	7067	1	2	3	3	4	5	6	7	8
51	.7076	7084	7093	7101	7110	7118	7126	7135	7143	7152	1	2	3	3	4	5	6	7	8
52	.7160	7168	7177	7185	7193	7202	7210	7218	7226	7235	1	2	2	3	4	5	6	7	7
53	.7243	7251	7259	7267	7275	7284	7292	7300	7308	7316	1	2	2	3	4	5	6	6	7

TABLE VI : LOGARITHMS

	0	1	2	3	4	5	6	7	8	9	1	2	3	4	5	6	7	8	9
54	.7324	7332	7340	7348	7356	7364	7372	7380	7388	7396	1	2	2	3	4	5	6	6	7
55	.7404	7412	7419	7427	7435	7443	7451	7459	7466	7474	1	2	2	3	4	5	5	6	7
56	.7482	7490	7497	7505	7513	7520	7528	7536	7543	7551	1	2	2	3	4	5	5	6	7
57	.7559	7566	7574	7582	7589	7597	7604	7612	7619	7627	1	2	2	3	4	5	5	6	7
58	.7634	7642	7649	7657	7664	7672	7679	7686	7694	7701	1	1	2	3	4	4	5	6	7
59	.7709	7716	7723	7731	7738	7745	7752	7760	7767	7774	1	1	2	3	4	4	5	6	7
60	.7782	7789	7796	7803	7810	7818	7825	7832	7839	7846	1	1	2	3	4	4	5	6	6
61	.7853	7860	7868	7875	7882	7889	7896	7903	7910	7917	1	1	2	3	4	4	5	6	6
62	.7924	7931	7938	7945	7952	7959	7966	7973	7980	7987	1	1	2	3	3	4	5	6	6
63	.7993	8000	8007	8014	8021	8028	8035	8041	8048	8055	1	1	2	3	3	4	5	5	6
64	.8062	8069	8075	8082	8989	8096	8102	8109	8116	8122	1	1	2	3	3	4	5	5	6
65	.8129	8136	8142	8149	8156	8162	8169	8176	8182	8189	1	1	2	3	3	4	5	5	6
66	.8195	8202	8209	8215	8222	8228	8235	8241	8248	8254	1	1	2	3	3	4	5	5	6
67	.8261	8267	8274	8280	8287	8293	8299	8306	8312	8319	1	1	2	3	3	4	5	5	6
68	.8325	8331	8338	8344	8351	8357	8363	8370	8376	8382	1	1	2	3	3	4	4	5	6
69	.8388	8395	8401	8407	8414	8420	8426	8432	8439	8445	1	1	2	2	3	4	4	5	6
70	.8451	8457	8463	8470	8476	8482	8488	8494	8500	8506	1	1	2	2	3	4	4	5	6
71	.8513	8519	8525	8531	8537	8543	8549	8555	8561	8567	1	1	2	2	3	4	4	5	5
72	.8573	8579	8585	8591	8597	8603	8609	8615	8621	8627	1	1	2	2	3	4	4	5	5
73	.8633	8639	8645	8651	8657	8663	8669	8675	8681	8686	1	1	2	2	3	4	4	5	5
74	.8692	8698	8704	8710	8716	8722	8727	8733	8739	8745	1	1	2	2	3	4	4	5	5
75	.8751	8756	8762	8768	8774	8779	8785	8791	8797	8802	1	1	2	2	3	3	4	5	5
76	.8808	8814	8820	8825	8831	8837	8842	8848	8854	8859	1	1	2	2	3	3	4	5	5
77	.8865	8871	8876	8882	8887	8893	8899	8904	8910	8915	1	1	2	2	3	3	4	4	5
78	.8921	8927	8932	8938	8943	8949	8954	8960	8965	8971	1	1	2	2	3	3	4	4	5
79	.8976	8982	8987	8993	8998	9004	9009	9015	9020	9025	1	1	2	2	3	3	4	4	5
80	.9031	9036	9042	9047	9053	9058	6063	9069	9074	9079	1	1	2	2	3	3	4	4	5
81	.9085	9090	9096	9101	9106	9112	9117	9122	9128	9133	1	1	2	2	3	3	4	4	5
82	.9138	9143	9149	9154	9159	9165	9170	9175	9180	9186	1	1	2	2	3	3	4	4	5
83	.9191	9196	9201	9206	9212	9217	9222	9227	9232	9238	1	1	2	2	3	3	4	4	5
84	.9243	9248	9253	9258	9263	9269	9274	9279	9284	9289	1	1	2	2	3	3	4	4	5
85	.9294	9299	9304	9309	9315	9320	9325	9330	9335	9340	1	1	2	2	3	3	4	4	5
86	.9345	9350	9355	9360	9365	9370	9375	9380	9385	9390	1	1	2	2	3	3	4	4	5
87	.9395	9400	9405	9410	9415	9420	9425	9430	9435	9440	0	1	1	2	2	3	3	4	4
88	.9445	9450	9455	9460	9465	9469	9474	9479	9484	9489	0	1	1	2	2	3	3	4	4
89	.9494	9499	9504	9509	9513	9518	9523	9528	9533	9538	0	1	1	2	2	3	3	4	4
90	.9542	9547	9552	9557	9562	9566	9571	9576	9581	9586	0	1	1	2	2	3	3	4	4
91	.9590	9595	9600	9605	9609	9614	9619	9624	9628	9933	0	1	1	2	2	3	3	4	4
92	.9638	9643	9647	9652	9657	9661	9666	9671	9675	9680	0	1	1	2	2	3	3	4	4
93	.9685	9689	9694	9699	9703	9708	9713	9717	9722	9727	0	1	1	2	2	3	3	4	4
94	.9731	9736	9741	9745	9750	9754	9759	9763	9768	9773	0	1	1	2	2	3	3	4	4
95	.9777	9782	9786	9791	9795	9800	9805	9809	9814	9818	0	1	1	2	2	3	3	4	4
96	.9823	9827	9832	9836	9841	9845	9850	9854	9859	9863	0	1	1	2	2	3	3	4	4
97	.9868	9872	9877	9881	9886	9890	9894	9899	9903	9908	0	1	1	2	2	3	3	4	4
98	.9912	9917	9921	9926	9930	9934	9939	9943	9948	9952	0	1	1	2	2	3	3	4	4
99	.9956	9961	9965	9969	9974	9978	9983	9987	9991	9996	0	1	1	2	2	3	3	3	4

TABLE VII : ANTILOGARITHMS

	0	1	2	3	4	5	6	7	8	9	1	2	3	4	5	6	7	8	9
.00	1000	1002	1005	1007	1009	1012	1014	1016	1019	1021	0	0	1	1	1	1	2	2	2
.01	1023	1026	1028	1030	1033	1035	1038	1040	1042	1045	0	0	1	1	1	1	2	2	2
.02	1047	1050	1052	1054	1057	1059	1062	1064	1067	1069	0	0	1	1	1	1	2	2	2
.03	1072	1074	1076	1079	1081	1084	1086	1089	1091	1094	0	0	1	1	1	1	2	2	2
.04	1096	1099	1102	1104	1107	1109	1112	1114	1117	1119	0	1	1	1	1	2	2	2	2
.05	1122	1125	1127	1130	1132	1135	1138	1140	1143	1146	0	1	1	1	1	2	2	2	2
.06	1148	1151	1153	1156	1159	1161	1164	1167	1169	1172	0	1	1	1	1	2	2	2	2
.07	1175	1178	1180	1183	1186	1189	1191	1194	1197	1199	0	1	1	1	1	2	2	2	2
.08	1202	1205	1208	1211	1213	1216	1219	1222	1225	1227	0	1	1	1	1	2	2	2	3
.09	1230	1233	1236	1239	1242	1245	1247	1250	1253	1256	0	1	1	1	1	2	2	2	3
.10	1259	1262	1265	1268	1271	1274	1276	1279	1282	1285	0	1	1	1	1	2	2	2	3
.11	1288	1291	1294	1297	1300	1303	1306	1309	1312	1315	0	1	1	1	2	2	2	2	3
.12	1318	1321	1324	1327	1330	1334	1337	1340	1343	1346	0	1	1	1	2	2	2	2	3
.13	1349	1352	1355	1358	1361	1365	1368	1371	1374	1377	0	1	1	1	2	2	2	3	3
.14	1380	1384	1387	1390	1393	1396	1400	1403	1406	1409	0	1	1	1	2	2	2	3	3
.15	1413	1416	1419	1422	1426	1429	1432	1435	1439	1442	0	1	1	1	2	2	2	3	3
.16	1445	1449	1452	1455	1459	1462	1466	1469	1472	1476	0	1	1	1	2	2	2	3	3
.17	1479	1483	1486	1489	1493	1496	1500	1503	1507	1510	0	1	1	1	2	2	2	3	3
.18	1514	1517	1521	1524	1528	1531	1535	1538	1542	1545	0	1	1	1	2	2	2	3	3
.19	1549	1552	1556	1560	1563	1567	1570	1574	1578	1581	0	1	1	1	2	2	3	3	3
.20	1585	1589	1592	1596	1600	1603	1607	1611	1614	1618	0	1	1	1	2	2	3	3	3
.21	1622	1626	1629	1633	1637	1641	1644	1648	1652	1656	0	1	1	2	2	2	3	3	3
.22	1660	1663	1667	1671	1675	1679	1683	1687	1690	1694	0	1	1	2	2	2	3	3	3
.23	1698	1702	1706	1710	1714	1718	1722	1726	1730	1734	0	1	1	2	2	2	3	3	4
.24	1738	1742	1746	1750	1754	1758	1762	1766	1770	1774	0	1	1	2	2	2	3	3	4
.25	1778	1782	1786	1791	1795	1799	1803	1807	1811	1816	0	1	1	2	2	2	3	3	4
.26	1820	1824	1828	1832	1837	1841	1845	1849	1854	1858	0	1	1	2	2	3	3	3	4
.27	1862	1866	1871	1875	1879	1884	1888	1892	1897	1901	0	1	1	2	2	3	3	3	4
.28	1905	1910	1914	1919	1923	1928	1932	1936	1941	1945	0	1	1	2	2	3	3	4	4
.29	1950	1954	1959	1963	1968	1972	1977	1982	1986	1991	0	1	1	2	2	3	3	4	4
.30	1995	2000	2004	2009	2014	2018	2023	2028	2032	2037	0	1	1	2	2	3	3	4	4
.31	2042	2046	2051	2056	2061	2065	2070	2075	2080	2084	0	1	1	2	2	3	3	4	4
.32	2089	2094	2099	2104	2109	2113	2118	2123	2128	2133	0	1	1	2	2	3	3	4	4
.33	2138	2143-	2148	2153	2158	2163	2168	2173	2178	2183	0	1	1	2	2	3	3	4	4
.34	2188	2193	2198	2203	2208	2213	2218	2223	2228	2234	1	1	2	2	3	3	4	4	5
.35	2239	2244	2249	2254	2259	2265	2270	2275	2280	2286	1	1	2	2	3	3	4	4	5
.36	2291	2296	2301	2307	2312	2317	2323	2328	2333	2339	1	1	2	2	3	3	4	4	5
.37	2344	2350	2355	2360	2366	2371	2377	2382	2388	2393	1	1	2	2	3	3	4	4	5
.38	2399	2404	2410	2415	2421	2427	2432	2438	2443	2449	1	1	2	2	3	3	4	4	5
.39	2455	2460	2466	2472	2477	2483	2489	2495	2500	2506	1	1	2	2	3	3	4	5	5
.40	2513	2518	2523	2529	2535	2541	2547	2553	2559	2564	1	1	2	2	3	4	4	5	5
.41	2570	2576	2582	2588	2594	2600	2606	2612	2618	2624	1	1	2	2	3	4	4	5	5
.42	2630	2636	2642	2649	2655	2661	2667	2673	2679	2685	1	1	2	2	3	4	4	5	6
.43	2692	2698	2704	2710	2716	2723	2729	2735	2742	2748	1	1	2	3	3	4	4	5	6
.44	2754	2761	2767	2773	2780	2786	2793	2799	2805	2812	1	1	2	3	3	4	4	5	6
.45	2818	2825	2831	2838	2844	2851	2858	2864	2871	2877	1	1	2	3	3	4	5	5	6
.46	2884	2891	2897	2904	2911	2917	2924	2931	2938	2944	1	1	2	3	3	4	5	5	6
.47	2951	2958	2965	2972	2979	2985	2992	2999	3006	3013	1	1	2	3	3	4	5	5	6
.48	3020	3027	3034	3041	3048	3055	3062	3069	3076	3083	1	1	2	3	4	4	5	6	6
.49	3090	3097	3105	3112	3119	3126	3133	3141	3148	3155	1	1	2	3	4	4	5	6	6

TABLE VII : ANTILOGARITHMS

	0	1	2	3	4	5	6	7	8	9	1	2	3	4	5	6	7	8	9
.50	3162	3170	3177	3184	3192	3199	3206	3228	3221	3228	1	1	2	3	4	4	5	6	7
.51	3236	3243	3251	3258	3266	3273	3281	3289	3296	3304	1	2	2	3	4	5	5	6	7
.52	3311	3319	3327	3334	3342	3350	3357	3365	3373	3381	1	2	2	3	4	5	5	6	7
.53	3388	3396	3404	3412	3420	3428	3436	3443	3451	3459	1	2	2	3	4	5	6	6	7
.54	3467	3475	3483	3491	3499	3508	3516	3524	3532	3540	1	2	2	3	4	5	6	6	7
.55	3548	3556	3565	3573	3581	3589	3597	3606	3614	3622	1	2	2	3	4	5	6	7	7
.56	3631	3639	3648	3656	3664	3673	3681	3690	3698	3707	1	2	3	3	4	5	6	7	8
.57	3715	3724	3733	3741	3750	3758	3767	3776	3784	3793	1	2	3	3	4	5	6	7	8
.58	3802	3811	3819	3828	3837	3846	3855	3864	3873	3882	1	2	3	4	4	5	6	7	8
.59	3890	3899	3908	3917	3926	3936	3945	3954	3963	3972	1	2	3	4	5	5	6	7	8
.60	3981	3990	3999	4009	4018	4027	4036	4046	4055	4064	1	2	3	4	5	6	6	7	8
.61	14074	4083	4093	4102	4111	4121	4130	4140	4150	4159	1	2	3	4	5	6	7	8	9
.62	4169	4178	4188	4198	4207	4217	4227	4236	4246	4256	1	2	3	4	5	6	7	8	9
.63	4266	4276	4285	4295	4305	4315	4325	4335	4345	4355	1	2	3	4	5	6	7	8	9
.64	4365	4375	4385	4395	4406	4416	4426	4436	4446	4457	1	2	3	4	5	6	7	8	9
.65	4467	4477	4487	4498	4508	4519	4529	4539	4550	4560	1	2	3	4	5	6	7	8	9
.66	4571	4581	4592	4603	4613	4624	4634	4645	4656	4667	1	2	3	4	5	6	7	9	10
.67	4677	4688	4699	4710	4721	4732	4742	4753	4764	4775	1	2	3	4	5	7	8	9	10
.68	4786	4797	4808	4819	4831	4842	4853	4864	4875	4887	1	2	3	4	6	7	8	9	10
.69	4898	4909	4920	4932	4943	4955	4966	4977	4989	5000	1	2	3	5	6	7	8	9	10
.70	5012	5023	5035	5047	5058	5070	5082	5093	5105	5117	1	2	4	5	6	7	8	9	11
.71	5129	5140	5152	5164	5176	5188	5200	5212	5224	5236	1	2	4	5	6	7	8	10	11
.72	5248	5260	5272	5284	5297	5309	5321	5333	5346	5358	1	2	4	5	6	7	9	10	11
.73	5370	5383	5395	5408	5420	5433	5445	5458	5470	5483	1	3	4	5	6	8	9	10	11
.74	5495	5508	5521	5534	5546	5559	5572	5585	5598	5610	1	3	4	5	6	8	9	10	12
.75	5623	5636	5649	5662	5675	5689	5702	5715	5728	5741	1	3	4	5	7	8	9	10	12
.76	5754	5768	5781	5794	5808	5821	5834	5848	5861	5875	1	3	4	5	7	8	9	11	12
.77	5888	5902	5916	5929	5943	5957	5970	5984	5998	6012	1	3	4	5	7	8	10	11	12
.78	6026	6039	6053	6067	6081	6095	6109	6124	6138	6152	1	3	4	6	7	8	10	11	13
.79	6166	6180	6194	6209	6223	6237	6252	6266	6281	6295	1	3	4	6	7	9	10	11	13
.80	6310	6324	6339	6353	6368	6383	6397	6412	6427	6442	1	3	4	6	7	9	10	12	13
.81	6457	6471	6486	6501	6516	6531	6546	6561	6577	6592	2	3	5	6	8	9	11	12	14
.82	6607	6622	6637	6653	6668	6683	6699	6714	6730	6745	2	3	5	6	8	9	11	12	14
.83	6761	6776	6792	6808	6823	6839	6855	6871	6887	6902	2	3	5	6	8	9	11	13	14
.84	6918	6934	6950	6966	6982	6998	7015	7031	7047	7063	2	3	5	6	8	10	11	13	15
.85	7079	7096	7112	7129	7145	7161	7178	7194	7211	7228	2	3	5	7	8	10	12	13	15
.86	7244	7261	7278	7295	7311	7328	7345	7362	7379	7396	2	3	5	7	8	10	12	13	15
.87	7413	7430	7447	7464	7482	7499	7516	7534	7551	7568	2	3	5	7	9	10	12	14	16
.88	7586	7603	7621	7638	7656	7674	7691	7709	7727	7745	2	4	5	7	9	11	12	14	16
.89	7762	7780	7798	7816	7834	7852	7870	7889	7907	7925	2	4	5	7	9	11	13	14	16
.90	7943	7962	7980	7998	8017	8035	8054	8072	8091	8110	2	4	6	7	9	11	13	15	17
.91	8128	8147	8166	8185	8204	8222	8241	8260	8279	8299	2	4	6	8	9	11	13	15	17
.92	8318	8337	8356	8375	8395	8414	8433	8453	8472	8492	2	4	6	8	10	12	14	15	17
.93	8511	8531	8551	8570	8590	8610	8630	8650	8670	8690	2	4	6	8	10	12	14	16	18
.94	8710	8730	8750	8770	8790	8810	8831	8851	8872	8892	2	4	6	8	10	12	14	16	18
.95	8913	8933	8954	8974	8995	9016	9036	9057	9078	9099	2	4	6	8	10	12	15	17	19
.96	9120	9141	9162	9183	9204	9226	9247	9268	9290	9311	2	4	6	8	11	13	15	17	19
.97	9333	9354	9376	9397	9419	9441	9462	9484	9506	9528	2	4	7	9	11	13	15	17	20
.98	9550	9572	9594	9616	9638	9661	9683	9705	9227	9750	2	4	7	9	11	13	16	18	20
.99	9772	9795	9817	9840	9863	9886	9908	9931	9954	9977	2	5	7	9	11	14	16	18	20

DELHI UNIV. B.COM.(HONS.)/ III SEM./2024
PAPER DSC- 3.1 : BUSINESS MATHEMATICS

***Time* : 3 Hours** ***Maximum Marks* : 90**

Attempt All questions. All questions carry equal marks. Use of Simple Calculator is allowed

1. (*a*) The total sales (S) in thousands of rupees of a firm selling two products *X* and *Y* is given by the relationship, $S = a + bX + cY$. Data for the first three months is given as under :

Month	*Total Sales*	*X*	*Y*
1	18	1	3
2	10	2	1
3	26	5	3

Using determinant method, determine the sales in the month when it sells 10 units of *X* and 20 units of *Y*. (6)

OR

(*a*) Mr. *Y* invested ₹50,000 divided into three investments. Part of the money is invested in a savings account with an annual return of 4%, partly in 9% annual yield bonds and the remainder in business. In 2019, when he lost 2% of the money that he invested in the business, his net income from all the three investments is ₹1,500. If he invested ₹10,000 more in the business than in the savings account, how much was invested in each. Use the matrix algebra. (6)

(*b*) Suppose the inter-industry flow of the products of two industries is given as under :

Production Sector	Consumption Sector		Domestic Demand	Total Output
	X	*Y*		
X	20	90	90	200
Y	30	15	105	150
Labour (hours)	50	30		

(*i*) Determine the technology matrix and test Simon-Hawkins conditions for the viability of the system.

(*ii*) If the domestic demand changes to 144 and 72 units respectively, what should be the gross output of each sector in order to meet the demands?

(*iii*) If the total labour available is 100 hours, is the solution feasible?

(*iv*) Find the value added in each sector.

(*v*) Find the equilibrium prices, if wage rate is ₹150. (12)

OR

(*b*) You are given the following transaction matrix for a two-sector economy :

Purchase Sector	*Sales Sector*		*Final Demand*	*Gross Output*
	I	*II*		
I	4	3	13	20
II	5	4	3	12
Primary Input	11	5		

(*i*) Determine the technology matrix and test Simon-Hawkins conditions for the viability of the system.

(*ii*) Rewrite the new transaction matrix when the final demand for the output of sector I increases to 23 units. (12)

2. (*a*) A publishing house purchases 9000 units of a particular item every year at a unit cost of ₹20. The ordering cost per order is ₹40 and the storage cost is 10% of the value of average inventory. Using calculus, find the optimal order quantity and the minimum total cost. Would you like to accept an offer by a supplier, who proposes to allow a discount of 10% on order size of 1000 units? (9)

OR

(*a*) The demand function for a certain commodity is $5q = 400 - 2p$ $(0 \leq p \leq 200)$.

(*i*) Determine where the demand is elastic, inelastic and of unit elasticity with respect to price.

(*ii*) Use the result of part (*i*) to determine the intervals of increase and decrease of the revenue function and the price at which revenue is maximized. (9)

(*b*) A cable service operator has a subscriber base of 1000 households, from each of whom he charges ₹500 per month. In order to increase his subscriber base, he proposes to decrease the rate of monthly subscriptions. His assessment is that for every decrease of ₹50, 200 additional households would seek connection of cable service. You are required to find out the decrease in service charge that enables him to maximize his total revenue. (9)

OR

(*b*) The cost of fuel in running an engine is proportional to the square of the speed in kms per hour, and is ₹48 per hour when the speed is 16 kms. Other costs amount to ₹300 per hour. Find the most economical speed. (9)

3. (*a*) Assume that a monopolist sells his product in two separate markets and the demand functions are: $p_1 = 12 - x_1$, $p_2 = 20 - 3x_2$. The total cost function for his product is given by: $C = 3 + 2x$, where $x = x_1 + x_2$. Determine the prices that the monopolist will charge and the quantities he will sell in each of the markets to maximize his profit. (9)

OR

(*a*) A production function is given by $Q = AL^{2/7}K^{2/7}$, where L is labour and K is capital.

(*i*) What is nature of returns to scale?

(*ii*) What is nature of returns to inputs?

(*iii*) What is the total reward of labour and capital if each factor is paid a price equal to its marginal product? Show that total product is not exhausted. (9)

(*b*) The elasticity of demand of a commodity with respect to price is calculated to be $\frac{p}{(p-1)(p-2)}$, where p is price. Find the demand function if it is known that the quantity demanded is 4 units at $p = 3$. (9)

OR

(*b*) The demand and supply laws for a commodity are $p_d = 56 - x^2$ and $p_s = 8 + (x^2/3)$. Find the consumer surplus and produce surplus at equilibrium price. (9)

4. Attempt **any three** questions. Each question carries 6 marks.

(*a*) A debt of ₹6000 due 8 years from now and ₹6000 due 12 years from now is to be repaid by a payment of ₹3000 in 3 years and a payment of ₹3000 in 5 years and a final payment at the end of 6 years. If the rate of interest is 9% compounded annually, how much is the final payment?

(*b*) Machine *A* costs ₹8,000 and has a useful life of 7 years. Machine *B* costs ₹6,000 and has a useful life of 5 years. Suppose Machine A generates an annual labour savings of ₹2,000 while Machine *B* generates an annual labour savings of ₹1,800. Assuming that the time value of money of the corporation is 10% per annum, which machine is preferable?

(*c*) Mr. Harshdeep makes 8 annual payments of ₹1,00,000 each, the first being made at the end of 5 years. Find the amount of the annuity if the money is worth 7% compounded annually.

(*d*) A machine depreciates at the rate of 8% of its value at the beginning of a year. The machine was purchased for ₹1,00,000 and the scrap value realized when sold was ₹43,440. Find the number of years the machine was used.

5. Given below are the objective function, the constraints and the final simplex tableau for a linear programming product-mix problem :

Max. $Z = 15x_1 + 12x_2 + 11x_3$

Subject to Constraints:

$3x_1 + 4x_2 + 5x_3 \leq 63$ (hours, Dept. *A*)

$4x_1 + x_2 + 4x_3 \leq 64$ (hours, Dept. *B*)

$x_1 + 4x_2 + 2x_3 \leq 40$ (hours, Dept. *C*)

$x_1, x_2, x_3 \geq 0$

Final Simplex Table :

C_j	Product Mix	15	12	11	0	0	0	Quantity
		x_1	x_2	x_3	S_1	S_2	S_3	
		0	1	8/13	4/13	–3/13	0	60/13
		1	0	11/13	–1/13	4/13	0	193/13
		0	0	–17/13	–15/13	8/13	1	87/13
	Z_j							
	$C_j - Z_j$							

(*i*) Complete the table and test whether the solution is optimal or not.

(*ii*) Write the optimal product mix and the profit contribution shown by the above solution.

(*ii*) Is the solution feasible? Give reason.

(*iv*) Does the problem have any alternative solution? If so, show one such solution.

(*v*) Indicate the shadow prices of three departments.

(*vi*) If the company wishes to expand the production capacity, which of the three departments should be given priority.

(*vii*) If the company produces 13 units of x_3, how many units of x_1 and x_2 shall have to be reduced, if any?

(*viii*) If a customer is prepared to pay higher prices for product x_3, how much should the price be increased so that the company's profit remains unchanged.

(*ix*) Indicate whether the solution given in the table is degenerate. If yes, which variable is degenerated? (18)

OR

(*a*) A 24-hour supermarket has the following minimal requirement for sales personnel :

Period	1	2	3	4	5	6
Time of day (24 hr clock)	04-08	08-12	12-16	16-20	20-00	00-04
Minimum no. required	8	18	16	25	8	4

Period 1 follows immediately after period 6. Sales personnel works for 8 consecutive hours, starting at the beginning of one of the six periods. To determine a daily employee worksheet which satisfies the requirement with the least number of personnel, formulate the problem as a linear programming problem. (6)

(*b*) The final product of a firm has a requirement that it must weigh exactly 5 kg. The two raw materials used in the manufacture of this product are *A* with a cost of ₹5 per kg and *B* with a cost of ₹8 per kg. At least 2 kgs of *B* and not more than 4 kg of *A* must be used. How much of each type of raw material should be used for each unit of the final product if the cost is to be minimized? Use simplex method. (12)

SOLUTIONS

1. (*a*) Similar to Example 26, Chapter 2.

Ans. 121

Or

1. (*a*) Similar to Example 21, Chapter 2.

Ans. Saving : ₹ 11,875 Bonds: ₹ 16,250, Business: ₹ 21,875

1. (*b*) Similar to Example 36, Chapter 2.

Ans. (*i*) Technology matrix : $A = \begin{bmatrix} \frac{1}{10} & \frac{3}{5} \\ \frac{3}{20} & \frac{1}{10} \end{bmatrix}$

(*ii*) $X = \begin{bmatrix} 240 \\ 120 \end{bmatrix}$

(*iii*) Yes, the solution is *feasible*.

(*iv*) Value added in sector X = ₹ 9000

Value added in sector Y = ₹ 3600

(*v*) Equilibrium prices : $\begin{bmatrix} p_1 \\ p_2 \end{bmatrix} = \begin{bmatrix} 53.125 \\ 68.75 \end{bmatrix}$

Or

1. (*b*) Similar to Examples 45 and 47, Chapter 2.

Ans. (*i*) Technology matrix : $A = \begin{bmatrix} \frac{1}{5} & \frac{1}{4} \\ \frac{1}{4} & \frac{1}{3} \end{bmatrix}$

(*ii*) The new transaction matrix is :

	Sector I	*Sector II*	*Final Demand*	*Gross Output*
Sector I	6.83	4.3275	23	34.16
Sector II	8.54	5.71	3	17.31
Primary Input	18.788	7.2125		

2. (*a*) Similar to Example 91, Chapter 6.

Ans. 600; ₹ 1,81,200; Yes

Or

2. (*a*) Similar to Example 45, Chapter 6.

Ans. (*i*) elastic on $100 < p \leq 200$; inelastic on $0 \leq p < 100$; and unitary at $p = 100$

(*ii*) increasing on $100 < p \leq 200$; decreasing on $0 \leq p < 100$; and maximum at $p = 100$.

2. (*b*) Similar to Example 61, Chapter 6.

Ans. Decrease in service charge = ₹ 125

R_{max} = ₹ 4,37,500

Or

2. (*b*) Similar to Example 70, Chapter 6.

Ans. 40 km/h

3. (*a*) Similar to Example 42, Chapter 8.

Ans. $x_1 = 5$, $x_2 = 3$; $p_1 = 7$; $p_2 = 11$

Or

3. (*a*) Similar to Example 22, Chapter 8.

Ans. (*i*) Decreasing returns to scale

(*ii*) Diminishing returns to inputs

(*iii*) $\frac{4}{7}Q$

3. (*b*) Similar to Example 25, Chapter 10.

Ans. $x = \frac{2(p-1)}{p-2}$

Or

3. (*b*) Similar to Example 41, Chapter 10.

Ans. $CS = 144$, $PS = 48$

4. (*a*) Similar to Example 33, Chapter 11.

Ans. ₹ 1472.22

4. (*b*) Similar to Example 75, Chapter 11.

Ans. Machine *A* is *preferable*.

4. (*c*) Similar to Example 84, Chapter 11.

Ans. ₹ 10,25,980

4. (*d*) Similar to Example 42, Chapter 11.

Ans. 10 years (app.)

5. Refer to answer no. 22, Chapter 4 (page 4.56).

Or

5. (*a*) Similar to Example 4, Chapter 3.

Ans. Minimize $Z = x_1 + x_2 + x_3 + x_4 + x_5 + x_6$

subject to : $x_1 + x_2 \geq 18$; $x_2 + x_3 \geq 16$; $x_3 + x_4 \geq 25$; $x_4 + x_5 \geq 8$; $x_5 + x_6 \geq 4$;
$x_6 + x_1 \geq 8$;
$x_1, x_2, x_3, x_4, x_5, x_6 \geq 0$

5. (*b*) Refer to Example 34, Chapter 4.